LASER PLASMAS AND NUCLEAR ENERGY

LASER PLASMAS AND NUCLEAR ENERGY

Heinrich Hora

Professor of Physics
University of New South Wales
Kensington–Sidney, Australia
and
Rensselaer Polytechnic Institute
Hartford, Connecticut

Edited by **Yuri Ksander**

Senior Staff Member
The Rand Corporation
Washington, D.C.

PLENUM PRESS • NEW YORK AND LONDON

Library of Congress Cataloging in Publication Data

Hora, Heinrich.
Laser plasmas and nuclear energy.

Includes bibliographical references and index.
1. Plasma (Ionized gases) 2. Laser radiation. 3. Nuclear fusion. 4. Atomic energy. I. Title.
QC718.H67 530.4'4 74-32287
ISBN-13: 978-1-4684-2087-6 e-ISBN-13: 978-1-4684-2085-2
DOI: 10.1007/978-1-4684-2085-2

The author and publisher gratefully acknowledge permission to reprint material in the Appendix, granted by the following:

American Institute of Aeronautics and Astronautics, for pp. 369–375 (from *AIAA Journal*, **11**, 1347–1349, 1973).

The American Institute of Physics, for pp. 123–124 (from *Applied Physics Letters*, **3**, 210–211, 1963); pp. 199–206, 273–282, 283–289 (from *The Physics of Fluids*, **7**, 981–987, 1964; **12**, 182–191, 1969; **14**, 371–377, 1971).

The American Physical Society, for pp. 383–385, 409–416, 425–431 (from *Physical Review*, **A1**, 821–824, 1970; **A6**, 2335–2342, 1972; **A8**, 1582–1588, 1973); pp. 377–381, 417–419, 421–423 (from *Physical Review Letters*, **30**, 89–92, 1973; **30**, 1116–1118, 1973; **31**, 1184–1187, 1973).

Chapman and Hall Ltd., for pp. 433–443 (from *Opto-Electronics*, **5**, 491–501, 1973).

Éditions Bordas, Dunod, Gauthier-Villars, for pp. 113–117 (from *3rd International Conference on Quantum Electronics, Paris, 1963*, Vol. 2, P. Grevet and N. Bloembergen (eds.), Dunod, Paris, 1964, pp. 1373–1377).

Institute of Electrical and Electronic Engineers, Inc., for pp. 119–122 (from *IEEE Journal of Quantum Electronics*, **QE-4**, 864–867, 1968).

International Atomic Energy Agency, for pp. 239–247 (from *Nuclear Fusion*, **10**, 111–116, 1970).

Verlag der Zeitschrift für Naturforschung, for pp. 225–238 (from *Zeitschrift für Naturforschung*, **25A**, 282–295, 1970).

Softcover reprint of the hardcover 1st edition 1975
A Division of Plenum Publishing Corporation
227 West 17th Street, New York, N.Y. 10011

United Kingdom edition published by Plenum Press, London
A Division of Plenum Publishing Company, Ltd.
4a Lower John Street, London W1R 3PD, England

TO MY WIFE ROSE, nee WEILER

PREFACE

Most of this book was written before October 1973. Thus the statements concerning the energy crisis are now dated, but remain valid nevertheless. However, the term "energy crisis" is no longer the unusual new concept it was when the material was written; it is, rather, a commonplace expression for a condition with which we are all only too familiar. The purpose of this book is to point out that the science and technology of laser-induced nuclear fusion are an extraordinary subject, which in some way not yet completely clear can solve the problem of gaining a pollution-free and really inexhaustible supply of inexpensive energy from the heavy hydrogen (deuterium) atoms found in all terrestrial waters.

The concept is very obvious and very simple: To heat solid deuterium or mixtures of deuterium and tritium (superheavy hydrogen) by laser pulses so rapidly that despite the resulting expansion and cooling there still take place so many nuclear fusion reactions that the energy produced is greater than the laser energy that had to be applied. Compression of the plasma by the laser radiation itself is a more sophisticated refinement of the process, but one which at the present stage of laser technology is needed for the rapid realization of a laser-fusion reactor for power generation. This concept of compression can also be applied to the development of completely safe reactors with controlled microexplosions of laser-compressed fissionable materials such as uranium and even boron, which fission completely safely into nonradioactive helium atoms.

The papers that have been published on this subject are widely scattered and quite different in content, occasionally providing contradictory and therefore rather confusing results. To undertake the preparation of an exhaustive monograph would therefore be akin to the labors of Sisyphus: when finished it would be obsolete. Many international conferences have been held which are turbulent mixtures of new and unexpected results with an accumulation of earlier unexplained results. With an eye to the necessity of resolving the energy crisis, some attempts have been made to provide introductory material to this subject. The biannual International Workshop Conferences on "Laser Interactions and Related Plasma Phenomena" organized by Helmut J. Schwarz and the author are held for this purpose. Both reviews and new results are presented and then -- to the extent that the material is suitable -- published in proceedings volumes of a fairly representative character. However, the level is necessarily advanced; and, therefore, both students and beginners have need of a quick survey and an introduction to plasma physics of such a nature as to enable them to follow the papers of these proceedings as well as the current literature. With this need in mind, this introductory volume was written to provide in combination with the appended collection of important papers something akin to a seminar in printed form, to be studied alone or to be used in conjunction with lectures and discussions.

The author has not been neutral in writing this introduction, but he has tried to avoid the shortcomings of many so-called "review articles." Such reviews often tend to be either muddled and undigested collections of key words and formulas from a large number of papers or extended discussions of very specific work by the reviewer himself to which he has added only results by others which are relevant to his own. However, the author is not free of human motivation -- otherwise he would not have been able to present the subject with the enthusiasm and conviction of an active participant in an exciting field of research -- and, therefore, this work may also suffer to some extent from the above-mentioned defects. If that is so and some aspects are emphasized more than they might have been by another writer, the reader may compensate for it by studying further in the papers of other authors.

Unavoidably, there is some mixing of highly elementary and introductory descriptions of the physical properties of a plasma with unresolved problems and with some new results. The yet unsolved questions are especially pointed out, particularly when different sources lead to confusing and conflicting results. For example, measurements of the reflectivity of laser-produced plasmas differ by a factor of 100 or more. There seem to be unexplained characteristic phenomena at moderate laser intensities in ruby and neodymium-glass laser pulses around 10^9 W/cm^2, such as the Linlor effect; and there are discrepancies between the models of homogeneous heating and of generation of compression waves at 10^{10} to 10^{13} W/cm^2. At higher intensities the scattering results are equally complicated. The inclusion of new results could not be avoided in view of the presentation of a new method of compression of laser-produced plasma which is much more efficient than the previous approach of gas-dynamic ablation. This method provides a new approach to the entire field, and the whole discussion is influenced by this new point of view.

It is a pleasure for the author to thank colleagues and friends for direct and indirect help in finishing this work, especially Vice-President Warren C. Stoker and Prof. Helmut Schwarz of Rensselaer Polytechnic Institute; Dr. Marcel Salvat, Director of Euratom at IPP Garching, for constant and fruitful cooperation over many years; and Prof. Moshe J. Lubin of the University of Rochester for helpful arrangements during the author's term at his laboratory while on leave from the Max Planck Institut für Plasmaphysik (Association Euratom) in Garching, Germany.

Ottobrunn (Germany)
Rotdornweg 4

Heinrich Hora

CONTENTS

1. INTRODUCTION

The problems of the interaction of high-intensity laser radiation with plasma represent a very new, fascinating, and fundamental field of research, which could be an important key for solving the energy crisis. Whatever other results may come from this field, the technique of inertial confinement of compressed pure hot deuterium by laser radiation may be the only clean, inexpensive, and inexhaustible nuclear energy source of the future. For the short-term, perhaps within 10 years under the most advantageous conditions, laser fusion may lead to a practicable reactor if compression of plasma by lasers to densities 10,000 times the solid state density can be achieved within the near future.

This unusual prognosis -- though based on a highly optimistic estimate -- is made with a somewhat simplistic point of view, without considering other nuclear reactor concepts being developed at present. If someone in 1908 had predicted the development of airplanes from knowledge of the more than one century of development of balloons and the most advanced "Zeppelin," he would have arrived at a completely incorrect picture of transoceanic travel in 1974.

This book is intended to give the reader some feeling of just how close the realization of laser-fusion reactors is; but even if it is further away than we believe -- if some as yet unknown difficulty should arise -- it will at the very least show the way to clean and cheap energy in the more distant future. The energy crisis[1] is "a real problem, not a problem that has been simply put together by some plot," as Governor Love, President Nixon's assistant for energy, said. H. G. Stever of the National Science Foundation sees an "urgent necessity for the U.S. to take aggressive action now" in terms of long-range priorities.[1]

To simplify the picture a little, we find from many references on the earth's energy resources that the United States' reserves of natural gas and oil can meet the national needs for the next 30 years, and coal supplies should last at least 500 years.[1] The use of fossil fuels may be limited by the tremendous increase in the CO_2 content of the atmosphere,[2] which can cause an increase in the average temperature of the atmosphere,[3] resulting in the melting of polar ice and elevation of the sea level.

The next possible source of energy from the earth (after solar energy or geothermal energy) is nuclear fission. The technology of liquid-metal fast-breeder reactors is highly advanced, and the design of a nearly 100 percent safe reactor will be available within the next few years. A disadvantage of the breeder reactor is that all of its reaction products are radioactive. A further problem is the limited amount of ore which is worth working for production of the fuel by present methods. While the earth's crust of 10 miles thickness contains an inexhaustible amount of uranium, the energy needed to separate the dispersed atoms is greater than that gained from their fission reactions. According to a favorable estimate,[4] the uranium and thorium ores worth working up would last for only 50 years supporting an energy consumption of 10^{22} J/yr, the estimated minimum energy requirement for the year 2050.

1.1. Nuclear Fusion

Another source of energy is the heavy hydrogen isotope deuterium D (or H^2), which constitutes 0.03% by weight of the hydrogen in all the water on earth. It may be used in a simple way to support energy needs for millions of years. Deuterium reacts at temperatures of approximately 100 million degrees centigrade, producing in roughly equal amounts stable (non-radioactive) helium-3 (He^3) and neutrons (n) or superheavy hydrogen (H^3), also known as tritium (T), and a proton

$$D + D \rightarrow \begin{cases} He^3 \text{ (0.82 MeV)} + n \text{ (2.45 MeV)} \cong \text{(3.27 MeV)} \\ T \text{ (1.01 MeV)} + H \text{ (3.02 MeV)} \cong \text{(4.03 MeV)} \end{cases} \tag{1.1}$$

where the energy of the resulting particle is given in parentheses. The relation between the different energy scales is

$$1 \text{ eV} = 1.602\text{x}10^{-19} \text{ Wsec} = 1.602\text{x}10^{-12} \text{ erg} = 1.16\text{x}10^4 \text{ K} \tag{1.2}$$

where the temperature in degrees Kelvin (usually denoted only by K) corresponds to a monoenergetic velocity distribution of the particles. For a nondegenerate equilibrium distribution (Maxwellian distribution) of the electron velocity, the maximum energy is 2/3 times that of the monoenergetic case. The unit Wsec (watt-seconds) is equal to the joule.

He^3 and T undergo the following additional reactions:

$$D + T \rightarrow He^4\ (3.5\ \text{MeV}) + n\ (14.1\ \text{MeV}) \cong (17.6\ \text{MeV}) \tag{1.3}$$

$$D + He^3 \rightarrow He^4 (3.6\ \text{MeV}) + p\ (14.7\ \text{MeV}) \cong (18.3\ \text{MeV}) \tag{1.4}$$

The neutrons generated can be thermalized by the lithium-6 isotope (Li^6), which constitutes 7.5% of natural lithium:

$$Li^6 + n \rightarrow He^4 + T \quad (\cong 4.6\ \text{MeV}) \tag{1.5}$$

The lithium has to surround the fusion reaction to absorb the energy of the neutrons and also to "breed" new tritium for the basic reaction.

Summarizing, deuterium is converted into stable helium He^4, and -- without the breeding reaction -- 7.05 MeV of energy is generated for each deuterium atom, corresponding to an energy production of 3.55×10^{11} J = 10^5 kW/hr (kilowatt-hours) from 1 g (gram) of deuterium.

The most interesting part of the reaction sequence shown above is Eq. (1.3), for which the cross-sections of the thermonuclear reactions have been measured.[5,6] Therefore, present research is directed to the study of fusion of mixtures of 50% deuterium and 50% tritium (D-T mixtures). To achieve an efficient fusion reaction it is necessary to contain a D-T plasma of ion density n_i and temperature $T \cong 5$ keV ($\cong 50 \times 10^6$ K) during a minimum time τ such that

$$n_i\tau \geqslant 10^{14}\ \text{cm}^{-3}\ \text{sec} \quad \text{(Lawson criterion)} \tag{1.6}$$

Confinement of plasmas is also under study using magnetic fields of toroidal configuration, such as those of the tokamak or stellarator; and these are sufficiently advanced to offer a realistic chance to achieve a fusion reactor within the next few years. The success of these attempts is based on the discovery[7] of the Pfirsch-Schlüter diffusion[8] of plasma across a magnetic field.

1.2. Laser-Produced Nuclear Fusion

Laser-produced nuclear fusion utilizes confinement without external magnetic fields. Confinement results simply from inertia in an expanding, very high-density plasma following sufficiently rapid heating by lasers. The prospect of early success with a D-T reactor by this method compared to magnetic field confinement is based on the expectation that laser radiation can produce a very

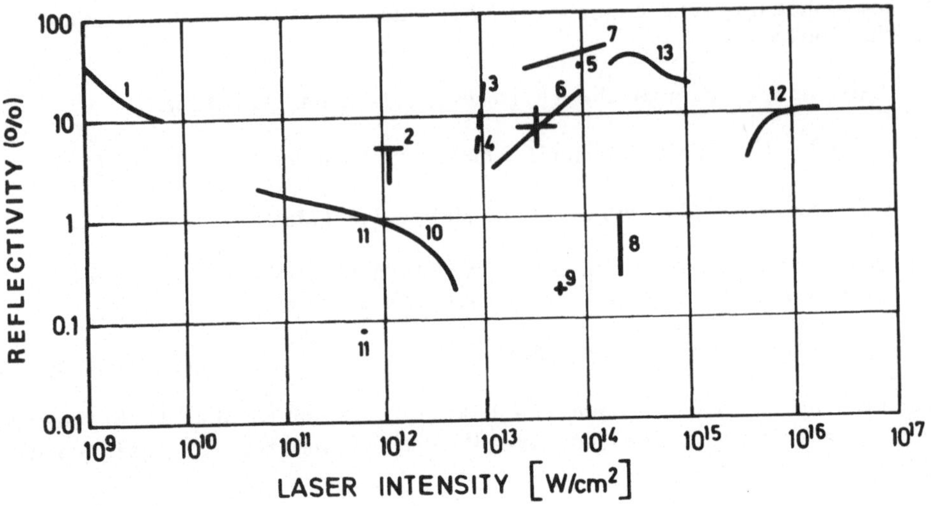

Fig. 1.1 Experimental reflectivity (%) from laser-produced plasma as a function of laser intensity. Data of various authors:[13]--1[14]; 2[15]; 3[16]; 4[17]; 5[18]; 6[19]; 7[20]; 8[21]; 9[22]; 10[13]; 11[23]; 12[24]; 13[25].

strong compression of the plasma.[9] The key goal for the immediate future is to achieve this compression. If it cannot be attained for some reason, the laser fusion concept may have to be relegated to a long-term program.

One advantage of the inertial confinement concept lies in the possibility of using pure deuterium as fuel for the fusion reactor, at least at the more advanced stages, while magnetic confinement can work efficiently only with D-T mixtures. The emission of cyclotron radiation from D-D due to confining magnetic fields is always stronger than the fusion energy gained -- even for reactors of impractically large size.[10] The use of T requires breeding with Li^6, which imposes a limitation owing to the restricted supply of lithium. Estimates of supplies of reasonably low-priced lithium indicate reserves sufficient to cover energy production at the minimum rate for the year 2050 for only 70 years.[4,11] Consideration of this problem[12] indicates the need for a pure D-D reactor, for which the laser concept, in principle, points the way.

The energy crisis has created an overpowering technological and economic need which may be met by study of the interaction of laser radiation, but the physical problems involved are new and very complicated. Just how confusing the experimental results can be is shown in Fig. 1.1. Measurements of the reflectivity of plasma irradiated by laser radiation of varying intensities are scattered over more than two orders of magnitude. Because there is as yet no theory or model for this property, the results are indicative of the very early stages of a pioneering field with nearly-statistical scattering of the measurements.

To illustrate the problem, we refer to the publication in 1905 of Einstein's[25] theory of the linear increase of the maximum energy of photoelectrons with the frequency of light. We now know that the curve intersects the abscissa at a value $\hbar\omega_0$ equal to the work function of the photoemitter. In the controversy that followed Einstein's paper, detailed measurements made in 1908[26] gave a curve passing through the origin of the coordinate system. Only the very careful experiments of Millikan[27] eventually confirmed Einstein's theory. The problem of measuring a quantity without prejudice is also reflected in Kircherner's publication of measurements of the photoelectric effect and the Compton effect on the direction of the polarization of light,[28] in which there is quite detailed philosophical consideration of the difference between what is to be measured and what one likes to measure, a problem that plagues even more recent measurements.[29] These facts should be taken into account when considering the very new field of laser-plasma interactions.

2. LASERS

This section gives a survey on the laser, its mechanism, properties, and present status, dealing with these subjects to the extent that they are relevant to laser-produced plasmas. The basic physical process involved is the stimulated emission of radiation. Consider an atom, molecule, or a solid, in which electrons can occupy two energy levels E_1 and E_2. We have $E = E_2 - E_1 = \hbar\omega$, where $\hbar$ is Planck's constant divided by 2π, and ω is the frequency of the light quantum (photon) which is emitted when the electron undergoes an optical transition from the upper level E_2 to the lower level E_1. Such an optical transition can occur spontaneously within a time τ, the spontaneous lifetime of the electron in the excited state E_2. For atoms and molecules these lifetimes are in the order of 10^{-8} seconds. Besides this spontaneous emission, the time in which the photon is emitted can be decreased by another photon of the same frequency ω which passes the excited atom and stimulates deexcitation by pulling out the photon from the atom. The two resulting photons have the same frequency and the same phase which results in a high coherence of the radiation. Stimulated emission was described in 1917 by Einstein in an ingenious derivation from Planck's law of quantized radiation.

2.1 Laser Condition

In common substances the stimulated emission of radiation is always less than the spontaneous processes or the usual absorption of the light. In thermodynamic equilibrium the number N_2 of the atoms in the state E_2 is always less than the number N_1 of the atoms in the state E_1. The necessary inversion $N_2 - N_1 > 0$ for producing predominantly stimulated emission can be obtained only by deviations from thermodynamic equilibrium (negative temperatures); e.g., by selecting atoms with forbidden optical transitions. This is possible with chromium ions in Al_2O_3 crystals (ruby) which exhibit a forbidden transition with E corresponding to an optical wavelength λ of 6943 Å with a spontaneous lifetime τ of 3 milliseconds. The population of the level E_2 is possible in an indirect way by optical excitation of atoms using green and blue light into the broad band of levels $E_n > E_2$, from which the electrons can fall within 10^{-7} sec into the level E_2. In a ruby crystal under intense irradiation with the blue light of a flashlamp, an inversion of the mentioned levels can occur and the emission of coherent plane wave modes due to stimulated emission of radiation of wavelength 6943 Å is observed leaving the crystal in various directions.

A laser is made from such a ruby crystal with plane ends. It is placed inside an optical cavity with a totally-reflecting mirror at one end and a mirror with a reflectivity R for the stimulated emission wavelength, so that wave modes along the crystal axis are reflected always parallel to this axis. Due to the repetitive reflections, the stimulated emission of the optically-inverted ruby occurs preferentially within these axial modes. A laser beam of very small beam divergence is emitted through the mirror with transmission $T = 1 - R$, with much narrower spectral width than the spectral width $\Delta\omega$ of the spontaneous emission line and with a very high degree of coherence. The only condition for laser emission (threshold)[30] is an inequality for the inversion $N_2 - N_1$ related to the volume V of the ruby

$$\frac{N_2 - N_1}{V} > 4\pi\left(\frac{2\pi n_1}{\lambda}\right)^2 \quad a\,\frac{1 - R}{\ell} \quad \Delta\omega \cdot \tau \tag{2.1}$$

where n_1 is the refractive index of the ruby, ℓ its length, and a is the factor[31] describing the ratio of the light intensity in the cavity without stimulated emission to the value after one pass in the cavity. This ratio thus describes all losses due to normal optical absorption, scattering, and diffraction in the system.

In addition to the early ruby laser[32,33] nowadays a large number of materials are known which can be used for the population inversion $N_2 > N_1$ and lasing. Instead of chromium atoms in Al_2O_3 one can use neodymium atoms to produce the YAG laser with emission at a wavelength of 1.06 microns. Instead of using the expensive sapphire crystal Al_2O_3 as a host material, one can use glass to get the neodymium-glass laser.[34] To reach other laser wavelengths, numerous other solid state lasers have been built using rare earth ions to provide the forbidden transitions. Dye lasers[35] use organic dyes with appropriate molecules in which forbidden transitions occur.

The preferred method of pumping these lasers, as in solid state lasers, is optical pumping; the light of flashlamps or even of other lasers is used to excite electrons to levels E_n from which they relax into the desired E_2 levels.

The excitation of semiconductor lasers is accomplished in yet another way. In a heavily doped GaAs junction diode[36] currents exceeding 20,000 Amp/cm^2 generate an inversion of electrons and holes in a thin layer at the junction from which laser emission can be detected as a highly-directional radiation with a narrower spectrum than that of spontaneous emission. Similar junction lasers of other substances, e.g., Ga(As,P) and others, have been discovered.

Emission from degenerate states of homogeneous semiconductors (E_2 in the conduction band and E_1 in the valence band) has been proposed, in which the inversion is produced by very strong electron beams.[31,37] After reasonable thresholds for the electron beam density and the necessary conditions of degeneracy became known,[31] electron-beam-pumped semiconductor lasers were demonstrated experimentally.[38]

To avoid damage in the solid materials and to obtain uniform pumping, liquid lasers were developed; e.g., the europium chelate laser[30] or solutions of neodymium ions in selenium oxychloride[40] in which pumping of the medium is provided by flashlamps as in the most common solid state lasers.

There are several reasons favoring use of a gas or a plasma as the active medium. The iodine laser with optical pumping is an example of such a laser.[41] The uv radiation from flashlamps causes photodissociation of the iodine atoms in molecules of CF_3I, or analogous aliphatic compounds. These are electronically excited after dissociation to the $5^2P_{3/2}$ state which can lase to the ground state $5^2P_{1/2}$. Other gas lasers are pumped by a gas discharge and are therefore much more efficient than optically-pumped lasers. The helium-neon laser[42] was the first known gas laser; it has still not produced sufficient power to be used for generation of plasmas. The same is true for the argon and xenon atomic lasers.[43] A very important system is found in the CO_2 laser.[44] It consists of a mixture of CO_2 with nitrogen and noble gases of pressures between 1 Torr and 60 atmospheres. A gas discharge is generated either along the cylindrical cavity containing the gas or by transverse excitation. In some cases irradiation by electron beams of some 100 keV or photoionization by a discharge in the laser medium are used to give homogeneous preexcitation of the gas to guarantee a homogeneous main discharge. The discharge excites the $v = 1$ level of the nitrogen molecule which has nearly the same energy as the vibrational transition of the CO_2 molecule from the state 00^01 to 00^00. This vibrational transition has a very long lifetime, 3.5 microseconds per atmosphere at 300 K and 1 microsecond per atmosphere at 500 K. The discharge in nitrogen, therefore, populates these vibrational levels of CO_2 molecules and results in a laser emission.

The long vibrational lifetime of the CO_2 molecule can also be used for a gas-dynamic method of population inversion. Consider CO_2 and N_2 mixtures heated up by a discharge or moving in a shock tube with a Laval nozzle. There the gas is cooled by transfer of the internal energy of the gas into kinetic energy. The vibrational levels still correspond to the higher initial temperature while the gas itself has a much lower temperature. The transition from the excited vibrational levels may then result in laser action.

The use of the energy of chemical reactions to obtain laser action[45] is one of the most promising ways to build highly-efficient lasers. Theoretically, an efficiency up to 80% can be expected. One system of this type is the HF laser, which, however, uses a chemical reaction occurring in several steps. The final goal would be a bootstrap laser, in which a reaction front in the chemically reacting gas converts the reaction energy into optical energy for a laser front which moves nearly with the speed of light through the gas, initiating the reaction and collecting laser energy. A laser illustrating this mode of operation, at least in part, has been demonstrated using a gas discharge for excitation instead of chemical reaction energy. In the nitrogen ultraviolet laser[46] the discharge front moves with the speed of light through the gas. The resulting laser pulses are of low quality, however, because of optical inhomogeneities in the moving discharge front.

2.2 Operation of Lasers

There are very different methods available for the operation of lasers. On the one hand, continuous wave (cw) operation is possible with more or less periodic substructure of emission profiles. Usually, gas lasers are used for cw operation, the helium-neon laser for intensities up to 100 W or the argon ion laser with green and blue emission lines and intensities up to 500 W. CO_2 lasers have attained intensities of 60 kW and more.[47] These laser beams have been used for welding steel up to 3 cm in thickness. The first ruby lasers operated in "spiking" regime.[32] Due to the complex sequence of population of the upper levels E_2 and the kinetics of the laser modes, the spiking laser emits an irregular sequence of pulses during its total emission time of up to one millisecond. An important advance was the discovery of the Q-switch;[48] the Q of the laser cavity is controlled by either of two methods: actively, by a rotating mirror at one end of the cavity or an electro-optical shutter (Pockels cell), or passively, by a dye placed in the cavity, the absorption and transmission of which is changed by the incident laser light. Initially, the reflection from one end of the cavity is very low; therefore, the laser action is suppressed and a high population of the upper level can accumulate. At the time of high reflectivity of the rotating mirror or opening of the Pockels cell to the highly-reflecting end of the cavity or the equivalent opening from bleaching of the dye, an intense laser pulse is built up in the cavity. The highly-converted states in the active medium are depleted and result in very intense laser pulses of some nanosecond duration. The discovery of the Q-switch in 1962 resulted in sharp ruby laser pulses of intensities exceeding 10^9 W/cm^2, which when applied to solids in vacuum immediately produced plasmas with up to keV ion energies.

Depending on the properties of the cavity, Q-switched laser pulses can have a temporal substructure with oscillations and subpulses of less than one picosecond duration. Using a passive Q-switch with a tuned dye concentration, nonlinear coupling of the laser modes, or their statistical correlation results in generation of a mode-locked pulse sequence consisting of a chain of pulses of few picosecond duration separated by a distance determined simply by the optical length between the mirrors of the cavity. The models of coupling of modes by DeMaria[49] and Letokhov[50] were investigated experimentally by Saltzmann.[51] One of the short pulses can be selected by an electro-optical switch and transmitted to a chain of laser amplifiers.

The properties of laser radiation are high coherence, small beam divergence, and extremely high power or intensity. Coherence means that the light beams after dividing and re-uniting can produce interferences, if the difference of the path of the beams is less than a certain length. This coherence length for noncoherent emitting atoms is of the order of meters. For lasers the length can exceed 10 km. This means that the phase fronts of the laser light are highly stabilized, which is the basic property needed in holography. For laser-produced plasmas this property seems of lesser importance, because the generation of high intensities is the primary requirement. However, if certain intensity distributions within the irradiated plasma are desired, the properties of coherence are needed.

The beam divergence is related to the mode structure of the laser beam and to the diffraction properties of the laser. It is obvious that not only is a wave mode parallel to the axis of the optical cavity of the laser generated but also a set of modes with a discrete sequence of angles off the axis. These transverse modes can be suppressed simply by focusing the laser beam to the aperture of a diaphragm transparent to the axial mode only. The higher transverse modes are absorbed by the material of the screen. The axial mode can also consist of several longitudinal modes separated by small frequency differences determined by wavelengths permitted within the optical length of the cavity, as far as the wavelengths are within the corresponding $\Delta\omega$ of the natural spectral width of the emitted radiation. The control or suppression of the longitudinal modes is possible with nonlinear materials arranged within the cavity. If a single-mode laser beam has been generated, it possesses a lateral structure of the intensity due to diffraction of the beam in the exit aperture of the optics involved. Transmission of the beam through subsequent laser amplifiers of varying thickness and optics to increase the cross section of the beam can produce further diffraction processes, resulting in a highly-complicated lateral distribution of the laser intensity with several Fresnel fringes.

Lasers can emit energy up to kJ within a few nanoseconds or tens of J within picoseconds. Because the beam divergence is in milliradians or less, power densities of 10^{16} W/cm^2 have been exceeded. The exact values of the intensity cannot be given with great accuracy because of the complicated spatial and temporal structure of the laser pulses arising in complex modes, and Fresnel fringes and filamentary structure of the laser pulses due to self-focusing within the optical media. Usually, time-integrated values of laser energy from calorimeters are used, or spatially-integrated intensities are measured by photodiodes with temporal resolution down to several picoseconds. One method of measuring the intensity of laser beams in vacuum with high spatial and temporal resolution was proposed by using the diffraction of electron beams by the laser beam (Kapitza-Dirac effect[52]). An example of the measurement of laser intensity in the focus of a lens is shown in Fig. 2.1; the result is far different from the expected interaction of an ideal plane wave with a plasma located at the focus. These markedly varying conditions have to be taken into account when widely-differing results of different experiments in laser-produced plasmas are discussed.

Several methods using nonlinear optics are available to modify the properties of a laser beam before it interacts with a plasma. The spatial distribution of laser beams can be corrected by lenses and Fresnel zone plates to provide desired intensity and phase distribution at a given surface. The time dependence of the beam can be modified by electro-optical shutters. The shutters operate through the rotation of polarization of nonlinear materials by electric fields, applied either with high voltages within one nanosecond or less or with the electric field of an additional laser pulse with a width down to picoseconds (Duguay's shutter[54]). A change in frequency of the laser beam is also accessible; e.g., frequency doubling which can be carried out with efficiencies of 20% or more[55] by passing the beam through a nonlinear crystal. The most advanced technique of generating higher harmonics is that discovered by Harris,[56] in which metal vapors produce the 5th harmonic. Another conversion of laser radiation to higher frequencies utilizes stimulated anti-Stokes Raman lines in liquid hydrogen or

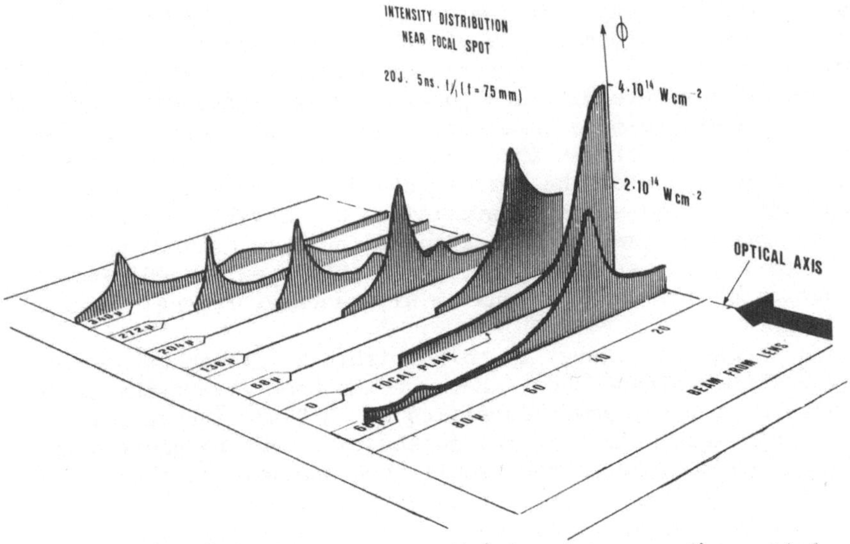

Fig. 2.1 Measurement of (time-integrated) spatial distribution of the laser intensity in a focus.[53]

nitrogen with conversion efficiencies of a few percent.[57] For change of a laser pulse into a desired profile with time, the technique of 2π, 4π, and higher pulses in passive nonlinear materials can be applied.[58]

Summarizing these results, it seems that in the future it will be feasible to generate very high intensity laser pulses and to transform them by passive nonlinear elements into the correct time, space, and spectral distributions especially suitable for laser-plasma interaction.

2.3 Available Lasers

The properties of currently available high-power lasers are of interest. The historic ruby laser with wavelength λ = 6943 Å has been developed for larger systems. The major limitation lies in the techniques of producing large ruby rods or disks. These lasers have been used in laser-plasma research at only a few laboratories. The maximum intensity obtained with ruby lasers is 10^{16} W/cm^2.[59] There are no known experiments which produced fusion neutrons with laser irradiation of pure deuterium or deuterium compounds. Laser systems featuring YAG oscillators (λ = 1.06 μ) followed by Nd-glass amplifiers have been reported as producing fusion neutrons at output powers of 10^{15} W/cm^2.[60]

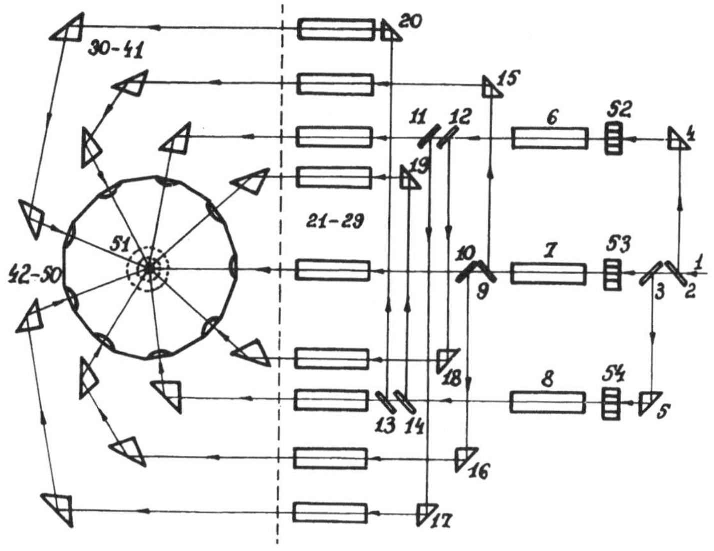

Fig. 2.2 Nine-beam, neodymium-glass laser system:[70] 2,3,9,10,11-14 = dividing elements; 4,5,15-20,30-41 = total reflection prisms; 6-9,21-29 = neodymium-glass rods; 42-50 = focusing objectives; 51 = spherical target; 52-59 = isolating shutters. At 1 a beam enters from a three-stage oscillator-amplifier with mode selection.

The major work on laser-produced fusion plasmas has so far utilized neodymium-glass lasers (wavelength $\lambda = 1.06\ \mu$). In several cases the traditional Schott glass and similar glasses of French, Soviet and Japanese origin were used. A special development at Owens-Illinois resulted in a glass with the preferred light atoms which has very high damage thresholds.[61] A neodymium-glass laser system with pulses of 100 picoseconds and very flexible control of the time profiles resulted in laser intensities exceeding 10^{16} W/cm^2.[62] Another system provides pulses of 800 J energy within 2 nanoseconds.[63] A Soviet system in operation since 1971 involves focusing nine parallel laser beams on the target with spherical geometry (Fig. 2.2). The laser pulses give 400 J within 2 nsec and 3 kJ within 16 nsec.

A very advanced laser system for producing fusion plasmas is the CO_2 laser ($\lambda = 10.6\ \mu$) with electron-beam preionization. The most advanced system that supplies several hundred joules within one nanosecond is that of Fenstermacher et al.[64] Preliminary results indicating generation of fusion neutrons were reported.[65]

The iodine laser ($\lambda = 1.3\ \mu$) is an interesting system which provides laser pulses of 20 J within few tenths of a nsec. The self-oscillating emission of oscillators results in pulses of 400 J within 1 microsecond. The goal of the development is a 1-kJ 1-nsec system.[41]

A very recent development is the electron-beam-pumped xenon laser with a very short wavelength in the ultraviolet ($\lambda = 1722$ Å).[66] Laser pulses of 200 J within 50 nsec have been achieved, limited at present only by the technology of the optical components, because the damage threshold is a few mJ/cm^2 at these wavelengths. The damage threshold for neodymium-glass laser radiation is from 5 J/cm^2 up to a few hundred J/cm^2.[67]

A very remarkable development of CO_2 lasers with electron-beam excitation is that of Basov et al. using pressures up to 60 atmospheres. Pulses on the order of tens of joules within ten nanoseconds have been obtained.[68]

The transverse-discharge HF laser[69] ($\lambda = 3.6\ \mu$) has been developed in a system giving pulses of the order of 100 J and pulse-lengths of 100 nsec (half width). This is also an interesting laser system for future research in fusion plasmas.

3. EARLY MEASUREMENTS AND GAS BREAKDOWN

Before discussing the physical basis of nuclear fusion in laser-produced plasmas and the present state of detailed problems, it is instructive to glean some important early experiments on plasma production by lasers, in which one can distinguish between laser-produced gas breakdown and plasma production from solids in vacuum.

3.1 Gas Breakdown

The first occurrence of breakdown and plasma generation by focusing of a laser beam in a gas was reported at the 3rd International Quantum Electronics Conference in Paris, February 1963, by Terhune and his group at the Ford laboratories.[71] Later studies were devoted to phenomenological details of the breakdown mechanism. The dependence of the breakdown threshold on the gas pressure followed a Paschen law, well known from the classical physics of gas discharges.[72] These relationships were measured by several other authors with only the exact value of the breakdown threshold varying.[73] For ruby lasers the minimum values required of the electric field strengths of the laser light were around 10^6 V/cm for pulses from 1 to 30 nsec in duration. Some exceptionally-low thresholds were probably due to the longer laser pulses. An increase of the breakdown threshold up to 10^8 V/cm for laser pulses of picosecond duration was measured some years later with neodymium-glass lasers.[74]

A quite remarkable phenomenon, first observed by Ramsden and Davies,[75] was the propagation of the breakdown front towards the incident laser light with velocities of 10^7 cm/sec or more, while the expansion velocity of the plasma at the rear side was 10^6 cm/sec or more. The unusual phenomenon with the "notorious"[76] 10^7 cm/sec velocity was known earlier from Linlor's measurement of the expansion of a plasma produced in vacuum by laser irradiation of compact targets.[77] A substantial motion of the nascent plasma in the direction of the aperture was confirmed from measurements with probes showing arrival of the expanding ions. The ions had surprisingly-high energies of 1 keV or more! In the case of gas breakdown, the existence of substantial motion could not be established as simply as in the case of the targets in vacuum because the surrounding gas prevented free motion of the generated ions. The measurement of the high-expansion velocity in gas breakdown was made by side-view image converter cameras using the velocity of the contour of the radiation front[75] and additionally by the Doppler shift of the emitted line radiation.[78] The spectra of Thomson scattering[75] and the Doppler broadening[78] indicated electron and ion temperatures of only a few eV. It is still an unexplained and remarkable phenomenon that the reflected laser light did not show a Doppler shift, according to very careful measurements by Minck and Rado.[79]

There were several different explanations of the "notorious" velocity based either on the assumption of a radiation-driven detonation wave[80] or on the assumption of a residual plasma and rapid motion of the phase of the breakdown.[81] Another model assumed substantial motion of the generated plasma against the laser light due to nonlinear forces of collisionless dielectric interaction of the laser light with the plasma.[82,83] The early experiments on the Doppler shift, however, could not distinguish between any of the proposed theories, except the observation of Minck and Rado.[79] Only the theory of substantial motion due to nonlinear forces resulted in a vanishing Doppler shift, because the laser light itself causes the motion and the Fizeau effect in plasmas (the motion of light with moving media) does not exist.[84] This force, however, is effective only at laser intensities exceeding 10^{14} W/cm^2 which can occur in gas breakdown only after self-focusing.[85,86] This has been confirmed in recent work.[87] The highly-sophisticated proof of the substantial motion of the plasma in gas breakdown was given in the work of Krasyuk, Prokhorov et al.[88] On this basis it seems quite probable that the mechanism occurs in the following way: the focused laser beam causes gas breakdown, and within a time of 10^{-10} seconds or less a self-focusing channel is built up. The laser intensities allow a nonlinear force mechanism for expansion of the plasma in the channel against the laser light with the high velocities which have been measured and without the need of heating (in accordance with measured low plasma temperatures[89]). Support of this model is given by the fact that the measured energy transferred into the plasma corresponds to averaged anomalous or nonlinear absorption constants along the whole plasma 10 to 100 times the highest classical values of inverse-bremsstrahlung absorption.[90] In the case of radiation-driven detonation fronts, these constants should be of even higher values than given classically. The nonlinear force mechanism, indeed, accounts for the measured nonlinear absorption constants.

A further basic question on gas breakdown is the mechanism of initiating the breakdown itself. There exist models of the classical avalanche breakdown mechanism presuming the existence of an initial electron, generated somehow, which is oscillating in the laser field and generates secondary electron-ion pairs by ionization.[91] Another model assumes that no electrons exist in the focus and the laser field causes ionization by multiphoton mechanisms.[92] There exist other theories with mixed mechanisms.[93] The multiphoton process seems to be essential in the experiments with picosecond laser pulses focused in gases of low density; for this case the breakdown fields of 10^8 V/cm are in good agreement with the theory. At higher pressures the breakdown threshold of about 10^5 V/cm for CO_2 laser radiation favors the avalanche mechanism, because the longer wavelength results in lower thresholds in this model. However, the observation of preluminosity from the laser-irradiated gases and the very strange behavior of several properties in gas breakdown, as observed by Papoular et al.,[93] indicate a large number of unresolved questions. The interest in this field has decreased during the last few years, because laser intensities grew to such high values that the details of gas breakdown at moderate intensities seem to be of little interest. However, many of the problems involved are still important for the whole field, and answers could be obtained by means of relatively inexpensive experiments.

If the very high laser intensities are actually interacting with atoms in such a way that the generation of plasma occurs within negligibly-short times, the mechanisms of self-focusing are also important in high-intensity interaction. A highly-critical question concerns whether satisfactory wavefronts of laser radiation can be produced to heat and compress plasma or whether self-focusing prevents uniform interaction with the irradiated plasma. It is quite important, therefore, to study the mechanisms of self-focusing in "transparent" materials like gases, in which diagnostics can provide direct information on the processes involved. It is remarkable that self-focusing occurs even at laser powers as low as a few megawatts and generates self-focusing channels of a few microns thickness.[94] Both the low thresholds and the diameter of the channels can be accounted for quantitatively from a model of the nonlinear forces.[86,95] The lateral decrease of laser intensity causes an electrodynamic pressure on electrons in the radial direction which is compensated by a gas-dynamic density gradient resulting in a thermokinetic pressure. The resulting depletion of the electrons (and thus plasma) in the center of the self-focusing filament has been demonstrated experimentally with interferograms.[96]

3.2 Plasmas Produced from Solids in Vacuum (Linlor Effect)

The interaction of laser radiation with solids in vacuum has been one of the topics of continuing interest since the discovery of the laser. Laser powers of less than one megawatt focused to intensities up to 10^8 W/cm^2 were available before 1962 with spiking ruby lasers. The results were all in good agreement with the classical theories:[97] the evaporated materials contained plasma with ions and electrons of energies of some electron volts. The low (milliampere) emitted electron currents corresponded to thermionic emission[98] in which the limit is given, not by the Richardson-Dushman[99] equation, but by Langmuir's space-charge limit applied to emission current densities.[100] There is still some interest in this work for drilling, welding,[101] removal or destruction of materials, and for evaporation; e.g., for producing thin films,[102] etc. Very intensive work for these purposes was done by Ready.[97] A consistent theory for the numerous classical processes involved was developed by Hans Zahn.[103]

The situation changed drastically, with the appearance of additional unexplained physical phenomena, when the Q-switched laser was discovered by Hellwarth[48] at Hughes Aircraft Labs and was used by Linlor[77] to generate plasmas at the surfaces of solids in vacuum.[104] The laser powers were in the range of 1 to 30 MW and the power densities (intensities) were around 10^9 W/cm^2. The plasma expanding against the incident laser radiation had velocities of 10^7 cm/sec, indicating ion energies of 1 keV or more, while at slightly lower laser intensities ions of only a few eV were observed. The emission of electrons suddenly changed from the classical values of less than 1 Amp/cm^2 at 100-kW laser powers up to thousand Amp/cm^2 at higher laser power.[105,106] These current densities are far greater than Langmuir's space-charge limitation.[100] The ion energy ε_i increases according to the relationship

$$\epsilon_i \sim I^\alpha \tag{3.1}$$

dependent on the laser intensity I, where the exponent α is between 1.5 and 2 (Fig. 3.1). It is, therefore, far greater than a linear function, while gas-dynamic heating of the plasma results in α's of 0.66, in the best cases.[107,108,109] The momentum transferred to the irradiated target increases at laser intensities around 10^9 W/cm^2 as the fourth power (Fig. 3.2) of the intensity.[110,111] The reflectivity of the plasma changes suddenly from the expected metallic values[14] of more than 60% to values less than 10%[15,17,109] (see Fig. 1.1).

This highly-anomalous behavior of the plasma changes into a much more reasonable mode amenable to plasma physical models at laser intensities exceeding 10^{11} W/cm^2. Then the exponent α in

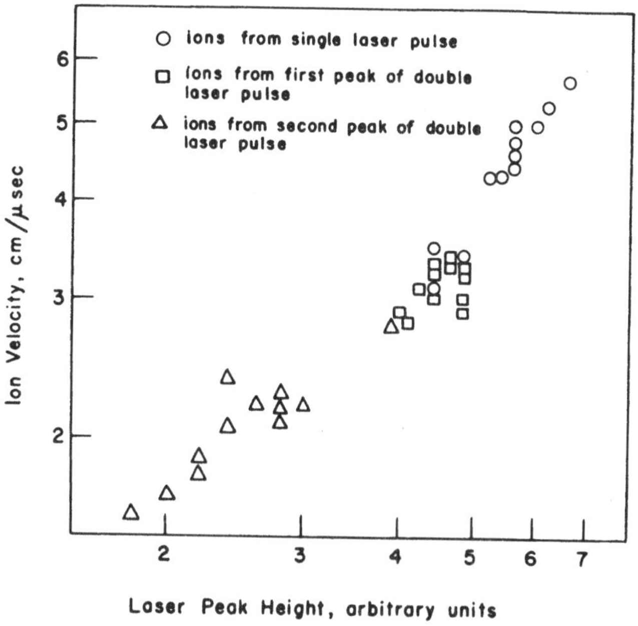

Fig. 3.1 Isenor's measurement[107] of nearly-linear increase of the ion velocity (and therefore a quadratic increase of the ion energy) with the laser intensity (to be identified with laser peak at constant focusing) at intensities close to 10^{10} W/cm^2.

Eq. (3.1) decreases to understandable values[112] around 0.5 to 0.7 (Fig. 3.3), and the measured recoil becomes reasonable.[113] The highly-pronounced anomaly at 10^9 W/cm^2 (Linlor effect) was obviously little noted in the world of physicists, because the reasonable behavior at higher intensities overshadowed the unusual behavior. It was even the case that reports of some anomalous results near 10^9 W/cm^2 were rejected from publication because of the existence of the reasonable results at higher intensity. It should be emphasized, that in this area much study can be done in small-scale experimental efforts. This can be very important and it may significantly influence major directions of the very expensive, big-science projects which have started in laser-produced nuclear fusion.

There still is no gas-dynamic or plasma-physical explanation for these anomalies. The assumption that the fast ions are due to electrostatic acceleration of the thin plasma in the Debye sheath at the surface does not hold, because the number of the fast ions exceeds 10^{15}, while a Debye-sheath mechanism can explain only 10^{12} ions or fewer. A sceptical physicist will be aware that the anomalies of the Linlor effect may be explained in the future by some

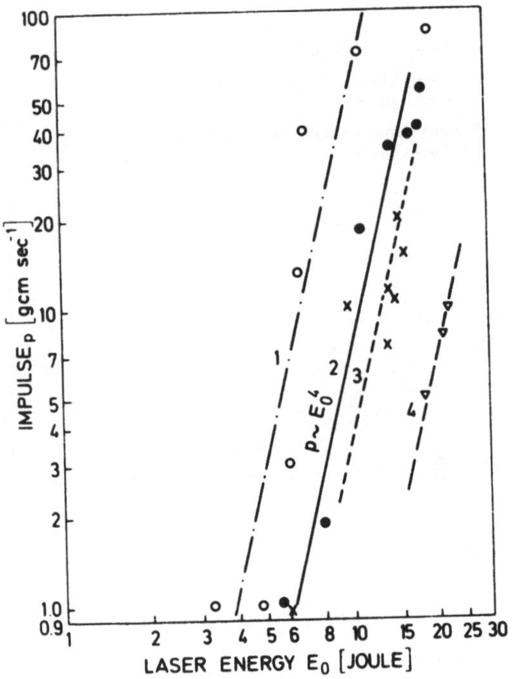

Fig. 3.2 Measurements of Metz[110] of an increase of the momentum p transferred to the irradiated target with the fourth power of the laser intensity I close to 10^{10} W/cm^2.

other, clearer theory; at present only one model seems to be applicable to understanding the anomalies: Starting with the threshold of 10^6 W for self-focusing, filaments are drilled into the targets, resulting in required laser intensities[95] of 10^{14} W/cm^2 for ruby or neodymium-glass lasers and 10^{12} W/cm^2 for CO_2 lasers, to induce the nonlinear-force acceleration mechanism. This mechanism explains without difficulty the keV ion energies,[86,87] the exponential increase of the ion energy,[107] and the recoil of the target as a function of the laser intensity. In addition, the anomalous electron emission can be explained by the fact that the electrons are accelerated primarily by the laser field within the space-charge-free interior of the plasma, from where they can be emitted without the difficulties presented by the Langmuir space-charge limitations.[106] At the higher intensities these mechanisms are thoroughly mixed with the expected gas-dynamic properties of high-temperature plasmas.[112,113] The nonlinear forces should dominate over the gas-dynamic processes again under conditions in which self-focusing is probably absent at intensities exceeding 10^{16} W/cm^2.

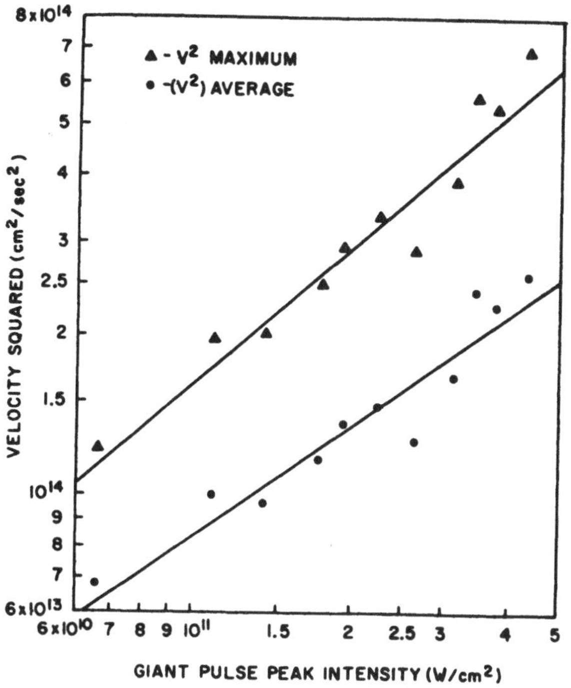

Fig. 3.3 "Normal" increase[112] of the ion energy (ion velocity squared) of about 0.7 as a function of the laser intensity at 10^{11} W/cm^2 and above.

In conclusion, one may say that neither the processes of gas breakdown nor the processes in laser-produced plasmas at intensities around 10^9 W/cm^2 (the anomalies of the Linlor effect and Ramsden's "notorious" expansion velocities of 10^7 cm/sec) are understood sufficiently to be considered a reliable foundation for future work on laser-produced nuclear fusion.

4. MICROSCOPIC PROPERTIES OF PLASMA

The physical description of a high-temperature plasma starts with the single particle motion of electrons and ions, as these are present in all gases at temperatures exceeding 1000° centigrade. The mechanism of ionization of molecules or atoms by collisions due to thermal motion or by radiation processes and interaction with neutral atoms is of little interest in laser-produced plasmas, which have plasma temperatures exceeding 100,000° centigrade (corresponding to 10 eV). Each particle has a charge Ze (e is the elementary charge, $e = 1.602 \times 10^{-19}$ Coulomb $= 4.807 \times 10^{-10}$ Gaussian cgs-units) where Z is an integer describing the number of elementary charges the ions of a mass m_i have. In nearly all of the following cases Z is one, because we shall consider hydrogen ions with a mass $M = 1.67 \times 10^{-24}$ grams, or of two or three times this value if deuterium or tritium ions are used. The mass of the electron is $m_e = 9.109 \times 10^{-28}$ grams.

The motion of the electrons (index e) and ions (index i) is described by their velocities v^0 determined by the equations (4.1) and (4.2), respectively, dependent on time t

$$m_e \frac{dv_e^0}{dt} = -eE - \frac{a}{c} v_e^0 \times H + \nabla\phi \tag{4.1}$$

$$m_i \frac{dv_i^0}{dt} = ZeE + \frac{Ze}{c} v_i^0 \times H + \nabla\phi \tag{4.2}$$

where E and H are the electric and magnetic field strengths at the spatial coordinate $r(dr/dt = v^0)$ and ϕ is a potential describing

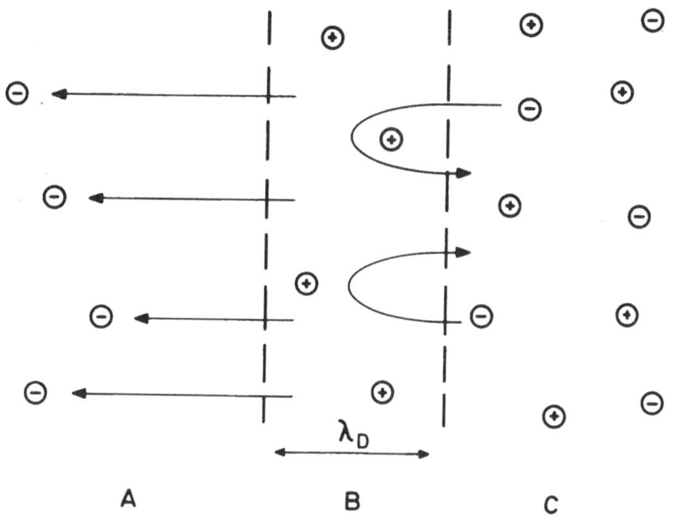

Fig. 4.1 Derivation of the Debye shielding distance λ_D from the depth of the positively-charged plasma boundary B which is depleted of high-velocity electrons and is of such charge as to reject the electrons from the space-charge-neutral interior C.

non-electromagnetic forces such as Coriolis forces. In the following, these forces will be neglected. E and H are due to externally-applied electromagnetic fields and fields generated by other particles of the plasma. In all cases to be discussed classical statistics are applicable, and the whole plasma can be described by a point-mechanical N-particle problem, where N is the total number of particles in the plasma.

4.1 Debye Length

One general property of a plasma is the space-charge neutrality with respect to some minimum distance or a minimum diameter λ_D called the Debye shielding distance. The value of this distance can be derived from the sheath at the surface region of a homogeneous plasma (Fig. 4.1). Its interior consists of a homogeneous plasma separated by a surface region B of thickness λ_D from the vacuum A. In addition to the gas-dynamic mechanism of expansion of the plasma against the vacuum, there is another process taking place. Although the energy of both electrons and ions is equal to kT (the Boltzmann constant) with both at the same temperature T, electrons, having much higher velocity than ions, leave the plasma very rapidly (arrows in A). This runaway process lasts only until the depletion of the electrons from region B produces such an electrostatic potential ϕ_D that further electrons of energy kT emitted from the plasma interior C are rejected electrostatically (curved arrows in B). From this we have the relation

$$kT = \phi_D = \lambda_D \; |E| \tag{4.3}$$

The electric field E in the region B is determined by the ion space charge of density n_i in B, given in cgs units by

$$4\pi\lambda_D n_i e = \oint n^2 E \cdot df = n^2 E \tag{4.4}$$

where the refractive index n usually is unity and the integral over the area is simplified by using an appropriate volume of thickness λ_D and a cross section of 1 cm^2. Rearranging Eqs. (4.3) and (4.4) one obtains

$$\lambda_D = \left(\frac{kT}{4\pi n_e e}\right)^{1/2} \text{(cgs)} = \left(\frac{\varepsilon_0 kT}{n_e e}\right)^{1/2} \text{(MKS)} \tag{4.5}$$

or in numerical quantities with λ_D in cm

$$\lambda_D = 6.9 \left(\frac{T(K)}{n_e(cm^{-3})}\right)^{1/2} = 7.43 \times 10^2 \left(\frac{T(eV)}{n_e(cm^{-3})}\right)^{1/2}$$

In addition to defining the plasma sheath at a plasma surface, the Debye length λ_D also defines the long-range potential ϕ_e of an excess charge within a plasma, given by a dependence on the radius r

$$\phi_e \sim \frac{1}{T} \exp\,(-r/\lambda_D) \tag{4.6}$$

where the amount of the excess charge determines the magnitude of ϕ_e.

4.2 Plasma Frequency

A further property of a space-charge-neutral plasma is the motion of the electrons and ions in specific electrostatic oscillations characterized by the plasma frequency ω_p, sometimes called the Langmuir frequency after its discoverer. Considering a plasma with a homogeneous electron density within a thickness Δx (upper part of Fig. 4.2), layers of electrons may be moved by varying differences $d\xi$ (lower part of Fig. 4.2) resulting in a change of the electron density

$$\frac{dn_e}{n_e} = \frac{d\xi}{dx} \tag{4.6a}$$

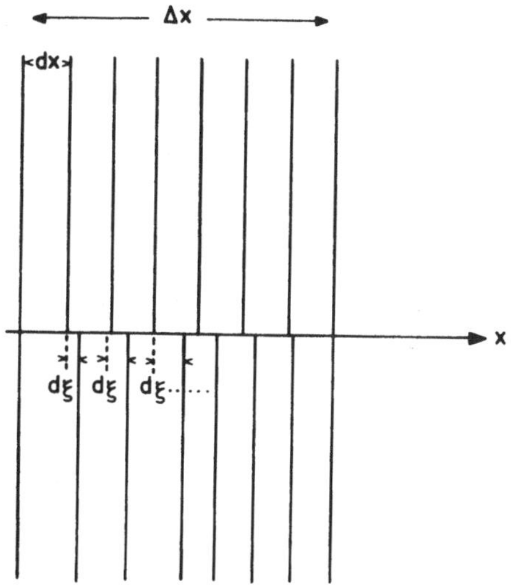

Fig. 4.2 Displacements of electrons of homogeneous density (upper part) by $d\xi$ to generate electrostatic oscillations with the plasma frequency ω_p.

The electrostatic potential ϕ_e generated is given by a Poisson equation

$$\frac{d^2\phi_e}{dx^2} \quad \frac{d}{dx}|E| = 4\pi n_e e \frac{d\xi}{dx} \tag{4.7}$$

resulting in an electric field strength

$$|E| = 4\pi n_e e\ \xi \tag{4.8}$$

and, therefore, in the equation of motion

$$m_e\xi = e|E| = 4\pi n_e e^2\xi \tag{4.9}$$

which defines the frequency ω_p of the electrostatic oscillation of the electrons

$$\omega_p{}^2 = \frac{4\pi e^2 n_e}{m_e}\ ;\ \omega_p = 5.65 \times 10^4 \sqrt{n_e(\text{cm}^{-3})} \tag{4.10}$$

The ion-plasma frequency is determined by the density n_i and the mass M of the ions, in place of the values for the electrons, n_e and m_e, respectively. The electrostatic oscillations cause a resonance of the plasma to incident electromagnetic waves of the same frequency which are then totally reflected. The observation of the total reflection of radio waves of about 30 meters wavelength at the plasma shell of the ionosphere surrounding the earth indicated an electron density of about $n_e = 10^6$ electrons/cm^3; this was in agreement with subsequent direct measurements by artificial satellites.

4.3 Collisions

The motion of single plasma particles can be changed by the electrostatic field of neighboring particles (Coulomb collision). The deviation in path of only small angles should be distinguished from wide-angle deviations. Only the latter type are called collisions and correspond to a considerable exchange of energy. According to the Lorentz model for Coulomb collision of electrons in a background of stationary ions modified by Spitzer's correction factor $\gamma_E(Z)$ for electron-electron collisions, the collision frequency ν for electron is

$$\nu = \frac{\omega_p^{\,2}\, \pi^{3/2}\, m_e^{\,1/2}\, Ze^2}{8\pi\gamma_E(Z)(2kT)^{3/2}} \ln\Lambda\,; \quad \Lambda = \frac{3}{2Ze^2}\left(\frac{k^3T^3}{\pi n_e}\right)^{1/2} \tag{4.11}$$

where $\ln\Lambda$ is the Coulomb logarithm. With T in eV and n_e in cm^{-3} we find

$$\nu = 1.483 \times 10^{-6}\, \frac{n_e}{T^{3/2}}\, \ln\left(1.555 \times 10^{10}\, \frac{T^{3/2}}{n_e^{\,1/2}}\right) \tag{4.12}$$

The value of Λ is the ratio of the Debye length λ_D (Eq. 4.5) to the impact parameter (defined by the maximum distance of the colliding ion from the path of the incident electron, which still gives a collision). The theory is limited to $\ln\Lambda$ larger than 2. This limits the electron densities n_e (in cm^{-3}) and the plasma temperatures T (in eV) to

$$n_e < 4.46 \times 10^{18}\, T^3 \tag{4.13}$$

A further restriction is that classical statistics is applicable, an assumption for the whole theory treated here as is usually the case in plasma physics. This is certainly the case if the electron temperatures are greater than ten times the Fermi temperature ξ_0

$$T > 10\,\zeta_o = 10\,\frac{h^3}{2m_e}\left(\frac{3n_e}{8\pi}\right)^{2/3} = 3.65 \times 10^{-14}\, n_e^{\,2/3} \tag{4.14}$$

where T is in eV, n_e in cm^{-3} and h is Planck's constant. For plasmas with solid state density this restriction gives plasma temperatures exceeding 5 eV, a limitation also for fully-ionized plasmas because recombination is important at lower temperatures at such high densities.

Using the collision frequency ν one can define the classical mean free path ℓ of electrons and ions neglecting in this estimate Spitzer's factor γ_E, which is approximately 0.5 in the most extreme case, and find from (Eq. 4.12)

$$\ell = \frac{v_e^0}{\nu} = \sqrt{\frac{2\,kT}{m_e}}\,/\,\nu \qquad (4.15)$$

This length is in most cases more than ten or one hundred times the mean ion distance. In laser-produced plasmas the desired mean free path is sometimes less than the laser wavelength to obtain special physical properties.

Besides this limitation given by quantum statistics, the conditions of classical statistics also give rise to a pronounced quantum mechanical property of the plasma which involves the plasmons. As mentioned in the discussion of the preceding equation, the mean free path in plasmas can be several times the ion separation, similar to electron motion in metals. If, however, the electron has energy ε_e

$$\varepsilon_e = \hbar\omega_p m \qquad (4.16)$$

where m is an integer, the plasma frequency is given by the electron density n_e (Eq. 4.10), and the quantized plasmon energy is $\hbar\omega_p$ with $\hbar = h/2\pi$, then a strong interaction occurs with the plasma oscillations of the electrons. The direct measurements in metals of the exit depth of photoelectrically-excited electrons show that the usual mean free path of some hundred ion distances is reduced to two or less if the electrons are emitted by photoelectric excitation with energies ε_e given by Eq. (4.16). Therefore, it can be expected in homogeneous plasmas at thermodynamic equilibrium that the well-known Maxwellian energy distribution (Fig. 4.3) will be modified by minima separated by the plasmon energy $\hbar\omega_p$. This modification of the Maxwellian energy distribution in a homogeneous and quiescent plasma at thermodynamic equilibrium has been demonstrated in the spectra of Thomson scattering from arcs.[115] The fact that there are energy differences measured corresponding to half of the plasmon energy is not in contradiction to the given explanation; this is related to the fact that only one-half of the oscillation energy of the electrons is effective as averaged kinetic energy of the electrons in the well-known formula for Thomson scattering.

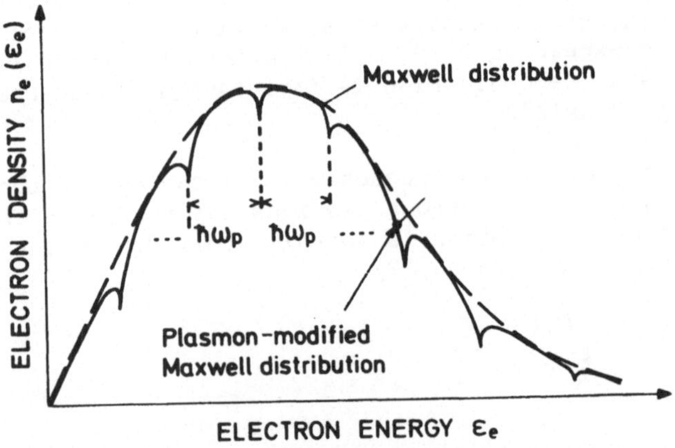

Fig. 4.3 The Maxwellian distribution will be modified due to strong interactions of electrons of an energy $\varepsilon_e = \hbar\omega_p m$ with plasmons.

The few basic properties of plasma in equilibrium which have been mentioned indicate a rather complicated state of matter and hint at highly-complex properties for inhomogeneous plasmas or for the nonequilibrium state. The mechanisms occurring in the latter conditions include several well-known types of micro-instabilities. Such mechanisms appear automatically if the plasma is described numerically as an N-particle problem, with N quite large. The behavior of the plasma under incident laser radiation in particular can be studied by this method of direct plasma simulation. In nearly all cases the necessarily long time intervals in the simulation require a description of collisions in some way other than simple Coulomb collisions, since for the latter very narrow time steps compared with the collision time are necessary. Even macroscopic effects can be demonstrated; examples of this are the increase of the electromagnetic energy density,[116] known from the macroscopic theory of the nonlinear force;[83] the generation of fast electrons and their interaction and heating of more dense regions; and the growth rate and saturation of parametric instabilities.[117]

5. MACROSCOPIC PLASMA PHYSICS

The macroscopic theory describes a plasma as a continuum with time-dependent spatially-varying functions of the density n(r,t), velocity v(r,t), temperature T(r,t), and energy exchange of the plasma. The plasma can be composed of two continuous fluids for the electrons (e) and for the ions (i), each of which follows Euler's equation of motion

$$n_i m_i \frac{dv_i}{dt} = eE + \frac{e}{c} v_i \times \mathbf{H} + n_e m_e \nu (v_i - v_e) + \nabla n_i k T_i + K_i \tag{5.1}$$

$$n_e m_e \frac{dv_e}{dt} = -eE - \frac{e}{c} v_e \times \mathbf{H} - n_e m_e \nu (v_i - v_e) + \nabla n_e k T_e + K_e \tag{5.2}$$

where equal particle densities for electrons and ions $n_i = n_e$ will be assumed in most cases. K_i and K_e represent external forces such as gravitation, and ∇nkT pressure gradients. v_i and v_e represent the continuous velocity fields of the two fluids, and the third terms on the right-hand side of the equations represent the interaction between the fluids (friction) given by the collision frequency ν, which is assumed to be known from microscopic theories.

5.1 Ohm's Law and Electromagnetic Waves

As has been shown by Schlüter,[118] subtraction of the equations results in a "diffusion equation" or an Ohm's law for plasmas, in which the difference of the velocities v_i and v_e represents the currents in the plasmas, described by the current density j. With the appropriate definition of j and using the gas-dynamic pressure

$p_e = n_e k T_e$,

$$\frac{m_e}{e^2 n_e}\left(\frac{\partial j}{\partial t} + \nu j\right) = E + v \times H + \frac{c}{en_e}\, j \times H + \frac{c}{en_e}\nabla p_e . \qquad (5.3)$$

The net plasma velocity v is given by addition of the velocities v_i and v_e. On the right-hand side of Eq. (5.3), the second term is the Lorentz term and the third term is the Hall term. In taking the differences only the electron pressure remains as a result of the much larger ion mass compared to the electron mass.

Equation (5.3) -- the difference between the Euler equations (5.1) and (5.2) for the two fluids -- can be used very simply to describe electromagnetic waves in a plasma. Neglecting all nonlinear terms and the electron pressure, and using the definition of the plasma frequencies ω_p, Eq. (4.10), we find

$$4\pi\left(\frac{\partial j}{\partial t} + \nu j\right) = \omega_p^2\, E \qquad (5.4)$$

If j and E exhibit a periodic time dependence of frequency ω,

$$j = j_0 \exp(-i\omega t + i\varphi);\quad E = E_0 \exp(-i\omega t) \qquad (5.5)$$

and

$$j = \frac{i}{\omega}\frac{\partial j}{\partial t};\quad E = -\frac{1}{\omega^2}\frac{\partial^2 E}{\partial t^2} . \qquad (5.6)$$

With the imaginary unit i, Eq. (5.4) becomes

$$4\pi\frac{\partial j}{\partial t} = \frac{\omega_p^2}{1+i\frac{\nu}{\omega}} E = -\frac{\omega_p^2}{\omega^2(1+i\frac{\nu}{\omega})}\frac{\partial^2 E}{\partial t^2} \qquad (5.7)$$

The phase φ in Eq. (5.5) has been discussed before;[119] it is neglected here. The Maxwell equations

$$\nabla \times E = -\frac{1}{c}\frac{\partial H}{\partial t} \qquad (5.8)$$

$$\nabla \times H = \frac{4\pi}{c} j + \frac{1}{c}\frac{\partial E}{\partial t} \qquad (5.9)$$

when reduced by differentiation and substitution in the wave equation for E

$$\Delta E + \frac{1}{c^2}\frac{\partial^2}{\partial t^2} E = -\frac{4\pi}{c^2}\frac{\partial j}{\partial t} \qquad (5.10)$$

can be rewritten using the result of Eq. (5.6)

$$\Delta E + \frac{1}{c^2}\left(1 - \frac{\omega_p^2}{\omega^2(1 + i\frac{\nu}{\omega})}\right)\frac{\partial^2}{\partial t^2} E = 0 \tag{5.11}$$

Here we have a wave equation for monochromatic fields E (Eq. (5.6)) and induced plasma current densities j, where the dispersion results in the complex refractive index $\tilde{n}$

$$\tilde{n}^2 = 1 - \frac{\omega_p^2}{\omega^2(1+i\frac{\nu}{\omega})} , \tag{5.12}$$

for which the letter η is used sometimes to avoid a confusion with n_i and n_e, the particle densities of plasmas.

Here the meaning of the plasma frequency can be seen immediately for a collisionless plasma ($\nu = 0$), where the refractive index vanishes for a frequency ω of the electromagnetic wave equal to the plasma frequency ω_p. The plasma is transparent only if $\omega > \omega_p$; otherwise the incident electromagnetic wave is totally reflected from the plasma. The frequency $\omega = \omega_p$ is called the critical frequency or the cut-off frequency and is related by Eq. (5.10) to a special electron density of the plasma n_{ec}, called critical or cut-off density. The cut-off densities for different lasers (in parentheses) are (in cm^{-3}):

$$n_{ec} = 2.3 \times 10^{21}(\text{ruby});\ 10^{21}(\text{Nd-glass});\ 10^{19}(CO_2);\ 3.8 \times 10^{22}(Xe_2^*) \tag{5.12a}$$

If the density of the plasma is higher than the cut-off density -- the solid state density of deuterium is 5.8×10^{22} cm^{-3} -- the plasma should then be totally reflecting like metals. As demonstrated in Fig. 1.1, however, the reflectivity is less than 1%. Only with laser intensities at or below the Linlor effect (10^9 W/cm^2 for ruby or neodymium-glass lasers) is the reflectivity of plasmas unusually high.

5.2 Equation of Motion and Equations of Conservation

The addition of the Euler equations (5.1) and (5.2) for the two fluids of the plasma, the electrons, and the ions, results in an equation of motion in terms of the velocity v obtained by adding v_1 and v_2. In Schlüter's original treatment the result was

$$m_i n_i \frac{dv}{dt} = -\nabla_p + \frac{1}{c}\, j \times H + j \cdot \nabla \frac{1}{\omega^2}\frac{\partial E}{\partial t} \tag{5.13}$$

The left-hand side of this equation is the force density f. If j and E oscillate with the frequency ω (using the real parts of Eq. (5.6)), the last term on the right-hand side is nonlinear. It should be mentioned that the derivation of the equation of motion by Spitzer[120] from microscopic theory using a kinetic equation does not give this nonlinear term. The assumption of space-charge neutrality when the lengths considered are larger than the Debye length has to be modified for the space charge of high-frequency oscillations. Such waves must be expected in a plasma from obliqe-ly-incident electromagnetic waves with a polarization of the E-vector parallel to the plane of incidence,[83] which generate a longitudinal component of the E-field. With this additional property, the more general equation of motion is given, expressing the plasma pressure p by $2n_i kT$,

$$\mathbf{f} = m_i n_i\left(\frac{\partial \mathbf{v}}{\partial t} + \mathbf{v}\cdot\nabla\mathbf{v}\right) = -\nabla 2n_i kT + \frac{1}{c}\mathbf{j}\times\mathbf{H} - \frac{1}{4\pi}\nabla\cdot\mathbf{EE}\,(\tilde{n}^2 - 1) \tag{5.14}$$

A formal transformation of this equation can be made by including the Maxwell equations and taking always the real parts of the quantities, E, H, γ whereby,

$$\mathbf{f} = -\nabla p + \nabla\cdot\left[\mathbf{U} + \frac{1}{4\pi}(\tilde{n}^2 - 1)\mathbf{EE}\right] - \frac{\partial}{\partial t}\frac{\mathbf{E}\times\mathbf{H}}{4\pi c} \tag{5.14a}$$

where U is the Maxwellian stress tensor

$$4\pi U_{rs} = \begin{bmatrix} \frac{1}{2}(E_x^2-E_y^2-E_z^2+H_x^2-H_y^2-H_z^2) & E_xE_y + H_xH_y & E_xE_z + H_xH_z \\ E_xE_y + H_xH_y & \frac{1}{2}(-E_x^2+E_y^2-E_x^2-H_x^2+H_y^2+H_z^2) & E_yE_z + H_yH_z \\ E_xE_y + H_xH_z & E_yE_z + H_yH_z & \frac{1}{2}(-E_x^2-E_y^2+E_z^2-H_x^2-H_y^2+H_z^2) \end{bmatrix}$$

From this formulation of the equation of motion one can immediately see a condition for equilibrium of a plasma (f = 0) at E = 0

$$\nabla(p - H^2/(8\pi)) = 0 \tag{5.14b}$$

in which the gas-dynamic pressure p is compensated by a magnetic field H. This is a basic relation for confining plasmas by magnetic fields.

In addition to the equation of motion, the state of a plasma is also determined by the equation of continuity

$$\frac{\partial n_i}{\partial t} + \nabla \cdot (n_i v) = 0 \tag{5.15}$$

and by the equation of conservation of energy

$$\frac{\partial}{\partial t}\left(\frac{n_i m_i}{2} v^2 + 2n_i kT\right) = -\chi \Delta T + W \tag{5.16}$$

in which the change of kinetic energy and of the internal energy is compensated by losses due to thermal conductivity with coefficient χ and the power W of emitted or absorbed energy. Equations (5.14) through (5.16) can be considered as the gas-dynamic equations of conservation of momentum (5.14), matter (5.15), and energy (5.16). The mathematical task is defined: The five plasma quantities $n_i = n_e$, T and the three components of v depending on r and t are determined by the five differential equations from (5.14) to (5.16). There are certain initial or boundary conditions if j, H, and W are given externally. Alternatively, the solution may utilize involved functions of the plasma parameters and further initial or boundary conditions. If the electrons and ions are treated separately, the quantities n_i, n_e, T_i, T_e, v_i, v_e are determined by the six equations of motion (5.1) and (5.2), two equations of continuity for electrons and ions (5.15), one equation of energy (5.10), and the Poisson equation (4.7).

The numerical solution of the gas-dynamic equations of conservation uses the properties of generated waves in an implicit way. The behavior of electromagnetic waves is determined by external, incident waves. The propagation follows according to the actual refractive index n (Eq. (5.12)) in the plasma, given by density n_i and temperature, as far as ω_p and ν are functions of these quantities (see the next Section). The generation of electron-plasma waves or ion-plasma waves is either given implicitly by the propagation of the electromagnetic waves or their generation as parametric Raman instabilities,[121,122] are treated separately. The waves generated by variations of the pressure, acoustic waves, are described implicitly by the gas-dynamic motion. The Alfven waves (hydromagnetic waves) are of importance only if constant magnetic fields are present. These have been neglected in most laser-produced plasmas but may become more important since the generation of spontaneous magnetic fields has been observed under nonuniform irradiation.[123,124] The Alfven waves result from a motion of the plasma perpendicular to a static magnetic field H_0 generating a current j perpendicular to both. Using the appropriate coordinates, the equation of motion (5.14) is

$$n_i m_i \frac{\partial v_y}{\partial t} = j_z H_0 \quad \rightarrow \frac{\partial}{\partial t} j_z = \frac{m_i n_i}{B_0} \frac{\partial^2 v_y}{\partial t^2} \tag{5.17}$$

and Ohm's law -- Eq. (5.3) -- neglecting the fast oscillations becomes

$$E_z - v_y B_0 = 0 \rightarrow \frac{\partial^2}{\partial t^2} v_y = \frac{1}{B_0} \frac{\partial^2 E_z}{\partial t^2} . \tag{5.18}$$

Using the two last equations we find the wave equation (5.10)

$$\Delta E_z = \frac{1}{c^2} \ddot{E}_z + 4\pi \frac{\partial j_z}{\partial t} = \frac{1}{c^2} \ddot{E}_z \left(1 + 4\pi \frac{m_i n_i c^2}{B_0^2}\right) \tag{5.19}$$

from which velocity of propagation of the Alfven waves v_A

$$v_A = c \,/\, (1 + 4\pi m_i n_i c^2 \,/\, B_0^2)^{1/2} \tag{5.20}$$

obtains directly. For densities of laser-produced plasmas of n_i exceeding 10^{18} cm^{-3}, the Alfven velocity v_A is in most cases negligibly small, if B_0 is less than 1 megagauss.

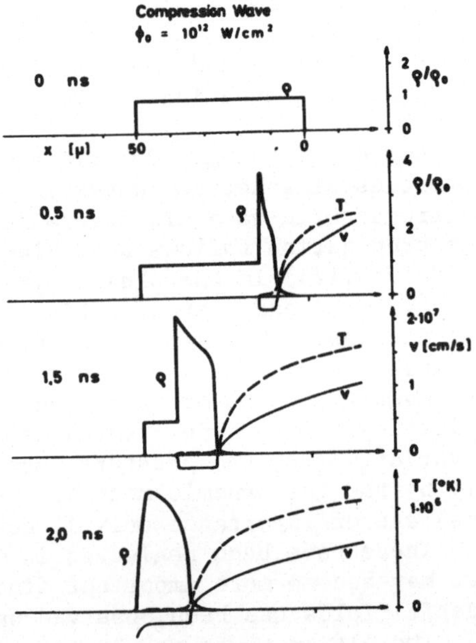

Fig. 5.1 One-dimensional numerical solution[125] of the hydrodynamic equations of conservation for laser light with a steplike intensity of 10^{12} W/cm^2 incident on a 50-micron-thick slab of solid hydrogen (density ρ_0). The resulting density $\rho = n_i m_i$, velocity v, and temperature T are shown for times t = 0, 0.5, 1.5, and 2.0 nanoseconds.

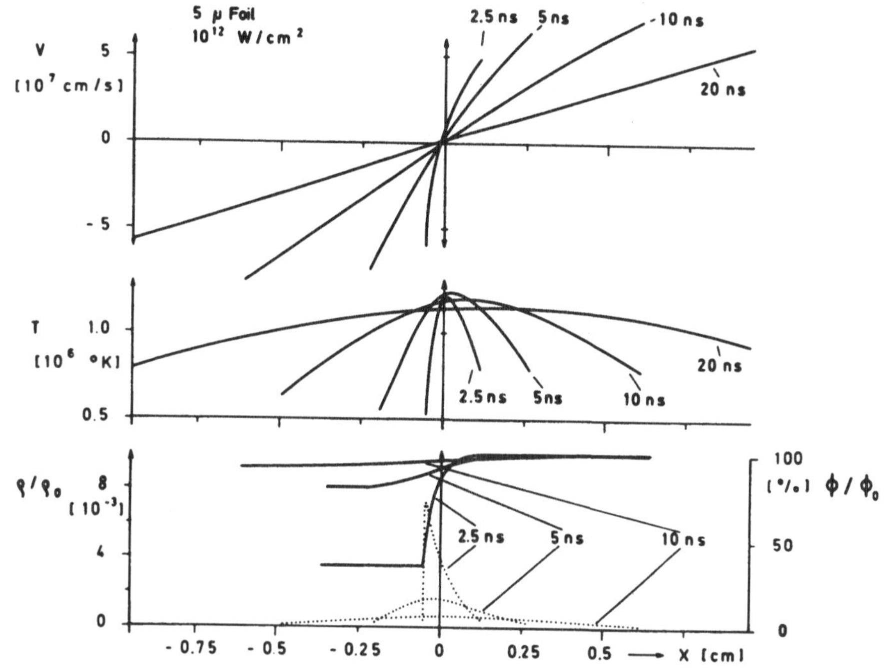

Fig. 5.2 One-dimensional numerical solution[125] for a foil of 5-micron thickness with linear velocity profile and Gaussian density profile at 5 nsec and later as found for irradiated spheres.[129]

A numerical solution of the equations of conservation has been treated by Mulser[125] on the basis of a single plasma velocity v based on a Lagrangian code. The results agree with those from an Eulerian code.[126] The treatment of electrons and ions as separate fluids was considered later by several authors.[127--130] The discussion of a one-fluid model of plasma demonstrates the basic behavior in a very clear way. If the irradiated slab of solid hydrogen is thick enough and if a classical value of thermal conductivity can be assumed for the highly-dense, low-temperature plasma generated, gas-dynamic ablation of the plasma surface causes a recoil to the interior where a shock-like compression of the plasma is produced (Fig. 5.1). If, however, the irradiated target is a very thin foil or if it has very high thermal conductivity, nearly homogeneous heating of the plasma occurs, resulting in a nearly Gaussian density profile and a linear velocity profile after some time (Fig. 5.2).

The shock-like interaction can also be described analytically on the basis of similarity laws, as shown by Krokhin and Afanas'yev,[131] Caruso and Gratton,[132] and as derived generally by Perth.[133]

5.3 Homogeneous Heating

The homogeneous heating of plasma can also be treated analytically (self-similarity model) assuming a sphere of plasma with homogeneous transfer of energy from the incident laser radiation. As derived[134] from the hydrodynamic equations of conservation, from (5.14) to (5.16), the expansion of such a sphere can be related to the analytic solution of simplified equations for a sphere with radius R and an averaged mass M.[134,135,136] The power supplied by the incident energy is given by W, and the temperature T is assumed constant within the whole sphere varying only with time t. For some constant value of $W = W_1$ and $g = W_1 t^3/N_1 m_1$

$$R^2 = (R_0^2 + 10\ g/9) \tag{5.21}$$

and

$$kT = \frac{W_1 t}{6N_i} \frac{2R_0^2 + 5g/9}{R_0^2 + 10g/9} \tag{5.22}$$

where N_i is the total number of ions in the sphere with an initial radius R_0. Taking into account the change of the cross section of the irradiated sphere in a laser beam of constant focal properties, Eq. (5.21) has to be corrected by a first approximation

$$R^2 = R_0^2 + \frac{10g}{9} [1 + G/(18\ R_0^2)] \tag{5.23}$$

as well as an appropriate expression for the time dependence of the plasma temperature. Following this model, Fig. 5.3 describes the laser energies and pulse lengths τ^* for neodymium-glass and CO_2 lasers, if a sphere of solid deuterium of a given initial radius is heated up to reach an averaged ion energy ε_0.

There has been extensive discussion of whether the compression-wave model or the homogeneous-heating self-similarity model better describes the experimental results. One criterion is the time at which the expanding plasma becomes transparent. The time calculated for penetration of the compression front through an irradiated solid target of a given thickness (or a sphere of a given diameter) is larger by factors of 20 to 100 than the time calculated for homogeneous heating to reach the cut-off density (transparency).[125,138] In all experiments on irradiation of spherical pellets of LiD,[139--142] aluminum[109] or other materials with diameters from 10 to 500 microns by ruby or neodymium-glass lasers at intensities between 10^{10} and 10^{13} W/cm^2 and pulse lengths of 5 to 30 nsec, the measurements of expansion velocity, transparency time, and energy transfer were in full agreement with the theory of

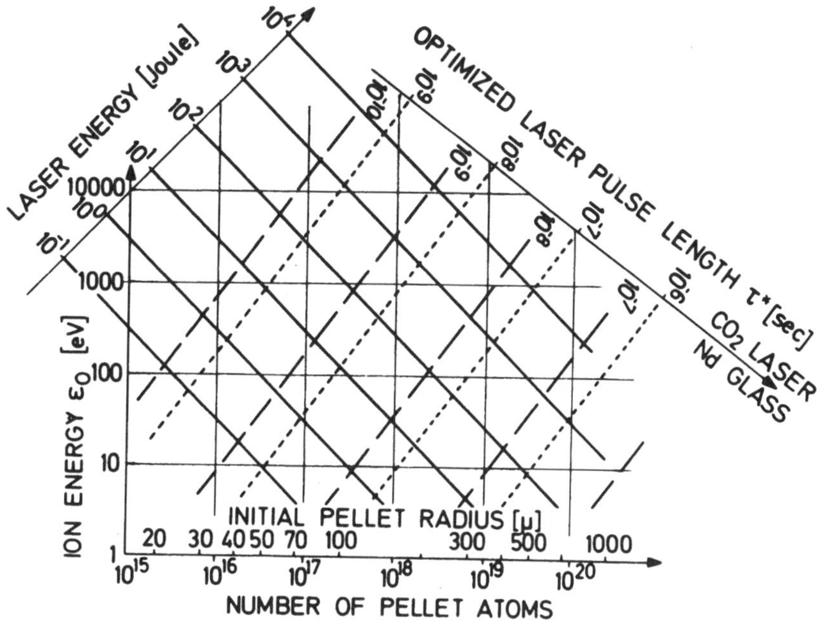

Fig. 5.3 Energy and pulse length τ* of neodymium-glass and CO_2 laser radiation for heating a solid deuterium pellet of given initial radius to averaged ion energies ε_0, derived from the self-similarity model of homogeneous heating.[137]

homogeneous heating. The transparency time of hydrogen foils[143] and, similarly, polyethylene foils[144] fully agree with the homogeneous heating model, while the compression-wave model gave thirty-fold longer times.[125] The compression-wave model based on one-dimensional interaction cannot be modified for lateral expansion mechanisms, because in most cases considered, the thickness of the foil is much less than the beam diameter. The observation of a small delay in plasma generated at the rear side of the foil compared to the front side does not contradict the homogeneous heating model. In some experiments with the same apparatus in which self-focusing in a transparent target was measured, the plasma at the rear side radiated earlier than at the front side.[145]

In the case of CO_2 laser-irradiated targets, some observations have been explained by the homogeneous heating model[146] and others by the compression-wave model.[133,147] However, the most critical quantity, the transparency time, cannot be used for a quantitative comparison,[147] because the power of the incident laser pulse varied too much during the time of passage of the light through the plasma.

The distinction between the two theories is not only an academic question of studying gas-dynamic models for laser-produced plasmas; the generation of a compression wave is an essential mechanism for reaching nuclear fusion with acceptable efficiencies. Thus, in the numerous experiments with neodymium-glass lasers at intensities from 10^{10} to 10^{13} W/cm^2 irradiating solids in vacuum, it is important to determine whether: (1) self-focusing occurred and a complicated nonhomogeneous transfer of energy to the plasma produced conditions for homogeneous heating without generation of compression fronts or (2) the compression-front model should be modified by including anomalous conditions of thermal conductivity in the compressed plasma to reproduce the very short transparency times which have been observed.

6. REFRACTIVE INDEX AND ABSORPTION

As derived from the macroscopic theory of a plasma, the complex optical refractive index $\tilde{n}$ is given by the dispersion relation of electromagnetic waves in a plasma, (Eq. (5.12)), where the real part n, and the imaginary part κ, are evaluated algebraically

$$\tilde{n} = n + i\kappa = \left(1 - \frac{\omega_p^2}{\omega^2(1 + i\nu/\omega)}\right)^{1/2} \tag{6.1}$$

$$n = \left[\frac{1}{2}\left\{\left[\left(1 - \frac{\omega_p^2}{\omega^2 + \nu^2}\right)^2 + \left(\frac{\nu}{\omega}\,\frac{\omega_p^2}{\omega^2 + \nu^2}\right)^2\right]^{1/2} + \left(1 - \frac{\omega_p^2}{\omega^2 + \nu^2}\right)\right\}\right]^{1/2} \tag{6.2}$$

$$\kappa = \left[\frac{1}{2}\left\{\left[\left(1 - \frac{\omega_p^2}{\omega^2 + \nu^2}\right) + \left(\frac{\nu}{\omega}\,\frac{\omega_p^2}{\omega^2 + \nu^2}\right)^2\right]^{1/2} - \left(1 - \frac{\omega_p^2}{\omega^2 + \nu^2}\right)\right\}\right]^{1/2} \tag{6.3}$$

The real part n is sometimes called the refractive index. Here the sum with the complex refractive index is denoted by the circumflex $\tilde{n}$. For a collisionless plasma ($\nu = 0$), both coefficients are clearly equivalent

$$n = \tilde{n} = (1 - \omega_p^2 / \omega^2)^{1/2} \quad (\text{if } \nu = 0) \tag{6.4}$$

The imaginary part of $\tilde{n}$, κ, is called the absorption coefficient. Its meaning is seen immediately from its relation to the absorption constant K, which determines the attenuation of a laser intensity I at some depth x; if I_0 is the intensity at x = 0

$$I = I_0 \exp(-Kx) \tag{6.5}$$

The absorption constant is then

$$K = \frac{2\omega}{c} \kappa \tag{6.6}$$

6.1 Linear Properties

As demonstrated by the preceding equations, the optical properties of a plasma depend on the plasma frequency ω_p (Eq. (4.10)) and, therefore, on the electron density n_e and the electron mass m_e (to the extent it can be changed relativistically), and they depend further on the collision frequency ν. Here it is important to recognize in what sense the collision frequency was defined. In Eqs. (5.1) and (5.2) the collision frequency ν was defined by changes in the motion of the electrons and ions due to Coulomb interaction in which energy is exchanged. This kind of collision leading to equipartition characterizes the friction of plasmas, and the thermal conductivity. It also characterizes the exchange of energy within one component (e.g., in the electrons if for some reason the velocity distribution is non-Maxwellian as shown in Fig. 4.3), or between electrons and ions if there is no thermal equilibrium for some reason, e.g., if the electrons are heated only by incident laser radiation. There is a question whether this collision frequency is to be identified with the high-frequency processes of the refractive index $\tilde{n}$. The formal derivation of $\tilde{n}$ from the difference of Euler's equations for the two fluids should prove the identity of the two definitions. However, there is some microscopic difference between these mechanisms. If the colliding particles were neutral molecules, their interaction would constitute simple two-particle interactions, and there would be no doubt of the identity. In the case of the charged plasma particles, the collision of an electron with an ion can cause in addition emission of an electromagnetic wave (photon), the well-known bremsstrahlung radiation.

As has been shown before,[148] the collisions for equipartition are indeed very close to the collisions of the bremsstrahlung mechanism. The absorption of radiation in a plasma is a mechanism of inverse bremsstrahlung: If an electron moves around the colliding particle within a Coulomb orbit, it can pick up energy from the high-frequency electromagnetic field which causes its oscillation.

Speaking in terms of quantum mechanics, there is the motion of an electron within a Coulomb field of an ion with continuous energy eigenvalues during which the electron changes to a higher energy state by absorbing the energy of one photon. The exact quantum mechanical description[149] results in an absorption constant (index B for bremsstrahlung)

$$K_B = \frac{64\pi^2}{3\sqrt{3}} \frac{Z^2 n_e n_i e^6 g(T)}{c\omega^2 (2\pi m_e kT)^{3/2}} \tag{6.7}$$

where T is the plasma temperature and g(T) is the "Gaunt factor,"[150] a value which corrects the point-mechanical description by a factor between 0.1 and 10. The justification for comparison of the quantum mechanical process of inverse bremsstrahlung and that defined by the collisions is given by the ratio of K_B to a value K derived from Eq. (6.5) using the plasma collision frequency ν (Eq. (4.11)),

$$\frac{K}{K_B} = 0.324 \frac{\ln \Lambda}{\gamma_E(Z) g} \tag{6.8}$$

in which the validity of K in

$$K = \frac{4\pi^2}{\gamma_E(Z)} \frac{Z^2 n_e n_i e^6 \ln \Lambda}{c(\omega^2 + \nu^2)(2\pi m_e kT)^{3/2} \left(1 - \frac{\omega_p^2}{\omega^2 + \nu^2}\right)^{1/2}} \tag{6.9}$$

is restricted to plasma densities below the cut-off density, $\omega \gg \omega_p$, $n_e \gg n_{e0}$ (Eq. (5.12a)) and to relatively-low collision frequency $\nu \ll \omega$. Using a special representation of the Gaunt factor[151] whose validity is limited in a similar way, with T in eV and n_e in cm^{-3}

$$g = 1.2695 \left(7.45 + \log T - \frac{1}{3} \log n_e\right) \tag{6.10}$$

and using the value of the Coulomb logarithm

$$\ln \Lambda = 3.45 \left(6.692 + \log T - \frac{1}{3} \log n_e - \frac{2}{3} \log Z\right) \tag{6.11}$$

we arrive at

$$\frac{K}{K_B} = \frac{0.883}{\gamma_E(Z)} \quad 1 - \frac{0.76 + \frac{2}{3}\log Z}{7.45 + \log T - \frac{1}{3}\log n_e} \tag{6.12}$$

The above equation shows once more the comparability of the two theories. There has been some discussion of whether Spitzer's correction $\gamma_E(Z)$ of the electron-electron collisions must be included. Although this correction is small ($\gamma_E(Z)$ varies from 0.582 to 1 when Z changes from 1 to ∞), it should be pointed out that electron-electron collisions can be accompanied by emission and absorption of photons. If the thermal energy kT is larger than the oscillation energy ε_α of the electrons, the coherent oscillation of two colliding electrons in the high-frequency field may be the source of another correction factor between the minimum value and unity. It should be mentioned that another theory for the absorption of laser radiation in a plasma exists, derived by Dawson and Oberman[153] from the description of the plasma by the collisionless Vlasov equation, including the interaction of the particles by phase mixing. The justification of this theory[148] was demonstrated by comparison with the quantum mechanical theory in the same way as shown previously for the theory of absorption due to Coulomb collisions. For the absorption constant of Dawson and Oberman K_{DO}, the ratio

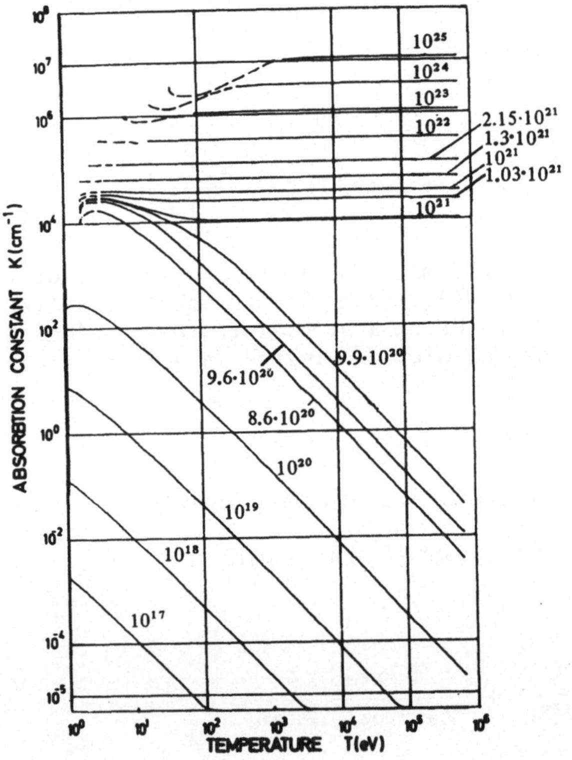

Fig. 6.1 Absorption constant $K(cm^{-1})$ from Eqs. (6.3) and (6.6) with Coulomb collision for neodymium-glass laser radiation in plasma as a function of the temperature T(eV) and density $n_e(cm^{-3})$.[148]

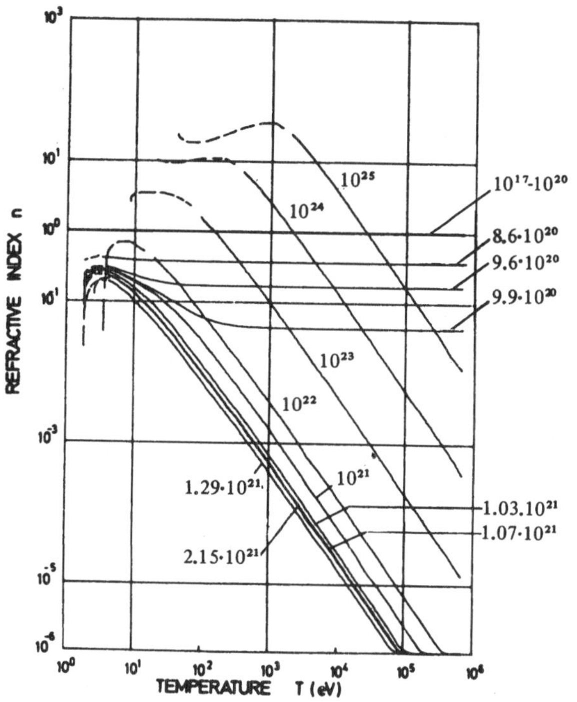

Fig. 6.2 Variation of refractive index n (real part) for neodymium-glass laser radiation in plasma with temperature T(eV) and density n (cm^{-3}).[148]

$$\frac{K_{DO}}{K_B} = 0.276 \frac{\ln \Lambda}{g} \quad (\omega \gg \omega_p ; \; \nu \ll \omega) \tag{6.13}$$

is quite close to unity. Comparing this result with Eq. (6.8), we find that the absorption theory based on Coulomb collisions is closer to the inverse bremsstrahlung theory (e.g., for deuterium by a factor of 2.02).

The quantum mechanical absorption constants may be considered as the most probable values. The disadvantage of the theory is its limitation to plasmas of low density and high temperature (low collision frequency), while the laser-plasma interaction is important for plasmas near the cut-off and higher densities. For these cases the refractive index n (real part) and the absorption constant K have been evaluated numerically at several interesting wavelengths (ruby, neodymium-glass, and CO_2 lasers and their second harmonics[148]). Figures 6.1 and 6.2 give examples for neodymium-glass lasers and Figs. 6.3 and 6.4 for CO_2 lasers.

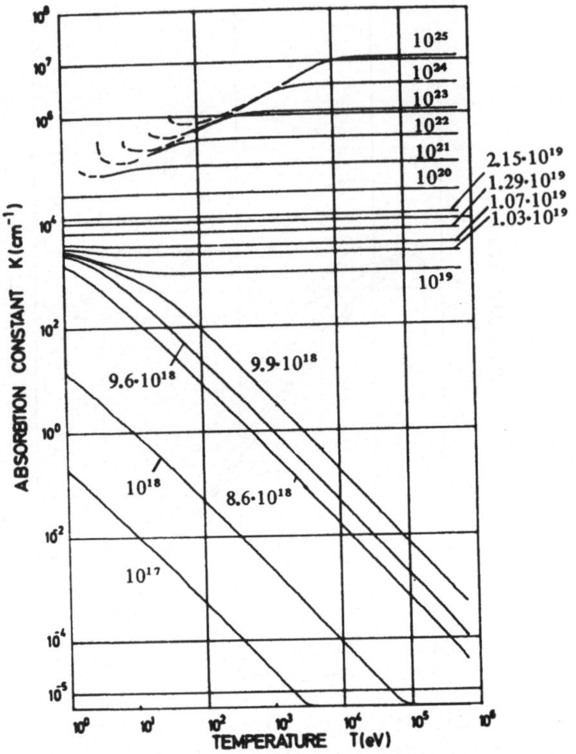

Fig. 6.3 Dependence of absorption constant $K(cm^{-1})$ for CO_2 laser radiation in plasma on temperature T(eV) and density n_e (cm^{-3}).[148]

For some purposes it is important to evaluate the absolute minimum value of the refractive index $|\tilde{n}|$, which can be much less than unity in a hot plasma near the cut-off density. From the exact value

$$|\tilde{n}| = \left[\left(1 - \frac{\omega_p^2}{\omega^2+\nu^2}\right)^2 + \left(\frac{\nu}{\omega}\,\frac{\omega_p^2}{\omega^2+\nu^2}\right)^2\right]^{1/4} \tag{6.14}$$

we find the minimum

$$|\tilde{n}|_{min} = (\nu/\omega)^{1/2} = a\,/\,T^{3/4} \quad (\text{at } \omega_p^2 = \omega^2 + \nu^2) \tag{6.15}$$

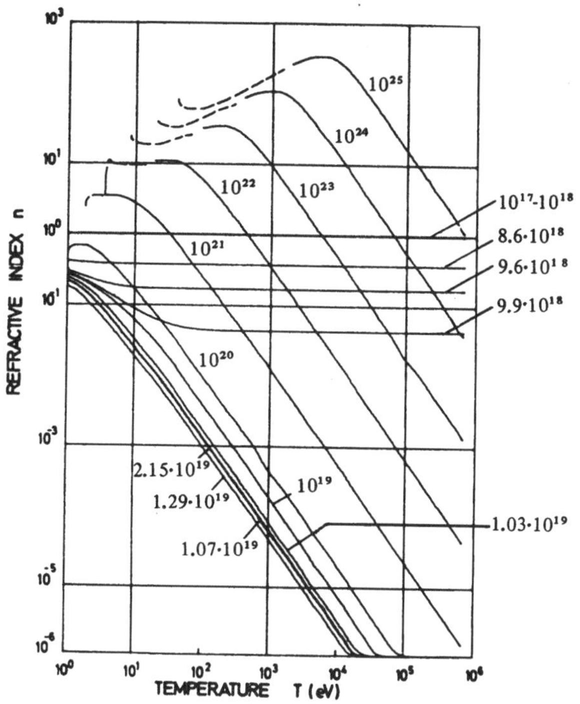

Fig. 6.4 Refractive index (real part) n for CO_2 laser radiation as a function of plasma temperature T(eV) and density n_e (cm^{-3}).[148]

where the constant

$$a = \left(\frac{\omega_p \pi^{3/4} m_e^{1/2} Z^2 \ln \Lambda}{8\pi \gamma_E (Z) (2k)^{3/2}} \right)^{1/2} \tag{6.16}$$

$$\begin{cases} 3.25\ (\text{eV})^{3/4} \text{ for neodymium-glass laser} \\ \text{for deuterium } (Z = 1) \text{ and } \ln \Lambda = 10, \\ 1.03\ (\text{eV})^{3/4} \text{ for } CO_2 \text{ lasers} \end{cases}$$

can be derived from Eq. (4.11).

6.2 Nonlinear Absorption

The assumption in the preceding discussion of the optical constants was that the laser intensities did not exceed the thresholds I* for which the energy of oscillation ε_{os} of the electrons in the laser field was larger than the thermal energy kT. For higher intensities the energy of the electrons is determined mainly by the intensity of the light. The electron kinetic energy (which determines the Coulomb collisions and, therefore, the optical constants) causes a nonlinear dependence of the collision frequency on laser intensity.

The energy of electrons is obtained from solution of the equation of motion for one electron (Eq. (4.1)) using only the first term on the right-hand side with E oscillating as given in Eq. (5.5) and taking only the real parts.

$$v_e = -\frac{e}{m}\frac{E_\alpha}{\omega}\sin\omega t \tag{6.17}$$

The oscillation energy ε_{os} is given by the maximum velocity v_e

$$\epsilon_{os} = \frac{e^2}{2m}\frac{E_a^2}{\omega^2} \quad ; \; E_a^2 = E v^2 / |\tilde{n}| \tag{6.18}$$

where the effective amplitude E_α can be expressed in a WKB approximation by the amplitude E_ν in vacuum and the absolute value of the refractive index. The WKB approximation for the solution of the wave equation (5.11)

$$\Delta E + \frac{\tilde{n}^2(x)}{c^2}\frac{\partial^2}{\partial t^2} E = 0 \tag{6.18a}$$

with linear polarization of E in the y-direction (i_2) and the propagation in the x-direction (i_1) becomes

$$E_a = \frac{i_2 E_\nu}{\sqrt{|\tilde{n}|}} \cos\left(\int^x \frac{\omega}{c}\tilde{n}(x)\,dx + \omega t\right) \tag{6.18b}$$

When one of the Maxwell equations (Eq. (5.8)) is included

$$\begin{aligned} H_0 &= -i_3 E_\nu \sqrt{|\tilde{n}|}\cos -\int^x \frac{\omega}{c}\tilde{n}(x)dx + \omega t \\ &+ i_3 \frac{c}{2\omega}\frac{E_\nu}{|\tilde{n}|^{3/2}}\frac{\partial|\tilde{n}|}{\partial x}\sin\left(-\int^x \frac{\omega}{c}\tilde{n}(x)dx + \omega t\right) \end{aligned} \tag{6.18c}$$

The WKB condition of restriction to small changes of $\tilde{n}(x)$ is

$$\Theta = \frac{e}{2\omega}\frac{1}{|\tilde{n}|^2}\frac{\partial|\tilde{n}|}{\partial x} \ll 1 \tag{6.18d}$$

where Θ can reach values as high as 0.3 without serious error in the approximation.

The effective temperature T which determines the optical constant is

$$T = T_{th} + \varepsilon_{os}/k \tag{6.19}$$

where T_{th} is the temperature given by the random motion of the electrons. For $\varepsilon_{os} = kT_{th}$, the threshold I* for neodymium-glass lasers and an electron temperature of 10 eV is 1.3×10^{14} W/cm^2. The relation between the amplitude E_a of the electric field strength and the laser intensity I is

$$E_a\,(\mathrm{V/cm})\sqrt{2I/\Omega_0} = 27.4\sqrt{I(\mathrm{Watts/cm^2})};\ E_a(\mathrm{cgs}) = \sqrt{8\pi I/c} = 2.91 \times 10^{-5}\sqrt{I(\mathrm{erg/cm^2\,sec})} \tag{6.20}$$

where Ω_0 is the resistivity of vacuum, equal to 277 ohms.

The absorption constant for I >> I* has been calculated by quantum mechanics methods from the theory of the inverse bremsstrahlung by Rand.[243] The result was a nonlinear absorption constant K_{NL}^{Rand}

$$K_{NL}^{Rand} = 0.25\pi^{1/2}\, c^{1/2} n_e^2 n_i^2 Z^2 e^3\, \omega m_e^{1/2} \ln\left[\frac{32e^2 I}{m_e k T_{th} c\omega^2}\right] \Big/ I^{3/2} \tag{6.21}$$

Starting from the theory of Coulomb collisions, the nonlinear absorption constant can be derived from Eq. (6.9) with $\nu \ll \omega$, using T from Eq. (6.19) and taking into account Eqs. (6.17) through (6.20),

$$K_{NL} = \frac{4\pi^4 e^6 n_e^2 n_i^2 Z^2 \ln \Lambda}{2^{1/2}\, 16\,(kT + \epsilon_{os})^{3/2}}, \quad \text{where } \nu = \frac{\pi^{1/2} \omega_p^2\, m_e^{1/2} e^2 \ln \Lambda}{2^{1/2}\, 16 (kT + \epsilon_{os})^{3/2}} \tag{6.22}$$

In this case the electron-electron collisions are negligible because the electron gas is oscillating coherently, and the optical constants are determined by electron collisions with ions only. The ratio of the two nonlinear absorption constants

$$\frac{K_{NL}^{Rand}}{K_{NL}} = \frac{4 \ln\left(\dfrac{16\,\epsilon_{os}}{\pi kT}\right)}{(2\pi)^{1/2}\pi \ln \Lambda} \tag{6.23}$$

is a constant which is close to unity within one order of magnitude.

6.3 Relativistic Absorption

The preceding discussion of the linear ($\varepsilon_{os} < kT_{th}$) and the nonlinear ($\varepsilon_{os} > kT_{th}$) absorption constant presumed that the oscillation energy ε_{os} of the electrons was less than the relativistic oscillation energy where ε_{os} exceeds $m_e c^2$. The relativistic thresholds of the intensities I^r for $\varepsilon_{os} = m_e c^2$ are

$$I^r = I_\nu^r / |\tilde{n}| = \frac{1}{|\tilde{n}| \ln \pi}\left(\frac{c\omega m_e}{2e}\right)^2 = \begin{cases} 3.7 \times 10^{18}\ \mathrm{W/cm^2}\ \text{(Nd-glass laser)} \\ 3.7 \times 10^{16}\ \mathrm{W/cm^2}\ (CO_2\ \text{laser}) \end{cases} \tag{6.24}$$

In the relativistic case, the exact solution of the equation of motion for the oscillation[154,155] (neglecting a possible longitudinal motion depending on initial conditions and the switching-on process of the laser field as shown by Nicholson-Florence[156]) results in an energy

$$\epsilon_{os} = m_e c^2 \left[\left(1 + \frac{e^2 E_\nu^2}{m_e^2 \omega^2 c^2 |\tilde{n}|}\right)^{1/2} - 1\right] = \begin{cases} \dfrac{e^2 E_\nu^2}{2 m_e \omega^2 |\tilde{n}|} & \text{if } \epsilon_{os} \ll m_e c^2 \\[2ex] ecE_\nu / (\omega |\tilde{n}|) & \text{if } \epsilon_{os} \gg m_e c^2 \end{cases} \tag{6.25}$$

Due to the relativistic change of the rest mass of the electron, the plasma frequency ω_p is

$$\omega_p^2 = \frac{4\pi e^2 n_e}{m_e} (1 - v^2 / c^2)^{1/2} \tag{6.26}$$

where use can be made of the exact solution of the relativistic equation of motion, yielding

$$(1 - v_{max}^2 / c^2) = 1 \Big/ \left(1 + \frac{e^2 E_\nu^2}{m_e^2 \omega^2 c^2 |\tilde{n}|}\right) \tag{6.27}$$

As a consequence of results given by Eq. (6.25) we find[157] $E_\nu >> E_\nu^r$

$$\omega_p^2 = \frac{4\pi e^2 n_e}{m_e} \frac{m_e c^2}{\epsilon_{os}} = 4\pi e n_e \omega \sqrt{|\tilde{n}|}\, c / E_\nu \tag{6.28}$$

From this the relativistic change of the cut-off density of the electrons n_{ec}^{rel}, presuming that the frequency ω of the incident laser radiation will not change, is

$$n_{ec}^{rel} = \frac{\omega^2 m_e}{4\pi e^2} \frac{\epsilon_{os}}{m_0 c^2}\ n_{ec} \frac{\epsilon_{kin}}{m_0 c^2} \tag{6.29}$$

where n_{ec} is the nonrelativistic cut-off density (see Eq. (5.12)).

The collision frequency ν of the electrons in a plasma at relativistic conditions changes from the value of Eq. (6.22) for nonrelativistic conditions ($I << I^r$)

$$\nu = \frac{\pi^{1/2} e^2 \omega^2 \ln \Lambda}{2^{9/2}\, c\, \epsilon_{os}} \quad \text{for } \epsilon_{os} \geqslant m_0 c^2 \tag{6.30}$$

These relativistic collision frequencies ν are remarkably small. As an example, for $\varepsilon = 1$ MeV, the value of ν is in the order of 100 kHz for CO_2 lasers and 10 MHz for Nd-glass lasers. Therefore, the plasma absorbs less energy at relativistic conditions and becomes more transparent. A similar result of increasing transparency was derived for relativistic instability.[158]

The Coulomb logarithm ln Λ used in Eqs. (6.29) and (6.30) must be corrected relativistically

$$\Lambda = \frac{c}{2e^2 \omega (\pi n_e)^{1/2}} \frac{E_\nu}{\sqrt{|\tilde{n}|}} \tag{6.31}$$

where n_e is always the cut-off value, if $|\tilde{n}|$ is far from unity. Using Eq. (6.28) we find

$$\Lambda = \begin{cases} 1.2 \times 10^9 \dfrac{\epsilon_{os}}{m_e c^2} & \text{for Nd-glass lasers} \\ 1.2 \times 10^7 \dfrac{\epsilon_{os}}{m_e c^2} & \text{for } CO_2 \text{ lasers} \end{cases} \tag{6.32}$$

Now we can evaluate the minimum absolute value of the refractive index $|\tilde{n}|$, (Eqs. (6.14) and (6.15)) for $kT \ll \varepsilon_{os}$ with $\omega_p{}^2 = \omega^2$

$$|\tilde{n}|_{min} = \begin{cases} \dfrac{\pi^{1/4} \omega^{1/2} e \ln \Lambda}{8\sqrt{2}\; c^{1/2}\, \epsilon_{os}^{1/2}} & \text{for } \epsilon_{os}{}^{3/4} \geqslant m_e c^2 \\ \dfrac{\pi^{1/4} \omega^{1/2} m_0{}^{1/4} e \ln \Lambda}{8\sqrt{2}\; \epsilon_{os}^{1/2}} & \text{for } \epsilon_{os} \leqslant m_e c^2 \end{cases} \tag{6.33}$$

The resulting transparency of the plasma at $\varepsilon_{os} > m_0 c^2$, avoiding loss of laser energy by collisions, is also favorable to the dielectric increase of the effective field E_a due to the denominator $\sqrt{|\tilde{n}|}$ in the WKB approximation. As an example, we find when $\varepsilon_{os} = 1$ MeV for CO_2 lasers a value $|\tilde{n}|_{min} = 1.2 \times 10^{-5}$, and for neodymium-glass lasers $|\tilde{n}|_{min} = 4.1 \times 10^{-5}$. Combining this result with Eq. (6.24) relativistic oscillation energies[157] for electrons are expected in plasmas near the cut-off density for neodymium-glass laser intensities of 10^{15} W/cm^2 if a WKB-like density profile can be assumed.

6.4 Anomalous Absorption and Instabilities

Besides the absorption due to Coulomb collisions described in the preceding subsections covering moderate laser intensities (linear case), high intensities causing a predominance of coherent oscillation (nonlinear absorption), and very high oscillation energies exceeding the rest energy of the electrons (relativistic case), an absorption of laser light in the plasma can occur from collective phenomena, instabilities, and nonlinear forces, which all together are often called anomalous absorption.

The first indication of anomalous absorption -- at least for the case of laser frequencies -- was reported by Dawson and Oberman[153] following the numerical evaluation of absorption constants calculated by phase mixing. The variation of K as a function of density displayed mysterious peaks near the cut-off density, and the slope was similar to those of Figs. 6.1 and 6.3. Because of the well-known numerical difficulties[148] which occur near the cut-off density in the region of poles of the complicated functions of K and κ, Eqs. (6.3) and (6.6), further study seemed worthwhile to find out the essential reason for the mysterious peak.[159] It was found, ultimately, that there is a physical basis which may be described as oscillating two-stream instability[160] and which manifests itself by increased reflectivity of high-intensity radiation from the plasma. The observation of anomalously-high reflectivity of microwaves was correlated with this result,[161] similar to microwave measurements in a Q-machine.[162] It was pointed out[163] that the intensity dependence of the reflectivity is in marked disagreement with the theory of anomalous absorption.

Parametric decay instabilities cause a transfer of laser energy into oscillations of electrons (Raman instabilities) characterized by oscillation with the plasma frequency ω_p or $2\omega_p$ of the electrons, or into acoustic oscillations (Brillouin instabilities[164,165]). The calculations for homogeneous plasmas must be distinguished from those for the more complex inhomogeneous plasmas.[166] One goal is to evaluate the threshold for the onset of the instability which varies under different conditions and for the different types of instabilities between 10^{11} and 10^{14} W/cm^2 for neodymium-glass laser radiation. Another aim is to study the (temporal) growth rate of the instabilities which may provide an experimental criterion for validation. The acoustic (Brillouin) instabilities especially have long growth rates which may help to exclude these processes from the interactions of laser radiation of very short pulse length. Another question is the saturation of instabilities;[167,168] above a certain intensity there is no further increase in the instability. Saturation is one of the major points of discussion on whether the instabilities create very high reflectivity of laser-produced plasmas[169,170] and prevent the necessary high input of laser energy into the interior of the plasma for purposes of thermonuclear fusion or whether saturation suppresses instabilities[171] and provides good input of laser energy.

The experimental results indicating relatively-low reflectivity from laser-produced plasmas seem to confirm the predominance of saturation,[23,24] although some experimental facts clearly indicate the mechanism of the instabilities from the spectrum of the reflected light, its shift to half, three-halves and double the frequency, and the spectra[23,172] showing red-shifted maxima which can be explained as stimulated Compton scattering.[172]

In addition to these instabilities, nonlinear forces in laser-produced plasmas (see the following Section) causing a net motion of plasma involve a kind of collisionless nonlinear absorption. Either laser energy may be transferred into mechanical energy of motion, or the ions may be heated adiabatically, preferentially over the electrons (dynamic absorption) or vice versa, as expected at the nodes of the standing waves generated. While the type of absorption could be determined directly from numerical calculations, a more analytic treatment of the instabilities and a systematic description was provided by Chen using nonlinear forces to calculate the microscopic mechanisms.[173]

While in the case of the macroscopic nonlinear forces it was possible to include the resonance-like increase of the mechanism, for the instabilities, the nonlinear feedback of instabilities leading to probable saturation was not included. The resonance mechanism may be the reason for the strong increase of the observed reflectivity of microwaves with intensity. The resulting strong saturation may be the reason for the observation of the unexpected--but for the purpose of the nuclear fusion highly desired--low reflectivity[23,24] of laser-produced plasmas at intensities from 10^{15} to 10^{16} W/cm^2 or more.

7. DIELECTRIC NONLINEAR FORCES AND DYNAMIC ABSORPTION

If an electromagnetic wave propagates into a plasma with a density very close to the cut-off density, the absolute value of the refractive index $|\tilde{n}|$ is much less than unity (Eq. 6.14)). The energy momentum flux density $E^2 + H^2$ is then greater by the factor $\vartheta = |\tilde{n}|^{-1}$, the effective wavelength of the laser light is increased, and the effective laser intensity $I = I_\nu/|\tilde{n}| = \vartheta I_\nu$ increases over its vacuum value I_ν by this factor. Spatial gradients of the time-averaged energy momentum flux density will cause forces corresponding to one hundred times the radiation pressure of the light or more. These forces can cause self-focusing and may be related to the Linlor effect and to the homogeneous heating of a laser-produced plasma. The same forces can, however, in determining the dynamics of the plasma, cause ablation of the plasma surface and result in recoil from compression of the plasma, and can cause disadvantageous energy transfer by dynamic anomalous nonlinear absorption. The nonlinear forces are mainly related to the fact that the swelling factor ϑ is larger than unity, but also arise if the factor is only a small quantity, as seen in dynamic absorption or in Chen's general derivation[173] of instabilities causing anomalous absorption. The essential property is the dielectric change of the refractive index from unity. This is the reason for the more specific expression "dielectric" nonlinear force.

7.1 Basic Properties of the Dielectric Nonlinear Force

As we have shown[83] the force density of the equation of motion (5.14a) can be written

$$\mathbf{f} = - \nabla \cdot \mathbf{p} + \nabla \cdot [\mathbf{U} - \frac{1}{4\pi} (\tilde{n}^2 - 1)\mathbf{EE}] - \frac{\partial}{\partial t} \frac{\mathbf{E} \times \mathbf{H}}{4\pi c} \tag{7.1}$$

where the gas-dynamic part is called the thermokinetic force, using a pressure tensor p

$$\mathbf{f}_{th} = - \nabla \cdot \mathbf{p} \tag{7.2}$$

and the force

$$\mathbf{f}_{NL} = \mathbf{f} - \mathbf{f}_{th} \tag{7.3}$$

is nonlinear owing to the quadratic form of the electromagnetic field components. In the following, the Poynting term can be neglected because of the much slower change of the averaged radiation intensities compared with the laser periods.

Neglecting the effects occurring within a laser period, we find the long-term forces by averaging Eq. (7.1) over the laser period. The nonlinear forces are then essentially governed by gradients of quadratic expressions for components E and H. Some forces of this type are well known. If a standing wave is produced in a plasma of such a density that collective effects are present, the plasma is driven toward the nodes of the standing wave. Another such force was reported by Schlüter[84] for a light beam penetrating a plasma when a certain lateral decrease of the intensity generates lateral gradients of E^2. These forces of static fields are well known as electrostrictive forces. The essential point in this consideration is that the fields are oscillating and, furthermore, that the magnetic field, too, effectively provides "electromagnetostriction." The collective effects, expressed by $|\tilde{n}| - 1$, are particularly important at plasma densities $n_e = n_{ec}$, where the plasma frequency ω_p is equal to the laser frequency.[82,83,174]

Using equations (7.1) and (7.3) to evaluate the time-averaged nonlinear force for a perpendicular incidence of plane waves on a stratified plasma, with the propagation in the x-direction (unit vector i_1) and linear polarization of E in the y-direction, gives[83]

$$\bar{\mathbf{f}}_{NL} = \frac{\mathbf{i}_1}{8\pi} \frac{\partial}{\partial x} (\bar{E}_y^2 + \bar{H}_z^2) \tag{7.4}$$

Taking the special case of reflection-free penetration as given by the WKB solution (Eqs. (6.18a) through (6.18d)) we find

$$f_{NL} = i_1 \frac{E_v^2}{16\pi} \frac{1-|\tilde{n}|^2}{|\tilde{n}|} \exp(-kx) \frac{\partial}{\partial x} |\tilde{n}| + i_1 \frac{E_v^2}{16\pi} \frac{1-|\tilde{n}|^2}{|\tilde{n}|} \frac{2\omega}{c} \mathrm{Im}(\tilde{n}) \exp(-kx) \tag{7.5}$$

with an amplitude of the electric field E_v in the vacuum and integral absorption

$$k(x) = \frac{1}{x} \frac{\omega}{c} \int^{x} \mathrm{Im}(\tilde{n}(\zeta)) d\zeta \tag{7.6}$$

The second term in equation (7.4) is the collision-induced radiation pressure. This term can be neglected compared to the first term of equation (7.5), which describes a nonlinear deconfining force if the electron density n_e exceeds the cut-off density n_{ec}, Eq. (5.12a).

In the case of a collisionless plasma we have

$$f_{NL} = i_1 \frac{E_v^2}{16\pi} \frac{\omega_p^2}{\omega^2 \tilde{n}^2} \frac{\partial n}{\partial x} \tag{7.7}$$

This expresses the purely-deconfining, collisionless nonlinear acceleration of the plasma towards lower electron density, giving collisionless absorption of radiation. Another derivation[82] of equation (7.7) is based on the two-fluid model[118] where nonlinear terms have to be added,[83] or alternatively on single-particle motion of the electron.[85]

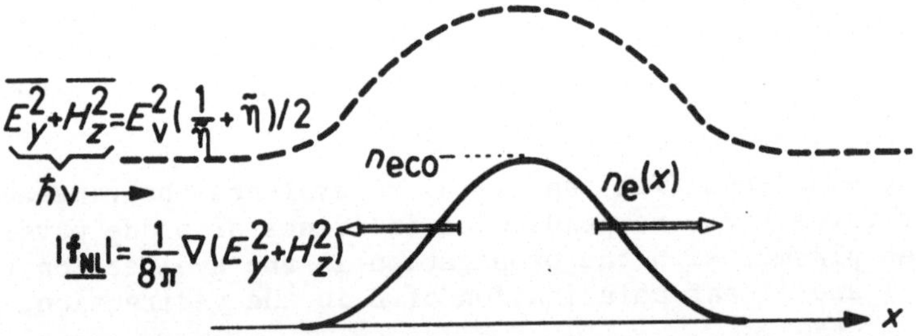

Fig. 7.1 A profile of the electron density $n_e(x)$ of a collisionless plasma with a penetrating laser beam ($\hbar\nu$) satisfying the WKB condition (n_e has to be at least a little less than the cut-off density n_{eco}). The variation of $E_y^2 + H_z^2$ causes gradients of this expression, resulting in nonlinear forces f_{NL}. η is used for $\tilde{n}$.

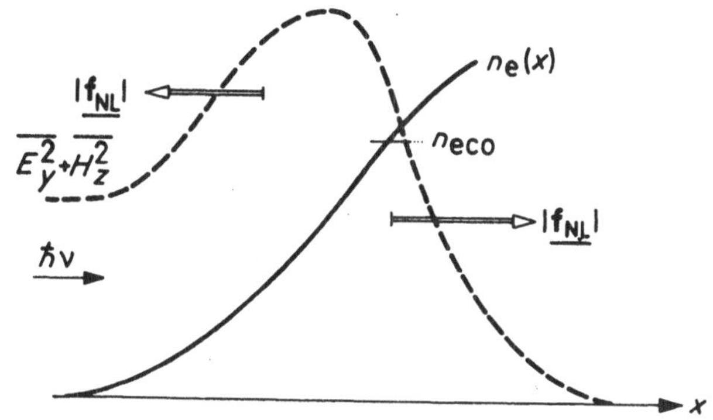

Fig. 7.2 Density $n_e(x)$ of a plasma with collisions exceeding the cut-off density n_{eco} with an incident, purely-penetrating laser beam ($\hbar\nu$) which can also be constructed in this case by the WKB approximation. The variation of the quantity $E_y^2 + H_z^2$ results in the nonlinear force f_{NL}.

The essential properties of the nonlinear deconfining force can be seen in the following schematic examples. Figure 7.1 demonstrates a collisionless plasma with a simply-propagating laser wave. The nonlinear force f_{NL} is caused by the variations of $E_y^2 + H_z^2$. The conservation of momentum is verified by direct compensation of the forces directed against the laser light on entering the plasma by the forces parallel to the light on leaving the plasma. This is also valid for nonsymmetric density profiles.

Figure 7.2 illustrates a plasma with collisions where the WKB approximation is possible even for densities exceeding the cut-off density n_{ec}. The quantity $E_y^2 + H_z^2$ decreases exponentially in the interior of the plasma according to the skin depth. The resulting forces are not compensated completely; there is an additional transfer of momentum according to the momentum of the absorbed photons.

7.2 Transferred Momentum and Ion Energies

The transfer of momenta can be seen by direct integration of the nonlinear force f_{NL}, equation (7.4) or (7.5). The total energy ε_L of a laser pulse between times t_1 and t_2 with a bundle having a perpendicular cross section K_1 in the y-z plane, is

$$\epsilon_L = c \int_{K_1} dxdz \int_{t_1}^{t_2} dt \, \frac{E_\nu^2(t_1 y_1 z)}{8\pi} = |\mathbf{P}_0| c \qquad (7.8)$$

which defines the total momentum P_0 of the photons. The momentum P transferred to the plasma between the vacuum, where the coordinate x_1 is chosen, and the depth x_2 in the plasma is

$$P = -\int_{K_1} dy\,dz \int_{x_1}^{x_2} dx \int_{t_1}^{t_2} f_{NL}\,dt = -\frac{P_0}{2}\left[\left(\frac{1}{|\tilde{n}(x_2)|} + |\tilde{n}(x_2)|\right) \exp(-k(x_2)x_2)\left(\frac{1}{|\tilde{n}(x_2)|} + |\tilde{n}(x_1)|\right)\exp(-k(x_2)x_2)\right] = 0 \tag{7.9}$$

This result was first obtained in a simplified way,[83 175] then rederived by Lindl and Kaw.[176] A more general discussion is presented here. The second term in the bracket is equal to 2 from the definition of the values $\tilde{n}$ and k for a vacuum. If x_2 is chosen far inside the plasma, we have $\exp(-k(x_2)x_2) = 0$, indicating that the total momentum transferred is that of the photons incident in the vacuum P_0, which is a trivial result. The function $P(x_2)$ can be monotonic[176] as indicated in Fig. 7.3, where no deconfining force (i.e., force in the plasma directed toward lower density in the negative x-direction) is expected. On the other hand, conditions for density and temperature gradients can be given, when $P(x_2)$ goes through a very low minimum, P describing the momentum P_{inh} at the minimum of the deconfining acceleration in the inhomogeneous plasma. The extreme value of P_{inh} is given by

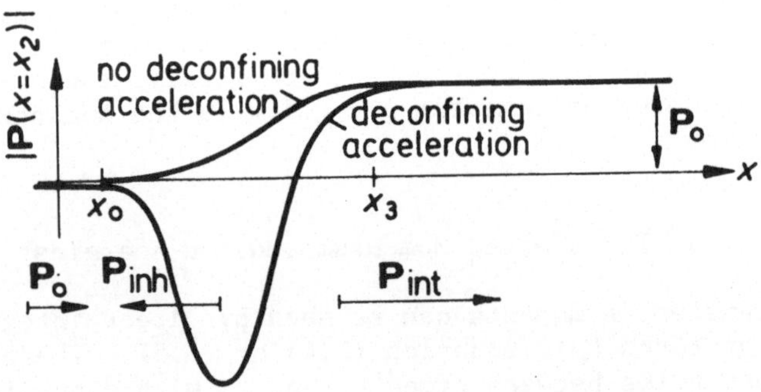

Fig. 7.3 Momentum P transferred to the plasma between the vacuum ($x_1 < x_0$) and a depth $x = x_2$ given by equation (7.9). The momentum of the incident photons P_0 is compared with the deconfining momentum P_{inh} of the plasma inhomogeneity and the final momentum P_{int} in the plasma interior.

equation (7.9) using $\exp(-k(x_2)x_2) = 1$

$$P_{inh} = -\frac{P_0}{2|\tilde{n}_2|}(1 - |\tilde{n}_2|)^2 \tag{7.10}$$

where $\tilde{n}_2$ is the value of $\tilde{n}$ at the depth x_2 and $|\tilde{n}|$ is the absolute minimum.

The deconfining momentum P_{inh} in the inhomogeneous layer can be approximated by using Eq. (6.5)

$$P_{inh} = -P_0 T^{3/4}(2a) \tag{7.11}$$

Therefore, the deconfining radiation pressure can exceed the normal radiation pressure by a factor of 100 or more at temperatures exceeding 1 keV. The deconfining momentum produces a recoil to the plasma interior where a momentum P_{int}

$$P_{int} = (|P_{inh}| + 1)P_0 \tag{7.12}$$

is to be expected. Thus, the plasma interior is confined by 100 times the normal radiation pressure or more. This nonlinearly-increased radiation pressure is essentially connected with the collective effects on which the nonlinear force f_{NL}, (Eq. (7.3)), is based. The plasma must therefore have a density n_e exceeding the cut-off density n_{ec} in its interior

$$n_e(x) \geqslant n_{ec} \; (x > x_3) \tag{7.13}$$

where x_3 is located in the skin layer (see Fig. 7.3), characterized by a strong exponential decrease of the laser intensity.

An additional general conclusion can be drawn concerning the magnitude of the energy of the ions after their acceleration in the inhomogeneous plasma layer.[85] Irrespective of the special density profile of the surface layer, we find that ion energy of directed motion can be expressed by the maximum oscillation energy of the electrons $\varepsilon_{os\ max}$ in the laser field near the cut-off density

$$\epsilon_i = Z\,\epsilon_{os\ max} \tag{7.14}$$

Because of the increase of the oscillation energy over the vacuum value $\varepsilon_{os\ vac}$ given by the WKB solution, we find

$$\epsilon_{os} = \frac{e^2}{m_e\omega^2}\frac{E^2}{4} = \frac{e^2}{m_e\omega^2}\frac{E_v^2}{4|\tilde{n}|} = \epsilon_{os\ vac}\frac{1}{|\tilde{n}|} \tag{7.15}$$

By means of equations (6.19) and (7.15) $T \gg T_{th}$

$$T = \epsilon_{os} \frac{T^{3/4}}{ka} \; ; \; T = \left(\frac{\epsilon_{os}}{ka}\right)^4 \tag{7.16}$$

Finally, equations (7.14) through (7.16) yield

$$\epsilon_i = Z\epsilon_{os} \frac{T^{3/4}}{a} = \begin{cases} Z \dfrac{E_v^2 T_{th}^{3/4}}{16\pi n_{ec} a}, & \text{if } \epsilon_{os} \ll T_{th} \\ Z \dfrac{E_v^8}{(16\pi a n_{ec} k^{3/4})^4}, & \text{if } \epsilon_{os} \gg T_{th} \end{cases} \tag{7.17}$$

The increase of a_i with the fourth power of the light intensity $I \sim E^2$ does not involve absorption due to the acceleration process itself. This mechanism limits equation (7.17) with respect to very high optical intensities.

The above considerations were based mainly on the WKB approximation. Its more extended, general validity was proved by Lindl and Kaw,[176] who used in equation (7.4) the well-known solution of E and H for a plasma with linear increase of the electron density. In this case the waves are totally reflected. Beyond the well-known acceleration of the plasma towards the nodes of the standing wave, we find an averaged force driving the plasma towards lower density which is exactly the same as in the WKB case, equation (7.7). This equation retains its validity also for increases of the electron density which are so steep that the WKB-condition ceases to be valid. A very important extension was made for the case of oblique incidence of radiation on a stratified plasma. The WKB case results in acceleration towards lower plasma density, the direction of which does not depend on the polarization of the light; only the magnitudes depend on third and higher orders of the polarization.[13] With linear density profiles[176] the results were the same as in the WKB case if the electric vector oscillates normal to the plane of incidence. When the E-vector is in the plane of incidence, there is a resonance effect causing a strong increase in the acceleration compared with the WKB case. This may be an experimental criterion for deriving information on the density profile to find whether it is WKB-like, linear, or otherwise. In the case of perpendicular incidence of the wave or standing waves, strong turbulence within the nodes of the wave was found[87] which could cause anomalous Brillouin scattering.

7.3 Predominance of the Nonlinear Force

The nonlinear force f_{NL} will be observed only when the

thermokinetic force f_{th} contained in the total ponderomotive force f (see Eq. (7.3)) is comparable with or less than f_{NL}. This question is connected with the nonlinear decrease of the collision frequency[159] caused by predominance of the coherent motion of the electrons in the laser field over the random motion in regions of minimum refractive index $|\tilde{n}|$ (see Eq. (6.22)). The temperature T, which determines the collision frequency, then consists in the limiting case of equal terms on the right-hand side of equation (6.19). This also distinguishes the two cases on the right-hand side of equation (7.17). Using the expression

$$T_{th} = \epsilon_{os}/k \, ; \, \epsilon_{os} = \epsilon_{os} T^{3/4} / a$$

we find a first approximation of the threshold value E^{1*} related to the intensity I^{1*}

$$E^{1*2} = 16\pi n_{ec} akT_{th}^{1/4} \, ; \, E_{\nu}^{1*} := MT_{th}^{1/8} \, ; \, I^{1*} := NT_{th}^{1/4} \qquad (7.18)$$

Neglecting the weak dependence of a on ln Λ, we find $M = 2.83 \times 10^8$ and $N = 2.08 \times 10^{14}$ W/cm^2 for ruby lasers, and $M = 1.67 \times 10^8$ and $N = 7.5 \times 10^{13}$ W/cm^2 for neodymium-glass lasers, where E_ν = V/cm, I = W/cm^2, and T_{th} = eV.

The limitation equation (7.18) of the nonlinear collision process also describes the predominance of the nonlinear force f_{NL} to some extent. The case in which the thermokinetic force $f_{th} = -n_e kT_{th}$ is equal to f_{NL} is calculated from equation (7.1) before differentiation, using equations (7.4), (7.5), and (6.15) for perpendicular incidence of the laser radiation on a stratified plasma. The difference in integration constants is taken into account by the term 1 in the brackets, which is the vacuum value of the momentum flux density of the radiation

$$n_e(1+1/Z)kT_{th} = \frac{E_{\nu}^{*2}}{16\pi} \quad \frac{T^{3/4}}{a} + \frac{a}{T^{3/4}} \exp_0 - 1 \qquad (7.19)$$

where "$\exp_0$" denotes an undefined value of the exponential function for the absorption which depends on the special profile of electron density and temperature in the inhomogeneous surface region of the plasma. In the following calculation we use $\exp_0 = 1, 0.5, 1/e$, and $1/e^2$. Starting from a special value of E_ν^* we obtain the associated temperature T_{th} by the following iteration: The first iteration of $T_{th} = T_{th}$ is found by putting $T = T_{th}^1$ in equation (7.19) and solving for $T_{th}^{(1)}$; the second iteration is found by solving $T_{th}^{(2)}$ from

$$n_e(1+1/Z)kT_{th}^{(2)} = \frac{E_\nu^{*2}}{16\pi}\left\{\left[\frac{\left(T_{th}^{(1)} - \frac{E_\nu^{*2}T_{th}^{(1)3/4}}{16\pi a n_{ec} k}\right)^{3/4}}{a} + \frac{a}{\left(T_{th}^{(1)} - \frac{E_\nu^{*2}T^{(1)3/4}}{16\pi a n_{ec} k}\right)^{3/4}}\right]\exp_0 - 1\right\}$$

(7.20)

the iteration being completed when $|T^{(1)} - T^{(3)}| < 0.5$ eV. The result for Z = 1 (hydrogen) and for neodymium-glass laser is shown in Fig. 7.4 for the specified values of $\exp_0$. The threshold for $\exp_0 = 1$ is equivalent to the limiting parameter in equation (7.18). From this it can readily be seen that the increase of ε_1 as the fourth power of the laser intensity shows the predominance of the nonlinear force. It should be noted that a special self-focusing process of laser-produced plasmas has an intensity threshold[177] which is similar to I^{1*}, being approximately 10^{14} W/cm^2 for solid state lasers.

Another estimate of the magnitude of the nonlinear force compared with the thermokinetic force was given by Steinhauer and Ahlstrom.[178] The pressures generated by thermalization of the radiation were compared to those produced by the nonlinear effects. The ratio

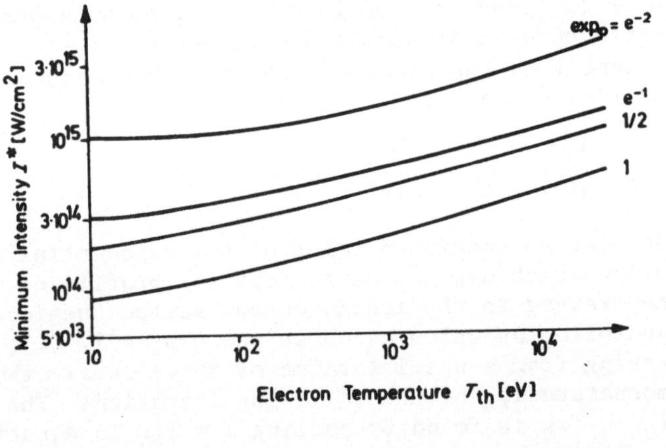

Fig. 7.4 Minimum intensity I* for neodymium-glass laser required to produce nonlinear de-confining force f_{NL} larger than thermokinetic force $f_{th} = - n_e kT_{th}$. The undefined exponential function $\exp_0$ refers to collision-induced attenuation.

$$\frac{|f_{th}|}{|f_{NL}|} = S \geqslant \frac{1.14 \times 10^5}{T^{5/4}} \left(\frac{Z}{\lambda_0}\right)^{1/2} \tag{7.21}$$

is a realistic upper bound, because the maximum values of temperature were used, decreases due to thermal conductivity being neglected. λ_0 is the laser wavelength (in μ) and T is that temperature of the electrons (in eV) which determines the averaged electron-ion collision time. The result is that even at intensities less than the threshold ($I \ll I^*$) the nonlinear force can exceed the thermokinetic force if $T > 10^4$ eV. The correlation with our result (Fig. 7.4) showing general predominance of the nonlinear force for $I > I^*$ can also be seen from equation (7.21). Because the Steinhauer-Ahlstrom temperature T is that of equation (6.19) by definition, we can use

$$T = T_{th} + \frac{\epsilon_{os}}{k} = T_{th} + \frac{\omega_p^2 E_\nu^2}{16\pi\omega^2 n_e k|\tilde{n}|} \approx \frac{E_\nu^8}{(16\pi a n_{ec} k^{3/4})^4} \tag{7.22}$$

For $I \gg I^{1*} \approx I^*$ we find in equation (7.21) at $Z = 1$; $\lambda_0 = 1.06$ μ

$$S \geqslant \left(\frac{7.2 \times 10}{E_\nu}\right)^8 \qquad [E_\nu] = \text{V/cm} \tag{7.23}$$

This immediately shows the strong increase of f_{NL} over f_{th} at E_ν of more than 7.2×10^8 V/cm, which correlates very well with the threshold I^* around 10^{14} W/cm^2 for neodymium-glass lasers. The high exponent in Eqs. (7.22) and (7.23) indicates a resonance mechanism, which is damped by the nonlinear absorption due to the nonlinear acceleration itself.

7.4 Self-Focusing of Laser Beams in Plasma

It was pointed out by Schlüter[84] that the lateral decrease of the intensity of a laser beam in a plasma should generate a ponderomotive force which causes a rarefaction of the plasma along the axis. The resulting density gradient produces a thermokinetic force in the antiradial direction. In equilibrium both forces will be compensated, as shown in Fig. 7.5. This condition was used to calculate the properties of self-focusing.[86,95] Starting from a Gaussian density distribution which varies only in the y-direction with the unit vector i_z of a beam penetrating in the x-direction and having a linear polarization of the E-vector in the y-direction

$$E_\nu^2 = \frac{E_{\nu o}^2}{2\tilde{n}} \exp(-y^2/y_0^2)\, ; \overline{H_z^2} = \tilde{n}^2\, \overline{E_y^2} \tag{7.24}$$

we find a maximum of the nonlinear force

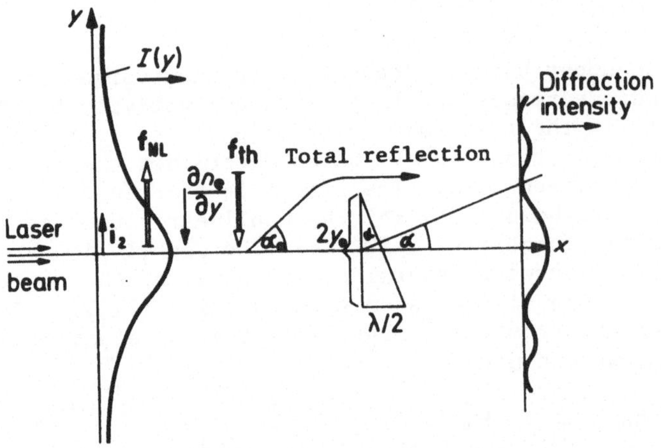

Fig. 7.5 Scheme of a laser beam in a plasma of lateral intensity decrease I(y), producing non-linear forces in the plasma f_{NL}, rarefying axial regions, compensated by thermokinetic forces f_{th}. The density gradient $\partial n/\partial y$ causes total reflection of a partial beam within a criticla angle α. The trapped light beam (filament) has to give a diameter resulting in a diffraction pattern where α_0 has to determine the first diffraction minimum. $2y_0$ is the effective diameter of the bundle and λ is the laser wavelength.

$$f_{NL} = -\frac{i_2}{8\pi}\frac{\partial}{\partial x}\left(\overline{E_y^2} + \overline{H_z^2}\right)_{max} = i_2 \frac{(1+\tilde{n}^2)E_{vo}^2}{16\pi\tilde{n}\,y_0}\sqrt{2}\exp(-0.5) \tag{7.24a}$$

where y_0 determines the width of the beam. Setting f_{NL} equal to the thermokinetic force $f_{th} = -i_2 kT(1+Z)\partial n_e/\partial y$, we find the relation

$$\frac{\partial n_e}{\partial y} = \frac{\sqrt{2/\exp(1)}}{16\pi kT(1+Z)}(1+\tilde{n}^2)\frac{E_0^2}{y_0} \tag{7.25}$$

If $\tilde{n}$ is the refractive index in the beam center and $\tilde{n}_{yo}$ is its value at $y = y_0$, total reflection of partial waves of the beam propagating at an angle α_0 to the axial direction is given by

$$\sin\left(\frac{\pi}{2} - \alpha_0\right) = n/n_{yo} \tag{7.26}$$

Using the expansion $\tilde{n}_{y_0} = \tilde{n} + (\partial\tilde{n}/\partial y)y_0$ and (in the case of negligible absorption) $\tilde{n} = 1 - \omega_p^2/\omega^2$, we find from equation (7.25)

$$\sin\alpha_0 = \sqrt{\frac{2}{\tilde{n}}\frac{\partial\tilde{n}}{\partial n_e}\frac{\partial n_e}{\partial y}\, y_0} \tag{7.27}$$

The resulting laser beam will have an effective thickness of $2y_0$, which determines an aperture for the first diffraction minimum at the laser wavelength $\lambda = 2\pi c/\omega$. We postulate that the trapped laser filament has a diffraction condition which does not exceed the condition of total reflection

$$\sin\alpha = \lambda/(4y_0) = \pi c/(2\omega y_0) \leqslant \sin\alpha_0 \tag{7.28}$$

From equations (7.25), (7.27), and (7.28) we find a threshold condition for the laser power $P = y_0^2 E_{vo}^2/C_1^2$ ($C_1 = 1.63 \times 10^{-5}$ cgs)

$$\bar{P} \geqslant \frac{(\pi c)^2 \tilde{n}^3 m_e kT(1+Z)}{e^2\sqrt{2/\exp(1)}\, c_1^2 (1+\tilde{n}^2)} \tag{7.29}$$

Using $\tilde{n} = 1$ for $\omega_p \ll \omega$ and $\tilde{n} = aT^{-3/4}$ from equation (6.15) for $\omega_p \leq \omega$ and $T \leq 30$ eV, we find for $a = 2.8$ and $Z = 1$

$$P \geqslant \begin{cases} 1.46 \times 10^6/T^{5/4}\ \text{W for } \omega_p \leqslant \omega \text{ and } T > 30\ \text{eV} \\ 11 \times 10^4\, T\ \text{W for } \omega_p \ll \omega \end{cases} \tag{7.30}$$

Therefore, a laser beam can start self-focusing by the nonlinear interaction force even at low power, such as 10^5 to 10^6 W, as in the case of self-focusing in dielectrics.[179] The model is limited by the time needed to form a filament. Starting from values of $y_0 \approx 10^{-4}$ cm and velocities of 10^7 cm/sec, the formation time for self-focusing is 10^{-11} sec. The second part of equation (7.30) is limited to lower densities by the condition that y_0 be larger than the Debye length. This treatment is based on self-focusing of a slab. The constants in Eq. (7.30) are reduced by the factor 0.8 because of the cylindrical symmetry of filaments.

The smallest possible radius of a filament (cylinder) y_0 is given when the center is completely empty. In this case $n_e kT$ is equal to $(E+H_z/8\pi)$. This condition occurs at intensities $I = I^* \approx I^{1*}$ (Eq. (7.18) or Fig. 7.4). Under cylindrical conditions we have

$$y_0 = (\bar{P}/\pi N T_{th}^{1/4})^{1/2} \tag{7.31}$$

The self-focusing filaments observed by Korobkin and Alcock,[94] who reported that laser power P of 3 MW produced filaments 2 μ in diameter, can then be readily explained. Assuming $NT_{th}^{1/4}$ = 2×10^{14} W/cm^2, we find $2y_0$ to be approximately 2 μ.

The described self-focusing mechanism of laser radiation in a plasma differs basically from the one presented by Kaw[177] which starts at a threshold intensity of I*. It also differs from the mechanism in stellar plasmas which explains very satisfactorily the phenomenon of quasars.[180]

7.5 Numerical Examples of Nonlinear Acceleration

There have been several studies of the dynamics and expansion of laser-produced plasma including the nonlinear force.[87,128,181] One example shown in Fig. 7.6 was intended to demonstrate the unimportance of the nonlinear force.[181] However, the intensities used were only a very small amount above the threshold I*. Below this threshold the nonlinear forces could not predominate over thermokinetic forces under any circumstances. But even in this case a net acceleration of the plasma resulted with values more than two times the thermokinetic force. Finally, a complete dynamic process has been investigated with similar moderated intensities and with a slow increase of the laser pulse.[181] When the nonlinear force was omitted, no change in the whole plasma expansion was observed. A detailed analysis of the result, however, indicated that the mechanism of gas-dynamic ablation with a driving force below the plasma corona was established before the necessary high laser intensity was present. Within the plasma corona there were finally larger nonlinear forces than thermokinetic, although the dynamics were determined at the later time only by the compressed material below the corona.

The following detailed numerical example based on a code of Kinsinger[182] for complete plasma dynamics using laser intensities exceeding 10^{16} W/cm^2 shows the important influence of the nonlinear force.[183]

The equation of motion for the plasma in a one-dimensional geometry was given by the force density, distinguishing between electrons and ions,

$$f = -\frac{d}{dx} n_e k T_e - \frac{d}{dx} n_i k T_i + \frac{1}{8\pi} \frac{d}{dx} (E^2 + H^2) \qquad (7.32)$$

where the electrons (e) and ions (i) of different temperatures T_e and T_i are coupled by a Poisson equation and by collisions. Besides

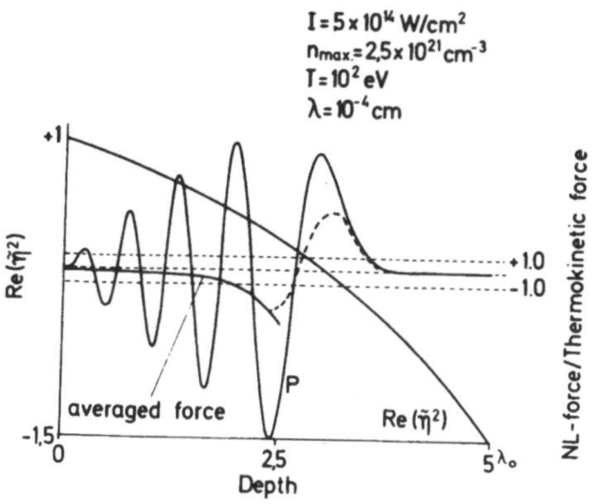

Fig. 7.6 Numerical calculation of the wave equations for a density profile with a parabolic real part of the refractive index.[181] The resulting standing wave at a plasma temperature $\tau = 100$ eV for neodymium-glass lasers of 5×10^{14} W/cm^2 incident intensity (from the left side) creates nonlinear forces whose value is given as the ratio to the actual thermokinetic force. The net force is given as a spatially-averaged value. It exceeds the thermokinetic force by a factor higher than 2.5

the equation of continuity (Eq. (5.15) for both components), the energy equation (Eq. 5.16) contains internal and external thermodynamic energies, thermal conductivity, and the incident laser radiation absorbed by collisions or by nonlinear electrodynamic motion as a source term.

The calculation begins at a time $t = 0$ with a distribution of plasma density starting from solid state density and a temperature profile. Without any laser field there would be gas-dynamic expansion and conservation of the total energy. At the successive time steps, a time-dependent incident laser intensity is prescribed, for which at each time step an exact stationary solution of the Maxwell equations is calculated including the actual plasma density and its refractive index. The retardation of the waves, the switching-on mechanism, and the development of the reflected wave are neglected.

The motion of the plasma within the next time step (with appropriately varying step size) is described generally according to Eq. (7.32), with the first two terms of the gas-dynamic force and the last term of the nonlinear electrodynamic force included. The following two examples from extensive series of computer runs illustrate characteristic results. In both cases the initial temperature of the electrons and ions (100 eV) was constant for the whole plasma. One may assume that this is produced by a prepulse from the laser. The laser pulse increases within 10^{-13} sec up to an intensity of 2×10^{16} W/cm^2, the increase is linear for 5×10^{-14} sec, and then follows smoothly a Gaussian profile. After reaching 2×10^{16} W/cm^2, the laser intensity remains constant.

The initial plasma density increased (for x < 0) quadratically above the cut-off density up to the solid state density of LiD. It then changed smoothly in Fig. 7.7a into a linear decrease up to the length x = 50 μ. The resulting electromagnetic momentum flux

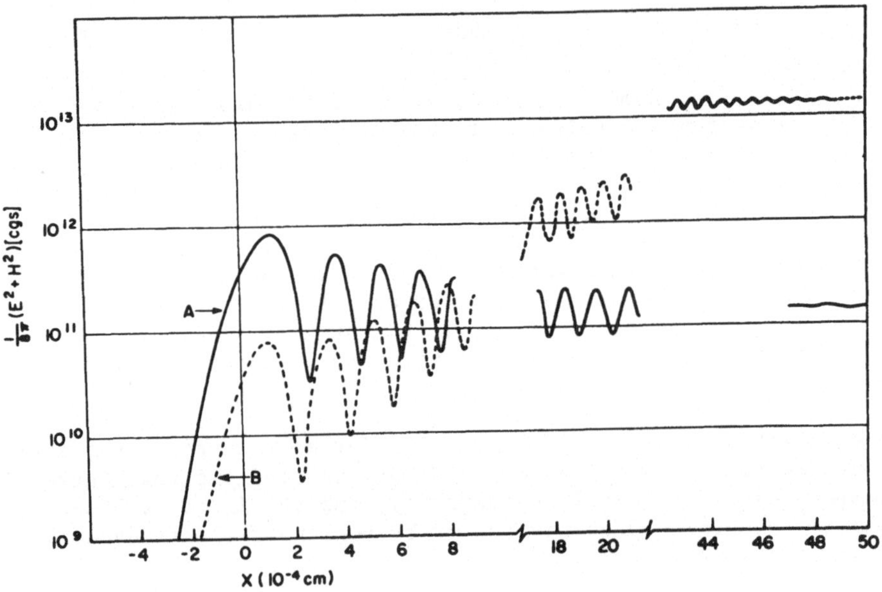

Fig. 7.7a A laser beam is incident from the right-hand side on a plasma of initial temperature of 100 eV and linear density increasing from zero at x = 50 μ to the cut-off density at x = 0 and then increasing more rapidly. The exact stationary (time-independent) solution without retardation of the Maxwell equations (5.8) and (5.9) with a nonlinear refractive index ñ Eq. (6.1) based on a collision frequency (Eq. (6.22)) results in an oscillation due to the standing wave and dielectric swelling of the amplitude (curve A). At a later time (2×10^{-13} sec) the laser intensity is 2×10^{16} W/cm^2 (curve B) where the relative swelling remains, but the intensity at x = 0 is attenuated by dynamic absorption.

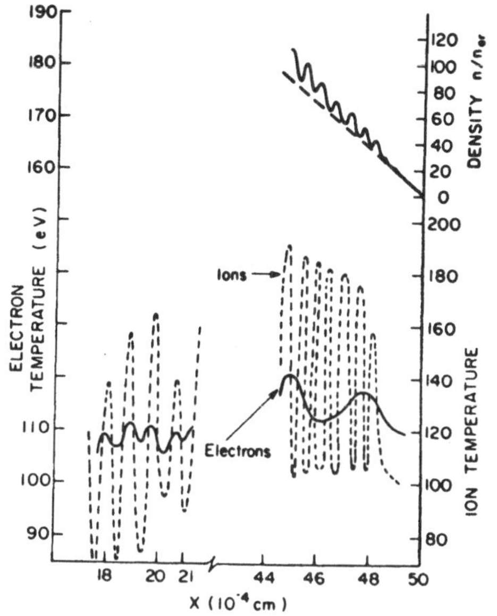

Fig. 7.7b The initial density (dashed line) and the density along curve B of Fig. 7.7a, where a ripple is created by the nonlinear force pushing plasma toward the nodes of the standing wave. The electron and ion temperatures are increased following the ripple by dynamic compression at conditions identical to curve B.

density $(E^2 + H^2)/8\pi$ is given in Fig. 7.7b. Curve A is taken at an early time, when the laser intensity is 2×10^{14} W/cm^2. In a very thin plasma ($x \approx 50\ \mu$) the value of $E^2 + H^2$ is constant, as is well known, e.g., from Chen's[173] general derivation of the instabilities from laser-plasma interaction involving the nonlinear force. At about 20 microns we find an oscillation of $E^2 + H^2$, which increases in amplitude and wavelength between 0 and 8 microns. This behavior is well known from the analytical work for the same linear density profile[176] (dealing, however, with temperature $T = 0$ and collisionless plasma). Near $x = 0$, the $E^2 + H^2$ value has increased due to the dielectric properties of the plasma.

The increase of the laser field E and of the wavelength λ over its vacuum values E_0 and λ_0 is given by a swelling factor $\vartheta = |\tilde{n}|^{-1} > 1$

$$E = \frac{E_0}{|\tilde{n}|} \quad ; \quad \lambda = \frac{\lambda_0}{|\tilde{n}|} \tag{7.33}$$

where ñ is identical with the optical (complex) refractive index of the plasma,[83,87] provided the electromagnetic field can be described by the WKB approximation. In the case of curve A in Fig. 7.7a, the plasma fits the WKB conditions at 1.3 micron quite well, and we find agreement between the swelling factor $|\tilde{n}|^{-1} = 6$, (taken from Fig. 7.7a) and the value calculated from the actual refractive index.

For later times, curve A moves in nearly parallel fashion to higher values. However, its upward shift is reduced near the cut-off density ($x = 0$). This decrease becomes so marked at $t = 2 \times 10^{-14}$ sec (and later even more pronounced) as shown by curve B that the intensity in the thin plasma ($x \approx 50$ μ) decreases by a factor of 10 up to $x \approx 20$ μ and to a thousandth part of the initial intensity at $x \approx 0$. The reason is very simple: The standing wave pushes the plasma toward its nodes with ion velocities as high as 10^7 cm/sec even at $x \approx 48$. The gas-dynamic velocities reached at that time are 10^3 cm/sec or less. The result is shown in Fig. 7.7b.

At $x = 46$ μ to 50 μ, the initial linear density (upper part of Fig. 7.7b) acquires a ripple. The slight total increase is due to net motion driven by the nonlinear force. The ion energy increases, oscillating up to 200 eV (from the initial value of 100 eV), out of phase with the ripple. There is a curious effect seen in the electrons, in which the temperature increase is only a fifth of that of the ions, as it should be for a collisionless shock; however, the periodicity is markedly different from that of the ions. At $x \approx 20$ μ the periodicity of the temperature of the electrons and the ions is nearly the same, but the electron temperature is again less by a factor than the ion temperature.

The density ripple explains the strong decrease of laser intensity with depth at later time: It causes strong reflection of light and a transfer of the optical energy into the ions by collisionless ripple shocks. Because it is not connected with decay of photons into microscopic acoustic modes, one may call this process of dynamic ion absorption a "collisionless ion heating by nonlinear-force-induced macroscopic dynamic ion decay (MDID)." It is to be noted that depression of the laser intensity in interior regions becomes marked only after the density acquires pronounced maxima and minima (ripple).

An obvious idea is to start with a modified density profile such that a change of the laser intensity I near cut-off results in a change of the actual nonlinear refractive index ñ, by virtue of the energy of the electron oscillation $\varepsilon_{os}(I)$ determining the electron temperature (Eq. (6.22)). In Fig. 7.7a at 1.3 microns the absolute value of the refractive index is determined only by

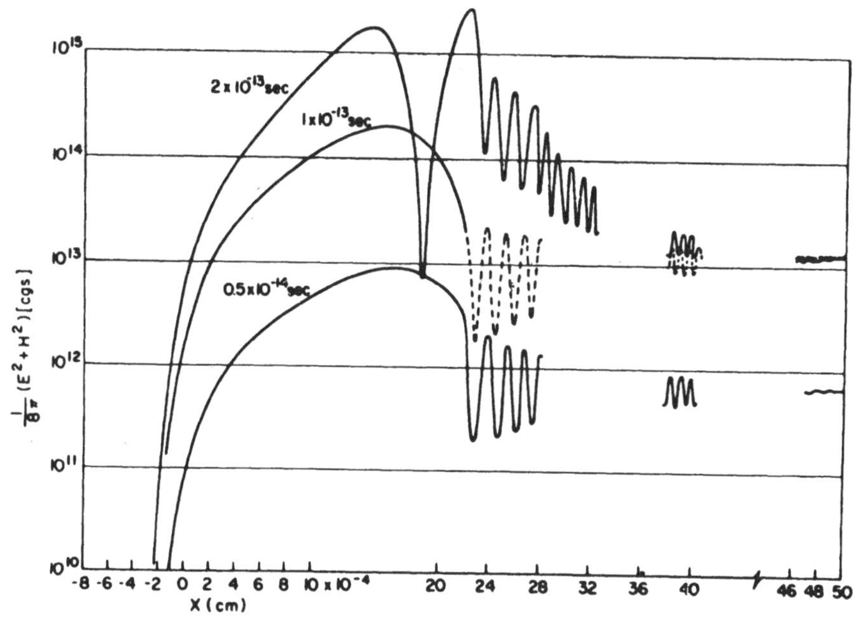

Fig. 7.8a Similar to Fig. 7.7a, but with an initial density decreasing linearly from n_{ec} at x = 0 to n_{ec} x 10^{-6} at x = 10 μ, and to n_{ec} x 10^{-2} at x = 20 μ, and then to n_{ec} = 0 at x = 50 μ . The $(E^2 + H^2)$ values show an increase by a factor of 31.2 at time 5 x 10^{-15} sec. At 10^{-13} sec the again-constant laser intensity of 2 x 10^{16} W/cm^2 was reached, with much less attenuation due to dynamic absorption at 2 x 10^{-13} sec. A swelling of $\vartheta \approx 400$ occurred.

its real part (given by the actual density), while the imaginary part (determined by the temperature) is too small. Therefore, we calculated a case with the same overdense density profile and a density varying linearly first from x = 0 to 10 microns from cut-off to 10^{-6} times cut-off, then from there to x = 20 microns to 10^{-2} times cut-off, and last linearly at x = 50 microns to zero cut-off. Figure 7.8a shows the $(E^2 + H^2)/8\pi$ values for an initial time (0.5 x 10^{-14} sec) at a laser intensity of 10^{15} W/cm^2 and at subsequent times of 10^{-13} sec and 2 x 10^{-13} sec. The swelling factor at the beginning was

$$\vartheta = |\tilde{n}|^{-1} = 31.2 \tag{7.34}$$

due to the dielectric properties of this density profile. At 10^{-13} sec, the little depression of the curve at x = 20 μ indicates some tendency to decrease due to density rippling. The maxima

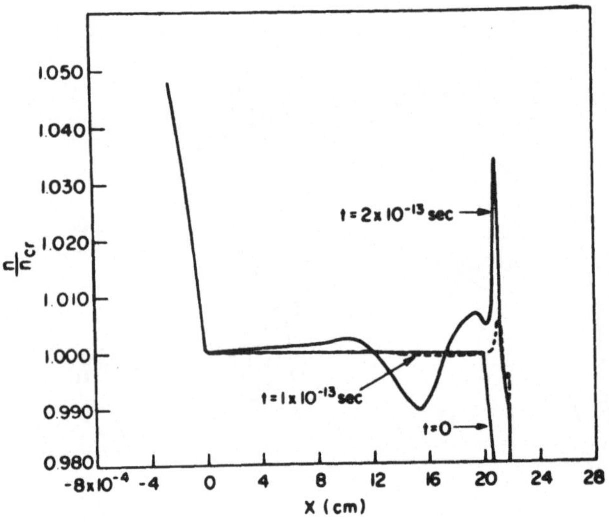

Fig. 7.8b Initial and subsequent density profiles indicating inward motion of plasma toward the interior for x less than 14 microns.

and minima of $E^2 + H^2$, however, change locally very quickly, and at 2 x 10^{-13} sec, we find a swelling of $\vartheta = |\tilde{n}|^{-1} = 400$. The dynamic change of the density near x = 20 μ is shown in Fig. 7.8b. A process resembling tunneling of the electromagnetic field through the overdense plasma near x = 20 microns at 2 x 10^{-13} sec is very strong, by virtue of the depth of the overdense plasma (≈ 1 micron) being small compared with the actual effective wavelength of more than 20 microns.

The resulting velocity profile at 2 x 10^{-13} sec is shown in Fig. 7.8c. The compression of plasma between x = 2 microns and x = 14 microns with velocities of 4 x 10^7 cm/sec does not change much within subsequent time intervals, while the velocities at x = 50 microns, for example, are still changing. The increased ion temperature from collisionless processes at x = 40 microns is again much larger than the electron temperature, as could be expected from macroscopic dynamic ion absorption. A curious effect occurs in the interval from 20 to 28 microns, where the electrons are much hotter than the ions and display an irregular oscillation. It can be assumed that a longitudinal acceleration of the electrons due to the nonlinear force occurs when the electrons have less interaction with the ions, comparable to hot electrons in solids at high electric fields. This phenomenon of

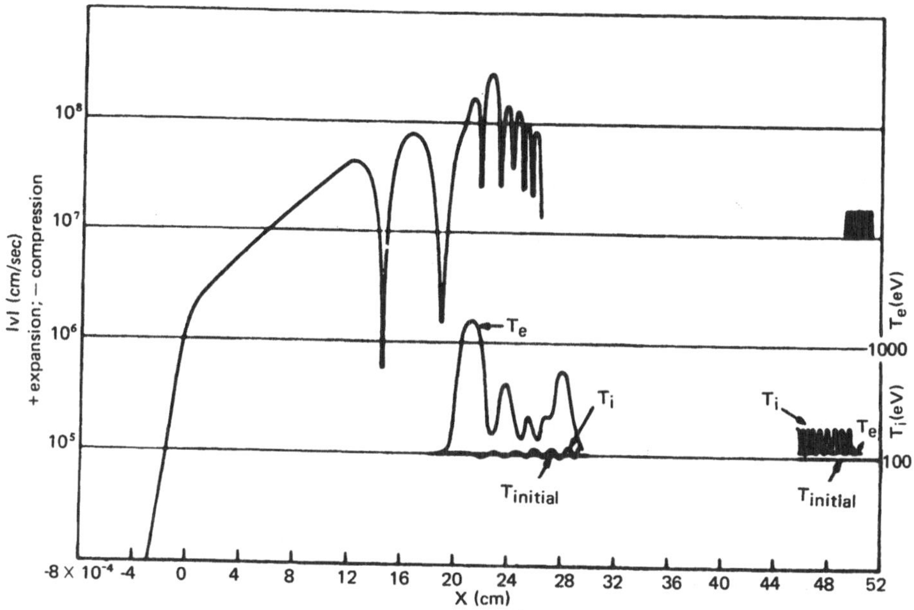

Fig. 7.8c Dynamic heating of ions (at x = 48 microns) and of electrons (x = 20 microns) occurring at time $t = 2 \times 10^{-13}$ sec. The resulting plasma velocity v shows a compressing motion of the whole plasma between - 4 and 14 microns with speeds up to 5×10^7 cm/sec, followed by expansion and alternating compression and expansion due to the standing-wave field.

dynamic electron absorption may be called "collisionless electron heating by nonlinear-force-induced macroscopic dynamic electron decay (MDED)."

The effects observed are highly sensitive to changes in the intensity I and the density profile. For an optimum case with $I = 4 \times 10^{16}$ W/cm^2 and the identical time and density profiles, the incident laser energy per cm^2 is 4.1 kJ up to 2×10^{-13} sec. The compression front between x = - 2 μ and x = 14 μ absorbs 0.96 kJ in kinetic energy, the ablating plasma takes 0.68 kJ in kinetic energy of net motion and 2.2 kJ for dynamic heating. The remainder of 0.16 kJ goes into reflection and collisional heating of electrons.

The dynamic description of a plasma by this numerical model for neodymium-glass laser intensities exceeding 10^{16} W/cm^2 indicates that

(a) A strong dielectric increase of the laser field and the effective wavelength occurs, described by a swelling factor $|\tilde{n}|^{-1}$ of up to 400 ($\tilde{n}$ is a parameter resulting from rigorous solution of the Maxwell equations and is identical with the complex refractive index in the WKB approximation);

(b) A compression of plasma driven by the nonlinear force occurs at a thickness of 15 microns with velocities exceeding 6×10^7 cm/sec. The mechanical energy in the compression front contains 23% of the incident laser radiation.

(c) A rippling of the density occurs in the whole plasma at densities below cut-off with nonlinear-force-generated velocities of 10^7 cm/sec or more. This may explain generation of laser-produced fusion neutrons from peripheral plasma coronas. The ripple decreases laser light penetration of a plasma down to the cut-off density;

(d) Rippling can be suppressed by selective profiling of the plasma density and laser intensity with time;

(e) A preferential transfer of laser energy to ions by collisionless shock occurs in the ripples (MDID);

(f) A preferential transfer of laser energy to electrons can occur at densities near cut-off (MDED).

8. THEORY OF LASER-INDUCED NUCLEAR FUSION

It is to the merit of Basov and Krokhin that they discussed[135] the possibilities of laser heating of solid deuterium or mixtures of deuterium with tritium for the purpose of nuclear fusion. Not only are the physical facts interesting but also theirs is the first discussion in the open literature. The earlier work done on a classified basis could never be dignified completely, and the step of Basov and Krokhin[135] was an act of authority without which the many attempts of other investigators would have been disqualified with the well-known expressions of "frivolous" and "fantastic" before any physical discussion. Indeed, this early phase of work was characterized by a flexible imagination -- as is usual in all new fields and especially in such an uncommon subject. The first subject studied was the gas dynamics of expansion of a laser-irradiated sphere under highly-simplified assumptions concerning the optical properties of absorption and radiation from laser-produced plasmas.

8.1 Inertial Confinement

Inertial confinement starts from a spherical plasma of radius R_0 and an initial ion density n_0 to which an energy E_0 has been transferred, giving an equilibrium temperature T_0. The initial expansion velocity $\dot{R}_0$ is assumed to be zero in the simplest case. How to realize such initial conditions is a very general question. Several methods are possible: Laser pulses with a duration of the order of 1 nsec[184] for the cases considered below, or electron beams,[185] particle beams, nuclear fission reactions, and annihilation radiation if antimatter[155] could be made available by efficient generation in laser fields. Many details of the transfer of laser energy have been discussed in which delays due to spatial energy transfer or to finite time of transfer of the energy from electrons to ions have been considered, especially with classical assumptions on the equipartition times.[186] These results may change under the conditions of high density in favor of shorter times to reach equilibrium. The initial parameters are related by

$$T_0 = \frac{E_0}{8\pi k n_0 R_0^{1/3}} \tag{8.1}$$

The temperature T during subsequent adiabatic expansion and the density in the plasma possessing a box-like density profile (the change from a box-like to Gaussian density profile affects numerical calculation of the gas-dynamic expansion only slightly) are given by

$$T = T_0 (R_0/R)^2 \; ; \; n_i = n_0 (R_0/R)^3 \tag{8.2}$$

Following the model of gas-dynamic expansion given in Eqs. (5.21) through (5.23), the variation of the radius R of the sphere is determined by

$$\dot{R} = \frac{5kT_0}{m_i} \left[1 - (R_0/R)^2 + \dot{R}_0^2\right]^{1/2} \tag{8.3}$$

where we assumed an initial expansion velocity $\dot{R}_0 = 0$.

The nuclear fusion energy gained without secondary processes such as heating and interaction with the alpha particles produced is given in relation to the incident laser energy E_0 by[184,187,188]

$$G = \frac{\epsilon_F}{E_0} \int_0^{\infty} dt \int_0^{R(t)} dxdydz \, \frac{n_i(R(t))^2}{A} <\sigma v> \tag{8.4}$$

where ε_p is the energy produced by one nuclear fusion process, (Eq. (1.1)) for pure deuterium (D-D) and (Eq. (1.3)) for deuterium tritium (D-T), and where A is 2 for D-D and 4 for D-T. The averaged value of the nuclear fusion cross-section σ with the velocity distribution v(T) of the Maxwell distribution for the actual temperatures T can be used from values in the literature.[189,191] Because these are empirical in nature, only numerical integration of Eq. (8.4) is possible.[188] Taking into account the constant total number of ions involved, Eq. (8.4) can be rewritten as

$$G = \frac{\epsilon_F}{E_0} n_0 \frac{4\pi}{3} R_0^3 \sqrt{\frac{m_0}{5kT_0}} \int_{R_0}^{\infty} \frac{n_i(R)}{A} < \sigma v(T(R)) > \frac{dR}{\frac{5kT_0}{m_i}\left[1-\left(\frac{R_0}{R}\right)^2\right]} \quad (8.5)$$

where a special numerical procedure is necessary around the pole of the integrand at $R = R_0$. Results are given in Figs. 8.1 and 8.2 for two cases of initial ion densities, n_0: the density of the solid state and that of one-tenth the solid state density. From the optimum gain G at the asymptotic line of a series of similar curves as given in Figs. 8.1 and 8.2 we derived the relationship for optimized initial volumes[184,188,192,193]

$$G = \left[\frac{E_0}{E_{BE}}\right]^{1/3} \left[\frac{n_0}{n_s}\right]^{2/3} \quad (8.6)$$

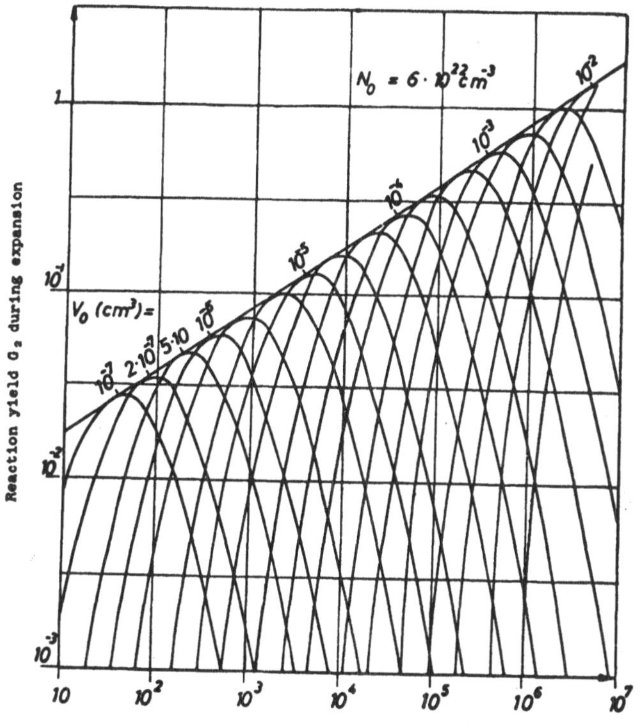

Fig. 8.1 Nuclear fusion reaction yield $G_2 = G$ (Eq. (8.5)) for D-T mixtures with initial density[110] of 6×10^{22} cm^{-3} (solid state) as a function of laser energy input E_0 for various initial spherical volumes[184] V_0

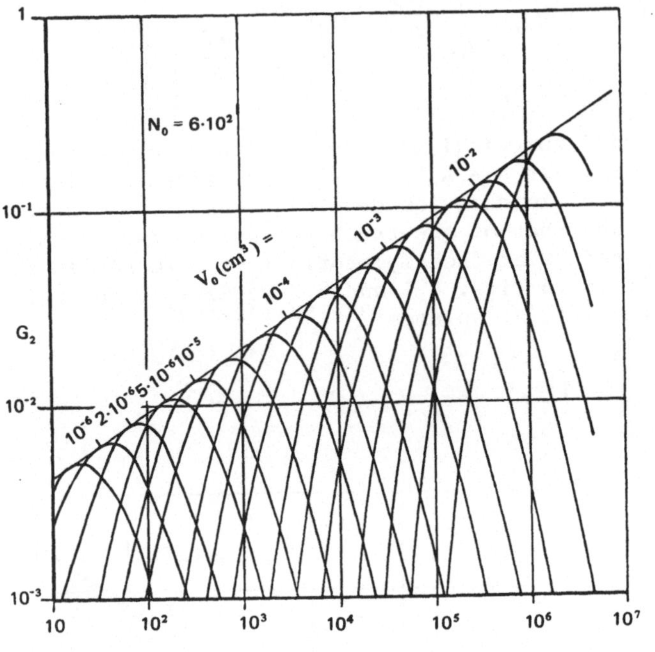

Fig. 8.2 Nuclear fusion reaction yield[184] as in Fig. 8.1 for an initial plasma density of $n_0 = 6 \times 10^{21}$ cm^{-3}.

where n_s is the ion density equal to the density of solid deuterium $n_s = 6 \times 10^{22}$ cm^{-3}, and the energies E_{BE} are the break-even energies for which G is 1 at an initial density of solid deuterium

$$E_{BE} = \begin{cases} 1.6 \times 10^6 \text{ Joules for DT} \\ 1.7 \times 10^{10} \text{ Joules for DD} \end{cases} \qquad (8.7)$$

Clearly, Eq. (8.6) is only correct for $G < 10^3$, since nearly all nuclei are burned at this limit. A further modification in evaluating Eq. (8.7) is possible considering the fact that the reacting ions may leave the plasma before a reaction takes place. These problems[187] are important only if a box-like density profile is assumed. It has been shown[134,188] that an initial Gaussian density profile (which is conserved exactly during the subsequent expansion) specifically avoids the problem of ion losses before reaction. In this case the break-even energy for D-T is higher by a factor $2^{3/2}$ (Eq. (8.7)) to reach $E_{BE}^{Gaussian} = 4.6$ MJ.[188] For D-D there is no essential difference because of the relatively-large plasma volume.[188] Taking into account the losses for a box-like profile, the numerical evaluation of Eq. (8.5) yields a break-even energy for D-T of 37 MJ. For D-D the effect of losses is again negligible.

It should be noted that the numerical calculations depend to a large extent on the empirical values of the cross section σ as a function of the ion velocity v if the averaging of (σ, v) is performed numerically. An error in v of 10% results in an error of 60% in σ at the important ion energies near 40 keV (where $\sigma \sim v^6$) and, therefore, results in an error in the break-even energy E_{BE} by a factor of 5. The high sensitivity of the results has to be taken into account when comparing several calculations. The situation becomes much more uncertain if much rougher assumptions are used. Approximations exist in which it is assumed that the plasma expands with the speed of sound to double its initial radius while holding constant its initial temperature, and the fusion reactions during this time are counted. With respect to our exact computation with Eq. (8.7) and its very high sensitivity to the empirical values of σ used, it is fortuitous that the break-even energies obtained in primitive approximations gave the same results as computations reported in 1964.[187]

8.2 Gas-Dynamic Compression

One significant feature of Eq. (8.6) -- a feature which is also seen directly from Figs. 8.1 and 8.2 -- is that it is necessary to have as high plasma densities as possible. A tenfold increase of n_0 permits a hundredfold decrease of the initial laser energy required for the same gain G. Taking into account necessary losses in the laser and in the conversion of the energy produced into usable electrical energy, G values of 100 or more and compression of the plasma up to densities of ten thousand times solid state density are required to permit use of feasible laser energies E_0 of magnitude from 10 to 100 joules.

The original basic concept for compression of plasma by the laser beam itself through the gas-dynamic ablation mechanism was first evaluated in the numerical studies of Nuckolls[194] and further authors,[195] followed by calculations made by other groups.[196] One motivation for this model is the fact that laser interaction with a plasma of density higher than the cut-off density causes compression in the overdense region, as seen, for example, in the calculations of Fig. 5.1. For a planar geometry the compression reaches densities four times the initial density. For a spherical geometry the compression in the center of the plasma increases to more than thirty times its initial value[197] if the laser pulse is rectangular in time. Computer solutions indicated that much higher densities could be produced if the laser pulse shape was not constant with time but instead increased monotonously or quasi-monotonously.

The formation of a steeply-sloped compression front was well known from Guderley's theory[198] of a sphere which considered a sequence of compression momenta applied at the surface.[199] Similar results were obtained by impact of a sequence of shock waves of increasing amplitude on a spherical plasma,[196] in which the speed of the later shock waves was larger by such an amount that all the waves converged in the center of the plasma.[200]

Figure 8.3 shows the mechanism schematically. Laser radiation enters from the left side and is absorbed in some way linearly, nonlinearly, or anomalously within the plasma corona, where the density is less than the cut-off density. It is assumed that efficient absorption takes place and that dynamic absorption does not occur in the very outermost, thin regions of the corona. The gas-dynamic motion of the plasma gives a radial dependence of the plasma density which is increased markedly if the laser pulse has the required monotonic increase of intensity. For spherical

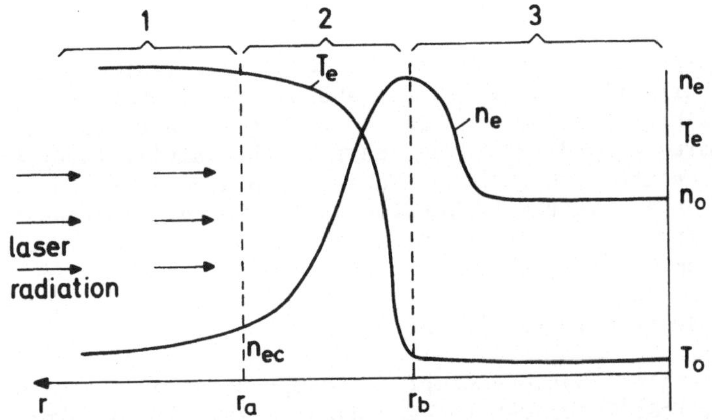

Fig. 8.3 Scheme of laser interaction with a spherically-irradiated pellet. The absorption of radiation takes place within area 1 (corona) down to a depth r_a, where the electron density reaches the cut-off value n_{ec}. Down to depth r_b (area 2) compression takes place increasing the density to a maximum value n_e above the initial density n_0 of the sphere. The temperature T_e decreases within area 2 and heats the compressed shell to cause an ablation of plasma in area 2 and a compression in area 3.

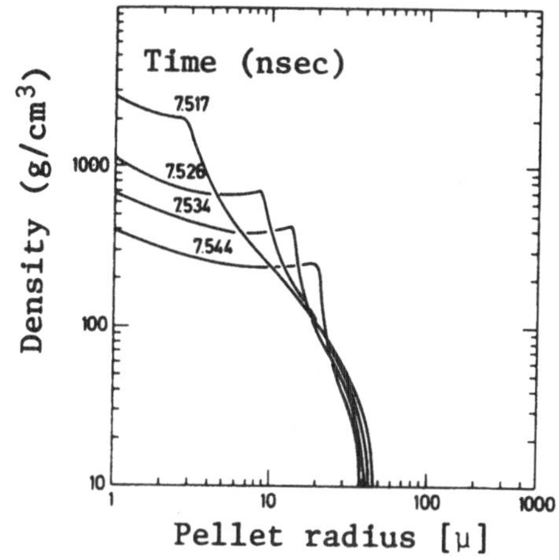

Fig. 8.4 D-T pellet densities as a function of radius at various times shortly before maximum compression[196] for a spherical laser pulse of 6.3 x 10^{11} W from 0 to 5.47 nsec, 6.3 x 10^{12} W to 7.21 nsec, followed by a linear increase to 4 x 10^{14} W up to 7.42 nsec (total laser energy E_0 = 60 kJ). The fusion energy gained is 510 kJ.

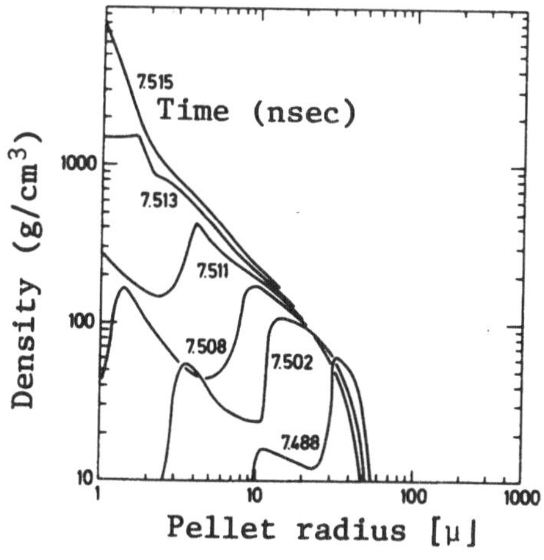

Fig. 8.5 The D-T pellet for the same case as Fig. 8.4[196] with density profiles immediately after the maximum compression.

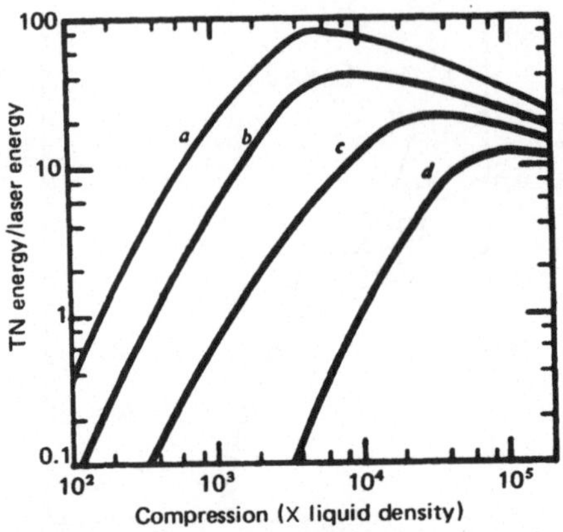

Fig. 8.6 Ratio of thermonuclear energy to incident laser energy E_0 as a function of the maximum compression for various values of E_0: a -- 1 MJ; b -- 100 kJ; c -- 10 kJ; d -- 1 kJ.

symmetry the final density in the center can reach ten thousand times the solid state density[194,196] (Figs. 8.4 to 8.6), thus yielding thermonuclear energy 40 times the incident laser energy E_0 = 100 kJ.

For the compression it is essential that the compressed core not be heated up too fast and so generate excessively-high final temperatures. On the other hand, the core must be heated quickly enough so that temperatures at least exceeding the Fermi temperature (Eq. (4.14)) are produced in the high-density plasma, otherwise the plasma undergoes a marked change in equation of state resulting in a much poorer compressibility. The inner side of the compressed plasma is set in motion by the ablation of plasma at the outer side of the compression shell.

The energy driving ablation is transferred from the laser energy absorbed within the corona. The following processes of thermalization are involved: The laser energy is transferred to electron energy in the corona by thermalization. The hot electrons with energies up to few hundred keV are then moving nearly without collisions in the plasma down to the compressed region where they are thermalized again to produce the energy for mechanical ablation. The mechanism of the energy transfer of the electrons has been analyzed quite extensively by Morse and Nielson.[201] The limiting energy conditions were derived for electrons involved in energy transfer to reach the compressed plasma yet not escape from it.

Due to the two steps of thermalization, at least two Carnot processes are involved which decrease the efficiency. The extremely-high speed of the processes will reduce the efficiency even more than the Carnot processes. This is the reason that in all known cases the numerical models have not shown energy transfer greater than about 7%; instead, 93% of the laser energy is dissipated in thermalization, ablation, and other losses. Nevertheless, the expected gains are singularly remarkable as shown in Fig. 8.6, and give reasonable hopes for building a thermonuclear fusion reactor.

There are a number of critical questions: How much energy is actually absorbed within the cut-off depth of the plasma; is reflectivity too high from the several types of instability -- a question which has been answered favorably by experiments;[23,24] is absorption too high in the peripheral corona due to dynamic absorption (Fig. 7.1). In addition there is a serious question of the stability of the spherical geometry of the plasma when temporal and spatial irregularies in the incident laser radiation are taken into account. Clearly, as mentioned in Section 2, there may be available (passive) correction elements to give the desired costant laser intensity at a specified spherical front within the irradiated plasma. However, inaccuracies and irreproducibilities in the production of the laser pulses may still occur, so that some tolerances may be necessary in the geometry. The effects of self-focusing at high laser intensities need further study. Spherical solutions have been made for the compression which show that some irregularities of a few percent in the geometry or in the intensity are tolerable.[202,203] The compression front has properties of self-stabilization, such that the plasma does not escape during the compression and still reaches rather spherical end states when compressed to ten thousand times the solid state density.

A further improvement of the calculations was made by including in the computer code not only the interaction of the alpha particles created[195,196,204] and their secondary heating of the high-density plasma but also the ignition of self-sustained nuclear fusion combustion fronts.[186,205,206] In this way it was possible to obtain nuclear fusion efficiencies exceeding 1000.[203]

8.3 Direct and More Efficient Transfer of Laser Energy into Mechanical Compression

The result described in Section 7.5 may be used to derive a more efficient type of compression than calculated by the gas-dynamic ablation model, in which 93% of the incident laser radiation is lost. Section 7.5 demonstrated that it is possible to transfer a much higher amount of laser energy into mechanical energy of a thick compression front by the mechanism of dielectric nonlinear forces. Because this mechanism always works better the smaller the amount of linear or nonlinear absorption described by an absorption constant due to collisions, it is possible to avoid thermalization and to reduce losses through transfer of thermal energy into mechanical energy, even by 100% under ideal conditions.

The only loss of laser energy is that due to the mechanical energy of the ablated plasma corona which is necessary to produce the compression momentum of the plasma layer at the cut-off density. In the specific example of Fig. 7.4, the amount of mechanical energy consumed in the corona blow-off was defined by the parameters of averaged velocity and mass of the compressed plasma and by the mass of the corona, which resulted in a 20% loss of the incident laser energy, while a 23% loss was sustained in the compressing plasma layer. The remainder of 57% was in reflected light and in thermal energy of the plasma, especially in energy due to dynamic absorption.

The example mentioned is only one of the first steps to optimize the necessary conditions for a high-efficiency compression scheme. Layers of various masses of the nuclei can be used in the region of the cut-off density in a way which supplies a much higher percentage of compression energy over ablation energy than in the above example. Further improvement is possible by more sophisticated variation of the laser intensity as shown by Nuckolls et al.[195] or by control of the laser frequency[207] to suppress dynamic absorption. It is not impossible to obtain by these means a compression of the plasma core to similar density as in gas-dynamic compression (ten thousand times the solid density) with more than 50% of the laser energy used effectively instead of 7% as in gas-dynamic ablation.

There is an additional question on how much energy is thermalized in the compression process which may decrease the efficiency. The optimum conditions would exist when the whole mechanical energy of the compressing plasma shell is transferred adiabatically into thermal energy. Deviation from this optimum may be caused, for example, by the impact of fast particles of the compressing shell with the cold stationary plasma which is to be compressed. The use of hollow spheres of the initial plasma (or of the irradiated target) and mixtures containing heavier atoms aid in optimizing conditions.

Preliminary and unpublished calculations of Goldman at Rochester deal with thermalization during the adiabatic compression of the sphere by the rapidly-compressing outer shell under realistic gas-dynamic conditions: Three-hundred joules in mechanical energy of compression in the outer shell of a sphere of deuterium at solid state density produced some 10^{13} fusion neutrons.

An analytic calculation of the optimized adiabatic compression can be based on a model in which the actual nonlinear pressure P_{NL} at the compressing layer is always larger than the gas-dynamic pressure $2n_i kT_i$, with equilibrium between electrons and the deuterium or tritium ions assumed for simplicity. The pressure difference defines a kinetic pressure P_{kin} when Eqs. (7.10) through (7.12) are used in the appropriate sense

$$P_{NL} = \frac{I}{2|\tilde{n}|} = 2\, n_i kT_i + P_{kin} \tag{8.8}$$

The following relations between final density n_{comp} and temperature T_{comp} and initial density n_1 and temperature T_1 hold for the adiabatic compression if the degenerate state is avoided

$$\frac{T_1}{T_{comp}} = \left[\frac{n_1}{n_{comp}}\right]^{2/3} ; \quad \frac{P_{comp}}{P_1} = \left[\frac{n_{comp}}{n_1}\right]^{5/3} \tag{8.9}$$

Using this relation and the one for the incident laser energy E_L, its intensity I, the laser pulse duration τ_L, and initial radius R_0 of the cut-off front, where $I = E_L/(4\pi R_0^2 \tau_L)$, we find

$$E_L/(4\pi R_0^2\, \tau_L) = I = c\,|\tilde{n}|\, n_1\, kT_{comp} \left(\frac{n_1}{n_{comp}}\right)^{2/3} \left[1 + \left(\frac{n_{comp}}{n_1}\right)^{5/3}\right] \tag{8.10}$$

$$\simeq c\,|\tilde{n}|\, n_1\, kT_{comp} \left(\frac{n_{comp}}{n_1}\right) \tag{8.11}$$

where $n_{comp} \geqslant 2n_1$

For the compression with nonlinear force the nuclear fusion efficiency G (Eq. (8.6)) is to be multiplied by a factor 2, because the fusion occurs during the phases of compression and expansion. This factor 2 enters as the factor 1/8 in the break-even energy E_{BE}, and with $n_0 = n_{comp}$ and $E_L = E_0$ in Eq. (8.6), we arrive at

$$G = \left(\frac{E_L}{E_{BE}/8}\right)^{1/3} \left(\frac{E_L}{4\pi r_0^{\,2}\, \tau_L\, c|\tilde{n}| n_0 kT_{comp}}\right)^{2/3} \tag{8.12}$$

The value of $|\tilde{n}|$ is to be understood as an average value from the detailed time-dependent compression process.

The laser pulse length τ_L can be determined from the adiabatic expansion or compression process (see Eq. (10) in Ref. 192)

$$\tau_L = \left(\frac{T_{comp}}{T_1} - 1\right)^{1/2} \left(\frac{m_i \tau_{min}^2}{2kT_{fusion}}\right)^{1/2} \tag{8.13}$$

Using Eq. (8.9) and the minimum radius $R_{min} = R_1 (n_0/n_{comp})^{1/3}$ we find

$$\tau_L = \left[\left(\frac{n_c}{n_0}\right)^{2/3} - 1\right]^{1/2} \left(\frac{n_0}{n_c}\right)^{1/3} R_0 \cdot 10^{-8}\ \text{sec} \simeq R_0\ 10^{-8}\ \text{sec} \tag{8.14}$$

and, therefore, from Eq. (8.12)

$$G = \frac{E_L}{R_0 |\tilde{n}|^{1/3}}\ 2.37 \times 10^{-4} \tag{8.15}$$

where E_L is in joules and R_0 in centimeters. R_0 and E_L are related by

$$R_0 = E_L^{1/3} \left(\frac{3}{4\pi n_0 kT_{comp}}\right)^{1/3} = 1.363 \times 10^{-3}\ E_L^{1/3} \tag{8.16}$$

which is independent of the adiabatic compression. Therefore, from Eqs. (8.15) and (8.16) we obtain

$$G = 0.167 \frac{E_L^{2/3}}{|\tilde{n}|^{1/3}} \tag{8.17}$$

with the laser energy E_L in joules. The necessary conditions are the initial radius R_0 (Eq. (8.16)), the total laser pulse length τ_L (Eq. (8.14)), and always

$$\frac{I}{c|\tilde{n}|} \gg (n_i kT_i) \tag{8.18}$$

with n_i below the cut-off density n_{co}.

The break-even energy is then near one joule. At E_L = 1 kJ and $|\tilde{n}| = 10^{-3}$, a gain of 167 would result.

8.4 New Concepts and Nuclear Fission

Advances in the compression of laser-produced plasmas by laser beams stimulated the concept of compressing fissionable materials such as uranium isotopes 235 and 233 (obtained from breeding thorium 232) or plutonium 239 (bred from uranium 238) by laser pulses. Much lower ion densities are found effective than in the case of hydrogen, because of the highly-stripped atoms, the accompanying electron density, and the limitations against degeneration. The detailed calculations of Winterberg,[208] as reproduced in later publications by other authors,[209] gave critical masses for fission chain reactions of some milligrams. The laser energies necessary to obtain positive yield were some 100 kJ. The process indicated by these calculations would thus be one of laser-triggered micro-explosions of nuclear fission reactions, which would be completely controllable. The fission reactors made possible with such laser-driven micro-explosions are the only known, absolutely safe reactors, since there is very fast control of the power generation. The reactor can be stopped immediately by interrupting the supply of uranium pellets. Each readily controllable reaction is a separate event without connection to a large amount of fissionable material, so there is little danger of a catastrophic explosion.

A further application of laser-induced compression for obtaining nuclear power is available in the use of boron 11, as reported by Weaver.[210] If a mixture of boron and hydrogen (the common light isotope) is compressed to 100,000 times the solid state density at a temperature of 30 keV, the reaction

$$H^1 + B^{11} \rightarrow 3\,He^4\ (+\ 8.78\ \text{MeV}) \tag{8.19}$$

takes place with the generation of a large amount of energy per reaction. The fission reaction products are stable helium atoms. Such a reaction would have the advantage over fission reactions of heavy atoms and also the D-T reactor since only stable and readily available atoms of hydrogen (H) and boron (B) are used. The reactor is absolutely free from danger of any catastrophic explosion (as in Winterberg's fission of uranium[208]) and the direct reaction product is clean, nonradioactive helium. The only problem is that higher compression of the pellets is necessary than for D-T fusion, which itself has been solved only theoretically by use of gas-dynamic ablation[197] or ablation driven by a non-thermalizing nonlinear force (Section 8.3).

It should be noted at this point that there is a much more fantastic concept of power generation with the laser than the previous concepts, the generation of antihydrogen. A rough estimation of the principle shows the possibility of efficient conversion of the energies involved.[155,157] It is possible to reduce the threshold intensities from 10^{19} W/cm^2 to 10^{16} W/cm^2 or less (Eqs. (6.24) and (6.33)) for relativistic cases to obtain electron-pair production with neodymium-glass lasers, in agreement with measurements of pair production at these laser intensities.[212] It should be possible to produce proton-antiproton pairs at intensities of 10^{19} W/cm^2 for neodymium-glass lasers or at 10^{17} W/cm^2 for CO_2 lasers with sufficiently-long pulses if a similar dielectric decrease in the relativistic threshold for getting GeV oscillation energies of protons can be obtained. Because of the necessarily long laser pulses (up to milliseconds), the laser powers required are not extremely high at 10^{11} W.[157]

Even without the details of the mechanisms, it appears that the efficiency of pair production should be higher by a factor of 10^{10} or more than in GeV accelerators, because the density in a plasma at solid state density is greater than a particle beam of an accelerator by this factor of 10^{10}. The reason is that the laser operates free of space-charge problems in a plasma and can generate extremely-high particle energies, which is completely impossible for particle beams due to the electrostatic forces. Taking into account the 10^{-7} grams of antihydrogen produced per year (which corresponds to 9 MJ of energy) in Serpukhov[211] where the accelerator may have consumed an energy of approximately 10^{17} J, we find for 10^{10} times higher efficiency a near balance between input energy and the energy stored in antihydrogen. A similar energy balance was estimated for antiproton generation by lasers,[157] where the Bhaba formula for the trident mechanism was simply used. A more realistic condition would be similar to that of accelerators with the exception of the oscillation of particles and the subsequent radiation losses. Energy generation is then a very easy procedure; the antihydrogen is stored at pressures below 10^{-15} torr (corresponding to temperatures of less than 0.5 K) by electromagnetic levitation. Manipulation and guiding, for example with laser beams, provides triggering of micro-fusion explosions by pellets of 10^{-8} grams of antihydrogen dropped into holes of solid D-T particles of appropriate size such that a controlled explosion is possible. This concept would also be workable with mini-explosions of deuterium, which would really provide a clean, inexhaustible, and inexpensive energy source if for any reason there were no way to compress plasmas by lasers to the densities necessary for D-T fusion or H-B fission.

Note added in proof: The concept of more efficient transfer of laser energy to a plasma (Section 8.3) was developed for a transfer rate of 50% (H. Hora, Atomkernenergie 24, 187, 1974). This efficiency, ten times higher than that of Nuckolls's gas-dynamic ablation,[202] leads by Eq. (8.6) to an energy requirement only one-thousandth as large for the same reaction efficiency G. The total efficiency $G_{tot} = 60$ based on incident laser energy E_i will be reached for $E_i = 246$ kJ (D-T), $E_i = 98$ kJ (D-D), and $E_i = 1.47$ MJ (H-B). Comparison with detailed computer studies (H. Hora, E. Goldman, and M. Lubin, University of Rochester, LLE-Rept. 21, 1974) confirms these values. Thus, this scheme of compression of plasma by nonlinear force-driven fast-imploding nonheated plasma gives for the first time the possibility of exothermic H-B fission (M. L. E. Oliphant and Lord Rutherford, Proc. Roy. Soc. A141, 259, 1933), which is absolutely clean, does not involve neutrons, and avoids heat pollution.

9. EXPERIMENTS FOR LASER-INDUCED NUCLEAR FUSION

The first observation of neutrons as a reaction product of the D-D reaction from a laser-irradiated target of lithium deuteride -- published by Basov, Kryukov et al. in 1968[213] -- was a milestone in the experiments for laser-induced nuclear fusion. The target was irradiated by laser pulses of energy from 10 to 20 joules and pulse duration of about 10 picoseconds. The overall number of fusion neutrons was between 30 and 100 and was very close to the noise level of the detectors. This experiment has been repeated by Gobeli et al.,[214] with similar pulses of a few picoseconds duration, but the number of neutrons generated was again not far above the noise level. Since these experiments no further results have been published on neutron production from deuterated targets with picosecond pulses. The breakthrough came in 1969, when Floux et al. irradiated solid deuterium with laser pulses of a few nanoseconds duration, producing 10^3 neutrons or more.[215] Similar results were reported by Lubin at a Gordon Conference one week before the disclosure of Floux, without, however, undergoing the normal processes involved in regular publication.

9.1 Irradiation of Spherical Targets

Some time before the first fusion neutrons were observed, several studies were made of irradiation of solid spherical targets with radius varying from 5 to 150 microns. The particle was normally suspended on a quartz fiber, a technique used also at present.[23] In other schemes, the particle was supported by an electromagnetic field[139--142,226] produced with three pairs of plates each crossed and carrying a hf-field, such that the pellet described a stable path according to the solutions of Mathieu's differential equation.[217] In yet different schemes, free-falling pellets have been used,[199] where the initially-fixed pedestal was removed downward with high acceleration.[219] The technique of preparing spherical pellets of solid deuterium, hollow spheres, or pellets of varying composition is still a major obstacle for laser-induced nuclear fusion or, indeed, for generation of clean, well-defined, high-temperature plasmas for other purposes.[220]

The early measurements were intended to analyze the process of generation of the plasma, in particular whether a spherically-symmetric plasma could be formed and whether the plasma parameters of density, temperature, and ion velocity agreed with the values expected from the theory of the self-similarity model[109,139--142,146] (Eqs. (5.21) and (5.22)). The highly spherically-symmetric expansion was established experimentally by Lubin et al.[141] only when a prepulse of about 10 nsec duration was incident on a target and having an intensity of less than 10% of the main laser pulse (several joules) which followed the prepulse within 10 nsec or more. In all other cases non-symmetries were observed. An unusual phenomenon was noted with pellets of aluminum and using a pedestial system,[109] and also for LiD pellets suspended electromagnetically.[140] A fast bunch of plasma with high ion energies (several keV) moving preferentially toward the laser aperture was recorded.[108,221] Analysis of the experimental data indicated that these fast ions were generated by a surface mechanism,[109,219] because on varying the pellet diameter (with other parameters constant) there was no change of the ion energy. A further anomaly was the superlinear increase in ion energy with incident laser energy or incident laser power, if the varying pulse length was readjusted. However, the inner core of the plasma,[109] which contained most of the generated plasma, expanded with high spherical symmetry, and the parameters of ion energy, temperature, absorbed energy, and transparency time corresponded exactly with the theory of the self-similarity model. Again this confirmed that there was homogeneous heating and no shock-wave heating. These thermokinetic properties found by irradiation of aluminum spheres of 50- to 150-micron radius were also observed under the much more severe conditions of small, electromagnetically-suspended pellets,[139--142] whose size had to be determined by a diffraction technique,[141] and measurement of the absorbed radiation was much more difficult.[141] Maximum electron temperatures of up to 1 keV have been measured[141] with 1-J 10-nsec ruby laser pulses using the prepulse technique.

9.2 Neutron Generation

The generation of fusion neutrons by irradiation of deuterated polyethylene or lithium deuteride targets by laser pulses of more than 100 psec duration involves the question of the risetime of the laser pulses. During the 1968-1969 period several laboratories studied neodymium-glass laser pulses of some ten joules and 30 nsec duration. However, the temperature generated, as measured from the x-ray spectrum, indicated values of only a few hundred eV, although there were some anomalies in evaluation of the measurements which occasionally gave temperatures of several keV.[213]

The systematic study of a carefully-measured electron temperature from the x-ray spectrum was performed by Floux et al,[215] who discovered that there is an increase of the temperature if the laser pulse exhibits a steeper increase with time. Floux used electro-optical shutters with risetimes of about 1 nsec. With this method conditions were found in which at high temperatures a large number of neutrons was emitted. Figure 9.1 shows the results attained with this method. The confirmation of neutrons produced by D-D reaction was made by measuring the time of flight of the neutrons, which corresponded to an energy of 2.45 MeV. In very recent experiments,[23] measurement of the fast protons generated (Eq. (1.1)) with 3.02 MeV energy was used to establish the occurrence of more than 10^5 fusion reactions per shot.

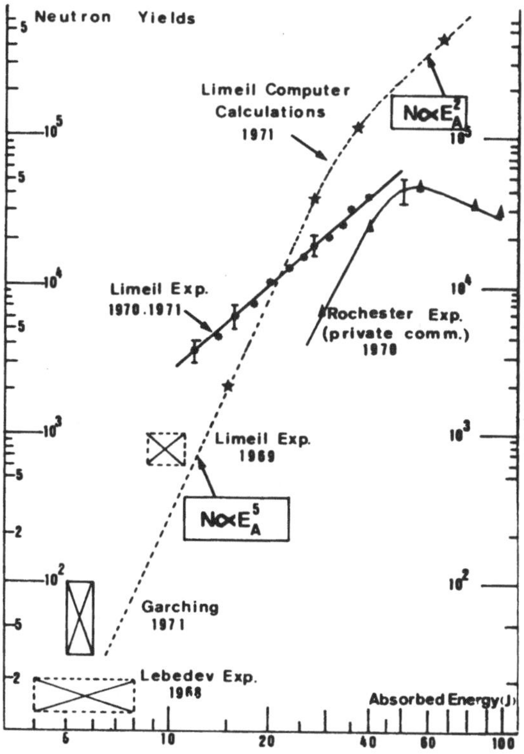

Fig. 9.1 Neutron yields from laser-produced plasmas.[222]

The next step in such experiments was the design of a laser system by Sklizkov[70] with eventually nine parallel beams operating simultaneously. A spherical $(CD_2)_n$ target was irradiated symmetrically by the beams with a time accuracy of better than 100 psec. These measurements resulted in the generation of 10^7 neutrons,[233] a large enough number so that an analysis of the thermal broadening of the signal could be made.[13] The conclusion was that the ion temperature of the plasma generating neutrons was less than 5 keV, a remarkable result proving the thermonuclear origin of the neutrons.[13]

There was some discussion of whether the neutrons came from the plasma, from the wall of the vessel (by collision of fast deuterons from the plasma), or from residual gas pressure in the experimental chamber. A clear answer was given by experiments in which after a prepulse laser pulses with energies from 2 to 50 J and 100-psec duration were incident[23] on spherical $(CD_2)_n$ or LiD targets in a chamber with vacuum better than 10^{-6} torr and walls at a large distance. By means of the geometry and the timing of the pulses, the neutrons could be definitely shown to originate only at the pellet.[23] This does not exclude, however, the mechanism of neutron generation at the walls under other conditions which favor generation of instabilities.

Some controversy has arisen in the theoretical interpretations of the neutrons generated by compression and thermalization of plasma and possibly also by instability mechanisms. While the desired neutron fluxes of 10^{14} might be expected for a sophisticated increasing laser pulse of 1 kJ spherically incident on a pellet and provided high compression can be achieved, it was pointed out that under the conditions of the nine-beam experiment the gas-dynamic theory could account for only 10^3 neutrons per pulse in contrast to the observed 10^7 neutrons.[13] The experimentalists were puzzled over the theoretical result, while interpreting their values for these high-neutron fluxes in terms of compressions up to factors of 40, presuming the measurements were related consistently to the analytic theory of generation of a compression wave.[13]

9.3 Anomalous Experimental Results

keV x-rays were measured from plasmas created by lasers in 1969, while more carefully executed measurements resulted in an order lower temperatures derived from the x-ray spectra. Similar reports were made of some ten keV x-ray signals[224] from plasmas with high-neutron efficiency, while other measurements indicated temperatures of only a few keV. Eidmann analyzed the x-ray spectra more rigorously and discovered that there is no spectrum which can be related to a definite temperature.[18] Indeed, the irradiation of carbon provided highly-distinct temperatures, clearly corresponding to the maximum temperature in the plasma. But in addition there were emission of a fast clump of peripheral plasma and radiation at times when lower temperatures were present. The latter processes are relatively of too little intensity. The markedly-anomalous spectrum indicated the generation of a strongly-non-Maxwellian temperature with an elevated higher value of several keV in addition to the regular temperature of about 1 to 3 keV. The elevated temperatures have been confirmed by several authors.[19] There was some correlation between the elevated temperature and the neutron signals. Polarization of the x-rays was detected in a direction parallel to the laser beam and, in evident contradiction to this, there was also an isotropic flux of x-rays.[53]

The anomalously-high temperatures extended the physics of laser-produced plasma very recently for the first time to the physics of elementary particles. X-ray signals from plasmas irradiated with laser pulses of intensities exceeding 10^{16} W/cm^2 were measurable through six centimeters of lead.[212] The spectrum corresponded to an electron temperature between 1 and 10 MeV. Although the intensity would have to exceed 10^{18} W/cm^2 to generate electron-positron pairs in a plasma with an electron density less than the cut-off density, it has been pointed out that the increase of the effective electric field strength E in a plasma near the cut-off density can significantly exceed the corresponding vacuum value due to the small values of the denominator $|\tilde{n}|$ (Eq. (6.24)).[155,157] Therefore, it is possible that the experiment in question resulted in the production of electron-positron pairs due to the dielectric effect or to another instability mechanism in the plasma.[212] Following this result it was pointed out[157] that laser intensities in vacuum may give rise to proton pair production at CO_2 laser intensities from 10^{17} to 10^{18} W/cm^2 (this is possible only with CO_2 lasers for solid state densities because of the relativistic shift of the cut-off density).

Another anomaly in neutron production -- correlated highly in some experiments -- is the generation of a group of fast ions.[18,19] This was not unknown in plasmas with laser intensities much below the thresholds for neutron generation.[108,109,140,221] While the major portion of the plasma had average ion energies not greatly exceeding 1 keV, a group of fast ions of 20 keV was discovered.[18,19] This fast group consisted of more than 10^{15} ions,[19] which excludes any electrostatic acceleration mechanism of the Debye sheath at the plasma surface since the sheath concentration is limited to 10^{12} ions/cm^3.

When the incident light was constant, the reflected light and the emission of x-rays were not constant but showed a pulsation with a frequency of about half a nanosecond.[226,227] This may indicate the generation of an instability in the plasma, or possibly, the generation of self-reflecting density ripples (see Section 7.7), their subsequent decay and re-establishment within the measured time. The spectra of the reflected light sometimes showed an anisotropy, depending on the polarization and the frequency of the reflected light. The spectra of the reflected light contained the second harmonics (2ω) or the half harmonics $(3/2)\omega$ and $(1/2)\omega$ which were in most cases red-shifted[228,229] (in few cases blue-shifted[230]) and broadened.[23] The likelihood of a stimulated Compton effect was deduced from the generation of a second line, a few Angstroms to the red of the second harmonic.[229] The pronounced emission of the half harmonics especially at 45° from the incident laser light indicated a two-plasmon decay instability (Raman instability).[165,173]

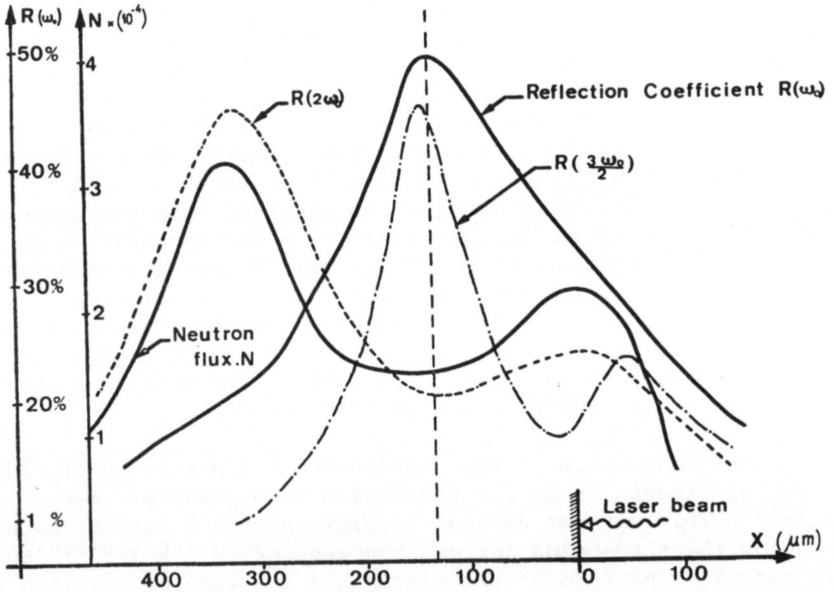

Fig. 9.2 Focusing of neodymium-glass laser radiation at various depths x in solid deuterium,[229,231] the accompanying neutron emission, and reflection of the basic frequency ω_0, and its harmonics $2\omega_0$ and $3\omega_0/2$.

A highly-important experiment on the correlation of the reflection and of neutron generation was performed by Floux, Bobin et al.,[229,231] in which the focal depth of the laser irradiating the solid deuterium was moved (Fig. 9.2) from 150 microns outside the target to 500 microns inside. At a depth of 130 microns the reflection of the incident light and of the 3/2 harmonics was extremely high, as was the emission of x-rays from the elevated high level, while the emission of neutrons was remarkably low. Only at very deep focusing did the neutron emission reach its greatest maximum paralleling closely the intensity of the reflected second harmonic. With such complexities, which still have not been analyzed completely, it is no wonder that such a high scatter of the experimental reflection coefficients has been observed as shown in Fig. 9.2

The design of a laser fusion experiment and the facilities at the Lawrence Livermore Laboratory in Livermore, California, is shown in Figs. 9.3 through 9.5. The essential fact of high compression as emphasized by Teller, of Lawrence Livermore Laboratory,[194] has not been demonstrated in a direct way. The observation of 10^7 neutrons from laser pulses of about 1 kJ was related to a compression of 40 times the solid state density;[13] only the 10^{14} neutrons expected in similar, but more sophisticated experiments shall be an indication of the desired high compression.[223] Measurements in laser-irradiated polyacrylate (Plexiglas) demonstrated directly a compression up to 3 times the solid state at a maximum pressure P_{max} of 10^6 atmospheres, using 10-J 6-nsec laser pulses.[234] From

Fig. 9.3 Model of a 12-beam neodymium-glass laser experiment at the Lawrence Livermore Laboratory in Livermore, California.

Fig. 9.4 Design of the building for the 12-beam laser of Fig. 9.3

these measurements a scaling law may be derived

$$P_{max} = a \sqrt{E_0} \ ; \ a = 10^6 \ \text{atm/joule}^{1/2}$$

The related compression is in distant agreement with the result of 40-fold compression for incident laser energies in the kilojoule range.[13]

The use of electron beams has been proposed for measuring the density attained in the compressed plasma by a direct method.[235] The energy of the beam, E_e = 100 keV or more, passing through the plasma will be reduced by a part of the electrons by discrete energy losses, $\Delta E_e = \hbar\omega_p$, where the plasma frequency ω_p is related to the actual electron density given by Eq. (4.10). These energy losses can be measured either directly by velocity analysis of the

Fig. 9.5 Overall view of laboratory containing the 12-beam experiment of Fig. 9.3.

electron beam or -- if the beam is highly monochromatic -- by its modulation, which causes an emission of photons at the plasmon frequency from targets hit by the electron beam. The modulation by plasmons has been evaluated[236] in analogy to the quantum-mechanical modulation of an electron beam by laser photons (Schwarz-Hora effect).[237] The necessarily high electron energy E_e for transmission through the plasma favors modulation. The modulation has the advantage of permitting time-integrated measurements, while a velocity analyzer of the electron beam requires high temporal resolution such as provided by the method of Siegbahn.[238]

Finally, a worldwide increase of the research effort in the production of fusion plasmas by laser should be underscored. Concepts have been advanced on the future design of power reactors.[235] The 50-million-dollar-plus annual budget for the activities in this field in the United States for 1973 may double in 1974.

10. CONCLUSIONS

The supply of inexpensive energy is one of the key problems of modern civilization. It can be accepted only as a compromise that such valuable resources as fossil oil or coal are used for power generation instead of as raw materials for organic chemistry.[240] This presumably temporary compromise is based on a comparison of the cost of the present type of nuclear fission reactor with the high cost of fossil fuel resources. Nuclear fission at present hardly reaches the level of competitivity[241] because the basic material -- uranium and thorium ore -- is extremely expensive. Economically-available ore can supply energy only for 50 years, when the minimum annual consumption of 10^{22} joules predicted for the year 2050 is taken into account. The problems of immense production of fissionable materials that are radioactive, the possibility of a catastrophic explosion of a fission power station, which cannot be completely ruled out, and other questions of security[242] basically categorize the present fission reactors only as a short-range energy source.

A truly inexpensive, inexhaustible, and clean energy source on earth -- as in the sun -- is hydrogen, especially its heavy isotope deuterium (present as 0.03 percent of the weight of the natural element), available in all water. Its thermonuclear reaction producing helium has been realized, at least by the method of uncontrolled large-scale reaction in the H-bomb, where inertial confinement is used. The deuterium compounds are heated up by small nuclear fission explosions to the desired temperatures exceeding 100 million degrees, and exothermic reactions occur in the subsequent expansion before adiabatic cooling takes place and the reactions are extinguished. Controlled nuclear fusion by magnetic confinement of the hot plasma may be considered in the best case only as a temporary concept even if such a reactor can be realized in the future. The cyclotron radiation that is necessarily generated limits the possibilities to reaction of deuterium with tritium, which can be bred only from lithium. However, lithium sources are expected to last only 70 years compared to the 50 years for uranium and thorium, as mentioned above. This comparison is still based on expensive fossil energy sources and not on the truly inexpensive energy source needed in the future.

The laser offers a means to produce controlled thermonuclear reaction by inertial confinement similar to the H-bomb. For lasers it is possible, at least in principle, to inject energies of several kilowatt-hours (or Megajoules) within 10^{-9} seconds into 1-cm^3 volumes of condensed heavy hydrogen isotopes, which causes an exothermic microexplosion. This task can be solved in a much more advantageous way if the laser-irradiated plasma is compressed by the interaction to ten thousand times the solid state density. Laser compression is presently aimed at fusion with deuterium-tritium mixtures, but in the future this concept should work with pure deuterium. Following realization of such a compression, the possibility of a tenfold greater compression may be considered for mixtures of hydrogen and boron; the greater compression would provide exothermic controllable burning directly and with absolute safety into clean helium. Laser compression could also give absolutely-safe fission reactions, e.g., for uranium, in which milligram critical masses suffice to produce controllable microexplosions. If for any reason the laser compression up to necessary densities is not effective, the laser may realize controllable microexplosions of solid deuterium by inertial confinement in another way: In the next few years very high laser intensities should produce antihydrogen efficiently, in contrast to accelerator methods. Pellets of some 10^{-8} g of antihydrogen should ignite microexplosions of deuterium under inertial confinement.

The physical problems of laser-plasma interaction are highly complex. The experiments of different research groups give results with a wide range of scatter, and numerous phenomena remain unexplained. Current empirical results are: Laser-irradiated plasmas have temperatures of several keV (tens of millions of degrees) in addition to "anomalous" temperatures of up to 10 MeV; neutrons are produced by nuclear fusion reactions of thermalized ions (10^7 neutrons from 200 joules of laser energy); reflected radiation indicates instabilities which suppress neutron production; at the desired intensities the reflectivity is low such that the laser fusion concept still appears to be feasible. The compression of plasma up to three times solid state density has been shown directly, while the possibility of compression up to a factor of

40 has been indicated indirectly. Self-focusing could be excluded in these cases, and the generation of a compression front was established.

Sophisticated experiments, however, can demonstrate the necessary compression, where yields of 10^{14} neutrons are expected from irradiation with 1 kilojoule laser pulses. Numerical calculations of the laser interaction, compression, and neutron production are based on gas-dynamic ablation in which 7% of the laser energy enters the compressed core. The compression is independent of some asymmetries when a spherical target is irradiated. Calculated fusion yields range up to 100 for incident laser energies of 100 kJ. A new type of compression by ablation due to non-thermalizing nonlinear forces of laser-plasma interaction gives a transfer of 23% of laser energy into mechanical compression energy and this may be improved at least in principle.

11. REFERENCES

1. Hudock, R. P., Astronautics & Aeronautics, 11, No. 8, 6 (1973).
2. Meadows, D. H. et al., *The Limits of Growth*, (Universe Books, New York, 1972).
3. Prokhorov, A. M., Paper presented to the opening session of "Laser '73," Munich, Sept. 4-7, 1973.
4. Häfele, W., Atomwirtschaft, 17, 384 (1972).
5. Glasstone, S., and R. H. Lovberg, *Controlled Thermonuclear Reactions*, (Van Nostrand, New York), 1960.
6. Tuck, J. L., Nucl. Fusion 1, 201 (1961).
7. Grieger, G., W. Ohlendorf, H. D. Pacher, H. Wobig, and G. H. Wolf, *Plasma Physics and Controlled Nuclear Fusion*, (Madison Conf.), TAEA, Vienna, 1971, Vol. III, p. 37.
8. Pfirsch, D., and A. Schlüter, Max-Planck-Institut für Physik (Munich), Rept. MPI/PA/7/1962.
9. Wood, L., and J. Nuckolls, Environment 14, No. 4, 29 (1972).
10. Trubnikov, B. A., and V. S. Kudryavtsev, U.N. Geneva Conference 1958, Vol. 32, p. 93.
11. Hora, H., Laser und Elektro-Optik 5, No. 4 (1972), p. 31.
12. Rocks, L., and R. P. Runyon, *The Energy Crisis*, (Crown Publishers, New York), 1972, p. 84; Committee on Resources and Man, *Resources and Man*, Chap. 8.
13. Rupasov, A. A., G. V. Sklizkov, V. P. Tsapenko, and A. A. Shikanov, ZhETF, 65, 1898 (1973).
14. Basov, N. G., V. A. Boyko , O. N. Krokhin, O. G. Semenov, and G. V. Sklizkov, ZhTF 38, 11 (1968).
15. Basov, N. G., V. A. Boyko, V. A. Gribkov, S. M. Sakharov, and G. V. Sklizkov, Preprint, Lebedev Institute, 3(1968).

16. Büchl, K., K. Eidmann, H. Salzmann, and R. Sigel, Appl. Phys. Letters 20, 231 (1972).
17. Sigel, R., S. Witkowski, H. Baumhacker, K. Büchl, K. Eidmann, H. Hora, H. Mennicke, and D. Pfirsch, Kvantovaya elektronika 2, No. 8, 37 (1971).
18. Büchl, K., K. Eidmann, P. Mulser, and R. Sigel, *Laser Interaction and Related Plasma Phenomena*, H. Schwarz and H. Hora (eds.), (Plenum, New York, 1972), Vol. 2, p. 503.
19. Yamanaka, C., ibid., p. 481.
20. Floux, F., J. F. Benard, D. Cognard, and A. Saleres, ibid., p. 409.
21. Shearer, J. W., S. W. Mead, J. Petruzzi, F. Rainer, J. E. Swain, and C. E. Violet. Phys. Rev. A6. 764 (1972).
22. Jones, E. G., G. W. Gobeli, J. N. Olsen, *Laser Interaction and Related Plasma Phenomena*, H. Schwarz and H. Hora (eds.), (Plenum, New York, 1972), Vol. 2, p. 469.
23. Goldman, L. M., J. Soures, and M. J. Lubin, Phys. Rev. Letters 31, 1184 (1973).
24. Brueckner, K. A., *Laser Interaction and Related Plasma Phenomena*, H. Schwarz and H. Hora (eds.), (Plenum, New York, 1974), Vol. 3B, p. 427.
25. Einstein, A., Ann. Physik, 17, 123 (1905).
26. Ladenburg, E., and K. Markov, Verh. Deut. Phys. Ges., 10, 564 (1908).
27. Millikan, R. A., Phys. Rev., 30, 287 (1910); *The Electron*, (New York, 1917).
28. Kirchner, F., Ann. Physik, 83, 521 (1927).
29. Hora, H., Jenaer Jahrbuch, 1960, II, p. 515.
30. Schawlow, A. L., and C. H. Townes, Phys. Rev., 112, (1940) (1958).
31. Hora, H., Z. Naturforsch., 20A, 543 (1965).
32. Maiman, T. H., Nature, 187, 493 (1960).
33. Collins, R. J., D. F. Nelson, A. L. Schawlow, W. Bond, C. G. B. Garrett, and W. Kaiser, Phys. Rev. Letters, 5, 303 (1960).
34. Snitzer, E., Proc. IEEE, 54, 1249 (1966); C. G. Young, Microwaves, 7, 69 (1968).
35. Sorokin, P. P., J. J. Luzzi, J. R. Lankard, and C. O. Pettit, IBM J. Res. Develop., 182 (1964); Schäfer, F. P. (ed.), *Dye Lasers* (Springer, Heidelberg, 1973).
36. Hall, R. N., Solid-State Electronics, 6, 409 (1963); Nathan, M. I., ibid, p. 425; Winstel, G., *Laser*, Kleen, W., and R. Müller (eds.), (Springer, Heidelberg, 1969), p. 360; Ivey, H. F., IEEE J. Quantum Electronics, QE-2, 713 (1966).
37. Hora, H., Phys. stat. solidi, 8, 197 (1965); J. Appl. Math. Phys., 16, 98 (1969).
38. Basov, N. G., O. V. Bogdankevich, and A. G. Devyatkov, Dokl. Akad. Nauk SSSR, 155, 783 (1964) (Sov. Phys. Doklady 9, 288 (1964); Horwitz, C. E., Appl. Phys. Letters, 8, 121 (1966).

39. Nehrich, R. B., E. J. Schimitschek, and J. A. Trias, Phys. Letters, 12, 198 (1964).
40. Lempicki, A., and H. Samelson, *Lasers*, A. K. Levine (ed.), (Marcel Dekker, New York, 1968), Vol. 1, p. 181.
41. Kasper, J. V. V., and G. C. Pimentel, Appl. Phys. Letters, 5, 231 (1964); Hohla, K., *Laser Interaction and Related Plasma Phenomena*, H. Schwarz and H. Hora (eds.), (Plenum, New York, 1974), Vol. 3A, p. 133.
42. Javan, A., Phys. Rev. Letters, 3, 87 (1959); McFarlane, R. A. et al., Proc. IRE 50, 2 III (1962).
43. Henry, L., and F. Legay, Onde Elect., 46, 410 (1966).
44. Patel, C. K. N., Appl. Phys. Letters, 7, 15 (1965); Patel, C. K. N., *Lasers*, A. L. Levine (ed.), (Marcel Dekker, New York, 1965), Vol. 2, p. 1.
45. Kompa, K. L., and G. C. Pimentel, J. Chem. Phys., 47, 857 (1967); Kompa, K. L., Topics in Current Chemistry, 37, 1 (1973).
46. Godard, B., IEEE J. Quantum Electronics, QE 9, 645 (1973).
47. Gerry, E. T., IEEE Spectrum, 7, 51 (1970).
48. McClung, F. J., and R. W. Hellwarth, J. Appl. Phys., 33, 828 (1962); Harvey, A. F., *Coherent Light* (Wiley-Interscience, London 1970).
49. DeMaria, A. J., and W. H. Glenn, *Laser Interaction and Related Plasma Phenomena*, H. Schwarz and H. Hora (eds.), (Plenum, New York, 1970), Vol. 1, p. 11.
50. Letokhov, V. S., Sov. Phys. JETP, 28, 562 (1969).
51. Salzmann, H., Phys. Letters, 32A, 113 (1970).
52. Schwarz, H., *Laser Interaction with Matter*, C. Yamanaka (ed.), (Japan Soc. Prom. Sci.), Tokyo, 1973.
53. Eidmann, K., and R. Sigel, *Laser Interaction and Related Plasma Phenomena*, H. Schwarz and H. Hora (eds.), (Plenum, New York, 1974), Vol. 3B, p. 667.
54. Duguay, M. A., and J. W. Hansen, Appl. Phys. Letters, 15, 192 (1969).
55. Harris, S. E., (Private communication, 1973).
56. Young, J. F., A. H. Kung, G. C. Bjorklund, and S. E. Harris, IEEE J. of Quantum Electr., QE-9, 700 (1973).
57. Mennicke, H., Phys. Letters, 37A, 381 (1971).
58. McCall, S. L., and E. L. Hahn, Phys. Rev., 183, 457 (1969).
59. Carman, R. L., W. H. Lowdermilk, J. Kysilka, M. Moran, and G. Kachen, IEEE J. Quantum Electronics, QE-9, 713 (1973).
60. Yamanaka, C., *Laser Interaction and Related Plasma Phenomena*, H. Schwarz and H. Hora (eds.), (Plenum, New York, 1974), Vol. 3B, p. 626.
61. Glass, A., and A. H. Guenther, ibid.
62. Lubin, M., ibid.
63. Brueckner, K. A., ibid.
64. Boyer, K., ibid; Fenstermacher, C., invited paper, CLEA, June 1973.

65. Godwin, R. G., (Private communication, 1973).
66. Gerardo, J. B., and A. W. Johnson, IEEE J. of Quantum Electronics, QE-9, 647 (1973); Rhodes, C. K., ibid., p. 647.
67. Yamanaka, C., *Laser Interaction and Related Plasma Phenomena*, H. Schwarz and H. Hora (eds.), (Plenum, New York, 1972), Vol. 2, p. 495.
68. Basov, N. G., 2nd Intern. Conf. Laser and Applications, Dresden, E. Germany, June 6-9, 1973, paper P1.
69. Pummer, H., W. Brietfeld, H. Wedler, G. Klement, and K. L. Kompa, Appl. Phys. Letters, 22, 319 (1973).
70. Basov, N. G., O. N. Krokhin, and G. V. Sklizkov, *Laser Interaction and Related Plasma Phenomena*, H. Schwarz and H. Hora (eds.), (Plenum, New York, 1972), Vol. 2, p. 389.
71. Maker, P. D., R. W. Terhune, and C. M. Savage, *Proc. 3rd Int. Quantum Electronics Conf., Paris*, February 1963, Bloembergen, N. and M. Grivet (eds.), (Dunod, Paris, 1964), Vol. I, p. 1559.
72. Meyerand, R. G., and A. F. Haught, Phys. Rev. Letters, 11, 401 (1963).
73. Gill, D. H., and A. A. Dougal, Phys. Rev. Letters, 15, 845 (1965).
74. Alcock, A. J., and M. C. Richardson, Phys. Rev. Letters, 21, 667 (1968).
75. Ramsden, S. A., and W. E. Davies, Phys. Rev. Letters, 13, 227 (1964).
76. Ramsden, S. A., (Private communication).
77. Linlor, W. I., Appl. Phys. Letters, 3, 210 (1963); Phys. Rev. Letters, 12, 383 (1964).
78. Mandel'shtam, S. L., P. P. Pashinin, A. V. Prokhind'yev, A. M. Prokhorov, and N. K. Sukhodrev, ZhETF, 47, 2003 (1964) (Sov. Phys. JETP, 20, 1344 (1965)).
79. Minck, R. W., and W. G. Rado, J. App. Phys., 37, 355 (1965).
80. Ramsden, S. A., and R. Savic, Nature, 203, 1217 (1964).
81. Rayzer, Yu. P., Sov. Phys.-Uspekhi, 8, 650 (1966).
82. Hora, H., D. Pfirsch, and A. Schlüter, Z. Naturforsch., 22A, 278 (1967).
83. Hora, H., Phys. Fluids, 12, 182 (1969).
84. Schlüter, A., Plasma Phys., 10, 471 (1968).
85. Hora, H., *Laser Interaction and Related Plasma Phenomena*, H. Schwarz and H. Hora (eds.), (Plenum, New York, 1970), Vol. 1, p. 383.
86. Hora, H., Opto-Electronics, 2, 201 (1970).
87. Hora, H., *Laser Interaction and Related Plasma Phenomena*, H. Schwarz and H. Hora (eds.), (Plenum, New York, 1972), Vol. 2, p. 341.
88. Kazakov, A. Ye., I. K. Krasyuk, P. P. Pashinin, and A. M. Prokhorov, JETP Letters, 14, 280 (1971).

89. The x-ray emission corresponding to some 100 eV may arise from bremsstrahlung of the fast-moving cold plasma; see A. J. Alcock, P. P. Pashinin, and S. A. Ramsden, Phys. Rev. Letters, 17, 528 (1966).
90. Litvak, M. M., and P. F. Edwards, IEEE J. Quantum Electronics, QE-2, 486 (1966); Hohla, K., Inst. f. Plasmaphysik, Garching, Rpt. 3/67 (1968).
91. Wright, J. P., Proc. Phys. Soc., 84, 41 (1964).
92. Bystrova, T. B., G. S. Voronov, G. A. Delone and N. B. Delone, JETP Letters, 5, 178 (1967).
93. Papoular, R., *Laser Interaction and Related Plasma Phenomena*, H. Schwarz and H. Hora (eds.), (Plenum, New York, 1972), Vol. 2, p. 79.
94. Korobkin, V. V., and A. J. Alcock, Phys. Rev. Letters, 21, 1433 (1968); Ahmad, N., B. C. Gale, and M. H. Key, J. Phys., B2, 403 (1969).
95. Hora, H., Z. Physik, 226, 156 (1969).
96. Alcock, A. J., *Laser Interaction and Related Plasma Phenomena*, H. Schwarz and H. Hora (eds.), (Plenum, New York, 1972), Vol. 2, p. 147.
97. Ready, J. F., *Effects of High-Power Laser Radiation* (Academic Press, New York, 1971).
98. Lichtman, D., and J. F. Ready, Phys. Rev. Letters, 10, 342 (1963).
99. Hora, H., and H. Müller, Z. Physik, 164, 359 (1961).
100. Ivey, H. F., *Advances in Electronics and Electron Physics*, L. Marton (ed.), (Academic Press, New York, 1954), Vol. 6, p. 137.
101. Locke, E. V., E. D. Hoag, and R. A. Hella, IEEE J. Quantum Electronics, QE-8, 132 (1972).
102. Schwarz, H. J., *Laser Interaction and Related Plasma Phenomena*, H. Schwarz and H. Hora (eds.), (Plenum, New York, 1970), Vol. 1, p. 71.
103. Zahn, H., and H. J. Dietz, Exp. Technik der Physik, 20, 401 (1972).
104. Linlor, W. I., *Laser Interaction and Related Plasma Phenomena*, H. Schwarz and H. Hora (eds.), (Plenum, New York, 1970), Vol. 1, p. 173.
105. Honig, R. E., Appl. Phys. Letters, 3, 8 (1963).
106. Siller, G., K. Büchl, and H. Hora, *Laser Interaction and Related Plasma Phenomena*, H. Schwarz and H. Hora (eds.), (Plenum, New York, 1972), Vol. 2, p. 253.
107. Isenor, N. R., Appl. Phys. Letters, 4, 152 (1964).
108. Namba, S., H. Schwarz, and P. H. Kim, *IEEE Symposium on Electron, Ion and Laser Beam Technology*, Berkeley, May 1967, p. 861.
109. Engelhardt, A. G., T. V. George, H. Hora, and J. L. Pack, Phys. Fluids, 13, 212 (1970).

110. Metz, S. A., Appl. Phys. Letters, 22, 211 (1973).
111. Hora, H., Appl. Phys. Letters, 23, 39 (1973).
112. Gregg, D. W., and S. J. Thomas, J. Appl. Phys., 37, 4313 (1966).
113. Gregg, D. W., and S. J. Thomas, J. Appl. Phys., 37, 2787 (1966).
114. Thomas, H., Z. Physik, 147, 395 (1957).
115. Ringler, H., and R. A. Nodwell, Phys. Letters, 29A, 151 (1969); Ludwig, D., and C. Mahn, Phys. Letters 35A, 191 (1971); Kronast, B., *Laser Interaction and Related Plasma Phenomena*, H. Schwarz and H. Hora (eds.), (Plenum, New York, 1972), Vol. 2, p. 181.
116. Biskamp, D., (Private communication, 1973).
117. Kruer, W. L., *Laser Interaction and Related Plasma Phenomena*, H. Schwarz and H. Hora (eds.), (Plenum, New York, 1974), Vol. 3A, p. 341.
118. Schlüter, A., Z. Naturforsch, 5A, 72 (1950).
119. Hora, H., *Laser Interaction and Related Plasma Phenomena*, H. Schwarz and H. Hora (eds.), (Plenum, New York, 1970), Vol. 1, p. 383.
120. Spitzer, L., *Physics of Fully Ionized Plasmas* (Wiley Interscience, New York, 1962).
121. Chen, F. F., *Laser Interaction and Related Plasma Phenomena*, H. Schwarz and H. Hora (eds.), (Plenum, New York, 1974), Vol. 3A, p. 291.
122. DuBois, D. F., ibid.
123. Korobkin, V. V., and R. V. Serov, JETP Letters, 4, 70 (1966).
124. Stamper, J. A., K. Papadopoulos, R. N. Sudan, S. O. Dean, E. A. McLean, and J. M. Dawson, Phys. Rev. Letters, 26, 1012 (1971).
125. Mulser, P., Z. Naturforsch., 25A, 282 (1970).
126. Rehm, R. G., Phys. Fluids, 13, 921 (1970).
127. Shearer, J. W., and W. S. Barnes, *Laser Interaction and Related Plasma Phenomena*, H. Schwarz and H. Hora (eds.), (Plenum, New York, 1970), Vol. 1, p. 307.
128. Shearer, J. W., R. E. Kidder, and J. W. Zink, Bull. Amer. Phys. Soc., 15, 1483 (1970).
129. Fader, W., Phys. Fluids, 11, 2200 (1968).
130. Goldman, E., Plasma Physics, 15, 289 (1973).
131. Krokhin, O. N., J. Appl. Math. Phys., 16, 123 (1965); Afanas'yev, Yu. V., O. N. Krokhin, and G. V. Sklizkov, IEEE J. Quantum Electronics, QE-2, 483 (1966).
132. Caruso, A., B. Bertotti, and P. Giupponi, Nuovo Cimento, 45B, 176 (1966); Caruso, A., and R. Gratton, Plasma Physics, 10, 867 (1968).
133. Dyer, P. E., D. J. James, G. J. Pert, S. A. Ramsden, and M. A. Skipper, *Laser Interaction and Related Plasma Phenomena*, H. Schwarz and H. Hora (eds.), (Plenum, New York, 1974), Vol. 3A, p. 191.

134. Hora, H., ibid., Vol. 1, p. 365.
135. Basov, N. G., and O. N. Krokhin, *3rd Quantum Electronics Conf., Paris, Feb., 1963*, N. Bloembergen and P. Grivet (eds.), (Dunod, Paris, 1964), Vol. 2, p. 1373; ZhETF, 46, 171 (1964)(Sov. Phys. JETP, 19, 123 (1964)).
136. Dawson, J. M., Phys. Fluids, 7, 981 (1964).
137. Hora, H., Inst. f. Plasmaphysik, Garching, Rept. 4/50 (1972).
138. Tanimoto, M. (Private communication, 1972); M. Salvat, Inst. für Plasmaphysik, Garching, Rept. (in preparation).
139. Haught, A. F., and D. H. Polk, Phys. Fluids, 9, 2047 (1966); Haught, A. F., D. H. Polk, and W. J. Fader, Phys. Fluids, 13, 2842 (1970).
140. Yamanaka, T., N. Tsuchimori, T. Sasaki, and C. Yamanaka, Tech. Progress Rept., Osaka University, 18, 155 (1968).
141. Lubin, M. J., H. S. Dunn, and W. Friedman, *Int. Conf. Contr. Thermonucl. Fusion, Novosibirsk, Aug. 1968* (IAEA, Vienna, 1969), Vol. 1, p. 945.
142. Sekiguchi, T., A. Kitsunezaki, and M. Tanimoto, Oyo Buturi (Japan), 40, 206 (1971), 498 (1968); Phys. Fluids (to be published).
143. Sigel, R., K. Büchl, P. Mulser, and S. Witkowski, Phys. Letters, 26A, 498 (1968).
144. Griffin, W. G., and J. Schlüter, Phys. Letters, 26A, 241 (1968).
145. Hohla, K., (Private communication).
146. Bonnier, A., and J. Martineau, Phys. Letters, 38A, 199 (1972).
147. Büchl, K. P., *Laser Interaction and Related Plasma Phenomena*, H. Schwarz and H. Hora (eds.), (Plenum, New York, 1972), Vol. 2, p. 245.
148. Hora, H., Inst. f. Plasmaphysik, Garching, Rept. 6/27 (1964); Hora, H., and H. Wilhelm, Nucl. Fusion, 10, 111 (1970).
149. Allen, C. W., *Astrophysical Quantities* (Athlon Press, London, 1955).
150. Gaunt, J. A., Proc. Roy. Soc. (London), A126, 654 (1930).
151. Smerd, S. F., and K. C. Westfold, Phil. Mag., 40, 831 (1949).
152. Lüst, R., Z. Astrophys., 37, 67 (1955); Fortsch. Phys., 7, 503 (1959).
153. Dawson, J. M., and C. Oberman, Phys. Fluids, 5, 517 (1962).
154. Hora, H., Nature (Physical Science), 243, 34 (1973).
155. Hora, H., Opto-Electronics, 5, 491 (1973).
156. Nicholson-Florence, M. B., Diss. Univ. of Essex (England), 1972.
157. Hora, H., *Laser Interaction and Related Plasma Phenomena*, H. Schwarz and H. Hora (eds.), (Plenum, New York, 1974), Vol. 3B, p. 803.
158. Kaw, P., and J. Dawson, Phys. Fluids, 13, 472 (1970).
159. Kaw, P., and A. Salat, Phys. Fluids, 11, 2223 (1968).
160. Valeo, E., and C. Oberman, Phys. Rev. Letters, 30, 1035 (1973).

161. Gekker, I. R., and O. V. Sizukhin, JETP Letters, 9, 243 (1969).
162. Chu, T. K., and H. W. Hendel, Bull. Amer. Phys. Soc., 16, 1297 (1971).
163. Hendel, H. W., Int. Conf. Waves and Instabilities in Plasmas, Innsbruck, Austria, April 1973.
164. Nishikawa, K., J. Phys. Soc. Japan, 24, 916 (1968).
165. DuBois, D. F., *Statistical Physics of Charged Particle Systems*, R. Kubo and T. Kohara (eds.), (Benjamin, New York, 1969), p. 87.
166. Rosenbluth, M. N., Phys. Rev. Letters, 29, 565 (1972).
167. Morse, R. L., and C. W. Nielson, Phys. Fluids, 14, 830 (1971).
168. Valeo, E., C. Oberman, and F. W. Perkins, Phys. Rev. Letters, 28, 340 (1972); Kruer, W. L., *Laser Interaction and Related Plasma Phenomena*, H. Schwarz and H. Hora (eds.), (Plenum, New York, 1974), Vol. 3A, p. 341.
169. Krasyuk, I. K., P. P. Pashinin, and A. M. Prokhorov, JETP Letters, 17, 93 (1973).
170. Galeyev, A. A., G. Laval, T. M. O'Neil, M. N. Rosenbluth, and R. Z. Sagdeyev, JETP Letters, 17, 35 (1973).
171. Vinogradov, A. V., B. Ya. Zel'dovich, and I. I. Sobel'man, ZhETF P, 17, 271 (1973).
172. Bobin, J. L., M. Decroisette, B. Meyer, and Y. Vitel, Phys. Rev. Letters, 30, 594 (1973).
173. Chen, F. F., *Laser Interaction and Related Plasma Phenomena*, H. Schwarz and H. Hora (eds.), (Plenum, New York, 1974), Vol. 3A, p. 291; DuBois, D. F., ibid., p. 267.
174. Klima, R., Plasma Physics, 12, 123 (1970); Javel, P., (Private communication, 1969).
175. Hora, H., Ann. Physik, 22, 402 (1969).
176. Lindl, J., and P. Kaw, Phys. Fluids, 11, 2223 (1968).
177. Kaw, P., Appl. Phys. Letters, 15, 16 (1969).
178. Steinhauer, L. C., and H. G. Ahlstrom, Phys. Fluids, 13, 1103 (1970).
179. Chiao, R. Y., E. Garmire, and C. H. Townes, Phys. Rev. Letters, 13, 479 (1964).
180. Shimoda, K., J. Phys. Soc. Japan, 24, 183 (1968).
181. Green, B., and P. Mulser, *Laser Interaction and Related Plasma Phenomena*, H. Schwarz and H. Hora (eds.), (Plenum, New York, 1972), Vol. 2, p. 381.
182. Kinsinger, R. E., Bull. Amer. Phys. Soc., 18, 1345 (1973).
183. Beaudry, G., and J. Martineau, Phys. Letters, 43A, 331 (1973).
184. Hora, H., Inst. für Plasmaphysik, Garching, Rept. 6/23 (July 1964).
185. Winterberg, F., Phys. Rev., 174, 212 (1968).
186. Chu, H. S., Phys. Fluids, 15, 413 (1972).

187. Hora, H., *Laser Interaction and Related Plasma Phenomena*, H. Schwarz and H. Hora (eds.), (Plenum, New York, 1970), Vol. 1, p. 427.
188. Hora, H., and D. Pfirsch, ibid., Vol. 2, p. 515.
189. Glasstone, S., and R. H. Lovberg, *Controlled Thermonuclear Reactions*, (Van Nostrand, New York, 1960).
190. Arnold, W. R., J. A. Phillips, G. A. Sawyer, J. E. Stovace, and J. L. Tuck, Phys. Rev., 93, 483 (1954).
191. Tuck, J. L., Nucl. Fusion, 1, 201 (1971).
192. Hora, H., and D. Pfirsch, *6th Int. Quantum Electronics Conf., Kyoto, Japan, 1970*, K. Shimoda (ed.), (Kyoto, 1970), p. 5.
193. Hora, H., ref. 11, Eq. (2).
194. Wood, L., and J. Nuckolls, ref. 9; and E. Teller invited papers, 7th Int. Quantum Electronics Conf., Montreal, Canada, May 1972.
195. Nuckolls, J., L. Wood, A. Thiessen, and G. Zimmerman, Nature, 239, 139 (1972).
196. Brueckner, K., KMS-Fusion Report, KMSF-NP 5, Ann Arbor, Mich., April 12, 1972.
197. Kaliski, S., Bull. Pol. Acad. Tech. Sciences, 5 (No. 2), 21 (1971).
198. Guderley, G., Luftfahrtforschung, 19, 302 (1942).
199. Daiber, J. W., A. Hertzberg, and C. E. Wittliff, Phys. Fluids, 9, 617 (1965).
200. Lengyel, L. L., AIAA Journal, 11, 1347 (1973).
201. Morse, R. L., and C. K. Nielson, Phys. Fluids, 16, 909 (1973).
202. Nuckolls, J., *Laser Interaction and Related Plasma Phenomena*, H. Schwarz and H. Hora (eds.), (Plenum, New York, 1974), Vol. 3B, p. 397.
203. Brueckner, K. A., ref. 24 and paper of the European Conference on Thermonuclear Fusion, Moscow, U.S.S.R., August 1973.
204. Kaliski, S., *Laser Interaction and Related Plasma Phenomena*, H. Schwarz and H. Hora (eds.), (Plenum, New York, 1974), Vol. 3B, p. 495.
205. Bobin, J. L., ibid.
206. Morse, R. L., ibid.
207. Hora, H., and B. Kronast, U.S. Patent 3,444,377.
208. Winterberg, F., Nature, 241, 449 (1973); Z. Naturforschung, 28A, 900 (1973).
209. Askar'yan, G. A., V. A. Namiot, and M. S. Rabinovich, JETP Letters, 17, 424 (1973).
210. Weaver, T. A., Lawrence Livermore Laboratory Laser Report 9 (No. 12), 1 (1973).
211. Prokoshkin, Yu. D., Umschau, 73, 153 (1973).
212. Shearer, J., J. Garrison, J. Wong, and J. E. Swain, *Laser Interaction and Related Plasma Phenomena*, H. Schwarz and H. Hora (eds.), (Plenum, New York, 1974), Vol. 3B, p. 803; Bull. Amer. Phys. Soc., 18, 1346 (1973).

213. Basov, N. G., P. G. Kryukov, S. D. Zakharov, Yu. V. Senatskiy, and S. V. Chekalin, IEEE J. Quantum Electronics, QE-4, 864 (1968).
214. Gobeli, G. W., J. C. Bushnell, P. S. Pearcy, and E. D. Jones, Phys. Rev., 188, 300 (1969), and ref. 22.
215. Floux, F., D. Cognard, J. L. Bobin, F. Delobeau, and C. Fauquignon, C. R. Acad., Sc., Paris, 269, 697 (1969); Phys. Rev., A1, 821 (1970); Floux, F., *Laser Interaction and Related Plasma Phenomena*, H. Schwarz and H. Hora (eds.), (Plenum, New York, 1970), Vol. 1, p. 469.
216. Faugeras, E., M. Mattioli, and R. Papoular, Plasma Physics, 10, 939 (1968).
217. Wuerkner, R. F., H. M. Goldenberg, and R. V. Langmuir, J. Appl. Phys., 30, 441 (1959).
218. Hora, H., and H. Schwarz, Japan. J. Appl. Phys. (1974).
219. Hora, H., *Laser Interaction and Related Plasma Phenomena*, H. Schwarz and H. Hora (eds.), (Plenum, New York, 1970), Vol. 1, p. 273.
220. Büchl, K., H. Hora, R. Riedmüller, M. Salvat, Instit. für Plasmaphysik, Garching, Rept. (Proposal) Oct. 1973.
221. Hirono, H., and T. Iwamoti, Japan J. Appl. Phys., 6, 1006 (1967).
222. See ref. 20, Fig. 11.
223. Basov, N. G., Yu. S. Ivanov, O. N. Krokhin, Yu. A. Mikhaylov, and S. I. Fedotov, JETP Letters. 15, 417 (1972).
224. Basov, N. G., Culham Conference, Sept. 1971.
225. McCall, G. H., F. Young, A. W. Ehler, J. F. Kephart, and R. P. Godwin, Phys. Rev. Letters, 30, 1116 (1973).
226. Sklizkov, G. V., *Laser Interaction and Related Plasma Phenomena*, H. Schwarz and H. Hora (eds.), (Plenum, New York, 1974), Vol. 3B, p. 553.
227. Korobkin, V. V., ibid.
228. Bobin, J. L., Erice Conference, June 1973.
229. Bobin, J. L., and F. Floux, *Laser Interaction and Related Plasma Phenomena*, H. Schwarz and H. Hora (eds.), (Plenum, New York, 1974), Vol. 3B, p. 591.
230. Belland, P., C. DeMichelis, M. Mattioli, and R. Papoular, Appl. Phys. Letters, 18, 542 (1971).
231. Saleres, A., F. Floux, D. Cognard, and J. L. Bobin, Phys. Letters, 45A, 451 (1973); Carion, A., J. Lancelot, J. deMetz, and A. Saleres, Phys. Letters, 45A, 439 (1973).
232. Teller, E., *Laser Interaction and Related Plasma Phenomena*, H. Schwarz and H. Hora (eds.), (Plenum, New York, 1974), Vol. 3A, p. 3.
233. Brueckner, K. A., ibid., Vol. 3B, p. 427; discussion, p. 543.
234. van Kessel, G. C. M., and R. Sigel, ibid.; Bull. Amer. Phys. Soc., 18, 1316 (1973).

235. Discussion, *Laser Interaction and Related Plasma Phenomena*, H. Schwarz and H. Hora (eds.), (Plenum, New York, 1972), Vol. 2, p. 271.
236. Hora, H., *Proc. Int. Conf. Light Scattering in Solids*, M. Balkanski (ed.), (Flamarion, Paris, 1971), p. 128.
237. Schwarz, H., and H. Hora, Appl. Phys. Letters, 15, 349 (1969). For more references see ref. 236.
238. Basov, N. G., (Private communication, June 1972).
239. Spalding, I. J., *Laser Interaction and Related Plasma Phenomena*, H. Schwarz and H. Hora (eds.), (Plenum, New York, 1974), Vol. 3; Bohn, F. H., H. Conrads, J. Darvas, and F. Förster, *5th Symp. Eng. Probl. Fusion Reactors*, Princeton, Nov. 5-9, 1973.
240. The Shah of Persia, M. Reza Pahlavi, Interview, Austrian Television, Dec. 30, 1973.
241. Teller, E., Die Situation der Modernen Physik, Arbeitsgemeinshaft für Forschung des Landes Nordrhein-Westphalen, Nr. 147 (Köln 1965; Westdeutscher Verlag).
242. Tuck, J. L., Nature 233, 593 (1971).
243. Rand, S., Phys. Rev., B136, 231 (1964).

12. APPENDIX

LIST OF REPRINTED PAPERS

1. Basov, N. G., and O. N. Krokhin. "The Conditions of Plasma Heating by the Optical Quantum Generator." *3rd Intern. Conf. Quantum Electronics, Paris, 1963.* Grivet, P., and N. Bloembergen (eds.), (Dunod, Paris, 1964), Vol. 2, p. 1373.
2. Basov, N. G., P. G. Kryukov, S. D. Zakharov, Yu. V. Senatskiy, and S. V. Chekalin. "Experiments on the Observation of Neutron Emission at the Focus of High-power Laser Radiation on a Lithium Deuteride Surface." IEEE J. Quantum Electronics, QE-4, 864 (1968).
3. Linlor, W. I. "Ion Energies Produced by Laser Giant Pulse." Appl. Phys. Letters, 3, 210 (1963).
4. Papoular, R. "The Initial Stage of the Laser-induced Gas Breakdown." *Laser Interaction and Related Plasma Phenomena*, H. Schwarz and H. Hora (eds.), (Plenum, New York, 1972), Vol. 2, p. 79.
5. Alcock, A. J. "Experiments on Self-focusing in Laser-produced Plasmas." Ibid., p. 155.
6. Siller, G., K. Büchl, and H. Hora. "Intense Electron Emission from Laser-produced Plasmas." Ibid., p. 253.
7. Hora, H. "Experimental Result of Free Targets." Ibid., Vol. 1, p. 273.
8. Dawson, J. M. "On the Production of a Plasma by Giant Laser Pulses." Phys. Fluids, 7, 981 (1964).
9. Hora, H. "Some Results of the Self-similarity Model." *Laser Interaction and Related Plasma Phenomena*, H. Schwarz and H. Hora (eds.), (Plenum, New York, 1971), Vol. 1, p. 365.
10. Mulser, P. "Hydrogen Plasma Production by Giant Pulse Lasers." Z. Naturforschung, 25A, 282 (1970).

11. Hora, H., and H. Wilhelm. "Optical Constants of Fully-ionized Hydrogen Plasma for Laser Radiation." Nucl. Fusion, 10, 111 (1970).

12. DuBois, D. F. "Laser-induced Instabilities and Anomalous Absorption in Dense Plasmas." *Laser Interaction and Related Plasma Phenomena*, H. Schwarz and H. Hora (eds.), (Plenum, New York, 1974), Vol. 3A, p. 267.

13. Hora, H. "Nonlinear Confining and Deconfining Forces Associated with the Interaction of Laser Radiation with a Plasma." Phys. Fluids, 12, 182 (1969).

14. Lindl, J. D., and P. Kaw. "Ponderomotive Forces on Laser-produced Plasmas." Phys. Fluids, 14, 371 (1971).

15. Hora, H. "Nonlinear Forces in Laser-produced Plasmas." *Laser Interaction and Related Plasma Phenomena*, H. Schwarz and H. Hora (eds.), (Plenum, New York, 1972), Vol. 2, p. 341.

16. Chen, F. F. "Physical Mechanisms for Laser-Plasma Parametric Instabilities," ibid., 1974, Vol. 3A, p. 291.

17. Nuckolls, J. H. "Laser-induced Implosion and Thermonuclear Burn," ibid., 1974, Vol. 3B, p. 397.

18. Lengyel, L. L. "Exact Steady-state Analogy of Transient Gas Compression by Coalescing Waves." AIAA Journal, 11, 1347 (1973).

19. Clarke, J. S., H. N. Fisher, and R. J. Mason. "Laser-driven Implosion of Spherical DT Targets to Thermonuclear Burn Conditions." Phys. Rev. Letters, 30, 89 (1973).

20. Floux, F., D. Cognard, L. G. Denoed, G. Piar, D. Parisot, J. L. Bobin, F. Delobeau, and C. Fauquignon. "Nuclear Fusion Reactions in Laser-produced Solid Deuterium Plasmas." Phys. Rev., A1, 821 (1970).

21. Basov, N. G., O. N. Krokhin, and G. V. Sklizkov. "Heating of Laser Plasmas for Thermonuclear Fusion." *Laser Interaction and Related Plasma Phenomena*, H. Schwarz and H. Hora (eds.), (Plenum, New York, 1972), Vol. 2, p. 389.

22. Basov, N. G. Yu. S. Ivanov, O. N. Krokhin, Yu. A. Mikhaylov, G. V. Sklizkov, and S. I. Feodotov, "Neutron Generation in Spherical Irradiation of a Target by High-power Laser Radiation," JETP Letters, 15, 417 (1972).

23. Yamanaka, C., T. Yamanaka, T. Sasaki, K. Yoshida, M. Waki, and H. B. Kang. "Anomalous Heating of a Plasma by Lasers." Phys. Rev., A6, 2335 (1972).

24. McCall, G. F., F. Young, A. W. Ehler, J. F. Kephardt, and R. P. Godwin. "Neutron Emission from Laser-produced Plasmas." Phys. Rev. Letters, 30, 1116 (1973).

25. Goldman, M., J. Soures, and M. J. Lubin. "Saturation of Stimulated Back-scattered Radiation in Laser Plasmas." Phys. Rev. Letters, 31, 1184 (1973).

26. Shearer, J. W., J. Garrison, J. Wong, and J. E. Swain. "Pair Production by Relativistic Electrons from an Intense Laser Focus." Phys. Rev., A8, 1582 (1973).

27. Hora, H. "Estimations for the Efficient Production of Antihydrogen by Lasers of Very High Intensities." Opto-Electronics, 5, 491 (1973).

The Conditions of Plasma Heating by the Optical Generator Radiation

N. G. BASOV and O. N. KROKHIN
Lebedev Institute. Academy of Sciences of the USSR

The possibility of concentrating energy of laser radiation in small volumes allows to put forward a question about heating small volumes of a dense hydrogen plasma up to the high temperatures at which thermonuclear reactions arise. The heating mechanism in its main stage consists of absorption of electromagnetic quanta by electrons in the field of ions transfer of some energy to ions (we do not consider the initial stage of heating to the temperatures of the order of several tens of thousands degrees, i.e. to the moment of formation of an essentially ionized plasma).

The final temperatures arising as a result of the plasma heating are determined mainly by the optical generator power, since the principal process limiting the heating is high thermoconductivity of the high temperature plasma. The energy of the plasma heated to high temperature was found out to be rather small comparing to the energy lost owing to thermo-conductivity and radiation.

1°/ The absorption coefficient of electromagnetic waves in plasma is connected with the imaginary part of the refraction index n'' by the ratio :

$$K = 2\frac{\omega}{c}n'' = 2\frac{\omega}{c}\left(\frac{-\varepsilon' + \sqrt{\varepsilon'^2 + \varepsilon''^2}}{2}\right)^{1/2} \qquad (1)$$

where ε' and ε''-real and imaginary parts of the plasma dielectric constant, respectively, and ω-frequency of the electromagnetic wave.

The plasma dielectric constant is equal to :

$$\varepsilon(\omega) = 1 - \frac{\omega_0^2}{\omega^2}\left(1 - i\frac{v}{\omega}\right) \qquad (2)$$

where ω -Langmuir's frequency equal to $\left(\frac{4\pi n e^2}{m}\right)^{1/2}$, v-collision frequency of electrons and ions in the plasma n-electron density.

Reprinted from: *3rd International Conference on Quantum Electronics, Paris, 1963,* Vol. 2, P. Grivet and N. Bloembergen (eds.), Dunod, Paris, 1964, pp. 1373–1377.

The maximum plasma density is limited by reduction of a real part of the dielectric constant (2) into zero, since at large plasma densities a complete reflection of the electromagnetic wave will take place.

If for the sake of simplicity one assumes that $\varepsilon'' \ll \varepsilon'$ takes place, then the expression for the absorption coefficient will take the form :

$$k \approx \frac{16\pi\sqrt{2\pi}\, n^2 e^6 . \max\left[\ln \frac{(kT)^{3/2}}{(4\pi n e^6)^{1/2}},\ \ln \frac{\sqrt{3}\, kT\, m^{3/2}}{(4\pi n e^2 \hbar^2)^{1/2}}\right]}{3\,(mkT)^{3/2}\, c\, \omega^2 \left(1 - \frac{\omega_0^2}{\omega^2}\right)^{1/2}} \tag{3}$$

One can see from (3) that the absorbtion coefficient decreases with the temperature growth. The numerical value of the absorption coefficient for radiation of a ruby optical generator $\omega = 3.10^{15}$ sec.$^{-1}$ in the hydrogen plasma with the maximum density $n = 3.10^{21}$ cm^{-3} heated to the temperature $T = 10^7$ degrees is equal, by the orders of magnitude, to 10^{+2} cm^{-1}. This shows that absorption of an optical radiation by the plasma is essential even at rather high temperatures.

2°/ Heating and gas-dynamic of small volume of plasma.

Let us assume that a small volume of plasma with the initial radius r_c is heated by the optical maser.

Variation of the temperature time and plasma motion can be determined from the hydrodynamics equations taking into account the thermo-conductivity processes. However, we shall not be interested in the distribution of temperatures and velocities inside the plasma volume, but consider only the average values dealing with the total volume as a whole.

The equation of conservation of momentum and energy give :

$$M\frac{dv}{dt} - 4\pi\, r^2 p = O \longleftarrow M\frac{dv}{dt} - 4\pi\, r^2 p = O \tag{4}$$

$$\frac{d}{dt}(M v^2/2 + E) = Q \longleftarrow \frac{d}{dt}\left[M\frac{v^2}{2} + E\right] = Q \tag{5}$$

where p is the averaged pressio, v radial velocity, E energy, M mass, Q power of laser.

The system of equation (4) and (5) has a simple solution for the case of ideal gas, when $E = 1{,}5\ pV = 1{,}5\ NkT$, where N number of particles in the system, V its volume.

If :

$$Q = \begin{cases} Q_o & o < t < \tau \\ O & t < o, \ t > \tau \end{cases}$$

the solution has the simplest form :

$$r^2 = r_o^2 + \frac{2}{3}\frac{Q_o t^3}{M} \tag{6}$$

$$T = \frac{Q_o t}{1,5\ Nk} \cdot \frac{Q_o t^3/6\ M + r_o^2}{2\,Q_o t^3/3M + r_o^2} \tag{7}$$

One can see from (7) that the divergence of the plasma decreases its temperature. The time, when the temperature decreases to one half of the temperature obtaining without divergence is about $t \sim r_o \left(\frac{3M}{\varepsilon}\right)^{1/2}$, where ε is energy of laser. If $r_o = 0,3.\,10^{-1}$ cm, $M = 0,2.\,10^{-4}$gr, $\varepsilon = 10^9$ erg, t is equal to 10^{-8} sec.

3°/ Energy processes in the high temperature plasma of large density.

a) Plasma radiations.

Radiation energy losses of the plasma are determined by Bremstrahlung and recombinaison radiations. The power of the plasma Bremstrahlung :

$$q_{ret.} = 1,6.\,10^{-27}\ n^2\ T^{1/2}\ \text{erg/cm}^3\ \text{sec} \tag{8}$$

The radiation power connected with recombination, respectively, is given by the formula :

$$q_{rec.} = 1,1.\,10^{-21}\ n^2\ T^{-1/2}\ \text{erg/cm}^3\ \text{sec} \tag{9}$$

One can see from (8) and (9) that for hydrogen plasma at $T_o = 4.\,10^5$ $q_{ret} = q_{rec}$. At the temperature $T = 10^7$ and density $n = 3.\,10^{21}$the radiation power is equal to $q_{ret} = 5.\,10^{19}$ erg/cm³ sec.

b) Thermoconductivity.

At high temperature, heat transfer in plasma is determined by the electron and radiative thermoconductivities.

In the first case, the thermoconductivity coefficient is equal to :

$$K_e = 1,24.\,10^{-6}\ T^{5/2}\ \text{erg/cm sec degr.} \tag{10}$$

takes into account electroconductivity. For the radiative thermoconductivity, the coefficient K_p is equal to :

$$K_p = \frac{16}{3} \sigma l T^3 \quad (11)$$

where σ Stephan-Boltzman constant, l Rosseland mean free path length of photons in plasma.

For the hydrogen plasma with the density $n = 3.10^{21}$ and at the temperature $T = 10^7$ linear dimensions of the plasma are considerably less than l, which, in that case, is equal by the order of magnitude to 10^2 cm. It means that the plasma is practically absolutely transparent and the total radiation flux from the plasma will be determined by the power of the Bremstrahlung radiations given by the formula (8).

c) The equation of state.

At high temperatures of the order of 10^7, one can consider the plasma as absolutely ionised and application the equation of ideal gas state. In this case the energy density is equal to $u = 3\ nkT$ and pressure $p = \frac{2}{3} u$. For $n = 3.10^{21}$ and $T = 10^7$, $u = 1{,}4.10^{13}$ erg/cm^3, hence one can see that, in principle, the radiation energy of laser of the order 100 joules will be enough for heating 10^{-4} cm^3 of dense plasma.

The corrections to the equations of state depending on the Coulomb interaction are of the order of $e^2/\lambda kT$, where $\lambda = \left(\frac{kT}{4\pi ne^2}\right)^{1/2}$ Debye screening length. At $n = 3.10^{21}$ cm^{-3} and $T = 10^7$ $\lambda = 3.10^{-7}$ cm and $e^2/\lambda kT = 10^{-3}$.

d) Ionization equilibrium.

According to the formula of ionisation equilibrium, the density of ions with the ionisation rate i, i + 1 is connected with the density of electrons by the relations :

$$\frac{n_{i+1}\, n}{n_i} = 0{,}58.10^{16}\ T^{3/2} \frac{g_{i+1}}{g_i}\ 1^{-I_i/kT} \quad (12)$$

where g_i and g_{i+1} statistical weights, I_i potential of ionisation. For more accurate calculations the values of g_i should be replaced by the statistical sum.

However, in the real plasma, it may happen that the relation (12) is not fulfilled owing to absence of the thermodynamic equilibrium. Indeed, if the photon density corresponded to the temperature T, the photoionization time would be equal to the time of recombination which can be estimated from (9) $\tau_{ip} \sim n_0 kT/q_{rec}$ where n_0 thermodynamically equilibrium density of neutral atoms. The time of ionizations by means of electron collisions is equal to $(6\,vn)^{-1}$ where σv average value of the ionization cross-section multiplied by the electron velocity.

It is obvious that in the case when the process of ionization by electrons is the fastest, i.e. $\tau_{i_p} \gg \tau_{i_e}$ the formula (12) will be valid as well in the absence of the thermodynamically equilibrium radiation in plasma. For instance, at the temperature $T = 10^7$ and concentration $n = 3.10^{21}$ the time of photoionization is of the order of magnitude of $\sim 10^{-10}$ sec while the time of electron ionization is 3.10^{-13} sec ($\sigma = 10^{-18}$).

The above analysis shows that development of the methods of generations of the optical radiations, in particular increase of power of the available quantum generation, apparently, leaves some hope for the experimental attempts to obtain small volumes of fairly dense plasma heated to high temperatures. For this propose it is nessesary to have at least 100 joules during 10^{-8} sec.

REFERENCES

L. SPITZER - "Physics of fully ionized gases" Interscience publishers, inc. N.Y.-L. 1956.

V. SILIN and A. RUKHADZE - "Electromagnetic properties of plasma and plasma like media" Atomizdat, Moscow 1961.

L.A. ARZIMOVITCH - "Controlled thermonuclear reaction" Phismatizdat, Moscow 1961.

O-11—Experiments on the Observation of Neutron Emission at a Focus of High-Power Laser Radiation on a Lithium Deuteride Surface

N. G. BASOV, P. G. KRIUKOV, S. D. ZAKHAROV, YU. V. SENATSKY AND S. V. TCHEKALIN

Abstract—Theoretical calculations indicate that laser radiation may be used to heat a deuterium plasma to temperatures at which thermonuclear neutron emission may be observed. Using a neodymium glass laser, producing a 20-joule pulse of approximately 10^{-11}-second pulse length, preliminary evidence of neutron emission has been obtained.

THE PRESENT paper reports preliminary results of experiments on the observation of neutron emission on focusing of high-power laser radiation on a lithium deuteride surface.

In 1962 it was shown in [1] that one may produce a high-density and high-temperature plasma by focusing high-power laser radiation on the surface of a solid target. From the calculations carried out in [1] it was evident that, with the help of laser radiation, it is possible to heat the deuterium plasma up to the temperature at which thermonuclear neutron emission may be observed. Laser powers of more than 10^9 watts were required for this purpose. There were no lasers of such power at that time, and extensive efforts were necessary to increase the output power of lasers.

The efficiency of the method of amplification for obtaining short light pulses with a high peak power was shown in our papers [2], [3]. Several gigawatts power was obtained for the ruby laser and more than ten gigawatts for the neodymium glass. The obtaining of such power pulses led to self-damage of the active medium. This fact made it difficult to carry out the experiments on plasma heating. From our investigations, the power damage threshold was found to increase with shorter pulse duration, and it was presumed that the use of still shorter pulses would permit the desired increase in giant pulse power. The remarkable opportunity to realize this was opened with the invention of ultrashort pulse lasers using the nonlinear absorber [4]. Using this laser as the oscillator, we managed to obtain (after amplification) the output light pulse with the energy of 20 joules in a very short time, not more, probably, than 10^{-11} second.[1]

Manuscript received May 17, 1968. This paper was presented at the 1968 International Quantum Electronics Conference, Miami, Fla.

The authors are with the P. N. Lebedev Physical Institute of the Academy of Sciences of the U.S.S.R., Moscow, U.S.S.R.

Fig. 1 illustrates the model of the device. A single pulse was separated from the train of ultrashort pulses of the oscillator with the help of the Kerr cell shutter, and was then amplified. Unlike [6], in our case the Kerr cell shutter to single out the pulse was placed outside the Fabry–Perot cavity, thereby making it possible to obtain the shorter pulses from the oscillator. The amplification of the singled out pulse was produced by five cascades of amplifiers, the length of the neodymium glass rods being 60 cm in each cascade. The diameter of rods in the two last cascades was 40 mm. The total amplification coefficient was 10^4.

Upon focusing of the laser radiation on a target surface, plasma formation takes place from a very thin surface layer. Since lithium deuteride is a very active material, there is the danger of contamination and substitution of deuterium by hydrogen on this target surface. Therefore an investigation of the composition of the plasma formed by the evaporation of the surface layer was carried out. Investigation of plasma composition evaporated from lithium deuteride by 10-MW ruby laser radiation was carried out using a time-of-flight mass spectrometer with electrostatic and magnetic analyzers. This investigation showed that the plasma formed at the first laser shot contains a great number of hydrogen and an insignificant quantity of deuterium ions; in addition, contaminants of oxygen and carbon were observed. Upon penetration into the target material with the successive laser shots, the deuterium ion portion significantly increased while the admixture portion decreased.

Additional experiments on the plasma composition were carried out using a laser neutron source. Fig. 2 illustrates the scheme of the experiment. 100-MW ruby laser radiation was focused on the lithium deuteride surface. Deuterium ions were then accelerated by a 50-kV electrostatic field and struck another target containing the deuterium. The output of neutron emission was evidence for the presence of deuterium in the surface layer of the target.

[1] The experiments carried out separately with the oscillator showed that the pulse duration was within the limits of 10^{-11} to 10^{-12} seconds. Pulse duration was measured using the method proposed in [5].

Reprinted from: *IEEE Journal of Quantum Electronics*, QE-4, pp. 864–867 (1968).

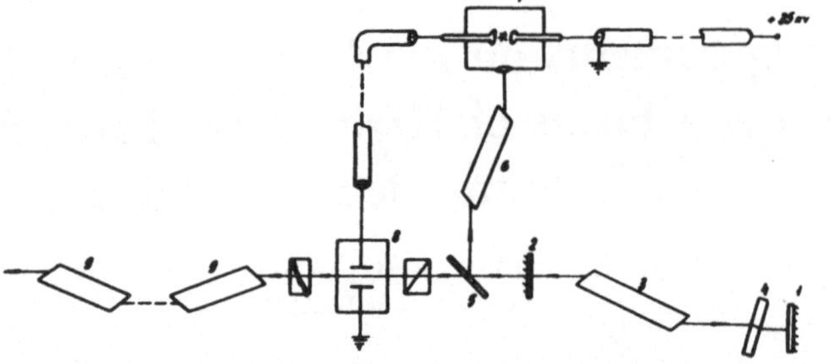

Fig. 1. Scheme of the experimental device: 1, 2—oscillator mirrors, 3—neodymium glass rod of the oscillator, 4—cell with the bleachable dye, 5—beam splitter, 6—cascade of the additional amplification, 7—spark-gap chamber, 8—Kerr cell shutter, 9—cascades of the main amplification. For the simplicity of the scheme the orientation of the Brewster angle rods of the main amplifier 9 after the shutter and the orientation of rods 6 and 3 before the shutter are shown in the same way.

The investigations carried out showed that special precautions must be taken to prevent the contamination of the target surface. The targets used in the following experiments on plasma heating were therefore prepared in an argon atmosphere.

Fig. 3 illustrates the scheme of the observation of neutron emission from plasma. Laser radiation was focused by a 60-mm-length plane-convex lens on the surface of a lithium deuteride sample in 10^{-3} torr vacuum. A large neutron scintillation counter was placed at a distance of 10 cm from the target. The plastic sample on the base of polystirol with the *p*-terphenyl and POPOP additions was used as a scintillator. The scintillator had a shape of a 30-cm-diameter cylinder turning to the truncated cone with the smaller base corresponding to the dimensions of 52 type photomultiplier photocathode. The total height of the scintillator was 20 cm. The counter was inserted inside the double duraluminum box with 16-cm-thick walls. The polished surface of the scintillator was surrounded by a magnesium oxide powder layer. The efficiency of the detection of the neutrons from the target in this arrangement was more than 10 percent. Electrical pulses from the photomultiplier were displayed on one of the two traces of the double-beam oscilloscope of C 1–17 type. The RC constant of the circuit in the PM anode was several tens of microseconds and the PM worked in the linear regime. An electrical signal from the Kerr cell shutter time coincident with the light pulse incident on the target was displayed on the second beam. The trace duration was 2×10^{-3} second, the signal from the Kerr cell shutter being roughly in the middle of the sweep. Fig. 4 presents one of the oscilloscope traces in which the pulse signal from the Kerr cell shutter may be seen on the upper trace and the background pulses from the counter on the lower trace. There is no coincidence of the pulses from the counter and from the Kerr cell shutter on this oscillogram.

Background pulses may be divided into two groups. The pulses of cosmic origin have their amplitudes more than 10 volts. The frequency of their appearance is small—

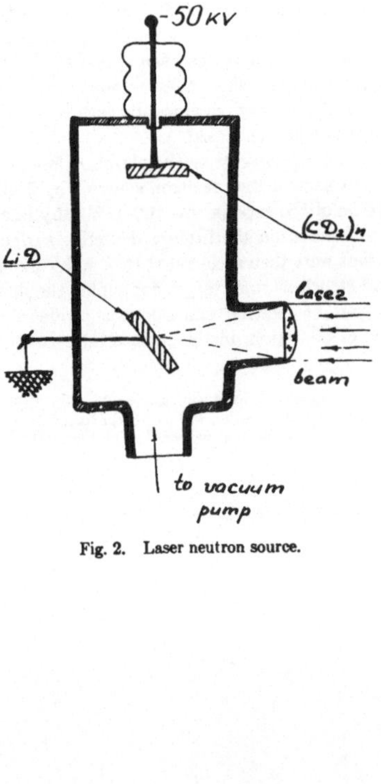

Fig. 2. Laser neutron source.

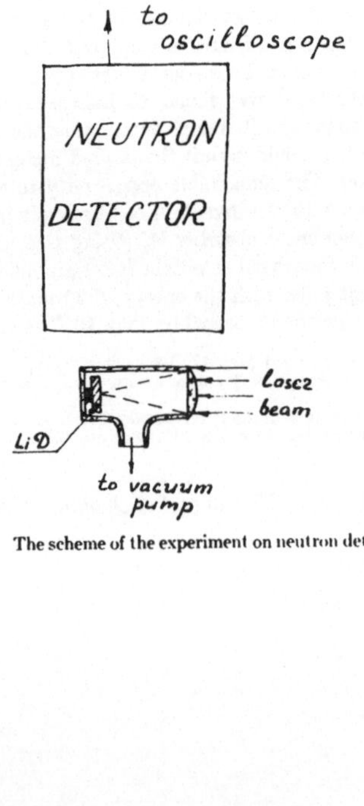

Fig. 3. The scheme of the experiment on neutron detection.

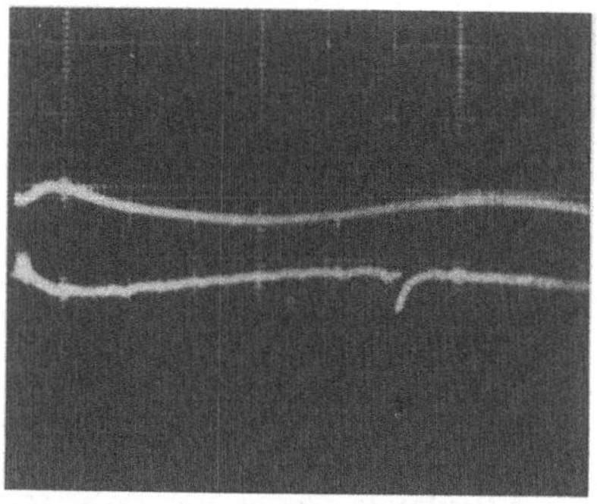

Fig. 4. Oscilloscopic traces of the pulses from the counter (lower trace) and from the Kerr cell shutter (upper trace); the case of no coincidence.

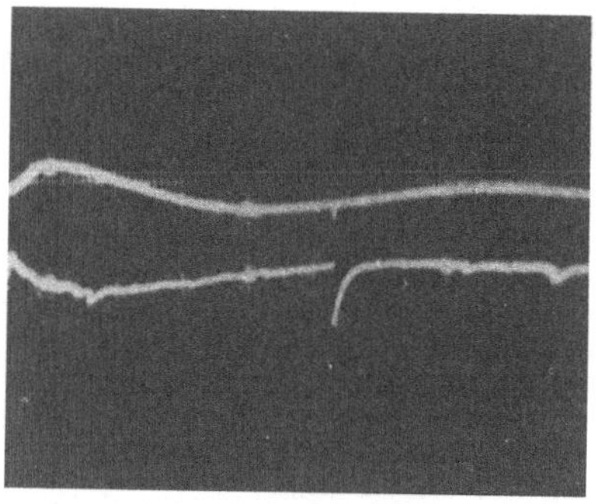

Fig. 5. Oscilloscopic traces of the pulses from the counter (lower trace) and from the Kerr cell shutter (upper trace); the case of no coincidence.

several pulses per second. The second group of pulses are from the natural radioactivity of the scintillator; their amplitude is about several volts, and the frequency of appearance is 10^3/second. The amplitude of the intrinsic PM noises was sufficiently less.

Neutron detection experiments were carried out in series of 5–10 shots. In each series the target used was moved in such a way that a new part of the target surface was irradiated at each shot. During the experiments the input-amplification cascade rods were found to be damaged. This damage took place only when the laser radiation was focused on the target. The plasma formed apparently reflects a sufficient portion of laser radiation back through the amplifier so that this radiation, upon being amplified, produces the damage. From this damage and also from probable changes in focusing and target quantity, the experimental conditions were not in general reproducible from series to series. The experimental results in the two series are presented in Tables I and II.

The oscilloscope traces relating to one of the cases of coincidence of the pulses from the counter and from the Kerr cell shutter is shown in Fig. 5. The amplitudes of the signals in cases of coincidence were approximately equal. The calibration of the counter with the help of Cs^{137} and Co^{59} sources showed that the pulses observed by us in the cases of coincidence may be related to single neutrons with the energies of 2.5 MeV.

TABLE I

Energy (joules)	6	4	17	6	5	6	10	8
Coincidence	no	no	yes	no	no	no	no	no

TABLE II

Energy (joules)	6	10	11	11	8	11
Coincidence	yes	yes	no	yes	no	no

The total number of coincidences (4) observed by us in the two experimental series (14 shots) is 20 times more than the probability of the accidental coincidence of a background pulse with the Kerr cell shutter pulse. Note that the cases of coincidence fall on the shots with energy more than 6 joules. Unfortunately, as we did not make simultaneous measurements of pulse duration, we cannot point out in which shot the power was at a maximum.

Let us make some evaluations of the average power on a particle and compare it with the experiment. Let the energy delivered to the plasma be 10 joules and pulse duration 10^{-11} second, light flux on a target being $J = 10^{16}$ W/cm^2. Then from the equations for an electron gathering energy in an electromagnetic field

$$\frac{\partial \epsilon}{\partial t} = \frac{e^2E^2}{2m\omega^2}\frac{\gamma}{\epsilon^{3/2}},$$

where ϵ is the electron energy, E the amplitude of the field, m the mass of the electron, ω the light frequency, and $\gamma/\epsilon^{3/2}$ the effective frequency of the collisions, we obtain

$$\epsilon = 0.5 \times 10^5 \text{ eV}.$$

From this one can calculate the thickness of the layer in which the absorption takes place. It equals 10^{-4} cm. The time between the electron–ion collisions is 3×10^{-13} seconds for electrons with the energy of 10 keV; therefore a free path length for the electron is 30 μ. Hence the volume of the heated plasma is 3×10^{-7} cm^3. Therefore the average particle energy is 2×10^3 eV. Considering that there were single neutrons in our experiments we may determine the average energy of the particles from the formula for the number of d–d reactions [7]:

$$\frac{VN^2}{2}\sigma v t \simeq 1,$$

where V is the volume, N the ion concentration of the plasma, σv the cross section and t the time of d–d reactions. From this formula it follows that the deuteron energy is 2×10^3 eV. The value for the diameter of the focal spot in this evaluation is 0.2 mm. These data were obtained from a photograph of a hot region of plasma made with the aid of an X-ray-obscure chamber. In this experiment the target was brass as the X-ray intensity from the lithium deuteride target was insufficient to produce any image.

Experiments in the heating of a gas by the focusing of ultrashort pulse radiation were also carried out by us. It was found that the plasma formed in the focus absorbs less of the energy of ultrashort pulses than of the energy of nanosecond pulses. Although we did not make neutron detection experiments in the case of gas, the preliminary tests carried out by us do not allow hope for reaching thermonuclear temperatures by ultrashort pulses in the case of gas. Addition studies of plasma production with ultrashort pulses are in progress.

ACKNOWLEDGMENT

The authors are very grateful to O. N. Krokhin for the help during the work and for the discussions of the results; to Academician L. A. Artsimovitch, V. I. Kogan, T. I. Philippova, and R. V. Lazarenko of the I. V. Kurtchatov Atomic Energy Institute for many discussions and help in the organization of the work; to Yu. A. Matveets for the assistance in the experiments; to Y. A. Bikovsky and V. I. Dymovitch for supplying the mass spectrometer device; to T. A. Romanova for help in the X-ray photograph preparations; to B. A. Benetsky for help in the calibration of the counter; to I. Ya. Barit and G. V. Belovitsky for the helpful discussions; to N. A. Phrolova for help during the work and supplying the isotopic sources; and to V. T. Yurov, L. M. Kyzmin, and D. B. Voronzov for assistance in the experiment.

REFERENCES

[1] N. G. Basov and O. N. Krokhin, *Zh. Eksperim. i Teor. Fiz.*, vol. 46, p. 171, 1964.
[2] R. V. Ambartzumian, N. G. Basov, V. S. Zuev, P. G. Kriukov, and V. S. Letokhov, *Zh. Eksperim. i Teor. Fiz., Pis'ma*, vol. 4, p. 19, 1966.
[3] N. G. Basov, V. S. Zuev, P. G. Kriukov, V. S. Letokhov, Yu. V. Senatsky, and S. V. Tchekalin, *Zh. Eksperim. i Teor. Fiz.*, vol. 54, p. 767, 1968.
[4] A. J. DeMaria, D. A. Stetser, H. A. Heynau, *Appl. Phys. Letters*, vol. 8, p. 174, 1966.
[5] W. H. Glenn and M. J. Brienza, *Appl. Phys. Letters*, vol. 10, p. 221, 1967.
[6] A. J. DeMaria, R. Gagosz, H. A. Heynau, A. W. Penney, and G. Wisnor, *J. Appl. Phys.*, vol. 38, p. 2693, 1967.
[7] L. A. Artsimovich, *Controlled Thermonuclear Fusion*. Moscow, 1963, p. 12.

ION ENERGIES PRODUCED BY LASER GIANT PULSE

William I. Linlor[1]
Hughes Research Laboratories
Malibu, California

Ion energies of ~1,000 eV have been produced by the action of a single "giant pulse" from a ruby laser.

The delivered energy was about 0.2 J in a pulse having full width at half maximum of about 40 nsec. The peak power was 5.4 MW, as measured with a Korad photodiode PDS-20-1C; all pulses were within 10% of this value. The area of the focal spot is not well known; approximate measurements indicate 10^{-3} cm^2. Energy was determined by the time of flight of the ions over a path of 4.3 cm, and ranged from 0.5 μsec for carbon to 1.2 μsec for lead.

The targets were in a system at 10^{-6} Torr prior to the laser burst. Within the vacuum system were mounted a lens of focal length 67 mm, a target plate on which the laser beam was focused, and a copper collector plate 3.1 cm o.d. with a 0.64 cm-diam hole to transmit the laser beam. Collector and target plate biases of −23 and −20 V added negligible energy to the ions in comparison with their thermal energy. As indicated in Fig. 1, resistors connected the collector and target to ground. The potentials produced by current flow, and the laser pulse signal from a 1P42 monitor, were displayed on oscilloscopes triggered by the laser electronics. Sweep synchronization was obtained by displaying the monitor signal on all sweeps.

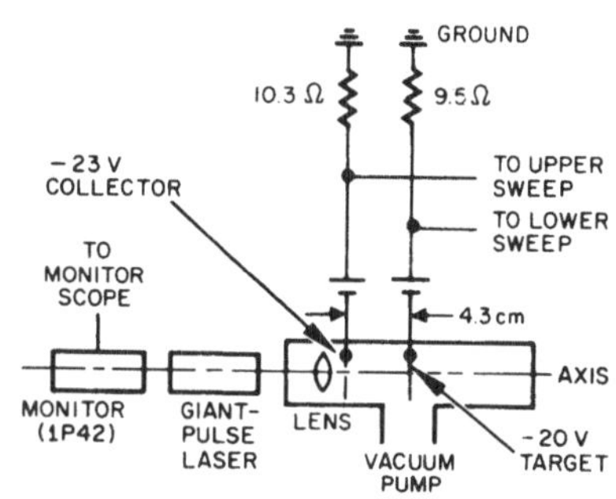

Fig. 1. Experimental system. Laser beam passed through hole in collector, and was focused on the target. Giant pulse was timed by monitor and by flow of electrons from target. System pressure before burst was 10^{-6} Torr.

The time of flight from the beginning of the laser burst to peak collector current is plotted vs the square root of the atomic number, in Fig. 2, for two laser power levels. If the initial temperatures of each plasma produced by the laser burst were the same, the points would theoretically lie on a straight line. Reasonably good agreement is obtained with the solid line corresponding to an ion energy of 1,000 eV for the 5.4-MW laser pulse – and with the dashed line corresponding to an ion energy of 420 eV for the 2.5-MW laser pulse. The latter pulse value is uncertain by about one megawatt because of a change in the monitor photodiode. The time-of-flight measurements at the higher power level have been repeated many times, always within 10%.

Oscillograms for high-Z materials showed positive pulses occurring very close to the time of the laser burst and lasting a few tenths of a microsecond. Such signals were absent for carbon and aluminum but began to appear for titanium and increased with the atomic number of the target material. These seem to be caused by photoelectrons ejected from the collector by photons from the energetic plasma.

A magnetic field of 1,200 G perpendicular to the axis of the system did not prevent passage of the plasma. Carbon plasma having ion energy of 400 eV exhibited no discernible bending.

Interpretation of the preceding results is complicated by "heat sink" problems and by uncertainty regarding the amount of material ionized. Some simplification was obtained by employing as targets

CONTENT ANALYSIS

A. C plasma	(ion)
A. Al plasma	C. giant-pulse laser
A. Au plasma	D. vacuum (10^{-6} Torr)
A. Ti plasma	E
B. thermionic emission	

Reprinted from: *Applied Physics Letters*, **3**, 210–211 (1963).

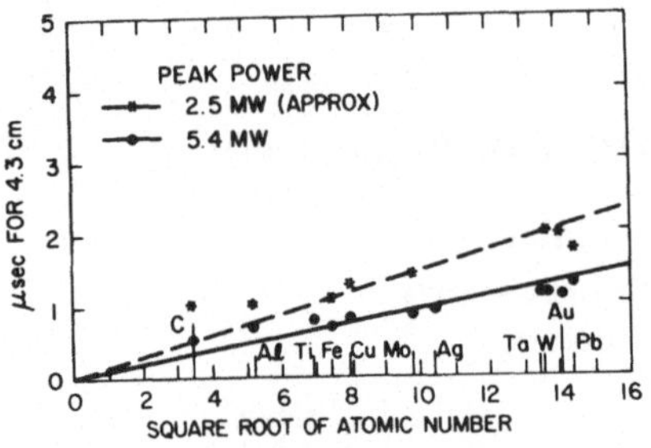

Fig. 2. Time of flight for various targets.

aluminum and gold foils whose thicknesses respectively were 1.8×10^{-4} and 1.6×10^{-4} g/cm^2. Ion speeds on both sides of single and multiple layers were measured for the same 5.4-MW laser pulse, by the addition of a collector on the "aft" side similar to the "fore" collector already described. Bias voltages were applied as in Fig. 1; for some measurements the bias voltages were removed and a magnetic field of 1,200 G was applied perpendicular to the axis of the system. The latter arrangement (no voltages applied) produced a net positive current flow to the collectors, such that the potential drop across the resistors was 30 V for aluminum and 15 V for gold foil targets.

From the foil investigation the following conclusions were obtained:

1. The ion energy measured by the fore collecto is very nearly the same whether a single foil or a thick sheet is the target.
2. After the laser burst, the single foils required about 0.3 μsec for the heat to penetrate the full foil thickness.
3. The ion energy on the aft side was about equal to that on the fore side, though sometimes a factor of two lower.
4. For multiple foils of gold, the product of thickness and ion energy on the aft side was approximately constant.

The ion energies obtained are considerably higher than indicated in previous publications dealing with the interaction of laser beams with various target materials.[2-10] Of these, only ref 8 involves a giant-pulse investigation, giving a carbon plume speed in air comparable to the present result.

I thank the directors of the Research Laboratories for providing a giant-pulse laser and other equipment, and for interested support. The use of Fred McClung's giant-pulse laser in the early phases of this work is appreciated, as well as the assistance of Larry Allum. Discussions of this material with A. Wayne Ehler were of considerable help.

---

[1]Author's present address: 14330 Blackmon Drive, Rockville, Maryland, and U. S. Atomic Energy Commission, Thermonuclear Branch, Washington 25, D. C.

[2]W. I. Linlor, "Plasma Produced by Laser Bursts," *Bull. Am. Phys. Soc.* **7**, 440 (1962).

[3]J. J. Muray, "Photoelectric Effect Induced by High-Intensity Laser Light Beam From Quartz and Borosilicate Glass," *Bull. Am. Phys. Soc.* **8**, 77 (1963).

[4]R. E. Honig and J. R. Woolston, "Laser-Induced Emission of Electrons, Ions, and Neutral Atoms From Solid Surfaces," *Appl. Phys. Letters* **2**, 138 (1963).

[5]D. Lichtman and J. F. Ready, "Laser Beam Induced Electron Emission," *Phys. Rev. Letters* **10**, 342 (1963).

[6]C. M. Verber and A. H. Adelman, "Laser-Induced Thermionic Emission," *Appl. Phys. Letters* **2**, 220 (1963).

[7]R. E. Honig, "Laser-Induced Emission of Electrons and Positive Ions From Metals and Semiconductors," *Appl. Phys. Letters* **3**, 8 (1963).

[8]J. F. Ready, "Development of Plume of Material Vaporized by Giant Pulse Laser," *Appl. Phys. Letters* **3**, 11 (1963).

[9]F. Giori, L. A. MacKenzie, and E. J. McKinney, "Laser-Induced Thermionic Emission," *Appl. Phys. Letters* **3**, 25 (1963).

[10]D. Lichtman and J. F. Ready, "Reverse Photoelectric Effect and Positive Ion Emission Caused by Nd-in-Glass Laser Radiation," *Appl. Phys. Letters* **3**, 115 (1963).

THE INITIAL STAGES OF LASER-INDUCED GAS BREAKDOWN*

Renaud Papoular

Association Euratom-CEA, DPh PFC

B.P. N°6,92-Fontenay-aux-Roses (France)

ABSTRACT

Instruments and methods are described which have been used to detect and count free electrons released in a gas by laser radiation at power levels below breakdown threshold. The experimental results are analyzed and their possible relations to known elementary phenomena are discussed.

INTRODUCTION

Gas breakdown is defined as the sudden onset of a high electrical conductivity in a normally non-conducting gas. This, of course, is due to the appearance of free electrons in the medium and is generally accompanied by the emission of a bright light and, in the case of laser-induced gas breakdown, by a strong absorption of the incident laser light.

In fact, breakdown is a composite and dynamic phenomenon, extending over a measurable time and going through different stages, starting from the absence of free electrical charges and ending possibly in complete

*Presented at the Second Workshop on "Laser Interaction and Related Plasma Phenomena" at Rensselaer Polytechnic Institute, Hartford Graduate Center, August 30-September 3, 1971.

Reprinted from: *Laser Interaction and Related Plasma Phenomena,* Vol. 2, H. Schwarz and H. Hora (eds.), Plenum, New York, 1972, pp. 79–96.

ionization of the gas, not to speak of the following after glow and recombination.

This paper is concerned with the first stages of gas-radiation interaction and with the mechanisms by which electrons are set free in the gas. Eight years after the discovery of laser-induced breakdown[1], these mechanisms are by no means completely clarified, probably because of the smallness of both space and time scales in the relevant experiments. The following pages must therefore only be considered as an attempt to describe experimental observations and their possible relations to already known elementary phenomena.

EXPERIMENTAL OBSERVATIONS

Fig. 1 represents the incident, unperturbed laser pulse (P_o), the light pulse (P) transmitted through the gas, and the light emission from the gas (E) in the visible band. The term "breakdown" is often used to designate the time when P decreases abruptly while e_λ increases rapidly. In this paper, we are interested in the events occurring before that time. Indeed, the laser

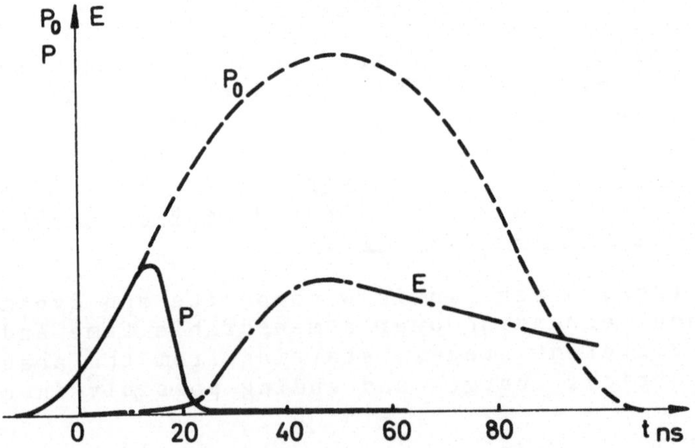

Fig. 1

Gas absorption and emission on breakdown.
P_o : incident laser light ; P : transmitted laser light,
E : light emitted by the gas in the visible band.

pulse itself has a bell-shaped time profile and starts much earlier than the instant of breakdown (a few tens of nanoseconds). As a consequence, and although nothing particular seems to occur, on Fig. 1, before breakdown, it is possible, even then, to detect and count free electrons, if only sensitive instruments are used, such as cloud chambers photomultipliers and proportional counters.

Cloud Chambers[2]

When free charges appear at some point in a supersaturated vapor with enough density, the vapor condenses locally at that point and may be detected photographically as a small cloud or droplet. Naiman et al. used a cloud chamber based on this principle to detect charges formed at the focus of a laser beam. They used methanol or dimethylmethylphosphonate as condensible target vapors. The sensitivity of their chamber was 5000 ion pairs/cm^3, enough to detect minimum ionization cosmic-ray tracks. No such tracks were normally observed passing through the focal volume when the laser was fired. A clearing electric field (100 v/cm) was applied to sweep away any electron originating from other parts of chamber. Thus, any charge detected on firing the laser could only be due to ionization of the vapor by the laser itself.

Using a Q-switched ruby laser which could deliver a maximum of 0,2 J in 30 ns (i.e. about 10MW), they could exceed the breakdown threshold. A bluish plasma was then formed at the focus, giving rise to profuse swirling clouds that expanded throughout the chamber and persisted for a few minutes.

At lower powers, however, the passage of the laser beam resulted in a cluster of droplets, ranging in diameter from 0.1 to 0.5 cm, which fell to the bottom of the chamber at about the same rate as the background mist.

This proves that the laser could release a number of charges without necessarily ending in complete breakdown (ionization) of the vapor : this is called prebreakdown ionization.

Photomultipliers[3]

Let the light beam from a Q-switched laser be focused by a lens onto the gas in a pressure chamber. In the particular experiment to be described, the laser had an Nd : glass rod and delivered 40 nsec pulses of 25 MW peak power, and the lens had a 100 mm focal length. The light emitted by the focal region at a mean angle of 90° relative to the laser beam direction was collected by a field lens of 25 mm focal length and detected by a photomultiplier type Radiotechnique 56 TVP (S20 cathode, 14 dynodes, 2-ns rise-time). The anode signal was displayed on a Tektronix oscilloscope type 585 A (4.5 nsec rise-time)[4]. The stray laser light is trapped by a system of black-paper stops and a high-pass filter in front of the photocathode.

For a given gas and a given set of pressure and laser power values, oscillograms are taken with different achromatic attenuators in front of the photocathode. The smaller the attenuation, the earlier one can follow the development of the discharge ; from the corresponding partial curves (shown in fig. 2), an overall curve of brightness (I) against time (t) is drawn for each set of physical conditions.

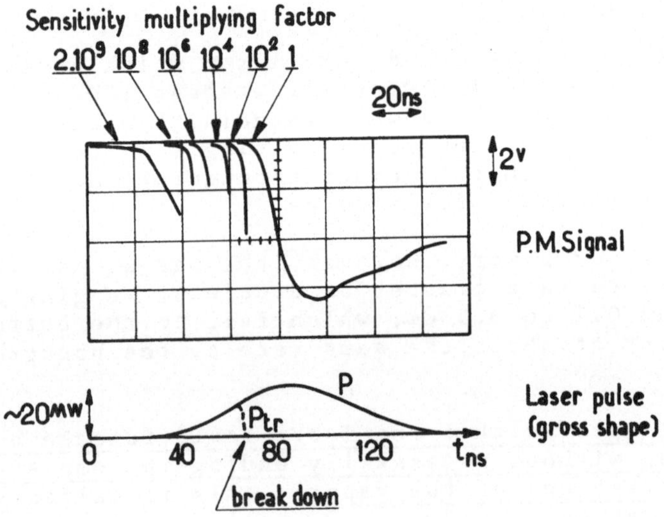

Fig.2

Raw photomultiplier data; gas: argon, 1 atm; laser energy, ε ≃ 1 J.

At low attenuator settings the photomultiplier is strongly saturated during most of the time, but this does not cause any permanent damage ; it is enough to overlook anode signals higher than about 100 mA.

The experimental results are summarized in the following figures. Fig. 3 shows the result of <u>spectral analysis</u> of I(t) by means of a premonochromator with a spectral resolution of about 200 Å at λ = 5000 Å.

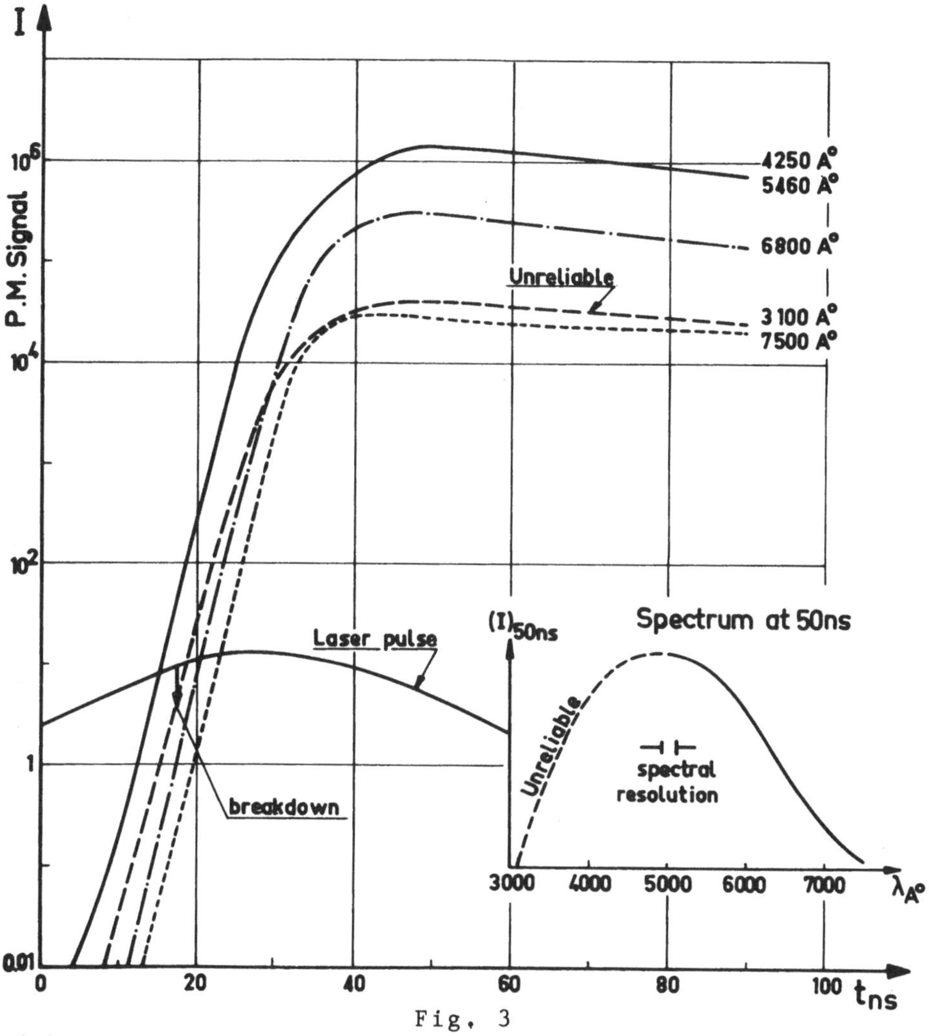

Fig. 3
I(t) for a few wavelengths (vertical scale arbitrary). Argon, 1 atm. ε ≃ 1 J. Laser pulse shape also shown ; the vertical arrow shows time of breakdown.

Each curve corresponds to one wavelength and the inset shows the spectral profile at one instant near maximum brightness. These very crude measurements are an indication of the predominance of the continuum over the line emission during the period of interest.

Fig. 4 shows I(t) for different pressures. Notice that, when the pressure increases, the curves are displaced towards earlier times, their gross shape remaining unchanged in that part which occurs before breakdown (the latter is represented by a vertical arrow for each pressure). Light emission from the gas is clearly detected as early as 20 to 30 nsec before breakdown. At such times, the laser power is less than 1/10 th of the threshold, i.e. the minimum peak power to cause breakdown, which is indicated for 1 atm. by the horizontal dashed line, marked $(P_{th})_{1atm}$.

Around the time of breakdown, light emission increases approximately exponentially during 20 to 30 nsec. In order to interpret this observation in terms of electron density, it should be noted that, before breakdown, the degree of ionization of the gas is still low and the light must be emitted mainly by free-free bremsstrahlung due to electron-neutral collisions ; I(t) should therefore be proportional to the free electron density in the observed volume. At breakdown and immediately after wards, electron-ion collisions are prevailing and I(t) should be proportional to the square of electron density.

Fig. 5 shows the influence of laser energy (ε) on the early gas emission. An increase of ε from 0,6 J up to 1J does not alter very much the gross shape of I(t) but shifts this curve to earlier times through about 60 ns. It has not been possible to draw a reproducible brightness curve for $\varepsilon < 0{,}6$ J.

Fig. 6 shows I(t) for different gases at 1 atm and $\varepsilon \simeq 1$ J.

Finally, Fig. 7 summarizes the general behaviour of the brightness I(t) in correspondance with the laser power P(t). The scale for I and P is arbitrary. If it is assumed that ionization is complete at maximum brightness, the free electron density, N, is then about 10^{19} cm^3 ; if, furthermore, I is taken to be proportional to N, as the case should be for emission by electron-neutral bremsstrahlung at constant temperature, then,

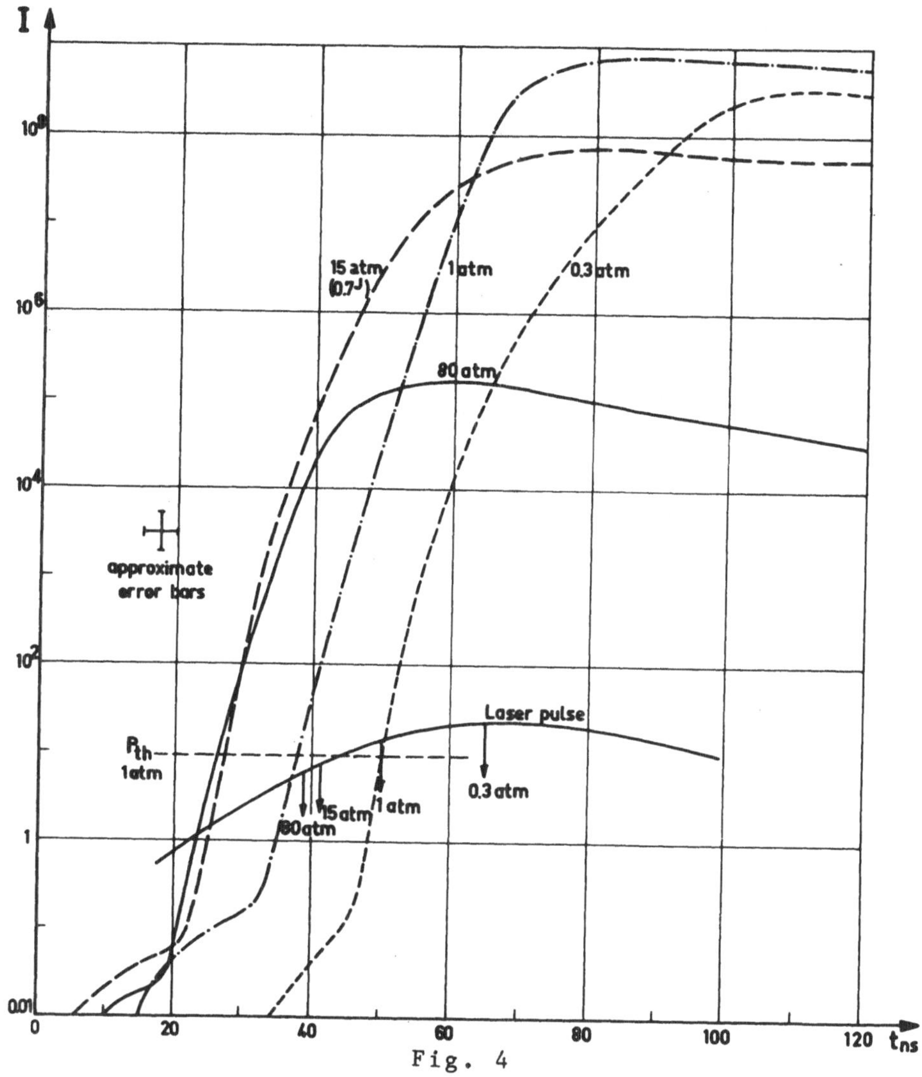

Fig. 4

I(t) for various argon pressures. ε ≃ 1 J.

at the foot of I(t), N should range between 10^9 and 10^{12} cm^{-3}, corresponding to 10^4 to 10^7 electrons in the focal volume ($10^{-5}cm^3$). If account were taken of electron-ion collisions beyond breakdown, and also of temperature variations, then still larger numbers would have been found for the minimum observable N. This is an indication that the photomultiplier method is not

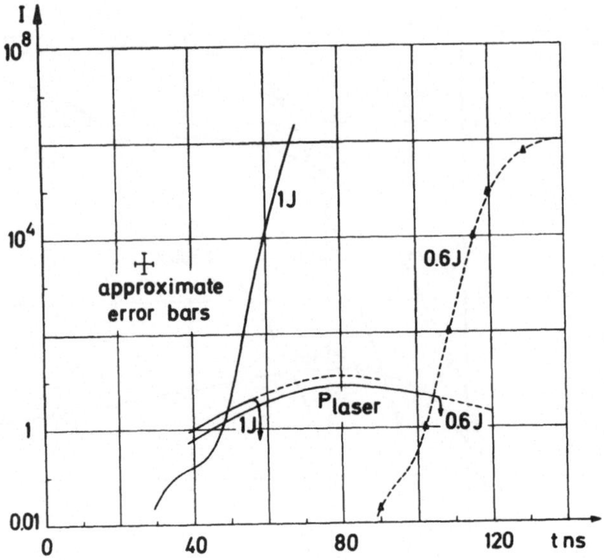

Fig. 5

I(t) for argon at 1 Atm. and two different laser energies (0,6 and 1 J) ; the corresponding laser pulses are also shown (P_{laser}).

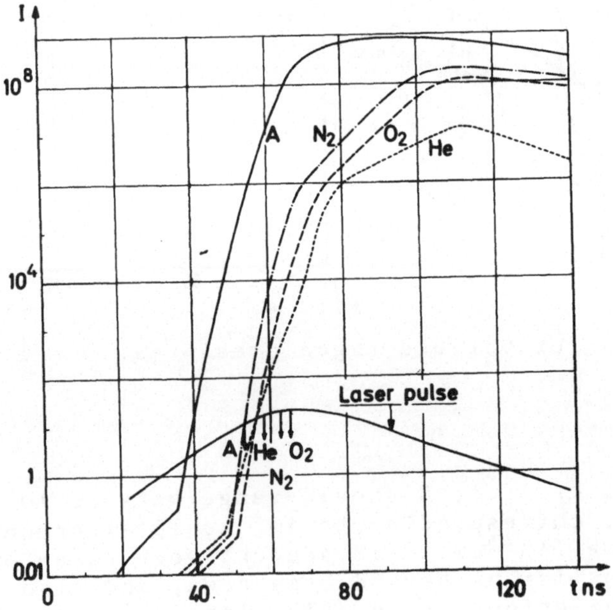

Fig. 6

I(t) for different gases at 1 atm and $\varepsilon \simeq 1$ J.

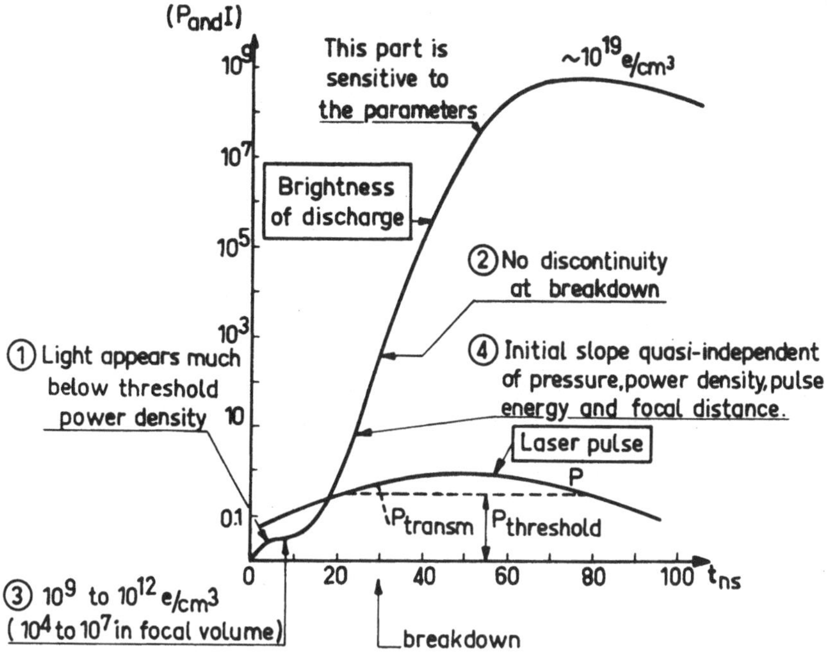

Fig 7

Summary of observations for a pressure of order 1 atm. Vertical scale for I and P is arbitrary.

enough to detect the <u>initial</u> or <u>priming</u> electrons for a definition, see "Discussion" below). Probably, when meaningful signals start to be detected by this method, the breakdown process has already developed into the <u>avalanche</u> stage, (which is treated in detail elsewhere). This could provide a method of studying the growth of <u>cascade ionization</u> ; there is, however, a difficulty here, in that it is not easy to account for the role of temperature in the light emission from the focal volume.

Proportional counters[5,6,7]

Fig. 8 is a sketch of a measuring device[6], showing the laser beam traversing a pyrex chamber containing two spherical bronze electrodes, 50 mm in diameter, to which is applied a dc potential difference varying between 0 and 40 Kv. The laser beam crosses the common axis of the electrodes, a few millimeters above the surface of the lower one (cathode). The n_t electrons which are released by the laser pulse in this region are accelerated by the dc electric field and thus can ionize,

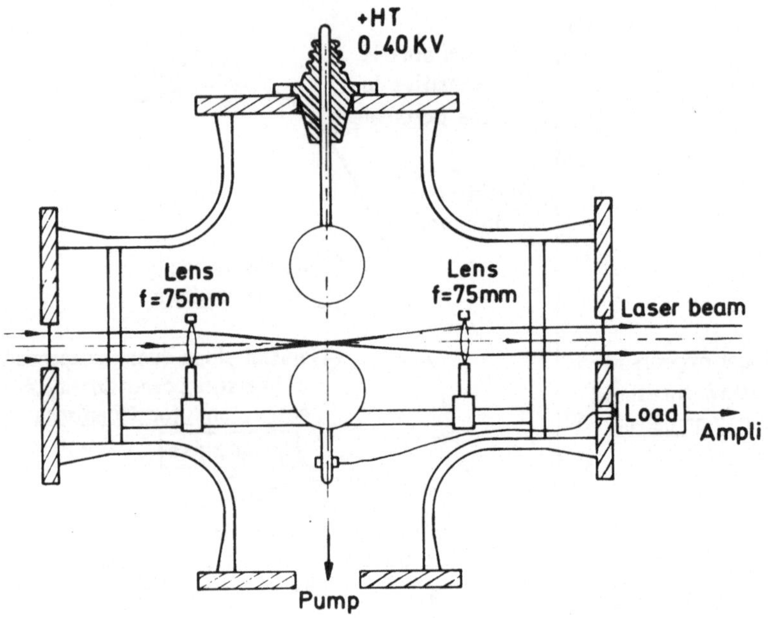

Fig. 8

Proportional counter with spherical electrode.

by collision, the molecules of the ambiant gas. The resulting free electrons are also accelerated and the cascade process goes on, giving rise to a growing avalanche of charges which can easily be detected when they reach the anode. If the total space charge, n, does not exceed the critical value (10^8 elementary charges) leading to the formation of a streamer[8], then the amplitude of the voltage pulse in the high voltage circuit is proportional to the initial number of free electrons in the focal region, assuming the time constant of the load is large enough (a few tens of μsec). It is clear that the maximum amplification factor $M = n/n_t$, is of the order of 10^8 ; but since 10^5 charges is already enough to give a detectable pulse of 1 mV (if the load capacity is of the order of 10 pF), an amplification of only 10^5 will allow one to detect a single initial electron.

Of course, the amplification factor, M, must be decreased as the expected number of initial electrons increases, so as not to exceed the streamer limit. This is done by dereasing the high voltage, for given gas and pressure. Because of the stochastic nature of the cascade process, M only represents the average value of a number of measurements, the dispersion of which decreases as n_t increases. M is determined either by direct calibration, using a known value of n_t, or by computation, using the known values of Townsend and attachment coefficients.

The analysis of experimental results often requires a knowledge of the volume in which every free electron gives rise to an avalanche. In the case of fig. 8, this is grossly the volume limited by the surface of the spheres and the circumscribed cylinder, and the amplification factor is function of the initial position of the priming electron. In practice, one can take M to be constant over about 20 mm along the laser beam axis. The cylindrical counter[7] with coaxial electrodes is superior in this respect, because M is not sensitive to the initial position inside the counter (see fig. 9)

Fig. 10 shows experimental results obtained with a counter of the type in fig. 8 ; curves of n_t (number of electrons released by the laser pulse) against Pm (peak laser power) are drawn for CO_2 at different pressures and a focusing lens of 75 mm focal length (f_1), using an Nd : glass laser with 40-nsec. pulses. The vertical bars define the dispersion of measurements. For each pressure, P_m has a minimum value, P_i, below which no electrons are detected. This value is to be distinguished from the laser-induced gas breakdown threshold, P_c, which is much higher (>20 MW for p<200 torr).

In between P_i and P_c, the curves exhibit three distinct regions :

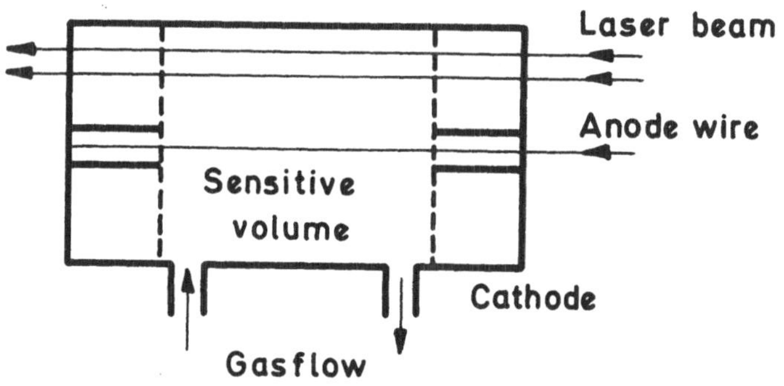

Fig. 9

Proportional counter with cylindrical electrodes.

a) an initial abrupt rise of n_t from zero to 10 or 100;

b) an intermediate region where $n_t \sim P_m^k$, with $k \simeq 2$;

c) an approximately exponential region, ending up in complete laser-induced gas breakdown.

The initial abrupt part is probably to be identified with results obtained with a cylindrical counter[7]; apparently, the range of powers used there was not sufficient to cover the three regions observed here (except, perhaps, in the case of CO_2).

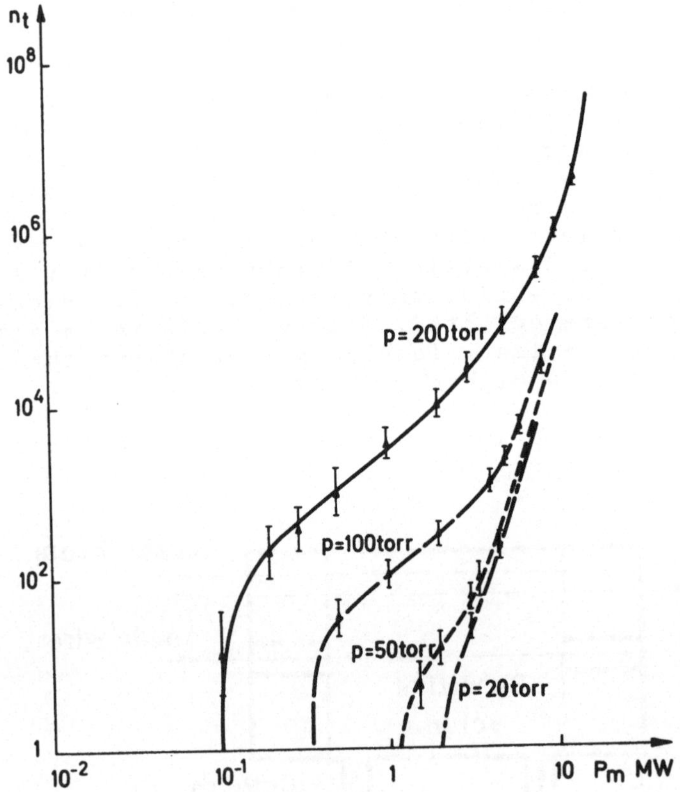

Fig. 10

Number of priming electrons, n_t, as a function of peak laser power, P_m ; f_1 = 75mm is the focal length of the focusing lens; gas:CO_2; laser wavelength : 1,06μ

As regards stage (C), one is tempted to link it up with the exponential part of the brightness curve (fig.4) in the sense that, in both cases, the breakdown process has developped to the point where the laser light gives rise to cascade ionization of the gas. It must be kept in mind, however, that the counter integrates the electron production over the pulse duration while the photomultipliers respondes to the instantaneous population of electrons.

Fig. 11 shows typical curves of the minimum power, P_i, as a function of pressure, p, for a couple of focal distances, f_1=75 and 2500 mm. Below a pressure of some tens of torrs, P_i scarcely depends on p while, at higher pressures, the slope of the curves lies between -1 and -2. Similar results have been obtained for molecular hydrogen and for a ruby laser[6].

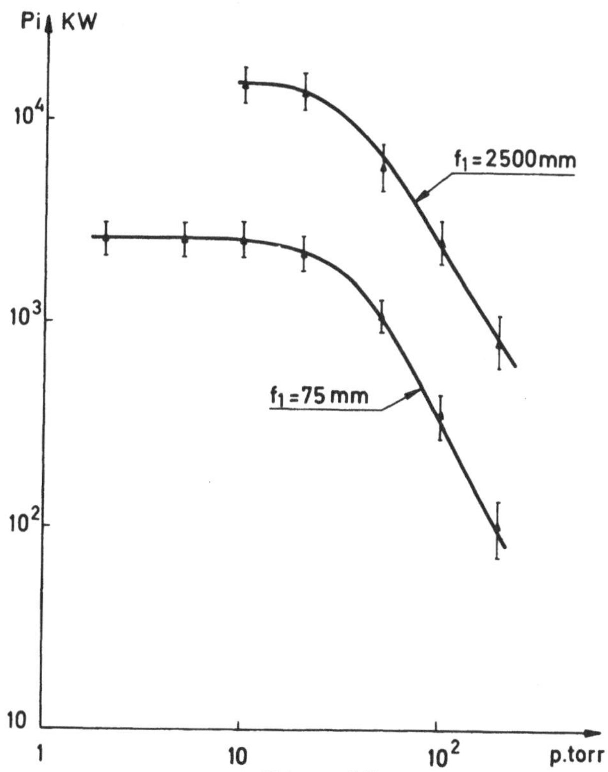

Fig. 11

Minimum power, P_i, as a function of pressure, p, for different focal distances ; gas : CO_2 ; laser wavelength 1,06μ.

As regards absolute values, P_i is of the order of a megawatt at pressures of about 100 torrs ; this value is not very sensitive to the nature of the gas (CO_2,H_2, N_2,A,He,Kr) nor to the difference in wavelength between ruby and neodymium glass ; it only increases by a factor of 10 when focal distance increases from 75 to 2500 mm.

Finally, the light flux corresponding to P_i is about 10^9 to 10^{10} W/cm^2, for f_1=75mm. This result is in order of magnitude agreement with results on CO_2, obtained in reference 7.

DISCUSSION[9,10,11]

In order to explain the production of free electrons during the early stages of breakdown, which we are interested in, one can think of two different mechanisms :

a) photoionization of neutral particles by one or more (say k) photons "acting together" so as to climb the potential barrier[12]: this is the photoelectric effect of order k and gives rise to the so-called initial or priming electrons ;

b) cascade ionization : this is due to free-free photon absorption by electron-neutral collisions, leading to an increase of electron energy, followed by neutrals ionization by electrons (like in microwave breakdown) ; at least one priming electron is required to start this process.

(It is generally admitted that cosmic or other high energy radiation do not produce more than 10 electron-ion pairs per cm^3 and sec, and that the equilibrium concentration of negative-positive ion pairs does not exceed 10^3cm^{-3} at NTP[13]. This, together with direct experimental evidence (e.g. reference 2), shows that background sources of electrons are not to be worried about).

The evolution of the free electron density N, in a gas of neutral particles of density N_o, under the action of a photon flux F, is then given by

$$\frac{dN}{dt} = a_k N_o F^k + bN_o NF \qquad (1)$$

where a_k and b are constant coefficients, and k is the smallest integer such that k times the photon energy exceeds the ionization potential (photoelectric effect of order k) ; for simplicity eq. (1) overlook loss terms (attachment, recombination and diffusion) and electron-ion collisions (the ion density being negligible in early stages).

The photons will be considered to be delivered in a pulse of duration τ and amplitude $F = Pm/s$, where τ is the half-power width of the laser pulse and P_m its peak amplitude, and s the area of the focal spot. By integration of (1) with suitable initial and boundary conditions, and neglecting the second term on the r.h.s, the minimum flux necessary to produce one free electron in the volume v under observation, is then found to be

$$Fi = Pi/_s = (a_k N_o \tau v)^{-1/k} \qquad (2)$$

which is proportional to $p^{-1/k}$, since $N_o \sim p$

When $F > F_i$, it is found that the total number of electrons set free in the volume v at the end of the pulse is :

$$n_t \simeq a_k F^k N_o \tau v \qquad (3)$$

When the photon flux is much larger than F_i,

$$n_t \simeq \frac{a_k}{b} F^{k-1} e^{bN_o F\tau} \cdot v \qquad (4)$$

which is exponential in F when $bN_oF\tau$ is large enough.

If the photon flux is not spatially homogeneous, the v in the above formulae is an "effective" volume over the flux distribution.

Let us now proceed to compare these equations with the experimental results, fig. 10 and 11. From (3) and (4), the slope $\alpha = d(\log n_t)/d(\log P_m)$ is equal to k or $k-1 + bN_o\tau F$, respectively. Thus, α should always be greater than k and increase with F steadily. Also, for a given F, α should increase with p (i.e N_o). Neither of these predictions is verified experimentally in the range of parameters considered here.

Also, according to ref. 2, the slope $\beta = d(\log F_i)/d(\log p) = d(\log F_i)/d(\log \cdot N_o)$ should be equal to $-1/k$. For the gases and photons employed here, k is about 10 to 20 , so that β should be very small, which fig. 11, for instance, shows not to be the case, except at pressure lower than a few tens of torrs.

Finally, one can compare experimental values of F_i with the values deduced from eq.2 by injecting the coefficients a_k given by measurements of the multiphoton effect[14], done at low pressures ($<2x10^{-3}$ torr) so that cascade ionization is absent. It appears that multiphoton ionization of the main gas (CO_2,H_2,N_e,H_e, etc.) is hardly probable in the present experiments (at p>50 torrs) because the photon fluxes involved are much too low.

One would then conjecture that impurities of low ionization potential are present among the main gas molecules and give rise to the priming electrons[11,15] However, the concentration of impurities should be independant of pressure, which make it difficult to explain why doubling the pressure can result in a increase of the electron yield by a factor larger than 10, at constant light flux (see fig. 10). The fact that the slope β (fig. 11) is of the order of -1 or less is also hardly accounted for by the presence of impurities. The above considerations apply all the more to breakdown induced by CO_2 lasers[16], since the photons, there, are ten times less energetic than those of Nd : glass lasers

Another conjecture could be that, at pressures higher than about 10 torrs, some sort of permanent, absorbing[17] or transient, "macromolecules" are formed with the molecules of the main gas, their size and/or concentration depending on pressure. For a given pulse duration, a minimum power would be required to ionize these particles (viz P_i). Would this critical value be reached, a large number of electrons could be liberated simultaneously, but once the particles were blown off, the only remaining mechanism for the production of free electrons would be cascade ionization.

This rough scheme is compatible a) with the sharp initial rise of $N_t(P)$ for low P_i's, followed by a bend going into an exponential, and b) with the strong inverse dependance of P_i on p.

On top of any such mechanism, which remains to be assessed, multiphoton ionization of impurities could also occur, and it would be useful to estimate the relative importance of the two phenomena. One way to do so is to analyze the products of ionization by means of a mass spectrograph and hence determine the concentration of impurities in the main gas.

In conclusion, it may be noted that a large amount of effort has been spent to build a theory of the avalanche process [10,11,18]. The present theories appear to be quite successful in predicting breakdown thresholds. This, however, is only a global test of the theory, and it is very much desirable that a direct experimental study of the cascade be carried out (e.g. the e-folding time as a function of pressure, gas and light flux). In principle, the proportional counter could be employed for this purpose, by measuring n_t as a function of the variables.

REFERENCES.

1. P.D. Maker, R.W. Terhune and C.M. Savage, Proc.3rd Int. Quantum Electr. Conf. Paris (1963) ; Dunod (Paris) vol. II, 1559 (1964).

   E. K. Damon and R.G.Tomlinson , Appl. Opt. 2, 546 (1963).

   R.G. Meyerand and A.F.Haught , Phys. Rev. Lett. 11, 401 (1963)

2. C.S. Naiman et al, Phys. Rev., 146, 133 (1966)

3. V. Chalmeton and P. Papoular, C.R. Acad. Sc. (Paris), 264, 213 (1967) ; Meeting of the Plasma Physics Division of the A.P.S., Boston 1966, N° 3C2 ;
   V. Chamelton, Thèse (Paris) 1969, Chap. II.

4. The time response of a chain comprising a photomultiplier and an oscilloscope has been studied by R.Papoular, in Rev. Phys. Appl., 3, 169 (1968) for different time-profiles of the light signal.

5. W.R. Pendelton and A.H. Guenther, Rev. Sc. Inst., 36 1546 (1965).

6. V. Chalmeton and R. Papoular, Phys. Lett. 26 A (11), 579 (1968) ,
   V. Chalmeton, J. de Phys. 30, 687 (1969) - Thèse (Paris, 1969), Chap. III-IV

7. A. Blanc and D. Guyot, Int. Conf. Phys. Ionized Gases Bucharest 1969, p.35, and private communications.

8. P. Raether. Electron Avalanches and Brakdown in Gases, London, Butterworth's, 1964.

9. V. Chalmeton, Thèse (Paris) 1969, Chap. V-VI

10. F.V. Bunkin and A.M. Prokhorov, Sov. Phys. JETP, 25, 1072, (1967)
Ya. B Zeldovich and Yu, P. Raïzer, Sov. Phys. JETP, 20, 772 (1965).

11. A.V. Phelps, in Physcs of Quantum Electronics, 538, Mc. Graw-Hill, (1966).
M. Young and M. Hercher, J. Appl. Phys. 38, 4393 (1967).

12. L.V. Keldysh, Sov. Phys. JETP, 20, 1307 (1965)

13. In Am. Inst. of Phys. Handbook, 5-278 (Mc Graw-Hill, 2nd ed.)

14. N.K. Berejetskaya et al, 9th Inter. Conf. on Phys. of Ionized Gases, Bucharest, 40 and 43 (1969).
P. Agostini et al., C.R. Acad. Sc., 270B, 1566(1970)

15. S.L. Chin, Canad. J. Phys., 48, 1314 (1970)

16. N.A. Generalov, et al. JETP Lett., 11 (7), 228(1970)

17. Non-resonant absorption of light by gases at medium pressures has been observed by C. Bordier et al., C.R. Acad. Sc., 262 B, 1389 (1966) ; 263B, 619 (1966) and N.R. Isenor and M.C. Richardson, Appl. Phys. Lett. (1971)

18. F. Morgan et al, J. Phys. D, 4, 225 (1971)

EXPERIMENTS ON SELF-FOCUSING IN LASER-PRODUCED PLASMAS *

A.J. Alcock

Division of Physics, National Research Council of Canada, Ottawa, Ontario, Canada

ABSTRACT

Recent investigations of laser induced gas breakdown have revealed a number of phenomena which are compatible with the occurrence of self-focusing at the time of breakdown. Some examples of the effects observed are: 90° scattering of laser light from regions having transverse dimensions as much as an order of magnitude less than the diameter of the focal volume, intense forward scattered radiation at a wavelength close to that of the laser, and plasma filaments of $\sim 10\mu$ diameter which have been detected by means of high spatial resolution, sub-nanosecond, Schlieren photography and interferometry. In this talk the large amount of experimental evidence for self-focusing is reviewed and a number of possible mechanisms discussed.

INTRODUCTION

The first indication that self-focusing might be associated with laser-produced plasmas was reported by Basov et al.[1] who observed the creation of long sparks, extending over a distance of ~ 2 meters, in air by means of weakly focused neodymium:glass laser radiation. Although this observation was explained in terms of the temporal variation of the laser beam divergence, the possibility of self-focusing leading to the observed effects was also introduced. Within a short time a similar suggestion was made by Korobkin et al.[2] who carried out a detailed study of the spark in air produced

* Talk presented at the 2nd Workshop on "Laser Interaction and Related Plasma Phenomena", Rensselaer Polytechnic Institute, Hartford Graduate Center, Aug. 30 - Sept. 3, 1971.

Reprinted from: *Laser Interaction and Related Plasma Phenomena*, Vol. 2, H. Schwarz and H. Hora (eds.), Plenum, New York, 1972, pp. 155–175.

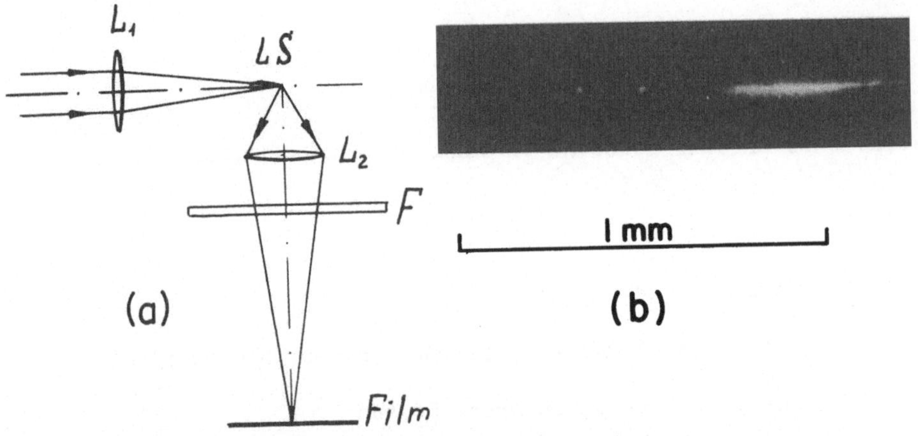

Fig. 1 (a) Experimental arrangement used to obtain high magnification time integrated photographs of air sparks by means of laser radiation scattered at 90°: L_1-focusing lens; L_2-imaging optics; LS-spark; F-narrow band filter. (b) Spark in air showing filamentary and discrete scattering regions. The direction of the laser beam is from right to left.

by means of a 100 MW, Q-switched ruby laser. From photographs of the spark, obtained with the scattered laser radiation emerging at 90° to the beam axis, and laser-illuminated shadowgraphs, the authors concluded that a number of discrete breakdown regions were formed during the development of the plasma. Although this phenomenon was discussed in terms of a radiation wave mechanism the authors also referred to earlier theoretical work by Askaryan[3] and Litvak[4] who discussed the possibility of intense electromagnetic radiation being trapped within a plasma.

EXPERIMENTAL STUDIES IN THE NANOSECOND REGION .

Shortly after the possibility of a connection between self-focusing and plasma production with lasers had been pointed out, additional studies of spark structure, carried out by means of scattered light photography, were reported.[5,6] In the case of (5) a multimode ruby laser, Q-switched by means of a rotating prism, was used to investigate sparks produced in hydrogen, nitrogen, oxygen, carbon dioxide, chlorine, methane and inert gases. The results obtained indicated a bead-like structure in molecular gases, when the laser power was slightly above threshold, and the formation of fork-like scattering regions as the laser intensity was increased. Further studies of the polarization of the scattered light were carried out and from these, and photographs taken simultaneously in two directions, it was concluded that a "surface" scattering process was responsible for the observed structure.

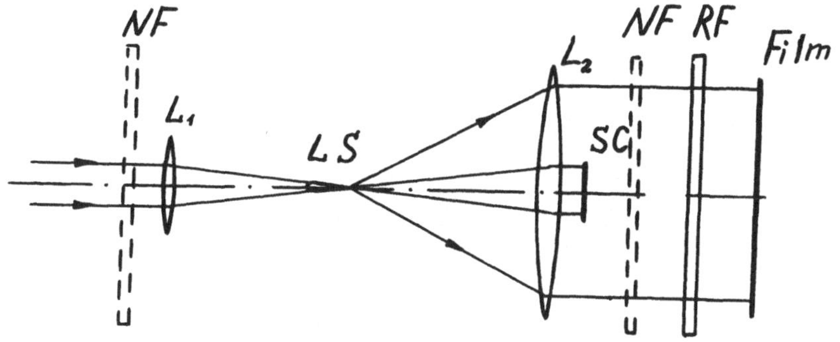

Fig. 2 Experimental arrangement used to detect forward scattered radiation emerging from a laser spark: L_1, L_2-lenses; LS-spark; NF-neutral density filters; RF-narrowband filter; SC-obstacle.

Some evidence for much weaker scattering was also obtained in the case of noble gases, and in particular, the authors noted the appearance of 3 to 5 widely separated scattering points in the case of argon.

Somewhat different results, obtained with a passively Q-switched ruby laser, operating in a single axial and transverse mode, were reported by Korobkin and Alcock[6] who investigated sparks in air by means of the scattered laser radiation emerging in the forward direction and at approximately 90° to the laser beam axis. Photographs of the sparks were obtained with a spatial resolution of ∿5μ using the type of experimental arrangement shown in Fig. 1(a). The diffraction limited laser beam which had a divergence of 0.5×10^{-3} radians was focused by means of a 10cm focal length lens and the highly magnified image of the breakdown region was recorded on infrared film using a narrow band filter to select only the scattered laser radiation. A typical result, obtained with the aid of this technique, is presented in Fig. 1(b) where it can be seen that the scattered radiation originated in filamentary regions having a diameter which did not exceed the resolution of the optical system. Thus the transverse dimensions of the scattering region were at least an order of magnitude smaller than the estimated focal spot diameter of 50μ.

The strong resemblance of the filamentary scattering region to the self-focusing of intense laser radiation in liquids prompted an investigation of the forward scattered radiation and, using the arrangement shown in Fig. 2, forward scattered light emerging from the focal region was detected both photographically and photo-

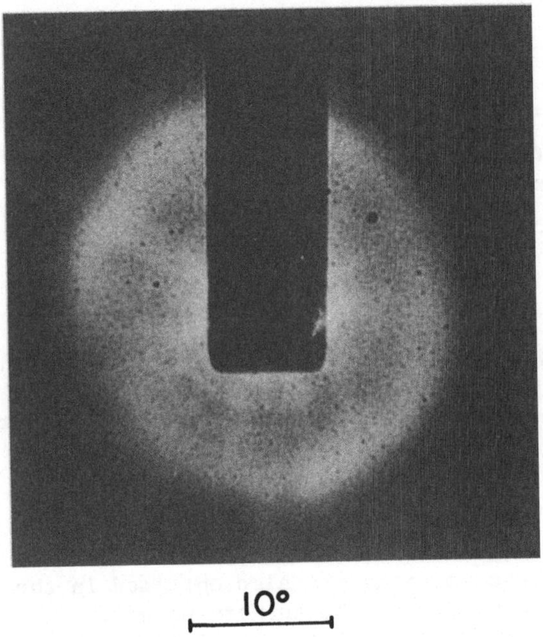

(a)

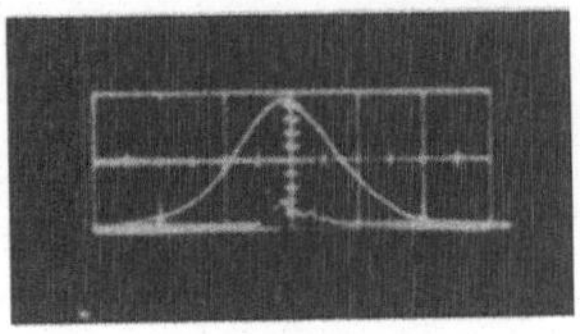

(b)

Fig. 3 (a) Photograph of forward scattered light. (b) Oscilloscope trace showing both the incident laser pulse and the forward-scattered pulse.

electrically. Photographs of the forward scattered radiation, such as that shown in Fig. 3(a), revealed that the scattered radiation from an air spark emerged in a well-defined cone having an included angle of $\sim 30^{o}$. When the camera was replaced by a planar photodiode coupled to a Tektronix 519 Oscilloscope it was found that, within the 0.5 nanosecond resolution of the detection system, the scattered light was produced at the instant of breakdown* and had a risetime which could not be resolved. As shown in the oscilloscope trace of Fig. 3(b), the peak power of the forward scattered light corresponded to approximately 30% of the incident 3 MW pulse.

Having confirmed that the polarization of the forward scattered radiation was always the same as that of the incident laser light, its coherence was also investigated by allowing two beams emerging at opposite sides of the cone to interfere with one another. The results of this test clearly demonstrated that the forward scattered light was coherent and on the basis of this observation, the spatial distribution of the scattered light, and its intensity relative to that of the incident beam, it was

*ie., the time at which the transmission of the non-scattered laser radiation decreased sharply.

considered unlikely that scattering from the plasma alone could account for the observed effects.

Although the above results, obtained in air, were initially interpreted in terms of self-focusing in the neutral gas it was clear that much additional evidence was needed to confirm the presence of self-focusing effects and to determine the processes involved. Valuable experimental data has been provided by more detailed studies of scattered laser radiation which were carried out by a number of authors.[7,8,9] Both Ahmad et al.[7] and Tomlinson[8] carried out quantitative investigations of the fraction of incident laser radiation scattered at 90°. On the basis of results obtained in helium[7] and a number of gases[8] it was concluded that the scattering was not Thomson scattering but was probably due to reflection from regions of high electron density. An important, additional, feature of both of these investigations was the emphasis given to the transverse dimensions of the source of scattered laser radiation. In the model proposed by Ahmad et al., the geometry of the optical system, used to collect the scattered light, was taken into account, and, on the assumption of reflection from a spherical shock front, it was predicted that the image of the reflecting region would have transverse dimensions corresponding to the diffraction limit. Although it was suggested that anomalous features of previously reported streak photographs[10] might be due to this 'apparent' reduction in the dimensions of the reflecting surface, no direct comparison was made with time integrated photographs obtained by means of scattered radiation. Such photographs were presented by Tomlinson[8] who applied both time-integrated and high speed streak photography to sparks produced in atmospheric pressure air and argon by the 5 MW, 30 nanosecond pulse from a rotating-mirror Q-switched ruby laser. When obtained with scattered laser radiation, both photographic techniques revealed the existence of discrete scattering centres having dimensions apparently less than the 14μ resolution limit of the recording optics. In the case of air these scattering centres were sufficiently close to one another, along the axis of the focal region, to form an almost continuous scattering 'filament', while it was noted that the centres in argon were clearly separated. The possibility that self-focusing might be occurring in the plasma sometime after the beginning of the ionization process was proposed as an explanation for the observed structure.

A similar suggestion was put forward by the author and co-workers[9] following a detailed study of sparks produced in a number of gases by means of the 4 MW peak power, 10 nanosecond, pulse from a passively Q-switched single-mode ruby laser having a measured linewidth of 0.003 cm^{-1} and a full angle beam divergence of 0.63 milliradians. The gases investigated were nitrogen, freon, methane, helium, argon, neon, krypton and xenon, and the pressure was varied in the range from 760 to 9000 Torr. Time-integrated

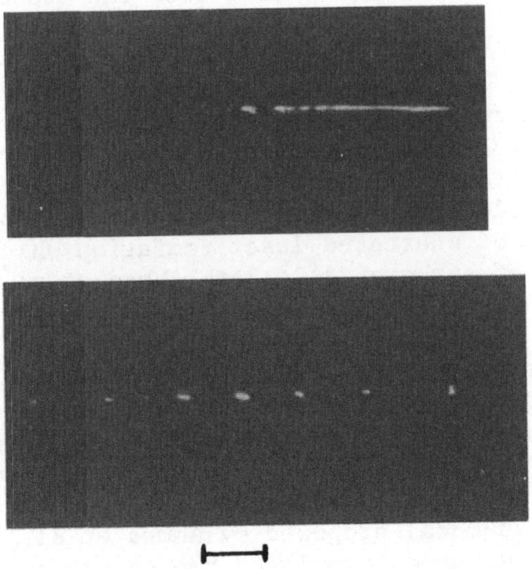

Fig. 4 Photographs of ruby laser light scattered at 90° from sparks in (a) nitrogen at 3800 torr and (b) argon at 1800 torr. The direction of the laser beam is from left to right.

photographs having a spatial resolution of $\sim 5\mu$ were obtained in nitrogen and argon and typical results are presented in Fig. 4. The two photographs in this figure demonstrate both the small transverse dimensions of the scattering regions and the sharp contrast between the spatial structure observed in the molecular and noble gases. In all cases the transverse dimensions of the scattering regions did not exceed the resolution of the optical system even though the Gaussian beam from the laser was focused by several different lenses to give estimated focal spot diameters as high as 80 microns. In addition it was found that the spacing of the scattering centres in noble gases was not only a function of the type of gas but decreased from a value of several hundred microns as the pressure was increased from 760 torr. This result showed very clearly that the observed structure did not result from intensity variations in the focal region, produced by spherical aberration of the focusing optics.[11]

Image-converter streak photographs of the scattering regions were also obtained and typical results, such as those presented in Fig. 5, show both the smoothly developing scattering filament characteristic of molecular gases (Fig. 5a) and the smaller discrete centres which develop at intervals of a few nanoseconds in a gas such as argon (Fig. 5b). Similar photographs (Fig. 5c), obtained with the visible radiation emitted by the argon plasma itself, confirmed that these small scattering points did in fact

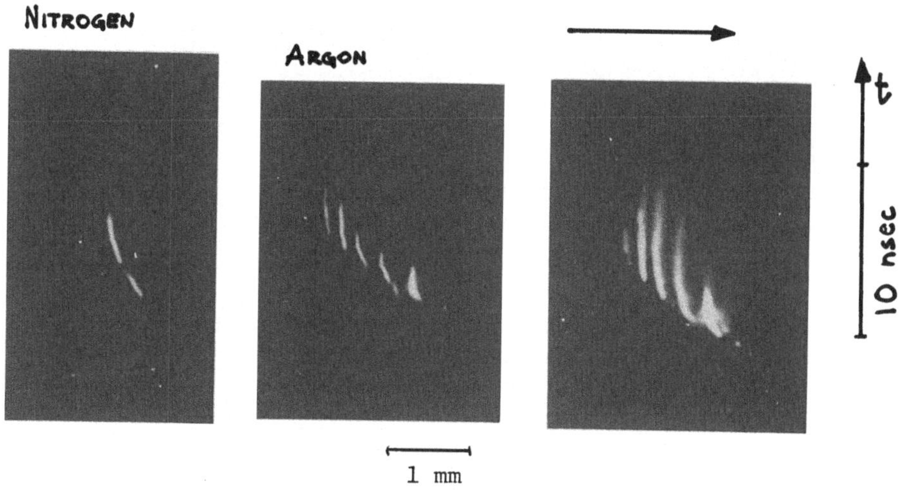

(a) (b) (c)

Fig. 5 90° streak photographs showing longitudinal development of sparks (a) scattered laser light from nitrogen at 3800 torr (b) scattered laser light from argon at 1800 torr (c) visible radiation emitted by spark in argon at 1800 torr. The arrow indicates the direction of the laser beam.

correspond to isolated plasma blobs.

An examination of the forward scattered radiation emerging from the sparks produced in noble gases yielded a number of unexpected results. The latter were obtained by detecting light emerging in the forward direction both photoelectrically and photographically, and oscilloscope traces showing the radiation emerging from a spark in argon at 8300 torr are presented in Fig. 6. On each oscilloscope trace the incident laser pulse is displayed as the first half while the second, delayed, photodiode signal shows the forward scattered light transmitted by a narrow band interference filter centred at the laser wavelength. In the absence of a spark equal signals were obtained from both detectors. Fig. 6(a) shows the effect observed when only radiation travelling in the same direction as the incident beam was permitted to reach the second photodiode. In this case the occurrence of breakdown was accompanied by the apparent absorption of approximately 80 percent of the remainder of the pulse. However, when all scattered light, emerging at angles up to 30° from the forward direction was collected, the output of the second detector corresponded to the trace displayed in Fig. 6 (b). This observation revealed that more than 80 percent of the incident pulse emerged from the focal region and that the true

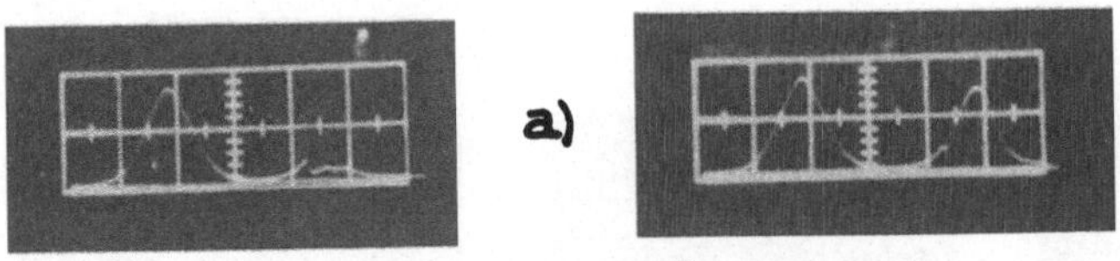

a) b)

$\sim$2.5MW/div" - vertical
10 nsec/div." - horizontal

Fig. 6 Oscilloscope traces showing forward scattered ruby laser light in argon at 8300 torr. The first signal on each trace corresponds to the incident laser pulse, while the second shows the forward scattered pulse. (a) signal obtained when only non-scattered laser radiation is permitted to reach the second photodiode. (b) signal obtained by collecting all light, at the laser wavelength, emerging from the focal region at angles up to 30° from the forward direction. The two detectors were normalized to give equal signals when no spark was produced.

absorption was in fact quite small. Although the largest signals were observed in the case of argon, similar results were obtained with all the gases investigated, and in all cases the polarization of the scattered radiation was found to be the same as that of the incident light. Both the incident power and the gas pressure determined the time at which breakdown occurred, however it was found that the ratio of the scattered radiation's maximum intensity to that of the incident light did not vary by more than $\sim$10%.

An additional property of the forward scattered radiation, revealed by recording it photographically, was the dependence of the angle of emission on gas pressure. This effect is illustrated by Fig. 7 which shows the scattered laser light emerging from argon sparks at pressures of 760 and 9000 torr. These photographs were obtained by using an obstacle to block out the direct laser radiation transmitted through the focal region and permitting only light that passed the obstacle to fall on the photographic film after passing through a narrow band filter. Fig. 7(a) shows scattered light emerging at an angle of $\sim 5^o$ from the forward direction when a spark was produced in high pressure argon. At lower pressures the higher power required for breakdown resulted in a noticeable leakage around the obstacle, however, in the case of atmospheric pressure argon the radiation emerging at an angle of $\sim 9^o$ was clearly visible, (Fig. 7b).

An investigation of the spectral characteristics of the scattered radiation was also carried out and this revealed a slight broadening of the spectrum towards longer wavelengths in the case of molecular gases, while both anti-stokes shifts and a

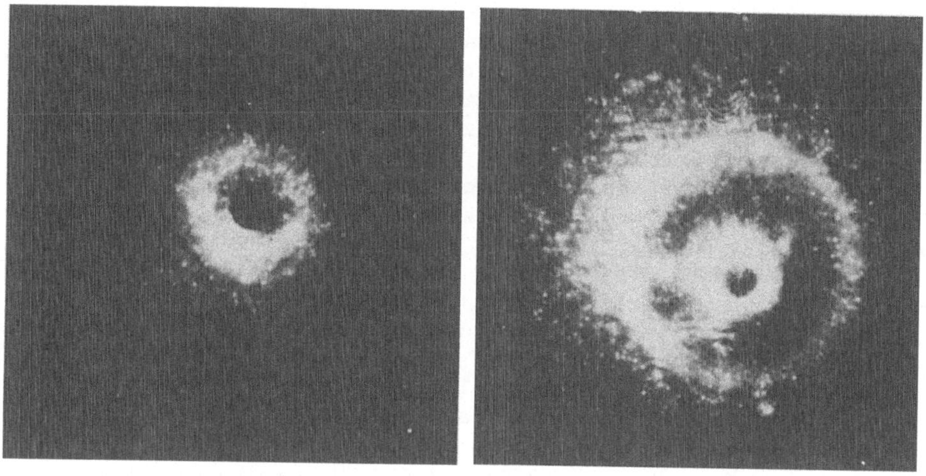

(a) (b)

Fig. 7 Forward scattered laser light in argon. a) Pressure of 9000 torr. b) Pressure of 760 torr.

spectral width of ∿0.1 cm^{-1} were observed in noble gases. In the latter case the intensity of scattered radiation permitted the pressure dependence of the shift to be measured, and it was found that it decreased smoothly from an initial value of 0.2 cm^{-1}, at a pressure of one atmosphere, and could no longer be resolved at a pressure of ∿7000 torr. Although it was expected that the non-linear effects associated with a self-focusing process might influence the spectral properties of the scattered radiation the shift observed in the case of noble gases strongly resembled a Doppler shift due to rapid motion of the 'scattering' source. However, it is by no means clear why a similar effect was not observed in molecular gases, and this discrepancy still lacks a satisfactory explanation.

Although, at the present time, the processes responsible for the spatial and spectral characteristics are not fully understood the high intensity of radiation emerging in the forward direction does provide a strong indication of self-focusing, particularly if it is assumed that the sources for 90° scattering and forward scattering are one and the same. However, the need for more direct evidence has prompted recent investigations where either high resolution Schlieren photography[12] or interferometry[13] have been applied to study the initial development of the spark plasma.

Key et al.[12] used a Nd:glass laser, operating in the pulse transmission mode to produce breakdown in atmospheric pressure argon. By passing part of the same laser beam through a gas breakdown cell, the 1 nanosecond rise time, 6 nanosecond duration, pulse was cut off to yield a 1 nanosecond pulse, which after frequency doubling and an appropriate optical delay, illuminated a Schlieren system. Photographs with 5μ resolution revealed filamentary regions, of high refractive index gradient, having a diameter less than 10μ and running between blobs of plasma separated by several hundred microns. The use of interferometric techniques by Richardson and Alcock[13] has permitted the identification of the observed filaments as extremely narrow channels of high density plasma. The same single-mode ruby laser, used in previous studies, was employed in the experimental arrangement shown in Fig. 8 and interferograms of sparks were obtained with a temporal resolution of ∿700 picoseconds and a spatial resolution of ∿10μ. As can be seen from the schematic diagram of the experiment, the oscillator beam was split into two parts, one of which passed through a subnanosecond electro-optical light gate, while the other passed twice through an amplifier rod to yield peak powers of ∿20 MW. The 1 MW peak power, 700 picosecond long, pulse transmitted by the polarizer-Pockels cell combination illuminated a conventional Mach-Zehnder interferometer while the amplified output from the oscillator was focused, with a 15 cm focal length lens, into a pressure cell situated in one arm of the interferometer. By means of appropriate optical delays the interferometer could be illuminated at various times within the first few nanoseconds after the initiation of breakdown, and, since the subnanosecond probe pulse was polarized orthogonally to the main laser beam, the orientation of a polarizer permitted either the interferogram or scattered light to be recorded.

The intensity distribution of the high power radiation within the focal region was investigated, by recording the light scattered from a dilute solution of milk, and revealed that the full width of the focal spot was ∿190μ at the half power points (Fig. 9a). An independent confirmation of this value was obtained by comparing the minimum power required for breakdown in argon with previously reported measurements of the breakdown threshold.[9]

Interferograms were obtained in a number of gases at pressures in the range of 0.5 to 8.0 atmospheres and examples of the results obtained are presented in Figs. 9 and 10. In Fig. 9 a series of interferograms, obtained at various times after breakdown in atmospheric pressure argon, provides a clear indication of how the plasma develops. As can be seen from the figure a filamentary region, with a fringe shift corresponding to a negative-going refractive index change, is created first at the time of breakdown (Fig. 9b). This filament, which initially has transverse dimensions of no more than 13μ, rapidly develops to a length of ∿300μ at which time a larger region of plasma is formed at the end of the filament farthest from the laser (Fig. 9c). As the filament continues to move rapidly towards

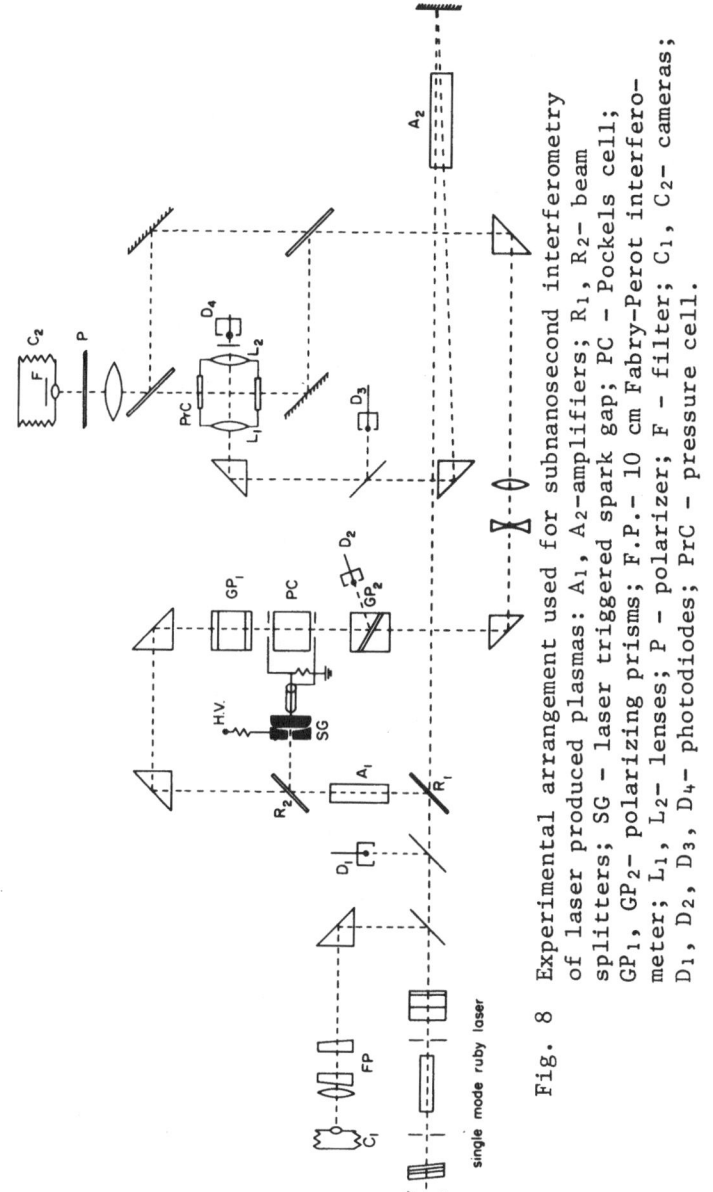

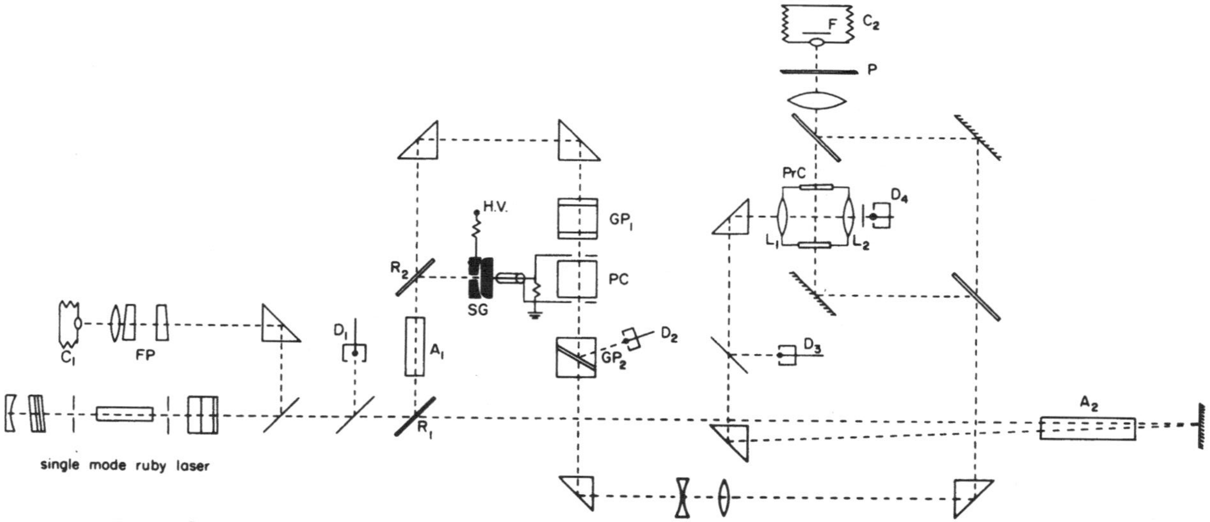

Fig. 8 Experimental arrangement used for subnanosecond interferometry of laser produced plasmas: A_1, A_2-amplifiers; R_1, R_2- beam splitters; SG - laser triggered spark gap; PC - Pockels cell; GP_1, GP_2- polarizing prisms; F.P.- 10 cm Fabry-Perot interferometer; L_1, L_2- lenses; P - polarizer; F - filter; C_1, C_2- cameras; D_1, D_2, D_3, D_4- photodiodes; PrC - pressure cell.

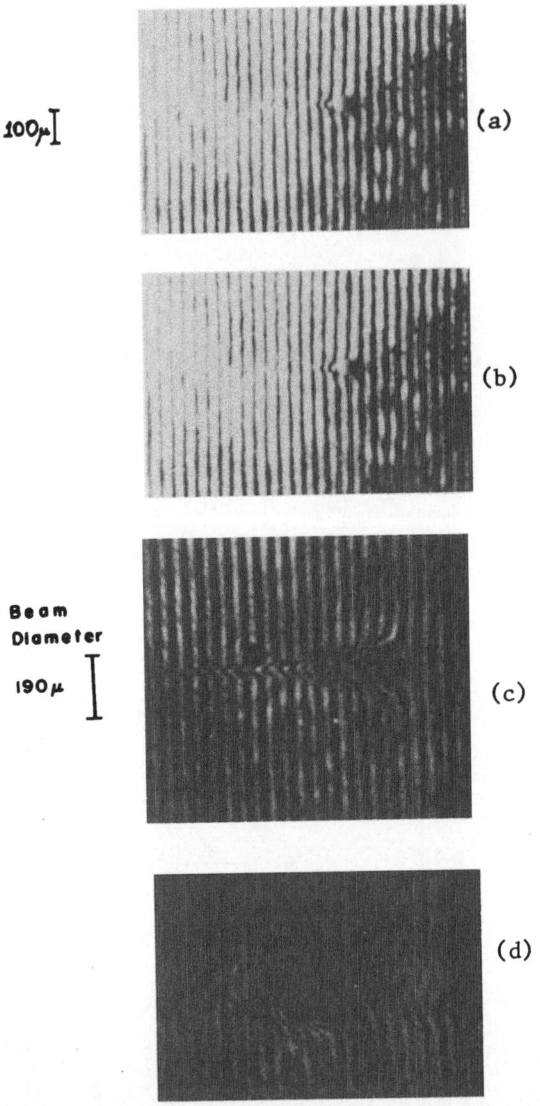

Fig. 9 Subnanosecond interferograms of breakdown in argon at 760 Torr; (a) laser light scattered by dilute solution of milk in focal region; (b) interferogram of spark obtained at time of breakdown; (c) and (d) show interferograms obtained 4 nsec and 10 nsec after breakdown respectively. The laser beam is incident from the left.

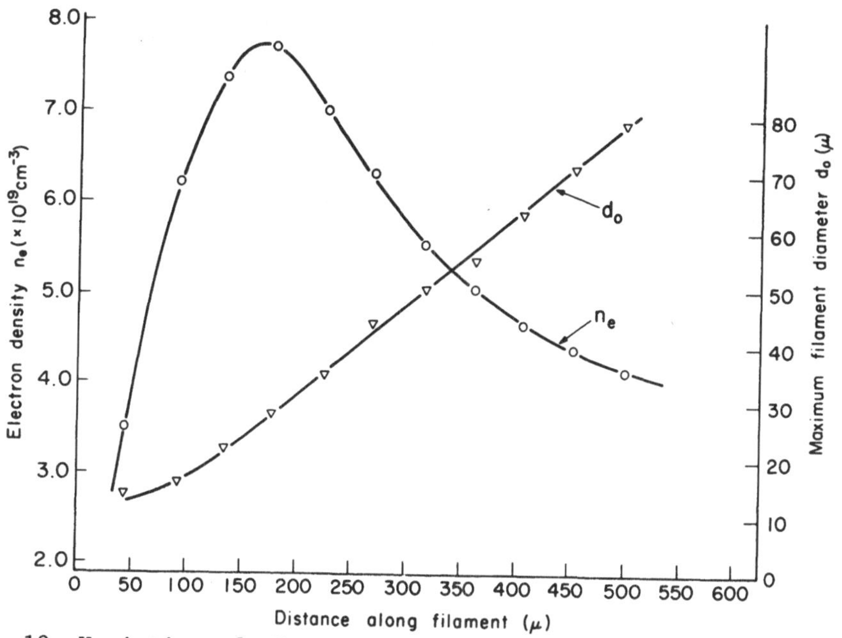

Fig. 10 Variation of electron density and maximum filament diameter along filament length, for the interferogram shown in Fig. 9c.

the laser additional blobs of plasma are generated at intervals of 200-400μ and following their formation the filament between them gradually decays (Fig. 9d).

From such interferograms an estimate of the average electron density in the direction of the illuminating beam can be obtained by assuming that the contribution of ions and neutral atoms to the refractive index change is negligible. Thus from the width, d, (cm) of the region where fringes are displaced, and the fringe shift, S, the average electron density, $\bar{n}_e$, is given by

$$\bar{n}_e = 3.25 \times 10^{17} \ Sd^{-1} \ cm^{-3}$$

For the interferogram of Fig. 9c the width of the plasma filament and the average electron density are plotted in Fig. 10. As can be seen from this figure the electron density has a maximum value of $\sim 8 \times 10^{19}$ cm^{-3} near the end of the filament closest to the laser.

In the case of the larger regions of plasma which develop along the length of the filament, regions of high electron density are observed within a few nanoseconds after the blob's formation. Such a high density core is clearly visible in the interogram

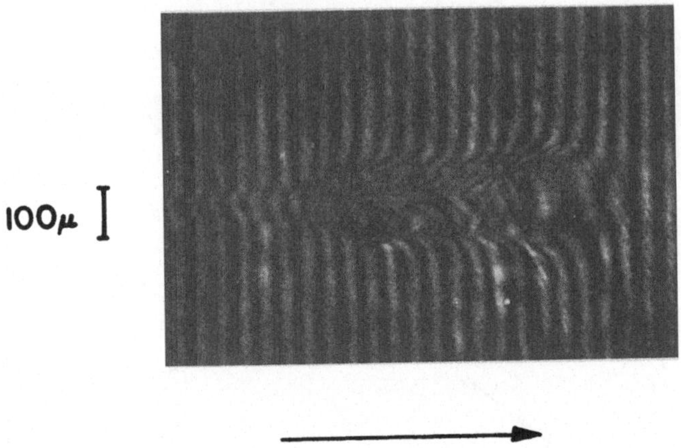

Fig. 11 Subnanosecond interferogram of a spark in argon, at a pressure of 760 Torr, obtained 5.5 nsec after breakdown and showing the internal structure of the plasma blob. The laser beam direction is indicated by the arrow.

of Fig. 11 and it appears likely that the discrete centres observed previously by means of 90° scattered light[5,8,9] have their origin in these regions.

In addition to the interferograms obtained in noble gases the development of sparks in molecular gases, such as hydrogen, has also been investigated. Although no extended filamentary regions of plasma were detected, and the plasma expanded smoothly during the laser pulse, the results indicated that the plasma developed initially in the form of small filament and then expanded rapidly behind the filament as it progressed towards the laser.

The direct observation of filamentary plasma regions, having transverse dimensions more than an order of magnitude smaller than those of the focal region, has provided the most significant evidence for self-focusing in laser-produced plasmas reported so far.

EXPERIMENTS INVOLVING SUBNANOSECOND LASER PULSES

Although all of the results described above were obtained with lasers generating pulses longer than a nanosecond, a number of similar observations were made on sparks produced by means of sub-nanosecond pulses. Both 90° scattering and forward-scattering from sparks produced by mode-locked ruby and neodymium:glass lasers have yielded preliminary evidence for self-focusing[14] as has an examination of the spectral structure of the forward

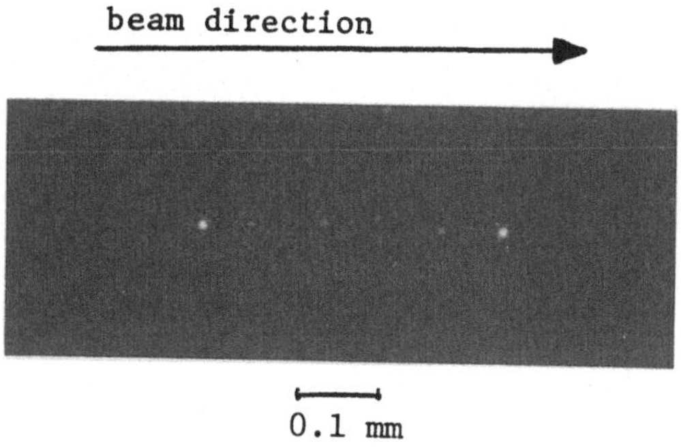

Fig. 12 Time-integrated photograph of spark produced in air by mode-locked Nd:glass laser. The photograph shows second harmonic laser radiation scattered at 90° to the incident beam. A 2 cm focal length lens was used.

scattered light.[15] These experiments were carried out with a mode-locked neodymium:glass laser and a mode-locked ruby laser generating pulses of 5 and 30 psecs respectively. Peak powers of ∿4 GW were obtained from the glass laser, in a beam which had a divergence of ∿2 milliradians, while the ruby system generated pulses of 100 MW peak power and operated in a single transverse mode to yield a diffraction limited beam divergence of 0.5 milliradians. Light scattered at 90° from sparks in air was detected photographically be means of the scattered ruby laser light or a second harmonic beam which was generated co-linearly with the neodymium laser radiation. In both cases it was found that the scattered light emanated from a number of discrete points distributed along the axis of the focal region (Fig. 12). The scattering centres bore a strong resemblance to those observed previously with nanosecond pulses and noble gas breakdown, however, it should be noted that in this case the sources of scattered light correspond to different breakdown regions formed by successive pulses in the mode-locked train. A more significant feature of the scattering points was their small diameter ($\lesssim$ 5μ) which, as in the nanosecond case, was found to be independent of the focusing lens used. Since the measured beam divergences indicated focal spot diameters varying from 25 to 200μ, and time-integrated measurements of the near and far field patterns provided no evidence for filamentary structure, it again appeared probable that some type of self-focusing mechanism was involved. Additional evidence was provided by observations of the forward scattered light and as can be seen from a typical result, obtained with the mode-locked ruby system, (Fig. 13), the scattered radiation emerged from the focal region within a cone of 15°-20° included angle. As described already in the preceding section an obstacle was used to block the laser light transmitted directly through

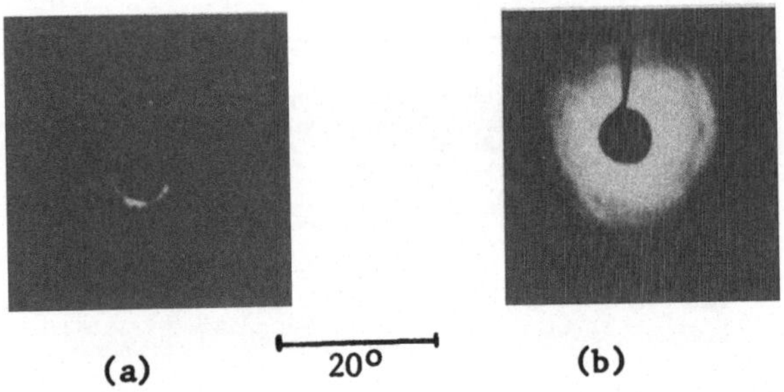

Fig. 13 Photographs of forward scattered radiation when the mode-locked ruby laser beam was focused by means of a 5 cm focal length lens. In (a) no spark was produced.

the focal region. Similar measurements made with either laser and a photodiode detection system revealed that approximately 10% of the incident power was transmitted beyond the obstacle when breakdown occurred. A study of the spectral width of this transmitted light was carried out with the neodymium:glass laser and indicated a broadening from the initial laser linewidth of ∿75 Å to a value of 200 Å. Unlike the results obtained with nanosecond pulses of single-mode laser radiation the broadening occurred towards both longer and shorter wavelengths with the shift to longer wavelengths being slightly greater. So far these observations have not been augmented by more detailed studies, however, the spectral broadening does indicate the presence of a non-linearity compatible with self-focusing effects.

Although similar observations of micron scale scattering centres have since been reported[16] only one experiment has indicated the occurrence of self-focusing when breakdown is produced by means of a single picosecond pulse. This investigation was carried out by Bunkin et al.[17] Who used a 2 GW ruby laser pulse, with a duration in the range of 20-100 psec., to produce breakdown in air, nitrogen and argon. With the laser intensity reduced well below the value required for gas breakdown, both the diameter of the beam and the intensity distribution in the focal region were measured by means of the fluorescence from a solution of Rhodamine 6G in ethanol. In addition to revealing a Gaussian intensity distribution in each section of the beam measurements performed with a 15 cm focal length lens yielded a minimum cross section of ∿200μ.

When sparks were produced in air and nitrogen, time-integrated and high speed streak photographs showed that a number of discrete breakdown regions separated by distances of 1.5-2 mm were formed in the vicinity of the focus. In some instances the individual breakdown points had dimensions less than 15μ and, from streak photographs, appeared to develop in the direction of propagation of the laser radiation with a velocity comparable to that of the actual pulse. Since the laser intensity decreased by a factor of 2 only 0.2 mm from the focus, neither of these observations could be accounted for without invoking a self-focusing process. Further evidence was provided by the observation of scattered laser radiation at power levels slightly below those at which the breakdown points appeared, and by measurements of the actual threshold intensity. The value of 3.5×10^{12} W/cm^2 was considerably smaller than that required to create a spark when the beam was focused to a 17μ diameter spot by a short focal length lens. In view of the fact that diffusion effects are negligible during a picosecond pulse such a dependence on the dimensions of the focal region would not be expected under linear focusing conditions. Although further experiments are required these results appear to confirm the earlier suggestion that self-focusing effects are associated with picosecond breakdown.

SELF-FOCUSING MECHANISMS

Although many of the experimental results described in the preceding sections indicate the presence of self-focusing effects, they do little to reveal the actual physical processes involved. However, during the last few years a number of possible mechanisms have been discussed and a brief review of these gives some insight into the more probable explanations.

Since all of the experimental evidence obtained so far applies to laser induced gas breakdown,self-focusing could occur during one or more of the following stages of spark formation:-

a) immediately prior to breakdown in the neutral gas
b) during the cascade ionization process
c) after a high density plasma has been produced.

During each of these stages it is possible to identify non-linear processes capable of producing refractive index changes of appropriate sign. In the case of (a), either electrostriction or the optical Kerr effect could provide the required nonlinearity and by using the expression for critical power derived by Kelley[18] threshold powers have been estimated.[9] The values obtained for molecular and noble gases, at atmospheric pressure, are of the order of tens of megawatts for electrostriction and gigawatts for the Kerr effect. Thus in the case of nanosecond pulses, at power levels of a few megawatts, a neutral gas process appears highly unlikely since the Kerr effect threshold is much too high and electrostriction requires both higher powers and longer pulse durations. In addition self-focusing in the neutral gas would inevitably lead to breakdown and a reasonable argument[8,9] can be based on the fact that there is little evidence to support the hypothesis that self-focusing rather than cascade ionization determines the breakdown threshold. However, processes occurring in the neutral gas cannot be ruled out very easily in the case of picosecond pulses where power levels may be well in excess of the Kerr effect self-focusing threshold.

Once the cascade ionization process, stage (b), has begun, the neutral gas in the focal region is replaced by a mixture of electrons, ions and neutral atoms with various degrees of excitation. The presence of free electrons results in a reduction of the refractive index which, in the absence of any compensating effect, will eventually lead to a defocusing effect. However, in many cases it appears likely that ionization occurs by photoionization of excited atoms and that the populations of ions, excited atoms and overexcited atoms (ie., atoms excited to within one or two photon energies of ionization) are approximately equal.[19] Since one would expect the absolute polarizabilities of the excited atoms to exceed those of unexcited or ionized atoms it is possible that the effect of the electrons might be cancelled and a net self-focusing effect result. Although there appears to be little experimental data on excited atom polarizabilities the possibility of such an effect can be illustrated in the case of argon where the observed polarizability of the $^3P_2$ state is $\sim 10^{-22}$ and exceeds by approximately two orders of magnitude that of the ground state.[20] The resulting refractivity at a wavelength of 7000Å is $\sim 6\times10^{-22}$ N_{AI} while the electronic contribution is $-2.3\times10^{-22}N_e$ (where N_e and N_{AI} are the densities of electrons and excited argon atoms respectively). Although from the above it is obvious that extremely high excited atom densities are required to produce a significant change in the refractive index, ie., of the order of 1%, it has been suggested that the effect of excited atoms could be considerably enhanced due to the resonant character of the nonlinear polarizability.[12] Furthermore the presence of excited molecules, at an early stage of the breakdown process might well initiate the self-focusing effect and, although this explanation has been proposed in the case of picosecond pulses,[17] it might also explain some of the observed discrepancies between molecular and noble gases.

Finally, the possibility of self-focusing occurring in the resulting plasma, stage (c), must be considered since the existence of numerous nonlinear plasma effects is well known. However, it appears that only two of these will be significant at the relatively low power levels where most of the experimental evidence for self-focusing has been obtained. Both of these processes involve localized changes in plasma density which arise either from thermal energy deposition or ponderomotive forces governed by the intensity gradient across the focal region. These mechanisms have been treated theoretically by Shimoda[21] and Hora[22] respectively and the minimum powers required to sustain a self-trapped filament have been estimated. In each case the power required is comparable to or less than the breakdown threshold and thus it appears possible that such a filament could exist within a laser produced plasma. However, in neither theory is any attempt made to determine the self-focusing length and from an order of magnitude estimate[12] it is difficult to see how the ponderomotive mechanism could account for the plasma filaments observed during the very early stage of spark development. In a more recent consideration of such effects Palmer[23] has applied the stimulated scattering theory of Herman and Gray[24] to a dense plasma and estimated the power densities required for the onset of several stimulated scattering processes. Although the values obtained indicate an extremely high threshold for thermal self-focusing effects the results for ponderomotive self-focusing in a high density plasma ($N_e \approx 10^{19}$) are compatible with observed breakdown thresholds.

CONCLUSION

Although investigations of laser-induced gas breakdown, carried out over a wide range of experimental conditions, have yielded a substantial amount of evidence for self-focusing, a satisfactory explanation for all of the observed effects is still lacking. Several mechanisms which could have an important role in such phenomena have been proposed and it appears that the effect of excited atoms and molecules during the breakdown process must be taken into account. However, before this can be done, additional experimental data obtained from more carefully controlled experiments, is essential.

ACKNOWLEDGEMENT

It is a pleasure to acknowledge the significant contributions of C. DeMichelis, V.V. Korobkin and M.C. Richardson to the investigations of self-focusing in laser-produced plasmas carried out at the National Research Council of Canada.

REFERENCES

1. N.G. Basov, V.A. Boiko, O.N. Krokhin and G. Sklizkov, Sov. Phys.-Doklady, 12, 248 (1967).

2. V.V. Korobkin, S.L. Mandel'shtam, P.P. Pashinin, A.V. Prokhindeev, A.M. Prokhorov, N.K. Sukhodrev, and M. Ya Shchelev, Sov. Phys. JETP, 26, 79 (1968).

3. G.A. Askaryan, Sov. Phys.-JETP, 15, 1088 (1962).

4. A.G. Litvak, Izv. Vuz. Radiofizika, 9, 675 (1966).

5. M.M. Savchenko and V.K. Stepanov, JETP Letters, 8, 281 (1968).

6. V.V. Korobkin and A.J. Alcock, Phys. Rev. Letters, 21, 1433 (1968).

7. N. Ahmad, B.C. Gale and M.H. Key, J. Phys. B. (Atom. Molec. Phys.) Ser. 2, Vol. 2, p. 403 (1969).

8. R.G. Tomlinson, Bull. American Phys. Soc. 14, 1021 (1969); IEEE J. Quant. Elect., QE-5, 591 (1969) .

9. A.J. Alcock, C. DeMichelis and M.C. Richardson, IEEE J. Quant. Elect., QE-6, 622 (1970).

10. N. Ahmad, B.C. Gale and M.H. Key, in "Advances in Electronics and Electron Physics" edited by J.D. McGee, D. McMullen, E. Kahan and B.L. Morgan, vol. 28B, p. 999 (1969).

11. L.R. Evans and C. Grey Morgan, Phys. Rev. Letters 22, 1099 (1969).

12. M.H. Key, D.A. Preston and T.P. Donaldson, J. Phys. B. (Atom. Molec. Phys.), 3, L88 (1970).

13. M.C. Richardson and A.J. Alcock, Appl. Phys. Letters, 18, 357 (1971); also Kvantoviya Electronica 1, no.5, p.37 (1971).

14. A.J. Alcock, C. DeMichelis, V.V. Korobkin and M.C. Richardson, Appl. Phys. Letters 14, 145 (1969).

15. A.J. Alcock, C. DeMichelis, V.V. Korobkin and M.C. Richardson, Phys. Letters 29A, 475 (1969).

16. Charles C. Wang and L.I. Davis Jr., Phys. Rev. Letters 26, 822 (1971).

17. F.V. Bunkin, I.K. Krasyuk, V.M. Marchenko, P.P. Pashinin and A.M. Prokhorov, ZhÉTF 60, 1326 (1971).

18. P.L. Kelley, Phys. Rev. Letters 15, 1005 (1965).

19. G.A. Askaryan and M.S. Rabinovich, Sov. Phys. JETP 21, 190 (1965).

20. R.H. Huddleston and S.L. Leonard, Eds., "Plasma Diagnostic Techniques". New York: Academic Press, 1965, p. 440.

21. K. Shimoda, J. Phys. Soc. (Japan), 24, 1380 (1968).

22. H. Hora, Z. Phys. 226, 156 (1969).

23. A.J. Palmer, these Proceedings - see page 367.

24. R.M. Herman and M.A. Gray, Phys. Rev. Letters 19, 824 (1967); R.M. Herman and M.A. Gray, Phys. Rev. 181, 374 (1969).

INTENSE ELECTRON EMISSION FROM LASER PRODUCED PLASMAS+

G. Siller, K. Büchl and H. Hora++

Max-Planck-Institut für Plasmaphysik/Euratom Association, Garching, Germany

ABSTRACT

Measurements are reported in which laser pulses are used to produce plasmas at tantalum targets biased as cathode against an anode by a voltage U_A = 3o to 2oo kV. The electron emission currents during 1oo nsec are a few hundred amperes up to 1 kA. The brightness of the emitted electron beams is more than a hundred times better than in beams generated from comparable field emission sources. The energy spread at 1oo kV anode voltage is less than 8oo eV. The measured high electron currents cannot be explained by classical thermionic emission because space charge effects permit currents a factor of $1o^{-4}$ less than observed. We suggest that self-focusing filaments are created by the laser at the ends of which the nonlinear force preaccelerates the electron in the space-charge free surface region of the plasma up to some keV energy. With such initial velocities the electrons in vacuum are not restricted by space-charge effects. Thus model also explains the independence of the emission current J of the focusing, the linear continuation of the measured dependence of J on U_A and the slope of J on the laser power P by a $P^{3/4}$ law.

+Presented at the Second Workshop on "Laser-Interaction and Related Plasma Phenomena" at Rensselaer Polytechnic Institute, Hartford Graduate Center, August 3o - September 3, 1971.

++ also from Rensselaer Polytechnic Institute.

Reprinted from: *Laser Interaction and Related Plasma Phenomena*, Vol. 2, H. Schwarz and H. Hora (eds.), Plenum, New York, 1972, pp. 253–269.

INTRODUCTION

One of the early applications of a laser was to heat solids in vacuum and measure the properties of the electron emission. While a completely regular behavior of the thermionic emission was established for laser powers up to a few megawatts[1-3] with emission currents of a few milliamperes, very unexpectedly high currents of up to 1oo amperes were observed with laser powers exceeding 1o MW[4,5,6]. Meanwhile emission currents of 1oo amperes and more have been measured[7,8].

Electron currents of such magnitude are technologically important, e.g. for relativistic electron ring accelerators (smokatron)[9], where about 10^{13} and more electrons of a beam of 10^{-8} sec duration are accelerated beyond a few MV and formed into a ring by a nearly homogeneous magnetic field. The compressed ring consists of a relativistically stabilized cloud of electrons which can finally be accelerated. High current electron sources are also very important for the techniques of nuclear fusion with relativistic megaampere electron beams [10]. This paper reports measurements of laser produced electron currents and their properties of beam divergence (brightness) and energy width, which are much better than the properties of field emission cathodes, used in smokatrons at present.

The problems entailed in explaining the anomaleously high electron currents are discussed more qualitatively. The classical description is not sufficient by orders of magnitudes. We suggest a mode where a laser induced high initial velocity of the electrons is concluded heuristically on the basis of a nonlinear acceleration. This mechanism should also explain why the field emission cathode cannot, in principle, achieve the properties of the laser produced electron emission principally.

EXPERIMENTAL SETUP

Two sets of apparatus were used, one for voltages between the cathode and anode up to 6o kV and another for voltages up to 2oo kV. Figure 1 describes the first apparatus. A Q-switched, two-stage ruby laser with an energy output up to 8 joule, a pulse half-width of 17 nsec, and a beam divergence of 5 mrad was focused on the cathode of a thick tantalum target. The focal lengths were 5 and 15 cm with focal radii of 2.5 and 7.5×10^{-2} cm

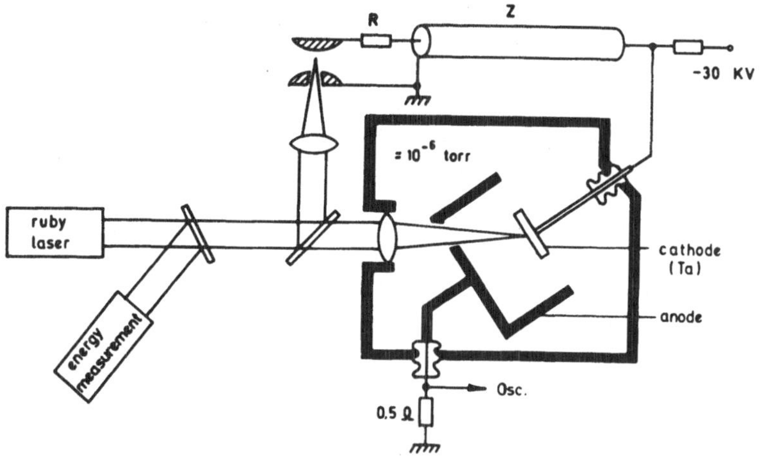

Fig. 1

Experimental setup for laser irradiation of a target biased by -3o to -5o kV against a collector anode during the time of electron emission. The switch-off after this time is performed by a laser switched line.

respectively, resulting in 10^{10} W/cm^2 intensities. The target was within a grounded vessel of 10^{-6} torr pressure. The total emission current was measured with an anode as described in Fig. 1, while the cathode was initially at a voltage of U_c = -3o to -6o kV. In order to measure at the anode only the primary electrons and not the following ions and to avoid a plasma discharge, the voltage U_c was switched off at the time of the arrival of the fastest ions at the anode, corresponding to the well known velocities[11] of 10^7 cm/sec, by means of a coaxial cable with a characteristic impedance Z of, for example, 3oΩ. This line was terminated at one end by a series connection of an ohmic load R equal to Z and a laser triggered spark gap (Fig. 1), the light of which was taken from the primary laser pulse.

The second apparatus (Fig. 2) used a Blümlein cable to keep the cathode at a voltage of -2oo kV for the times before arrival of the ions at the anode. The spark gap was triggered by a separate laser, while the main laser was obliquely incident on a cathode designed with a Pierce profile. The grounded anode had holes and only the electron current passing through the central hole was measured.

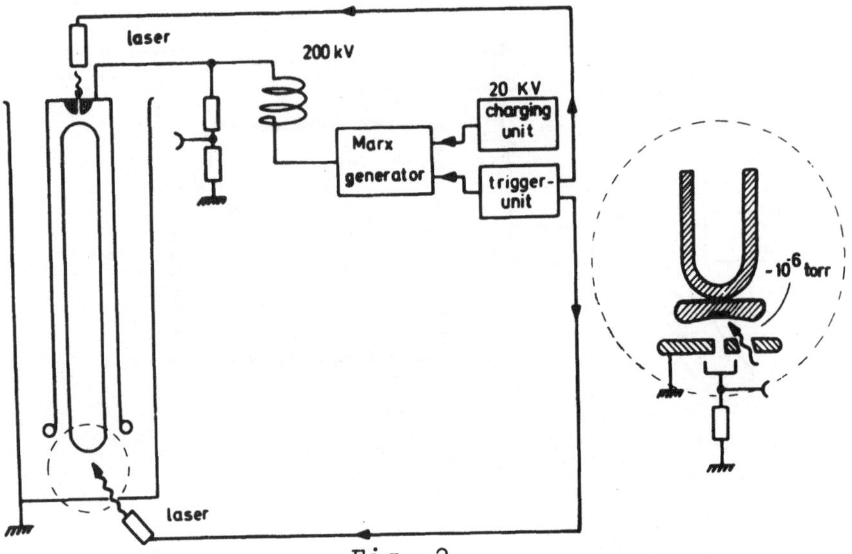

Fig. 2

Experimental setup similar to that in Fig. 1 with a Blümlein cable biasing the cathode up to -2oo kV. The wavy arrow indicates the incident laser pulse.

RESULTS OF THE ELECTRON EMISSION CURRENT

Identification of the Current Carriers

It is not evident from the very beginning that the measured negative currents are due to a free electron beam from the cathode to the anode. Therefore, it is necessary to identify the current carriers. The first indirect argument that the current carriers are electrons is as follows: It was found that the rise of the current pulse coincided with the laser pulse within the resolving time of the Tektronix 519. The cathode-to-anode distance is 4 cm. For electrons of 3o keV this means a transit time of about o.8 nsec. The expansion velocity of the plasma cloud is at least two orders of magnitude smaller than the velocity of the electrons. One method of direct identification of the current carriers is the deflection in electric fields we used.

Figure 3 shows a sketch of the experimental arrangement. The charge carriers fly through a hole (3 mm diameter) in the anode into the electric deflection field. Preference was given to identification of the deflection on a photographic film. The film was wrapped in a

2o μ aluminum foil. The range of 3o keV electrons in an aluminum foil is about 5 μ. The X-rays produced by the decelerated electrons in the foil will expose the film.

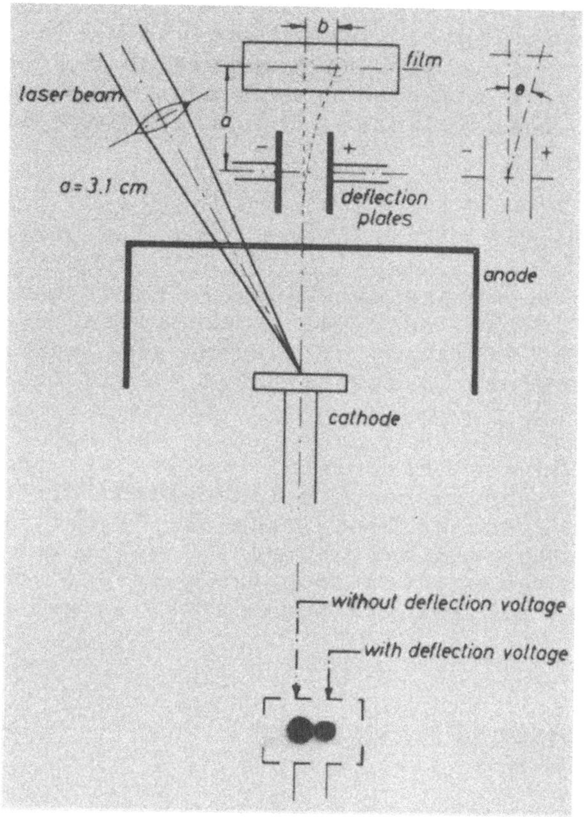

Fig. 3

Reflection of the emission current in an electric field for giving evidence of electrons. Scheme of the apparatus (upper part) and photographic film exposed by the electrons (lower part).

The deflection angle θ (see Fig. 3) is given by

$$\text{tg}\ \theta = \frac{\ell\ E}{2\ U_B} \tag{1}$$

where ℓ is the length of the deflection plates (2.5 cm), E the field strength (6x1o^3 Vcm^{-1}), and U_B the acceleration voltage of the electrons. In the case of Fig. 3 we used a coaxial line with a characteristic impedance of 5 Ω , and the amplitudes of the current pulses were of the order of 5oo amperes. So the voltage drop across the inner impedance was about 2.5 kV. The accelerating voltage U_B of the electrons between the cathode and the anode was therefore about $2.75 \times 1o^4$ V. This yields tg θ=o.273=b/a (see Fig. 3) and b$\approx$8.5 mm. From the exposed film in Fig. 3 (lower part) it can be seen that the deflection is that of a negative charge. The measured distance b is in good agreement with the computed value. This indicates that the charge carriers are electrons.

Properties of the Total Emission Current

Figure 4 shows the time dependence of the emission current and of the laser pulse for a laser energy of 1.3 joule. The variation of the emission current with various parameters is demonstrated by the following figures.

One of the most surprising results was that the electron emission current is independent of the distance of the focusing lens from the target. Figure 5 demonstrates the constant current J while the distance of the lens was shifted from the point of complete focusing five millimeters towards the target and ten

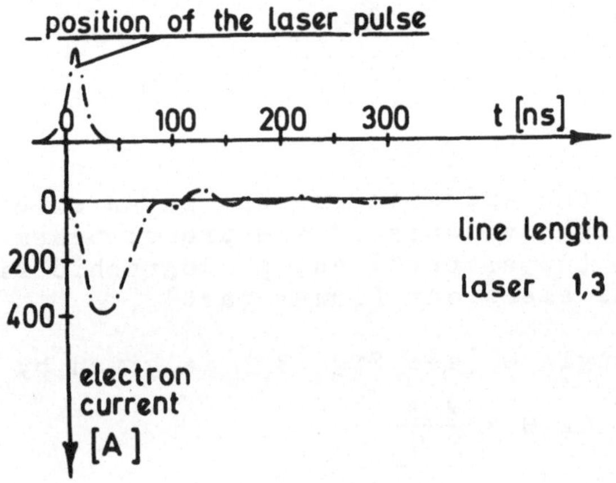

Fig. 4

Time dependence of the laser pulse and the electron emission current for a line length of 5.5 m and a laser pulse of 1.3 joule. The cathode voltage U_c was -6o kV.

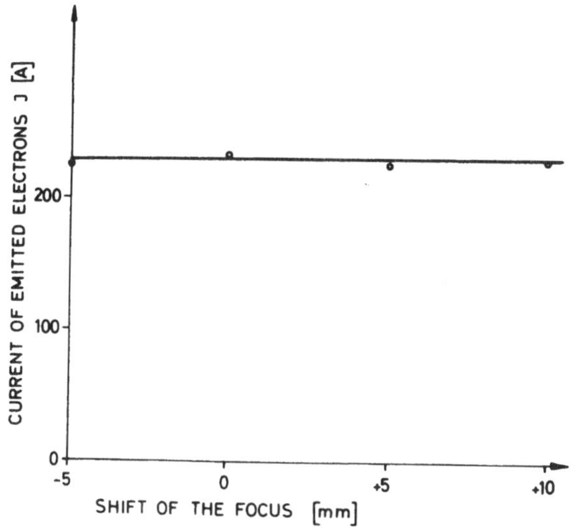

Fig. 5

Variation of the focus distance for a 15 cm lens shows no variation of the emission current.

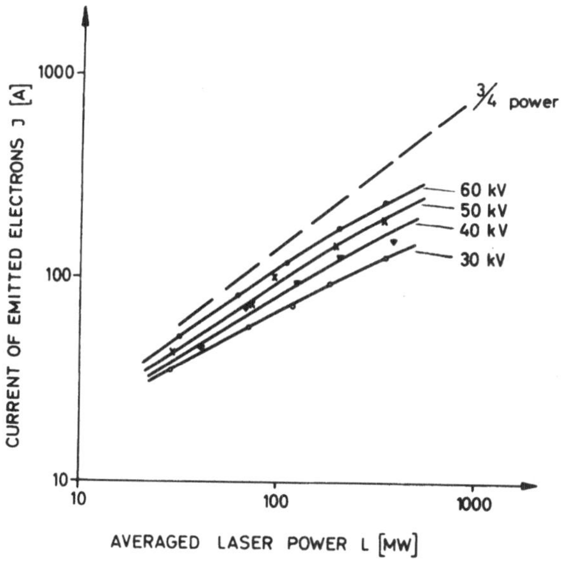

Fig. 6

The current of the emitted electrons as a function of the laser power for various voltages (3o to 6o kV) between anode and cathode.

millimeters in the reverse distance, where a lens of 15 cm focal length was used. The laser pulse length was 13 nsec and the power was 35o MW ± .15 % for these measurements.

Figure 6 demonstrates the measurements of the emission current J at varying laser power P when the cathode voltage U_c had values between 3o and 6o kV. A theoretical line of the power $\alpha = 3/4$

$$J \sim P^{\alpha} \tag{2}$$

was drawn for the following discussion.

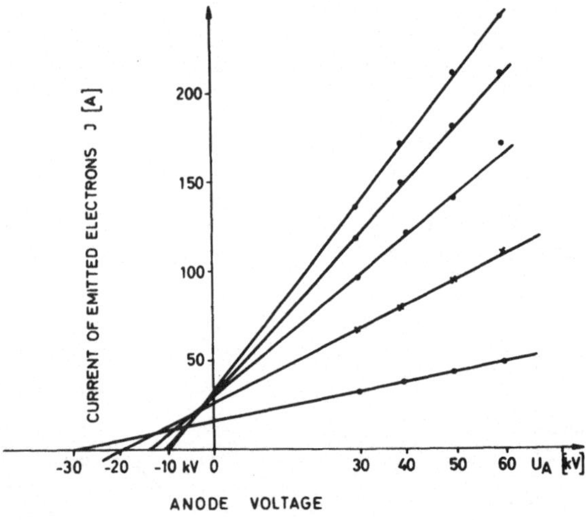

Fig. 7

Reversed diagram of Fig. 6 where the interpolated currents are used for laser powers of 4oo, 3oo, 2oo, 1oo, and 3o MW for lines from up to down, indicating a nearly linear increase of the emission current on the anode voltage.

Figure 7 is a reversed diagram of Fig. 6, where the dependence of J on $U_A = -U_c$ is demonstrated with the laser power as a parameter.

Emittance Measurements

The most interesting quantity of the electron source is the current that can be obtained in a certain emittance. The experimental setup used can be seen in Fig.8.

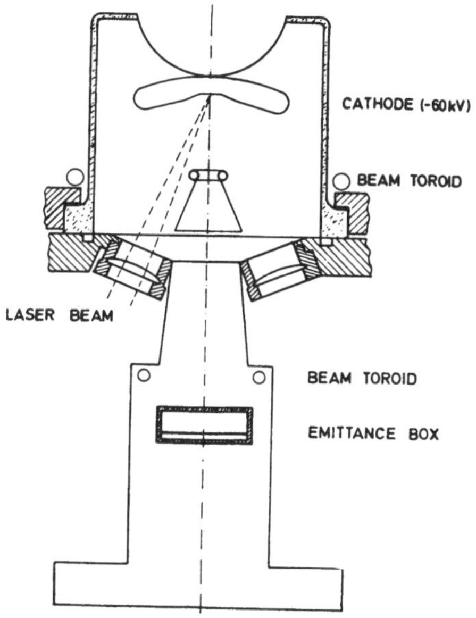

Fig. 8

Experimental setup for measuring the emittance

The tantalum plate of the cathode was surrounded by an electrode to shape the electric field lines (Pierce optical system). The anode was a grounded ring with an inner diameter of 1 cm. This ring was continued on one side by a cone so that only electrons passing through the ring can reach the emittance box. The cathode-to-ring-distance was 4 cm. The cathode was charged to -60 kV (see Fig. 1). The total current emitted by the cathode was measured by the upper beam toroid (Rogowski coil) and the current to the emittance box was measured by the lower one.

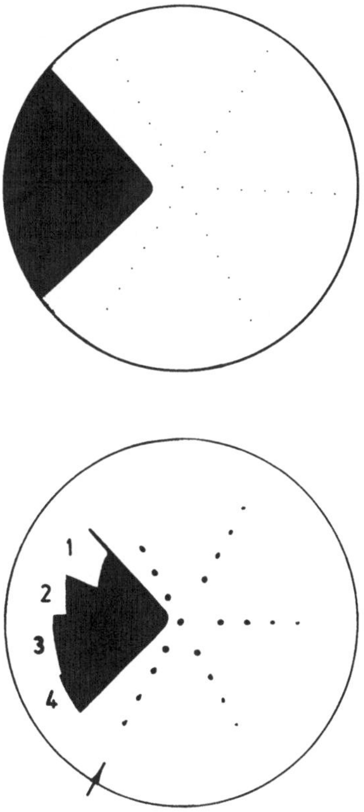

Fig. 9

Array of the holes in the emittance box (upper part) and photograph of the irradiated film (lower part).

We determined the emittance in the central part of the electron beam. The emittance ε is defined as an integral value of partial divergence of the electron beam (measured by diaphragms) over the radial coordinate of the cylindrical electron beam. This can be explained by the product of the averaged beam divergence of all parts of the electron beam times the cross section of the beam. A good electron source should have a small value ε , i.e. a small beam divergence at a small beam diameter. The divergences were measured with an emittance box consisting of a 1.5 mm thick metal plate with an array of apertures o.3 mm in diameter. The array of the holes in the box is shown in Fig. 9 (upper part). A film (Agfa Gevaert N 33) wrapped in 1o μ aluminum foil was mounted at a distance of 2 cm behind the plate. The range of 6o kV electrons in aluminum is 2o μ. A photograph of the irradiated film can be seen in Fig. 9 (lower part). The distance of the holes in the emittance box must be so large that the density curves of the holes on the film do not overlap. The black sectors in Fig. 9 are for calibrating the density of the film. In this area there is a piece cut out of the metal plate of the emittance box and the electrons hit the wrapped film directly. The film is exposed to four electron pulses. In Fig. 9 (lower part) the black sector is divided into four subsectors. The subsector 1 is hit by one electron pulse, the next by two, the next by three and the darkest one by four equal electron pulses. The difference in the density of the four subsectors cannot be seen very well in this reproduction, but is clear on the exposed film. The photometric curves along these four subsectors in the radial direction are evaluated and calibrated, as described in detail in a separate report[12], where the emittance ε is evaluated by integration over the measured two-dimensional phase-space diagram (Fig. 1o).

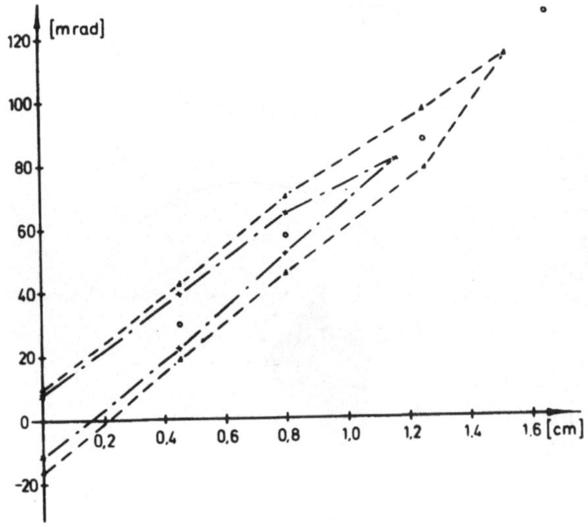

Fig. 1o

Measurement of the angle (mrad) at which the electron density of a particular beam passing through the holes of o.3 mm at a distance from the center of the beam (cm) has decreased to 1/2 (x) or 1/4 (▲) of the central intensity. The half-value of emittance ε is given by the area surrounded by curves.

The total current emitted by the tantalum plate was of the order of several hundred amperes. The current within the area where the emittance was measured was J = 17 A. We measured a current of J = 9.12 A in an emittance of ε = 31.4 mrad.cm or a current of J = 12.3 A in an emittance of ε = 63.6 mrad.cm. With these values of the emittance and the current one can compute the normalized brightness. This quantity is given by the formula

$$B = \frac{J}{\varepsilon^2 \, eU \, (1+eU/2m_oc^2)}$$

where eU is the kinetic energy and m_oc^2 the rest energy of the electrons. With the evaluated values one gets

$$B_{1/2} \approx 1.5 \; 10^{-1} \; \frac{A}{\text{sterad.cm}^2 \; eV}$$

and

$$B_{1/4} \approx 4.8 \; 10^{-2} \; \frac{A}{\text{sterad.cm}^2 \; eV}$$

We can compare these values with the electron emission system, where a brightness[8] of the order of 10^{-3} A/sterad cm^2 eV is measured, and with hot cathodes (Ardenne), where a brightness of 10^{-4} A/sterad.cm^2.eV is measured.

In our opinion the values evaluated for the emittance are only an upper limit and the real values are smaller. Therefore, the values of the real brightness are larger than those given here, because of the photometric method and with respect to our time integrated measurement[12].

Measurements of the Energy Spread

The next quantity in which we were interested was the energy spread of the electrons in the pulse. The energy of the electrons was measured by deflecting them in a static magnetic field. The chosen radius of curvature of the electrons in the magnetic field was about 10 cm. This corresponds to a field strength of about 110 gauss for 100 keV electrons. The measuring device for the electrons was an arrangement of 14 collectors, each in the form of a strip with a width of 1 mm, a height of 14 mm, and a distance of 1 mm from the next strip. A change of 1 % in the energy of the particles resulted in a deflection of about 1 mm at the place where the 14 collectors were mounted. For these measurements the voltage U_c at the cathode was of the order of -100 kV. Evaluating the time dependence of the currents from the strips we immediately found[12] an energy spread of less than 800 V.

THE DIFFICULTIES OF A CLASSICAL INTERPRETATION

Two mechanisms have to be considered for a theoretical interpretation of the electron emission current on the basis of classical theories: the thermionic electron emission laws and space charge effects. We present here a generalized equation of electron emission from a body (plasma) of temperature T including the Richardson-Dushman equation and the revised Richardson equation of the photoelectric emission[13] which becomes effective for plasmas at high temperature[14] for thermodynamic equilibrium[15] with respect to the black-body radiation.

In the case of no degeneration

$$n_e h^3 \ll (2m_e kT)^{2/3} \quad , \qquad (3)$$

the emission current density is

$$j = \begin{cases} A_{1/2} \cdot T^{1/2} \cdot \exp(-h\nu_o/kT) & \text{if} \quad (kT)^{5/2} < \frac{c^2h^3}{2\sqrt{2\pi m_e}} n_e \qquad (4a) \\ M \cdot T^3 \cdot \exp(-h\nu_o/kT) & \text{if} \quad (kT)^{5/2} > \frac{c^2h^3}{2\sqrt{2\pi m_e}} n_e \qquad (4b) \end{cases}$$

where[16]

$$A_{1/2} = n_e e \sqrt{k/2\pi m_e} \qquad (4c)$$

and the photoelectric efficiency constant[14]

$$M = \frac{2}{c^2} \left(\frac{k}{h}\right)^3 . \qquad (4d)$$

h is Planck's constant, k is Boltzmann's constant, c is the velocity of light, e is the charge, m_e the mass, and n_e the density of electrons, T is the temperature and $h\nu_o$ is the work function for electrons. In the case of degeneration

$$n_e h^3 \gg (2m_e kT)^{2/3}$$

the emission current density is

$$j = \begin{cases} A_2 \cdot T^2 \cdot \exp(-h\nu_o/kT) & \text{if} \quad kT < 2\pi m_e c^2 \qquad (4e) \\ M \cdot T^3 \cdot \exp(-h\nu_o/kT) & \text{if} \quad kT > 2\pi m_e c^2 \qquad (4f) \end{cases}$$

where

$$A_2 = 4\pi mk^2e/h^3 \qquad (4g)$$

is Sommerfeld's "universal constant".

The work function $h\nu_o$ is a well defined value for the electron emission from solids or liquid metals. The problems of semiconductors and insulators can be neglected in this discussion because of the considered high temperatures. In the case of a plasma located in vacuum and subjected to inertial or magnetic or radiation pressure confinement[17], the Debye sheath at the surface produces a work function $h\nu_o$ which has the value kT. Therefore, when the electrons of the Debye sheath are separated, a thermionic emission of the maximum value of equations (4) with an exponential factor exp(-1) can be expected.

The current densities (4) are the saturation values if no space charge processes reduce the emission. These saturation currents can indeed account for the measured high electron emission currents from laser produced plasmas even at very unfavorable conditions. If a plasma surface of an electron density of $n_e=10^{19}$ cm^{-3} is assumed, and if the electron temperature T is 10^4 °K - these values being lower bounds of those of plasmas produced by lasers from solids -, the electron emission current from a surface of 10^{-3} cm^2 cross section is found from Eq.(4a) to be

$$J = 9.2 \times 10^3 \text{ amp} \qquad (5)$$

where $h\nu_o$=kT was used.

Such high saturation currents are usually prevented by the space charge laws[18]. The electron emission current J_s from a surface F(cm^2) of a parallel plate cathode to an anode at a distance d_1(cm) biased by a voltage U_L (volts) is

$$J_s = 2.33 \times 10^{-6} \frac{F}{d_1^2} U_L^{2/3} \text{ amp} \qquad (6)$$

which gives for the case F = 10^{-3} cm^2; d_1 = 4 cm; U_2 = 60 kV

$$J_s = 2.14 \times 10^{-3} \text{ amp} \qquad (7)$$

A case more comparable with our experiment is a cylindrical geometry with a cathode wire of length l, radius r_1 of a temperature T within a coaxial anode of radius r_2. The maximum current J_M in the case of minimumfree potential is for r_2 = 6 cm, r_1 = 2 x 10^{-2} cm, l = 4x10^{-2}cm, and the extreme temperature $T = 10^7\ {}^oK$

$$J_M = 2.39 \times 10^{-3} \text{ amp} \qquad (8)$$

which is much too small to explain the measured currents of kA.

The electron emission currents at the moderate laser intensities of Ready[3] of about mA are somehow within the values given by space charge effects and are describable in terms of the usual classical theory of electron emission. In the case of the experiment of Siller et al.[7], the measured 10^3 amp was five magnitudes as large as the space charge limitation (eq.(7)), though the pure thermionic emission from plasmas (eq.(5)) would permit the measured values. Another necessary condition for the high electron currents is a sufficient electric conductivity of the plasma or of the metal below the emitting surface. The aim is to reach a current density of 10^6 amp/cm^2. With an electric conductivity of almost 10^5 $^{-1}$cm^{-1} for tantalum, we find a necessary field strength of E = 10 V/cm, which is low enough - also if it were increased by orders of magnitudes owing to the heating of the material - to explain the observed[5-7] currents up to 1 kA at the applied voltages.

NONLINEAR THEORY

The impossibility of explaining the high emission currents because of space charge effects is a serious difficulty. Other results for the space charge properties can only be expected by postulating boundary conditions for the electrons other than those used[18], e.g. by assuming a certain initial energy of the electrons of some keV which they must have received within the space charge free interior of the laser produced plasma. The fact that such accelerated electrons enter the vacuum within the cross section of the laser irradiated spot of the cathode can indeed result in other space charge limited currents, as could be shown from the change of the boundary conditions of the differential equations involved.

One mechanism of the preacceleration of the electrons within the space charge free interior of the inhomogeneous surface region of the laser created plasma is well known from the nonlinear force of the collisionless interaction of the laser radiation with the plasma due to the spatial change of the complex refractive index η [20]. For a reflectionless penetration of the light into the plasma with negligible collision induced absorption, the nonlinear force density in the plasma for perpendicular incidence of the light along a coordinate x

$$f_{NL} = - \frac{E_v^2}{8\pi} \frac{\partial}{\partial x} \left(\frac{1}{|\eta|} + |\eta|\right) \quad (9)$$

where E_v is the amplitude of the electrical vector of the light in vacuum. It was proved[21] that this nonlinear force exceeds the gasdynamic forces, if for ruby laser radiation the intensity I exceeds a value I^*

$$I > I^* = 2.08 \times 10^{14}\, T^{1/4}\ W/cm^2 \quad (T > 10^2 eV) \quad (10)$$

or if for lower intensities[22] T exceeds 10^4 eV. These intensities do not occur in our experiments, where I was near 10^{10} W/cm^2. However, it is well known from the application of the nonlinear theory[20] to self focusing[21] in plasma that laser powers P exceeding one megawatt can create filaments the cross section F_{SF} of which is simply given by

$$F_{SF} = \frac{P}{I^*} \quad (11)$$

because then the gasdynamic pressure of the filaments is equal to the thermokinetic pressure. The diameter measured of these filaments on gas breakdown[23] agree very well with the calculated values of a few μm. At the beginning of the filaments at the plasma surface the nonlinear acceleration can be effective because the high intensity I reaches I^* there.

This model gives the following explanation of the measurements:

(1) The high electron emission currents are explainable, in principle, in terms of the nonlinear space-charge free preacceleration in the plasma surface to avoid the well known space-charge limitations.

(2) The linear continuation of the measurement of Fig. 7 to vanishing emission currents J indicate an initial energy of the electrons of a few keV. If a J were to be measured between 0 and +30 volts of the anode voltage $U_A = -U_c$ to show a deviation from the linear relation, this may not upset our conclusion because at low values of U_A one can expect a mixing up of several mechanisms which overlap the very linear behavior measured for U_A exceeding 30 kV.

(3) The fact that the emission current J is independent of the laser focusing (Fig. 5) indicates once more the self-focusing mechanism, because the diameters of the resulting filaments are independent of the incident laser intensity.

(4) The nonlinear theory also provides a fair explanation of the slope of the curves of Fig. 6. It can be assumed that the emission current J is proportional to the cross section F_{SF} of the self-focusing areas.

$$J \sim F = \frac{P}{I^*} \qquad (12)$$

If the temperature T in the filaments is assumed to be proportional to the laser power P^α by an exponent α, we find from eq.(1o) and (12)

$$J \sim P^{1-\frac{\alpha}{4}} \qquad (13)$$

The slope drawn in Fig. 5 is a $P^{3/4}$-line according to $\alpha=1$, as can be expected from heating by thermokinetic processes.

The authors gratefully acknowledge the encouragement and stimulating remarks of Prof. A. Schlüter on the work presented.

REFERENCES

1. D. Lichtman and J.F. Ready, Phys.Rev.Lett. 1o, 342 (1963).
2. C.M. Verber and A.M. Adelman, Battelle Techn. Rev. 14, (7), 3 (1965)
3. J.F. Ready, J. Appl. Phys. 36, 462 (1965); Phys. Rev. 137A, 62o (1965); A.J. Alcock, H. Motz and D. Walsh, Quantum Electronics, 3rd Int. Congr. Paris 1963 (P. Grivet and N. Bloembergen Eds.) Dunod Paris 1964, Vol. II, p. 1687.
4. R.E. Honig, Appl. Phys. Lett. 3, 8 (1963).
5. R.E. Honig, Laser Interaction and Related Plasma Phenomena, (H. Schwarz and H. Hora Eds.) Plenum Press New York 1971, p. 85.
6. G. Bourrabier, T. Consoli and L. Slama, Phys. Lett. 23, 236 (1966).
7. G. Siller, K. Büchl and E. Buchelt, Max-Planck-Institut für Plasmaphysik, Rpt. o12, 3/1oo (1969).
8. C. Andelfinger, K. Büchl, E. Buchelt, W. Ott, G. Siller, Proceedings Electron, Ion and Laser Beam

Technology, 4th Int. Conf. Los Angeles, May 197o (R. Bakish, Ed.) The Electrochem. Soc., New York 197o, p. 3.

9. V. I. Veksler, G.I. Budker and Ya.B. Fainberg, CERN Symp. on Accelerators, 1956; see Symp. Electron Ring Accelerators, Febr. 1968, UCLR-181o3.

1o. F. Winterberg, Phys. Rev. 174, 212 (1968); M.V. Babykin, E.K. Zavoiskii, A.A. Ivanov, L.I. Rudakov, Proc. 4th Int. Conf. Controlled Thermonuclear Fusion, Madison, June 1971, IAEA, Vienna 1971.

11. W.I. Linlor, Appl. Phys. Lett. 3, 21o (1963); Energetic Ions Produced by Laser Pulses, Laser Interaction and Related Plasma Phenomena, (H. Schwarz and H. Hora Eds.) Plenum 1971, p. 173.

12. G. Siller, E. Buchelt and H.B. Schilling, Max-Planck-Institut für Plasmaphysik, Report 0/7 (1971).

13. P. Görlich, H. Hora and W. Macke, Exp. Tech. Phys. 5, 217 (1957); Jenaer Jahrb. 1957, p. 91; P. Görlich, and H. Hora, Optik 15, 116 (1958).

14. H. Hora and H. Müller, Z. Physik 164, 359 (1964).

15. W. Klose (private communication) see ref. 14.

16. A. Sommerfeld and H. Bethe, Handbuch der Physik, (A. Scheel Ed.) Springer Berlin 1953, Vol. 24/2; C. Herring and M.H. Nicols, Rev. Mod. Phys. 21, 266 (1949).

17. H. Hora, Application of Laser Produced Plasmas for Controlled Thermonuclear Fusion, in Laser Interaction and Related Plasma Phenomena (H. Schwarz and H. Hora Eds.) Plenum New York 1971, p. 437.

18. H. Rothe and W. Kleen, Grundlagen und Kennlinien der Elektronenröhren, Akad. Verlagsges. Leipzig 1953, p. 21; Henry F. Ivey, Advances in Electronics and Electron Physics (L. Marton, Ed.) Acad. Press New York 1954, Vol. 6, p. 137.

19. H.G. Möller and F. Detlefs, Jb. drahtl. Telegraphie und Telephonie 27, 74 (1926).

2o. H. Hora, D. Pfirsch and A. Schlüter, Z. Naturforschg. 22a, 278 (1967); A. Schlüter, Plasma Physics 1o, 471 (1968); H. Hora, Phys. Fluids 12, 182 (1969).

21. H. Hora, Nonlinear Effect of Expansion of Laser Produced Plasmas, Laser Interaction and Related Plasma Phenomena, (H. Schwarz and H. Hora, Eds.) Plenum 1971, p. 383; Opto-Electronics 2, 2o1 (197o).

22. L.C. Steinhauer and H.G. Ahlstrom, Phys. Fluids 13, 11o3 (197o).

23. V.V. Korobkin and A.J. Alcock, Phys. Rev. Lett. 21, 1433 (1968).

EXPERIMENTAL RESULTS OF FREE TARGETS*

Heinrich Hora

Rensselaer Polytechnic Institute and

Institut für Plasmaphysik, Garching, Germany

Abstract

Plasmas produced from free small (5 to 10μ radius) electromagnetically suspended LiH targets showed in the first measurements a quite symmetric expansion but finally an asymmetry with a preferential direction against the laser. Preheating by tailored laser pulses re-established conditions of the self-similarity model of a symmetric expansion and heating up to temperatures of few hundred eV. Aluminum balls of 50 to 150 μ radius showed two groups of plasma, a slow spherical inner core fulfilling thermokinetic properties describable by the self-similarity model, and an asymmetric outer shell expanding preferentially against the laser due to a nonlinear surface process. Free hydrogen targets of the same larger size expand symmetrically and drift slowly into the direction of the laser radiation, indicating a recoil by a surface acceleration against the laser. Magnetic fields decrease the expansion velocity and increase the duration of luminosity in some experiments. Emission of microwave radiation can be explained, but many effects are unsolved (increase of ion energy, increase of electric resistivity, instabilities etc.).

Introduction

When plasmas are produced from solids by incident laser radiation, the observable processes are highly complicated. To reduce some essential influences, the solids were arranged in vacuum[1], though many applications need the presence of surrounding gases[2,3], or vice versa - the processes in the

*Talk presented at the Workshop "Laser Interaction and Related Plasma Phenomena", Rensselaer Polytechnic Institute, Hartford Graduate Center, June 9-13, 1969.

Reprinted from: *Laser Interaction and Related Plasma Phenomena,* Vol. 1, H. Schwarz and H. Hora (eds.), Plenum, New York, 1971, pp. 273–288.

gases, e.g. the blast waves in very thin gases, could be studied starting from plasma produced at solids[4]. But also the arrangement in vacuum still seemed to be too complicated with respect to the energy transport into the cold surrounding material, if the targets were compact material with a plane surface and varying thickness[1,5]. So, many experiments were performed to study plasmas produced by lasers from small free targets suspended in vacuum.

In the preceding paper of S. Witkowski on "Free Targets" [6], the methods of arranging free targets in vacuum were reviewed. Here we shall describe the experimental results. One remarkable difference to the results of compact targets seems to be that very small free targets expand very well in agreement with a model of spherically symmetric self-similarity expansion[7], while larger free targets show successively the result observed at thick compact targets, namely, a fast group of ions of keV energy and a second group of slower expanding plasma.

Many experiments were reported with compact targets, where the keV ions have been observed but no different groups of plasma could be detected[1,8]. So, the groups were discovered only where the diagnostic method allowed the sensing of these details. The density of plasma in dependence on the velocity showed a minimum which separated obviously two groups[9]. Different groups of varying ionization were observed in carbon plasma[10]. The interpretation of time-of-flight-signals to discriminate two groups was only possible after a thorough treatment of the target[5,11]. Targets not carefully degased showed - as well known from other experiments - very confusing charge collector signals. The two different groups were observed at very different conditions with special charge collectors screened by grids[12]. Another experiment performed to look for the two groups, analyzed the properties of each group by an HF probe[13] and demonstrated the thermal property of the slow thermal group and a non-Maxwellian velocity distribution of the electrons of the fast group. This result of the groups at compact targets caused us to start this review with small free targets and to succeed with targets of increasing size. We shall discuss first the experiments without a surrounding magnetic field.

Small Free Targets

The first measurements with free targets suspended by a three dimensional ac charge collector system of Wuerker et al.[6,14] were performed by Haught and Polk[7]. The diagnostics were concentrated on charge collecting probes to detect the ions and electrons of the expanding plasma. The targets were specks of lithium hydride of 5 to 10 μ radius. The incident laser radiation was primarily restricted to peak power of 20 MW of some 10 nsec length focused down to a

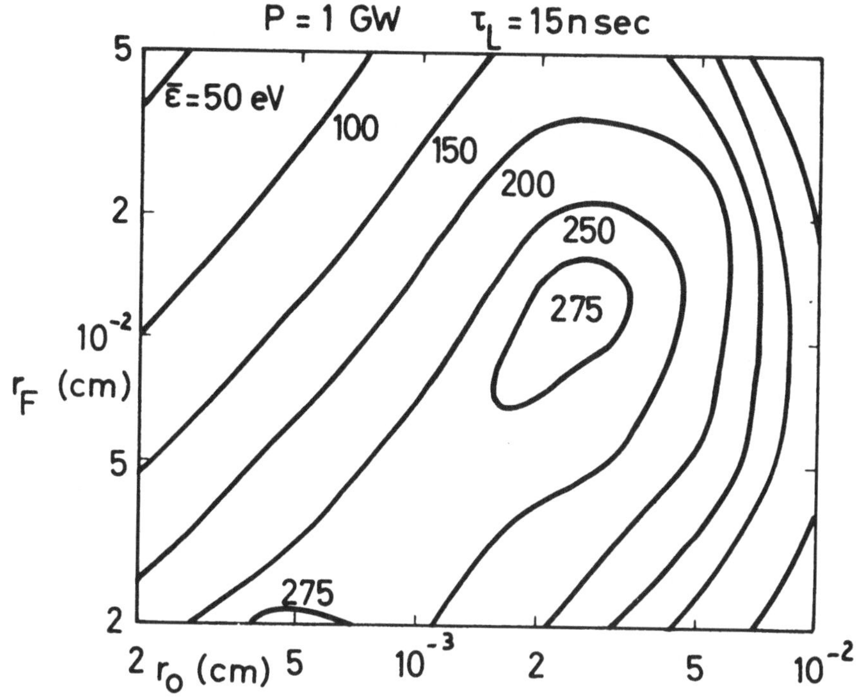

Fig. 1

Calculation of averaged energy ε from the self-similarity expansion model in dependence on the focus radius r_F, target radius r_O at given values of the laser power P and pulse length τ_L as evaluated by Haught, Polk and Fader[7].

diameter of 300 μ and less. One type of probes had a slit behind which charged plates, parallelly oriented to the plasma ribbon, collected electrons or ions. The result was that all atoms of the target were ionized. Times of flight of the ions were measured with open electric probes, consisting of two parallel plates with radially oriented surfaces. From measurements in different directions, it was concluded that the plasma expands spherically symmetric within 30% and that there was no evidence of any separation of the expanding plasma into a fast or slower group of ions. There was no measurable charge separation. These results allowed the application of the theory of the self-similarity expansion of the created plasma [15,16] from which it could be concluded that the measured maximum ion energies indicated an averaged ion energy of 100 to 200 eV. To get results of the plasma properties without measuring the target size (radius r_O), only starting from the maximum laser power P, laser pulse length τ_L (half width of intensity maximum), the similarity model was used to calculate the averaged particle energy $\bar{\varepsilon}$ at varying r_O and varying focus radius r_F

for given P and τ_L. A maximum of ε was found at a special value r_F (see Fig. 1). As one can understand, the procedure that from a large number of experiments with varying target size r_o at constant P, τ_L and r_F, the time of flight measurements showed a maximum of ε which could be compared with the calculated $\bar{\varepsilon}$ from Fig. 1. The agreement was primarily (Haught and Polk[7]) within a factor of two and later[7] within a few percent when powers of ruby lasers up to 300 MW were used.

The same method of suspension of free lithium hydride targets os similar size was used where an irradiation of 1.5GW neodymium glass lasers with τ_L = 30 nsec was applied[17]. From measurement of the density profile, the applicability of a self-similarity expansion with Gaussian profiles, which are of significance for this mechanism[16], was proven. The initial mean energy $\bar{\varepsilon}$ per particle was around 100 eV. Measurements of the emission of radiation of the H_β line indicated an increase of recombination at a time 200 nsec after the interaction of the laser beam. One interesting result was the occurrence of two groups of plasma according to the differentiated charge collector signals of the arriving plasma (Fig. 2). The reality of the two peaks of the curve of Fig. 2 was verified very carefully[17,18].

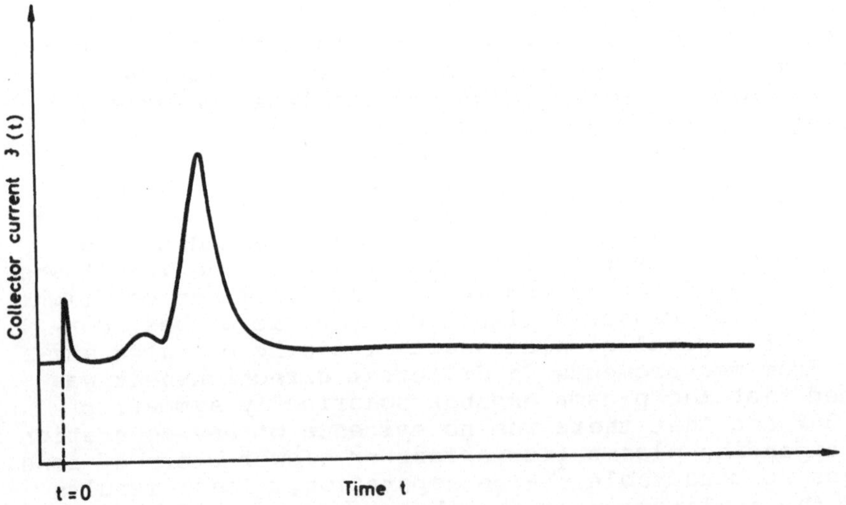

Fig. 2
Ion current I(t) by the time t of the carefully screened charge collector of a lithium hydride plasma produced by neodymium glass laser radiation from a target of about 10 μ radius, according to measurements of Papoular et al.[17] The first peak at the time of t= 0 of laser interaction is that of electron emission from the collector due to ultraviolet emission of the plasma of higher temperatures than 10 eV.

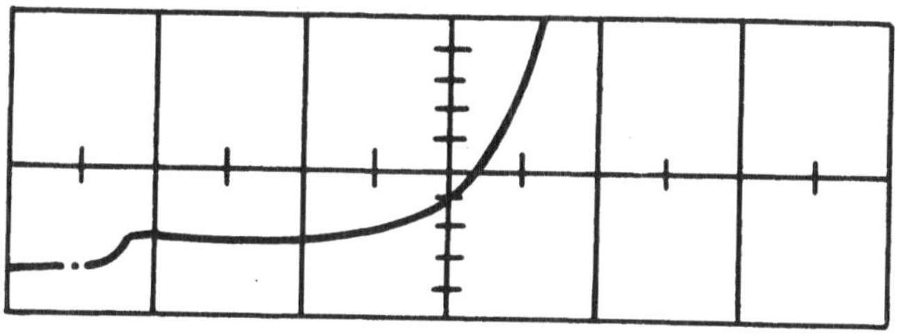

Fig. 3
Oscilloscope trace of the leading edge of a tailored laser pulse with an initially vaporization tail. 1.5 nsec/div. according to Lubin et al.[19].

Many details of the properties of plasmas produced from lithium hydride specks suspended electromagnetically[14] in vacuum were found by Lubin et al.[19]. The diagnostics were extended by measuring the size of the suspended targets, where an evaluation of the diffraction patterns of light by the target was necessary. The radius r_o of the targets was around 5 to 20 μ. Measurements were performed with charge collectors of different design with respect to the cases mentioned before[7,17] and the time of flight was measured at such a close distance as 5 mm. The result was that a laser pulse of P = 600 MW and τ_L = 9 nsec produced a plasma with an asymmetry of expansion. The plasma expanded in the forward direction against the plasma light 1.8 times faster, corresponding to a 3.2 times higher ion energy, than in the backward direction. It was suggested that a symmetric self-similarity expansion can be verified, if the plasma is preheated by a laser pulse before the main laser pulse is incident. The properties of the preheating pulse were derived heuristically from the values of energy for vaporization and ionization of the target and a time not larger than that required for a strong acoustic disturbance by a shock wave to traverse the pellet. The prepulse had to be 2 nsec and 5 MW for a lithium hydride target of 10 μ radius r_o, if a coupling efficiency of one percent was assumed. Fig. 3 shows the oscilloscope trace of the leading edge of such a tailored laser pulse. Measurements with laser pulses of P = 1.5 GW and τ_L= 4 nsec showed indeed the far field symmetry of the expanding plasma and a complete ionization of the target. The applicability of the self-similarity model[15, 16] for the heating and expansion of the plasma was then certain. Time of flight measurements resulted in the maximum ion energy and, by applying the model, in the average energy $\bar{\varepsilon}$ per particle near 800 eV. These measurements are shown in Fig. 4 for varying radius r_o. The calculation of $\bar{\varepsilon}$ from the self-similarity model[15,16] was possible immediately starting from the measured values of P, τ_L, r_o, and focus radius r_F (Fig. 4). In the same way, the calculation of the maximum temperature T_{max} of the plasma was possible. The measurement of T_{max} was tried

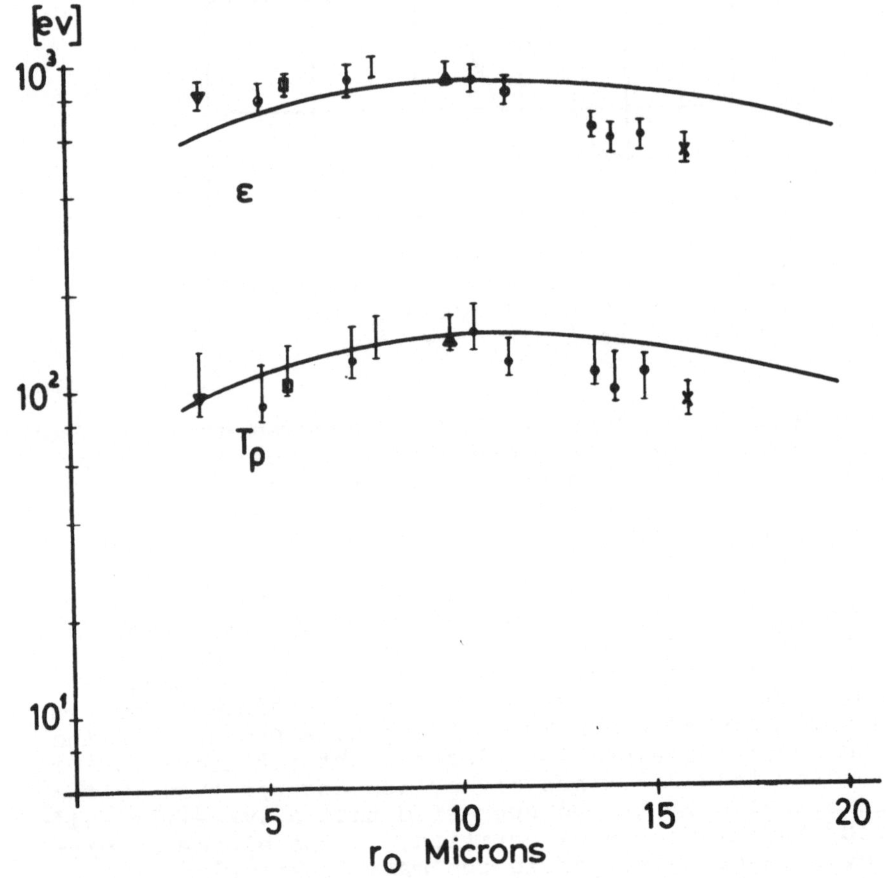

Fig. 4
Averaged energy per particle $\bar{\varepsilon}$ measured by time of flight and calculated from the self-similarity model for varying target radius r_o. Maximum temperature T_{max} measured by Thomson scattering and calculated from the self-similarity model according to Lubin et al.[19].

by means of Thomson scattering of an additional laser beam of $P = 100$ MW and $\tau_L = 15$ nsec. It was reconsidered whether the scattering takes place in a region of too strong spatial in-homogeneity of density and temperature, as it could be calculated[20,21]. The focusing[22] of the scattering beam that was reached down to 10μ seems to solve these problems. The agreement of the temperatures T_{max} from the evaluation of the wings of the spectrum of scattered radiation showed a considerable agreement with the average energy $\bar{\varepsilon}$, determined by time-of-flight-measurements, and to the calculated values of T_{max} (Fig. 4).

Yamanaka et al.[12] used the same method of electromagnetic suspension[14] of lithium hydride targets to produce plasmas by lasers. Side-on framing camera pictures demonstrated an asymmetric expansion with higher velocities against the laser light when the target had a radius r_o of 20 μ. This result is similar to that of Lubin et al[19]. The difference to the symmetric properties of the expanding plasma of Haught et al.[7] may be due to the fact that the targets of Yamanaka et al.[12] were of relatively large size if the question of preheating is neglected[19].

A further result[12] was the observation of two groups of plasma by the differentiated signals of grid-screened charge collectors. The maxima of the two groups corresponded to ion energies of 330 eV and 13 eV when the input laser power was 20 MW. The signals were very different to those of Papoular et al.[17] (Fig. 2), taken at much higher laser powers. Whatever the reasons for these features are, it is established that some interesting details of the plasma properties are to be found from such collector signals. The fact that these groups have not been observed by Haught et al.[7] and Lubin et al.[19] may be due to their method of integrated charge collector signals instead of differentiated signals.

Larger Free Targets

Another technique for irradiating free targets in the focus of a laser than that of electromagnetic suspension is the free fall of the particle. The techniques of free fall and the problems of guiding were reviewed in Witkowski's talk on "Free Targets"[16]. The size of the targets is then larger than in the case of the electromagnetic suspension and, therefore, the resulting plasma due to laser irradiation shows some other properties than in the case of small targets.

The work at Westinghouse started from irradiation of thin disks of aluminum of the size of the focus cross section held by thin files of quartz[23]. The results are very similar to those gained from aluminum spheres crossing the laser focus by free fall[24] after having been released from a retracing pedestal system[25]. The aluminum ball had a radius r_o of 50 to 150 μ, which could be measured by a microscope on the center of the pedestal before the ball was to fall down. The laser energy of reflection at the front side was measured. The transmitted light was registered time-resolved at an oscilloscope in comparison to the incident light, and integrated measurements of incident and transmitted light were taken by calorimeters. The focal diameter (half of maximum intensity) was about 400 μ. The total absorbed energy depended on the ball radius and varied between 30% and 50% at laser powers P between 50 and 400 MW and pulse lengths τ_L between 15 and 35 nsec. Framing camera pictures were taken side-on and end-on from 3 stages of the created plasma. Many differences in the exposure time had to be taken into consideration and it was not certain from the beginning whether it is right to relate the contours of the luminous plasma on the photographs with the surface front of the plasma. The following properties,however, of the created plasma could be figured out, the consistency of which proves once more its content of verity. Further, the side-on and end-on framing camera pictures allowed the identification of the outermost front of the pictures of the plasma with its surface. This could be checked by recalculating the radii of expansion of a set of exposures to the starting time, which was identified with the time of the laser pulse maximum. A further check was that the maximum energy ε_{max} of the fast ions expanding towards the laser was the same within 20% compared with the

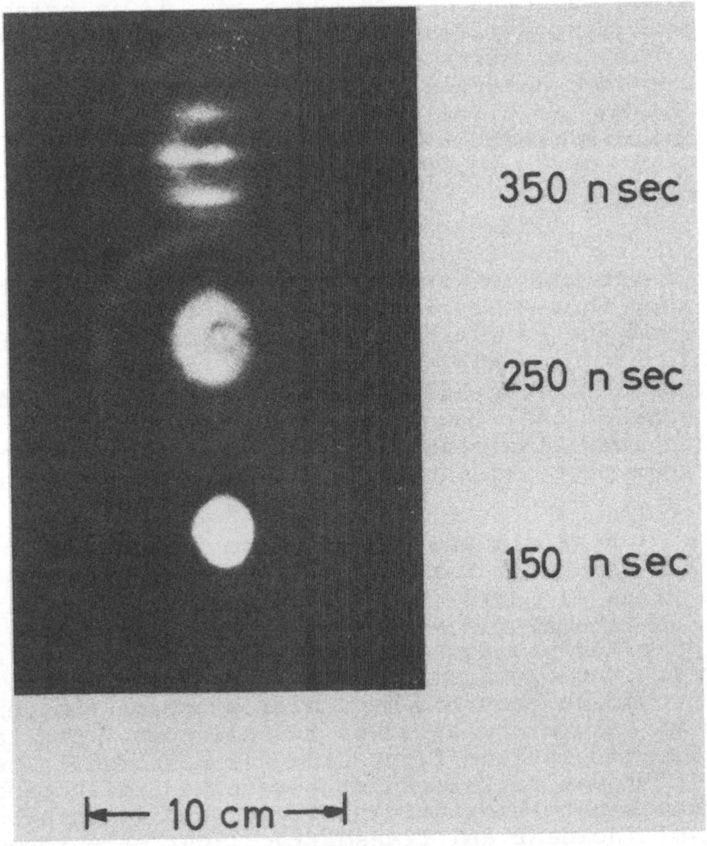

Fig. 5
Side-on framing camera picture of a plasma produced from an aluminum ball of 80μ radius by a 100 MW, 30 nsec laser pulse focused to 400 μ diameter at the times marked after the laser pulse maximum exposure times were 50 nsec. The second fram shows the outer plasma part expanding fast with a preferential direction towards the laser, and an inner spherical part[24].

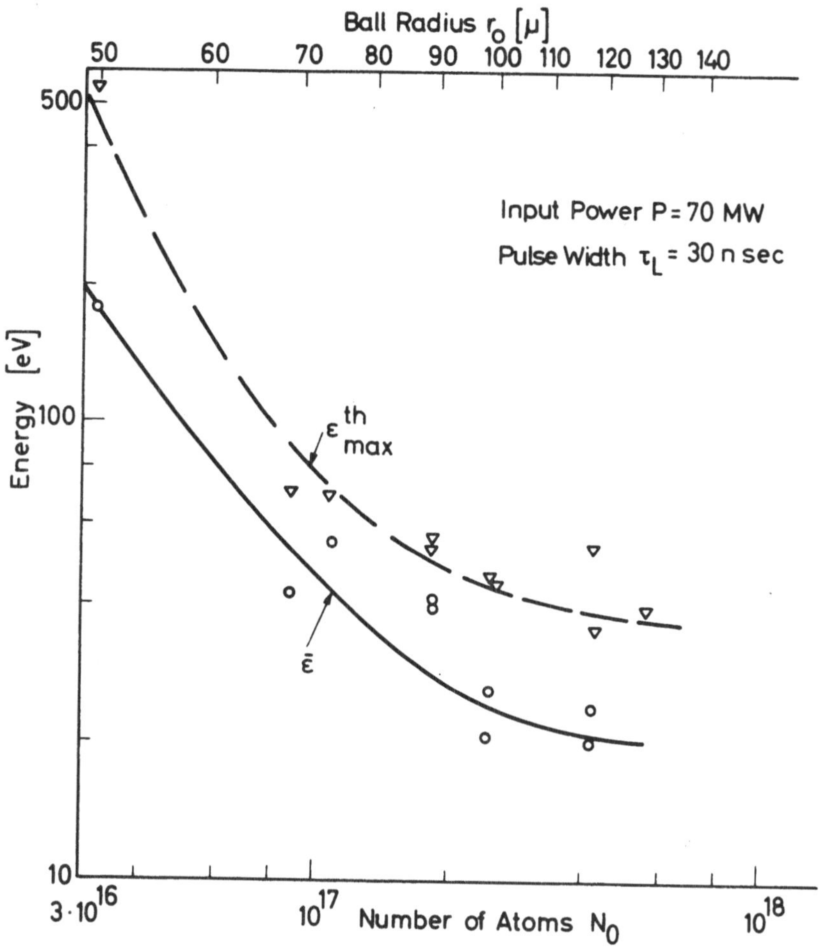

Fig. 6

Maximum ion energy ε_{max}^{th} (∇) taken from the velocity of the surface of the inner spherical part of the plasma (Fig. 3) and averaged energy per atom ε(o) calculated by distribution of all measured absorbed energy to all atoms of the aluminum ball. The dependence on the ball radius r_o was evaluable for cases of constant laser power P and pulse length τ_L only. In the case shown we have P ≈ 70 MW; $\tau_L \approx 30$ nsec[24].

measurement of plasma produced from compact aluminum targets at conditions of the same laser intensity P and the same laser pulse length τ_L, assuming the quadratic relation $\varepsilon_{max} \sim I^2$, as described in the following.

The results are then:

1) The expanding plasma consists of two parts, an inner thermokinetic and an outer part related to nonlinear surface

mechanism. Both parts are separated by a zone of lower brightness. (Fig. 5)

2) The thermokinetic properties of the inner part are proved by the maximum ion energy ε_{max}^{th}, taken from the photographs, and the averaged energy per particle $\bar{\varepsilon}$, derived from a distribution of the measured absorbed energy to all atoms of the ball (Fig. 6). ε_{max}^{th} is very close to values calculated from the self-similarity model of a thermokinetic expansion process[16]. The ratio $\varepsilon_{max}^{th}/\bar{\varepsilon}$ is of about two and agrees with the self-similarity model[15]. The brightness of the thermokinetic part decreases in the center, indicating some recombination[17,26,27]. The thermal core contains nearly all of the created plasma.

3) The properties of the outer part of the plasma, having a preferred expansion towards the laser light, could be seen from the properties of the maximum ion energy ε_{max}^{th}, taken from the velocity of the front of the expanding plasma. This energy ε_{max} had the usual values of some keV and showed no variation at varying ball radius r_o if the laser power P and pulse length τ_L were constant (Fig. 7). This confirmed a surface mechanism. The values of ε_{max} agreed with that of the case of a free disc[23] and compact targets (Gregg and Thomas[8]) as mentioned before. The property of the nonlinearity had to start from the fact that at compact targets the relation between ε_{max} and the laser intensity I was nearly quadratic[11,28] if the laser pulse length τ_L was constant. So we have to not simply plot ε_{max} against the measured laser power P because we had varying τ_L, but we had to reduce the measured values ε_{max} by the laser pulse length τ_L

$$\varepsilon_{max}^{1} = \frac{\varepsilon_{max}}{\tau_L^2} \qquad (1)$$

to plot these against the laser power P. Fig. 8 shows the result. The measured values ε_{max} are very close to a line with a power of 1.6 with little scattering, while the diagram of ε_{max} against P was sublinear with strongly scattering points of the measurements. An application of a theory of a nonlinear surface acceleration (involving self-focusing[19], shows the nonlinear absorption process and momentum transfer, and confirms that the fast part of the plasma contains less than a few percent of all plasma.

4) At large times (10^{-6} sec) after the laser was incident, an inner hole of the ion density of the expanding plasma was determined from reflection and transmission measurements of microwaves[30]. The concluded velocities are related to the thermal core. The thermokinetic theory does still not describe these recombination processes in details, at present; also the measured electron temperatures of few eV at

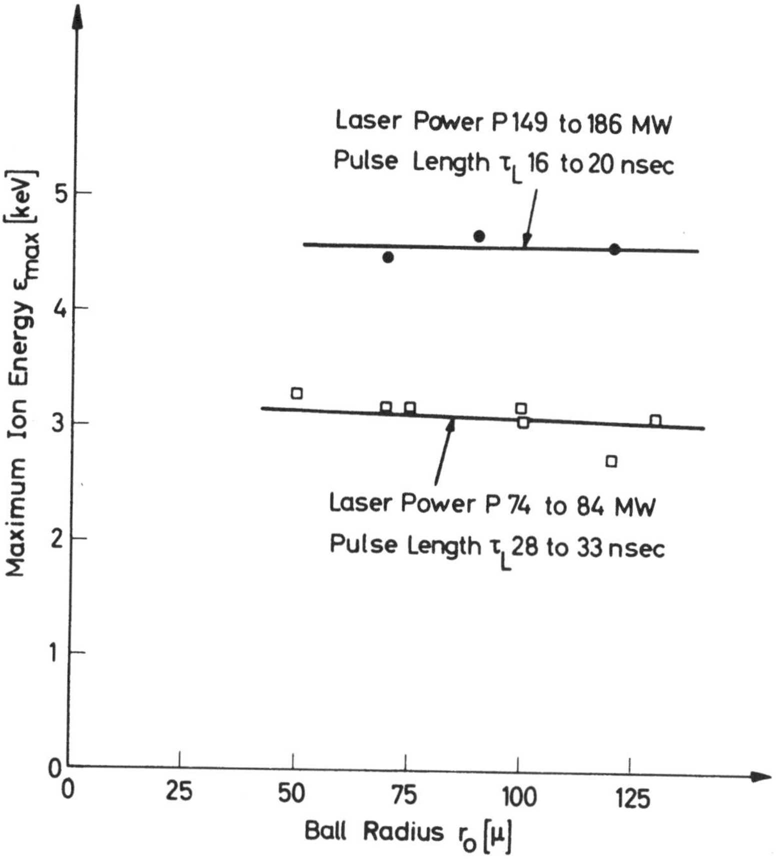

Fig. 7
Maximum ion energy ε_{max} of the fast outer front of the plasma showing no variation at varying ball radius r_o (in contrast to the thermokinetic core, Fig. 6, if the laser power P and the pulse length τ_L was constant[24].

times 90 or 190 nsec after the laser pulse[31] are still not understood theoretically.

Other measurements with larger free targets were performed using solid hydrogen. The first measurements used tips of hydrogen having a diameter of few 100μ, comparable with the diameter of the laser focus. Side-on framing camera pictures[32] demonstrated a spherically symmetric expansion with velocities understandable by a thermokinetic model. Fast ions have not been detected by the pictures. Later experiments started from free falling cylindrical specks of 200μ diameter and 250μ length of hydrogen as described before[6]. The plasma produced by a laser was measured by probes. It was concluded that the target had been fully ionized[33]. In another case[34] hydrogen of similar size was irradiated by a laser and the expansion of the

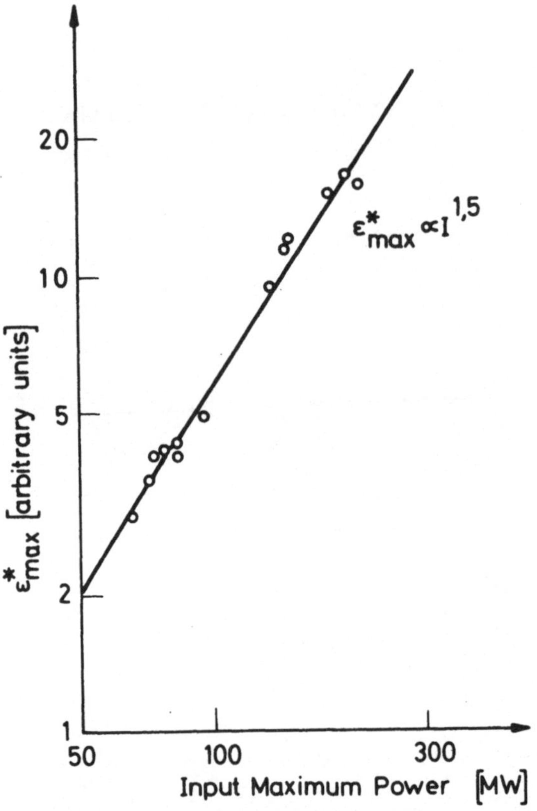

Fig. 8

Maximum ion energy ε^{1}_{max} reduced by Eq. (1) to constant laser pulse lengths τ_L, in dependence of the laser power P[24].

luminous plasma was measured by streak photographs. The result was a motion of the whole plasma along the direction of the laser radiation with an average ion energy of such drift of 4 eV. The magnitude of the drift could not be figured out from the measurements at free aluminum targets[24]. An estimation of the momentum transfer from the asymmetric expanding fast nonlinear surface allows the conclusion of a drift of the same magnitude as for hydrogen[34].

Magnetic Field Interaction

The question of interaction of laser produced plasmas with a magnetic field was connected with the first measurements using Q-switched laser pulses. The observations of plasmas produced from compact targets[1] were quite complex taking into consideration that the observed two groups showed various behaviour[11]. So it is not simple to understand that the ions were accelerated by the magnetic field both for field lines oriented perpendicularly or parallelly to the target surface. At special conditions an increase of the velocity of the fast group and a decrease of the slow group was measured[12]. Another curiosity was a jet-like narrowing and acceleration of the plasma expanding against the laser, due to a magnetic field[35].

In the case of free targets, initially the situations did not seem to be so complicated as in the mentioned case of compact targets. The suggestion of a confinement of the plasma by the magnetic field was proved when the longer time of emission of radiation (lines and continuum) from the plasma in a magnetic field was measured in contrast to the case without magnetic field[23]. The behavior of the plasma produced from small, electromagnetically suspended lithium hydride targets in magnetic fields[7, 17, 19] was quite well understandable[36]. The expansion in a minimum B-field or in a casp or bottle field against the field lines was slower than without the field, as concluded from time-of-flight-signals of charge collectors. In addition to such studies[19], the microwave radiation was calculated and measured, being created by the interaction of the plasma expanding into the magnetic field. In contrast to these cases of a decrease of plasma expansion by magnetic fields, also an increase was reported. The energy of expanding protons was larger by a factor of six by a magnetic field of only 7 kGauss[37]. This result has a similarity to the observations with compact targets[11] and indicates many unsolved problems.

The increase of the time of luminosity of the plasma by a magnetic field was also observed with plasmas produced from free solid hydrogen targets[34], where the magnetic field had values of 450 kGauss. But no final results could be reached. The compression of the plasma by the magnetic field could be calculated[38]. A more detailed theory of the interaction of the laser produced plasma was given by Bhadra[39]. The experiments performed parallelly indicated an anomalous increase of the electric resistivity and instabilities due to collective effects[40]. The measurements of Tuckfield and Schwirzke[40] were performed with paraffin targets of 10μ and larger radius suspended by a 1 μ diameter quartz fiber. In addition to the reviewed results, the plasma produced by laser pulses of P = 150 MW and τ_L = 200 nsec showed a spherical expansion. Using targets of 50 μ radius, a preferential direction of expansion towards the laser was observed. In the presence of a magnetic mirror field of 1200 Gauss, the expansion was stopped and an oscillation of the plasma diameter, synchronous with an oscillation of the signals of a diamagnetic probe was measured. The resulting shell structure in the luminescence of the streak photographs of the plasma are expectable by the model of Bhadra[39]. This observed oscillation is a very remarkable result in addition to all problems of increased ion energies[37], instabilities, etc.[40]. The oscillation was also calculated from a self-similarity expansion of plasma in a uniform magnetic field [41].

Peferences

1. W.I. Linlor, Appl. Phys. Lett. 3, 210 (1963); Phys. Rev. Lett. 12, 388 (1964); Laser Production of High Ion Energies, these proceedings, p. 173.

2. S. Panzer, J.Appl.Math Mech. 16, 138 (1965); Laser Angew. Strahltech. 1, No. 3, 23 (1969). A. W. Ehler, and G. L. Weissler, Appl. Phys. Lett. 8, 89 (1966).

3. D. Röss, Laser, Akad. Verl. Ges. Frankfurt/M., 1966; P. Görlich, and W. Wrede, Nova Acta Leopoldina, 25 (1965).

4. R. B. Hall, J. Appl. Phys. 40, 1941 (1969).

5. H. Schwarz, Linear and Nonlinear Laser Induced Emission of Ions from Solid Targets With and Without Magnetic Field, these Proceedings, p. 207.

6. S. Witkowski, Free Targets, these Proceedings, p. 259.

7. A. F. Haught, and D. H. Polk, Conf. Proced. Culham, Sept. 1965, IAEA, Vienna, 1966, Vol. II, P. 953; Phys. Fluids 9, 2047 (1966); A. F. Haught, D. H. Polk, and W. J. Fader, Conf. Proceedings, Novosibirsk, Aug. 1968, IAEA, Vienna, 1969 Vol. I, p. 925.

8. H. Opower, and E. Burlefinger, Phys. Lett. 16, 37 (1965); H. Opower, W. Kaiser, H. Puell, and W. Heinicke, Z. Naturforschg. 22a, 1392 (1967); D. W. Gregg, and S. J. Thomas, J. Appl. Phys. 37, 4313 (1966); B. C. Fawcett, A. H. Gabriel, F. E. Irons, N.J. Peacock, and P. A. H. Saunders, Proc. Phys. Soc. (London) 88, 1051 (1966); J. L. Bobin, F. Floux, P. Langer, and H. Pignera, Phys. Lett. 28A, 398 (1968).

9. G. Tonon, Compt. Rend. 262B, 706 (1966).

10. B. C. Boland, E. E. Irons, and R. W. McWhirter, J. Phys. B1 1180, (1968).

11. S. Namba, H. Schwarz, Proc. IEEE Symp. on Electron, Ion and Laser Beam Technology, Berkeley, May 1967, p. 86; S. Namba, P. H. Kim, and H. Schwarz, Transactions 8th Int. Conf. Phenomena in Ionized Gases, Vienna, Aug. 1967, p. 59; Sc. Papers Inst. Phys. Chem. Res., Tokyo, 60, 101 (1966); H. Schwarz,

Linear and Nonlinear Laser Induced Emission of Ions from Solid Targets With and Without Magnetic Field, these Proceedings, p. 207.

12. T. Yamanaka, N. Tsuchimori, T. Sasaki, and Ch. Yamanaka, Technol. Progress Rept. Osaka Univ. 18, 155 (1968).

13. H. Hirono, and J. Iwamoto, Japan J. Appl. Phys. 6, 1006 (1967).

14. R. F. Wuerker, H. M. Goldenbaum, and R. V. Langmuir, J. Appl. Phys. 30, 441 (1959).

15. J. M. Dawson, Phys. Fluids 7, 981 (1964); Thermo-kinetic Expansion Theory, these Proceedings, p. J. M. Dawson, P. Kaw, and B. Green, Phys. Fluids 12, 875 (1969).

16. H. Hora, Some Results of the Self-Similarity Model, these Proceedings, p. 365.

17. P. E. Faugeras, M. Mattioli, and R. Papoular, Euratom-CEA, Fontenay, Rep. FC-465 (1968).

18. R. Papoular, Euratom Conf. on Laser-Produced Plasmas, Paris, Febr. 1967.

19. M. J. Lubin, H. S. Dunn, and W. Friedman, Conf. Proceedings, Novosibirsk, Aug. 1968, IAEA Vienna, 1969, Vol. I, p. 945.

20. M. Decroisette, Ann. Physique, 2, 203 (1967).

21. A. F. Haught, D. H. Polk, and W. J. Fader, UARL-Rep. July 1968.

22. M. Lubin, private communication (1969).

23. E. W. Sucov., J.L. Pack, A. V. Phelps, and A. G. Engelhardt, Phys. Fluids 10, 2035 (1967).

24. A. G. Engelhardt, T. V. George, H. Hora, and J. L. Pack, Westinghouse Res. Lab. Rept. WERL-3472-9 (1968); Phys. Fluids (to be published); Bull. Amer. Phys. Soc. 13, 887 (1968).

25. J. L. Pack, T. V. George, and A. G. Engelhardt, Rev. Sci. Instr. 39, 1697 (1968).

26. A. G. Engelhardt, J. L. Pack, and T. V. George, Bull. Amer. Phys. Soc. 13, 320 (1968).

27. A. G. Engelhardt, and H. Hora, Bull. Amer. Phys. Soc. 13, 920 (1968).

28. N. R. Isenor, Appl. Phys. Lett. 4, 152 (1964); Can. J. Phys. 42, 1413 (1964).

29. H. Hora, D. Pfirsch, and A. Schlüter, Z. Naturforsch. 22a, 278 (1967); A. Schlüter, Plasma Phys. 10, 471 (1968); H. Hora, Phys. Fluids 13, 182 (1969); Ann. Physik 22, 402 (1969); Z. Physik 226, 156 (1969).

30. J. L. Pack, T. V. George, and A. G. Engelhardt, Phys. Fluids 13, 469 (1969).

31. A. G. Engelhardt, T. V. George, J. L. Pack, H. Hora, and G. Cox, Bull. Amer. Phys. Soc. 13, 1553 (1968).

32. P. A. H. Saunders, P. Avivi, and W. Millar, Phys. Lett. 24A, 290 (1967).

33. G. Francis, D. W. Atkinson, P. Avivi, J. E. Bradley, C. D.King, W. Millar, P.A.H. Saunders, and A. F. Taylor, Phys. Lett. 25A,486 (1967).

34. U. Ascoli-Bartoli, B. Brunelli, A. Caruso, A. DeAngelis, G. Gratton, F. Parlange, and H. Salzmann, Conf. Proc. Novosibirsk, Aug. 1968, IAEA, Vienna, 1969.

35. J. Brunteneau, E. Fabre, H. Lamain, and P. Vasseur, Phys. Lett. 28A, 777 (1969).

36. I. B. Bernstein, and W. J. Fader, Phys. Fluids 11, 2209 (1968)

37. M. Mattioli, and D. Véron, LiH Laser Produced Plasmas, Euratom-CEA- Fontenay, to be published (1969).

38. A. Cavaliere, P. Giupponi, and P. Gratton, Phys. Lett. 25A, 636 (1967).

39. D. K. Bhadra, Phys. Fluids 11, 234 (1960).

40. R. Tuckfield, and F. Schwirzke, Plasma Physics 11, 11 (1969); F. Schwirzke, and R. Tuckfield, Phys. Rev. Rev. Letters 22, 1284 (1969); F. Schwirzke, III. Europ. Conf. Contr. Fusion, Utrecht, June 1969, Wolters Publish. Groningen, 1969, p. 114.

41. P. E. Faugeras, M. Mattioli, and R. Papoular, AIAA Fluid and Plasma Dynamics Conf., Los Angeles, Calif., June 1968; M. Mattioli, Euraton-CEA-Fontenay Rep. FC-477 (1968).

On the Production of Plasma by Giant Pulse Lasers

JOHN M. DAWSON
Plasma Physics Laboratory, Princeton University, Princeton, New Jersey

Calculations are presented which show that a laser pulse delivering powers of the order of 10^{10} W to a liquid or solid particle with dimensions of the order of 10^{-2} cm will produce a hot plasma with temperatures in the range of several hundred eV. To a large extent the plasma temperature is held down by its rapid expansion and cooling. This converts much of the energy supplied into ordered energy of expansion. This ordered expansion energy can amount to several keV per ion. If the expanding plasma can be caught in a magnetic field and its ordered motion converted to random motion this might be utilized as a means for filling controlled thermonuclear fusion devices with hot plasma. Further, it should also be possible to do many interesting plasma experiments on such plasmas.

1. INTRODUCTION

ONE of the basic problems of controlled thermonuclear fusion is the filling of the various fusion devices with high-temperature, low-density plasma. One method which has been proposed is to vaporize a small speck of material with a strong laser pulse of short duration. Recently Basov and Krokhin[1] presented some calculations which indicate that this approach is feasible. It is the purpose of this paper to present some further calculations on this possibility. A number of the calculations given here are also contained in the work of Basov and Krokhin. Some similar calculations have also been made by Engelhart.[2]

A typical plasma which might be produced by this means would contain from 10^{16} to 10^{19} electrons and nuclei with energies ranging from a few hundred eV to a few keV per particle. To achieve such plasmas would require a laser pulse delivering powers of the order of 10^9–10^{12} W for times of the order of a few nanoseconds. Published figures[3] indicate that such pulses are close to the state of the art and may be achieved in the near future. It should be possible to do interesting experiments with pulses now available.

2. REQUIREMENTS ON THE TOTAL ENERGY AND POWER REQUIRED

If for the moment we assume that we can neglect radiation, then the energy required to give N_i ions and N_e electrons each an energy ϵ is

$$E = \epsilon(N_i + N_e), \tag{1}$$

or if we measure E in joules and ϵ in eV

$$E = 1.6 \times 10^{-19}\epsilon(N_i + N_e). \tag{1a}$$

For an ϵ of 1 keV and $N_i + N_e \approx 10^{17}$ we see we require 16 J. We would thus require energies in the range of 1 to 10^3 J. Now if this material is initially in solid or liquid form where typical densities are of the order of 5×10^{22} atoms/cm^3, then we would require the speck to have a volume of between 2×10^{-7} cm^3 to 2×10^{-4} cm^3. The linear dimensions of the speck will consequently be of the order of 6×10^{-3} to 6×10^{-2} cm. If we assume that the nuclei are protons and that they have 1 keV energy, then their velocity is 4×10^7 cm/sec. The smaller speck doubles its size in approximately 10^{-10} sec, while the larger one requires 10^{-9} sec. As we see later, the plasma rapidly becomes transparent to the light pulse upon expanding, and so for the energy to be effective, it must be delivered in the order of this time. The small speck which requires 1 J to reach the desired energy requires a power of 10^{10} W while the larger speck requiring 10^3 J requires 10^{12} W. Lower energies, of course, require less power both because less energy is needed and the gas expands more slowly. A detailed calculation will be given later.

[1] N. G. Basov and O. N. Krokhin, in *Proceedings of the Conference on Quantum Electronics, Paris, 1963*.
[2] A. G. Engelhardt, Westinghouse Research Laboratories Report 63-128-113-R2.
[3] A. L. Schawlow, Sci. Am. 209 [7], 34 (1963).

Reprinted from: *The Physics of Fluids*, 7, 981–987 (1964).

3. PENETRATION

In order for the light pulse to penetrate the plasma drop, the plasma frequency must be lower than the light frequency. The frequency put out by a ruby laser is 4.35×10^{14} sec^{-1}. The plasma frequency is given by[4]

$$\nu_p = 8.9 \times 10^3 n_e^{1/2}, \quad (2)$$

where n_e is the electron density. Equating these two frequencies gives

$$n_e = 2.4 \times 10^{21} \text{ electrons/cm}^3. \quad (3)$$

This is somewhat lower than the electron density in a solid or liquid (electron density in liquid hydrogen 4×10^{22}). Thus as soon as a high degree of ionization is reached at around 10 eV, the plasma will reflect the radiation. However, once the volume had increased by a factor of 10 or so (this would require 10^{-8} sec or so for hydrogen at 10 eV), the light will again penetrate the drop. Once the plasma frequency drops below the light frequency the fraction of the light entering the drop rapidly increases. If we assume the drop has a sharp boundry which the light is normal to, then the fraction entering the drop is given by

$$f = \frac{[\nu + (\nu^2 - \nu_p^2)^{1/2}]^2 - [\nu - (\nu^2 - \nu_p^2)^{1/2}]^2}{[\nu + (\nu^2 - \nu_p^2)^{1/2}]^2}, \quad \nu > \nu_p \quad (4)$$

where ν_p and ν are the plasma and light frequencies. For a ν of $1.05\, \nu_p$, this formula predicts an f of 0.7. If the boundary of the drop is diffuse on the scale of the lightwave length ($\lambda = 6.9 \times 10^{-5}$ cm), the penetration is even more efficient. On the other hand, if the light strikes the drop at an oblique angle, this penetration is hindered. Here penetration is obtained only when

$$\omega^2 - k_\parallel^2 c^2 > \omega_p^2, \qquad \omega = 2\pi\nu, \quad (5)$$

where $k_\parallel$ is the wavenumber parallel to the surface. For 45° incidence the critical density is reduced by a factor of 2.

4. ABSORPTION

In order for the light to heat the drop, it must be absorbed before it passes through. The primary absorption process for a fully ionized plasma is due to electron–ion collisions, i.e., resistance to the induced currents. The inverse of the absorption length due to such collisions for waves of frequency ν is given by[5]

$$K = \frac{8\pi Z^2 n_e n_i e^6 \ln \Lambda(\nu)}{3c\nu^2 (2\pi m_e kT)^{3/2} (1 - \nu_p^2/\nu^2)^{1/2}}, \quad (6)$$

$$K = \frac{1.17 \times 10^{-8} Z n_e^2 \ln \Lambda(\nu)}{3\nu^2 (kT)^{3/2}} \frac{1}{(1 - \nu_p^2/\nu^2)^{1/2}}, \quad (6a)$$

kT in eV$_0$, $\Lambda = v_T/\omega_p p_{min}$. Here Ze and n_i are the ionic charge and density, $-e$, m_e, n_e are the electronic charge, mass, and density, c is the velocity of light, v_T is the thermal velocity of the electrons, p_{min} is the minimum impact parameter for electron ion collisions [$p_{min} \cong$ maximum of $Z e^2/kT$ or $\hbar(m_e kT)^{1/2}$]. If ν is equal to ν_p, Λ is Spitzer's[4] Λ.

The term $(1 - \nu_p^2/\nu^2)^{-1/2}$ is important only for $\nu \simeq \nu_p$ and hence we will neglect it. In any case this term only increases K. For hydrogen at 1 keV, $n_e = 2.4 \times 10^{21}$ $\nu = 4.35 \times 10^{14}$ sec^{-1}, $\ln \Lambda = 10$, one finds $K = 40$ cm^{-1}. Thus the light will be absorbed in a drop of dimensions 2.5×10^{-2} cm.

5. RADIATION

The radiation from the plasma is black body radiation at those frequencies for which the absorption length is less than the plasma radius, while for frequencies where the absorption length is large compared with the radius, the radiation is bremsstrahlung. The maximum of the Planck distribution occurs for a frequency

$$h\nu_{max} \cong 3kT. \quad (7)$$

For $kT = 1$ keV, ν_{max} is equal to 7.2×10^{17} sec^{-1} or about 2000 times the laser frequency. Thus by Eq. (7)[6] the plasma is transparent to these frequencies and the radiation is primarily bremsstrahlung. The bremsstrahlung emitted is given by[4]

$$\epsilon = 4.86 \times 10^{-31} Z n_e^2 T^{1/2} \text{ W/cm}^3, \quad (8)$$

where T is in keV.

For hydrogen at 1 keV and a density of 2.4×10^{21} this formula gives

$$\epsilon = 2.8 \times 10^{12} \text{ W/cm}^3. \quad (9)$$

A volume of 10^{-4} cm^3 containing 2.4×10^{17} protons radiates at the rate of 2.8×10^8 W, which is small compared to the estimated required powers of the order of 10^{10} W. For higher Z materials the bremsstrahlung is more serious. The bremsstrahlung calculation also ignores radiation by bound electrons and from recombination. These processes become more important with increasing Z, but appear to be unimportant for Z's less than or the order of 3 or 4 at plasma temperatures in the hundreds of eV. For

[4] L. Spitzer, Jr., *Physics of Fully Ionized Gases* (Interscience Publishers, Inc., New York, 1956).
[5] J. M. Dawson and C. R. Oberman, Phys. Fluids 5, 517, 1962.

[6] Equation (7) is not strictly correct for $h\nu \cong kT$, nevertheless it gives a rough estimate of the absorption.

temperatures below 100 eV even black body radiation from the drop would be negligible.

6. ION ELECTRON THERMALIZATION TIME

The thermalization time for ions and electrons is given by[4]

$$t_{ie} = [1.05 \times 10^{13} A/n_e Z \ln \Lambda] T^{\frac{3}{2}}, \qquad (10)$$

$$\Lambda = v_T/\omega_p p_{min},$$

where T is in keV and A is the atomic weight. For hydrogen at a density of 2.4×10^{21} at 1 keV with $\ln \Lambda$ set equal to 10 we find

$$t_{ie} = 4.4 \times 10^{-10} \text{ sec}, \qquad (11)$$

which is of the order of the duration of the heating. At lower temperature t_{ie} is shorter so it appears that up to this temperature the ions and electrons remain in thermal equilibrium. If one tries to heat the plasma to higher temperatures the electrons are heated but the ions are not. If one had higher frequency lasers so that one could heat at higher densities, then equilibrium could be maintained at larger temperatures.

7. THERMAL CONDUCTIVITY AND ENERGY TRANSPORT

The primary process for temperature equalization in the plasma is by electronic heat conduction. Since the drop is transparent to its emitted radiation this contributes to an energy loss, but not to the heat conduction. The equation for the diffusion of heat is (see Ref. 4)

$$C_v \, \partial T/\partial t = \eta \, \partial^2 T/\partial r^2, \qquad (12)$$

$$\eta = \frac{5.85 \times 10^{12}}{Z \ln \Lambda} T^{5/2} \quad \frac{\text{ergs}}{\text{sec deg cm}},$$

where T is in keV,

$$C_v = 2(n_e + n_i) \times 10^{-16} \quad \text{ergs/cm}^3 \text{ deg}.$$

From this equation setting $\ln \Lambda = 10$ we find that the distance l that the heat will diffuse in time t is

$$l^2 = \frac{\eta}{C_v} t = \frac{2.93 \times 10^{27}}{Z(n_e + n_i)} T^{5/2} t. \qquad (13)$$

For hydrogen at a density of 2.4×10^{21} and at a temperature of 1 keV we find for l^2

$$l^2 = 3.05 \times 10^5 t. \qquad (14)$$

For $t = 10^{-9}$ sec, $l = 1.75 \times 10^{-2}$ cm, which is of the order of the plasma radius. Lower temperatures and higher Z's and higher densities decrease l.

In the early stages of the pulse, strong local heating may occur at the surface of the plasma. Since the absorption coefficient is a strong function of the temperature such regions should rapidly become transparent and the light will penetrate deeper into the plasma. The expansion of the surface layer also enhances the decrease in absorption coefficient. In addition to the advancing region of light penetration, shocks are set up in the plasma due to the local high temperature and pressure. Such shocks are quite effective in carrying energy away from the hot spots and in heating the plasma through which they pass and thus provide another means for equalizing the temperature. The details of all these complicated processes depend on how the drop is illuminated, i.e., uniformly over its surface or locally, and a complete analysis of them would be very complicated. Since the purpose of this paper is merely to show the feasibility of using this method to produce interesting plasmas we do not attempt such an analysis here. Instead we simply assume that the drop is heated uniformly. Towards the end of the time when the drop absorbs, this should be a reasonable approximation since the light penetrates the whole plasma at that time and the electronic heat conduction is fairly large.

A much more detailed analysis of the temperature profile within the plasma appears called for. However, here we assume that the whole drop is at a uniform temperature.

8. EXPANSION AND COOLING

For temperatures greater than 100 eV and number densities of the order of 10^{21} cm^{-3} any low Z material ($Z \leq 5$) is essentially completely stripped of electrons since the thermal energy is of the order of the binding energy for the most tightly bound electron. As a consequence the plasma behaves like an ideal gas of electrons and ions and, as already pointed out, rapidly expands. We may approximate this process as was done by Basov and Krokhin[1] by assuming a uniform density and temperature and equate the work done by the pressure to the increase in the kinetic energy associated with the radial expansion. We have

$$P\Delta V = \Delta \tfrac{1}{2} \bar{M} \dot{r}^2$$

or

$$P 4\pi r^2 \frac{dr}{dt} = \tfrac{1}{2} \bar{M} \frac{d}{dt} \left(\frac{dr}{dt}\right)^2, \qquad (15)$$

where P is the pressure and $\bar{M}$ is some suitable average total mass for the plasma. If we take the density of the plasma to be uniform and assume the velocity increases linearly from the center to the edge, then $\bar{M}$ is $\frac{3}{5}M$ where M is the actual mass (we use this value of $\bar{M}$).

The pressure is given by

$$P = 3(N_i + N_e)kT/4\pi r^3, \qquad (16)$$

where N_i and N_e are the total number of electrons and ions contained in the drop. Finally the temperature is determined by the equation for conservation of energy

$$\tfrac{3}{2}(N_e + N_i)k\frac{dT}{dt} = -4\pi Pr^2\frac{dr}{dt} + W, \qquad (17)$$

where W is the rate at which the laser is supplying energy. The left-hand side is the thermal energy of the plasma, while the first term on the right-hand side is the rate at which energy is going into the expansion. Here we have neglected the radiation though it could be subtracted from W. If one assumes W is constant and that the initial expansion velocity is zero, then these equations can be integrated to give

$$r = [r_0^2 + \tfrac{10}{9}Wt^3/N_i m_i]^{\frac{1}{2}}, \qquad (18)$$

$$kT = \frac{Wt}{3(N_e + N_i)}\left[\frac{2r_0^2 + 5Wt^3/9N_i m_i}{r_0^2 + 10Wt^3/9N_i m_i}\right], \qquad (19)$$

where r_0 is the initial radius of the plasma and m_i is the mass of the nuclei.

For times sufficiently large so that $Wt^3/N_i m_i \gg r_0^2$, we can approximate these expressions by

$$r = [\tfrac{10}{9}Wt^3/N_i m_i]^{\frac{1}{2}}, \qquad (20)$$

$$kT = Wt/6(N_e + N_i). \qquad (21)$$

For short times when the expansion is unimportant, kT goes like

$$kT = \tfrac{2}{3}Wt/(N_e + N_i). \qquad (22)$$

Comparing (21) and (22) shows that for long times $\frac{3}{4}$ of the energy goes into expansion and $\frac{1}{4}$ into thermal energy.

As the plasma expands it rapidly becomes transparent to the laser light because by Eq. (6a), K is proportional to n_e^2. We can find the time when the drop becomes transparent by equating Kr to 1. In doing this let us find the dependence on the rate of power input and the initial electron density and drop radius by making use of Eqs. (20) and (21) and the relations

$$N_e = n_{e0}\tfrac{4}{3}\pi r_0^3, \qquad n_e = n_{e0}(r_0/r)^3, \qquad (23)$$
$$N_i = N_e/Z, \qquad n_i = n_e/Z.$$

Substituting all these expressions in $Kr = 1$ gives

$$\frac{52.5 \ln \Lambda n_{e0}^6(1 + 1/Z)^{3/2} r_0^{18} m_i^{5/2}}{3Z^{3/2}\nu^2 W^4 t^9} = 1. \qquad (24)$$

Setting $\nu = 4.35 \times 10^{14}$, $m_i = 1.66 \times 10^{-24}A$, $\ln \Lambda = 10$, Eq. (24) becomes

$$\frac{1.45 \times 10^{-82} r_0^{18} n_{e0}^6}{3Z^{3/2}W^4 t^9} A^{5/2}\left(1 + \frac{1}{Z}\right)^{3/2} = 1, \qquad (25)$$

where r_0 is in cm, and W is in watts. Solving for t we find

$$t = 7.3 \times 10^{-10} n_{e0}^{2/3} \frac{r_0^2 A^{5/18}(1 + 1/Z)^{1/6}}{Z^{1/6}W^{4/9}} \text{ sec}. \qquad (26)$$

The energy delivered per particle is

$$\epsilon = \frac{Wt}{n_{e0}\frac{4}{3}\pi r_0^3(1 + 1/Z)}$$
$$= 6.5 \times 10^2 \frac{A^{5/8}W^{5/9}}{n_{e0}^{1/3}Z^{1/6}r_0(1 + 1/Z)^{5/6}}, \qquad (27)$$

where ϵ is in eV and W is in watts.

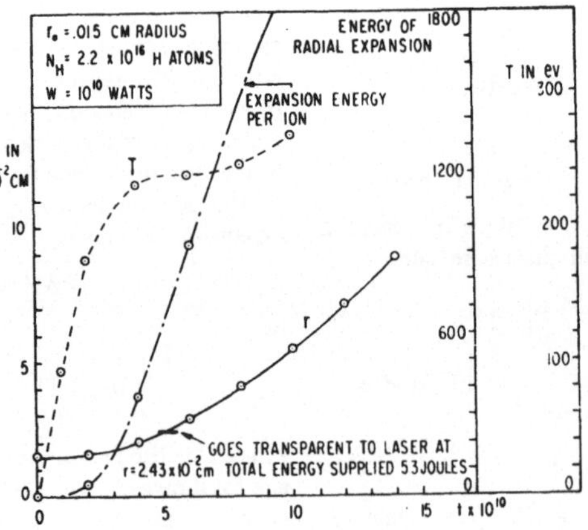

FIG. 1. Heating and expansion of hydrogen plasma drop.

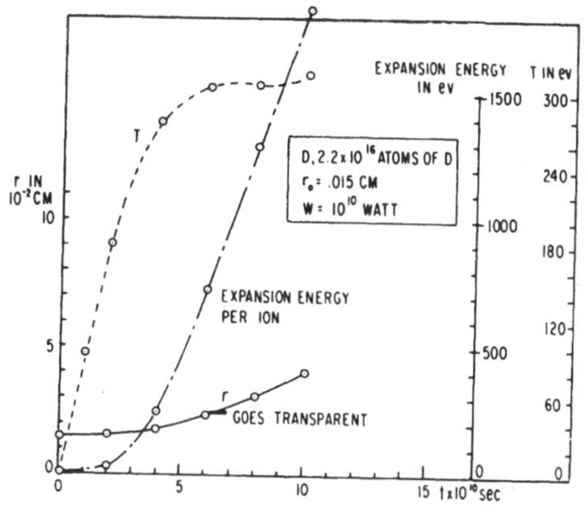

FIG. 2. Heating and expansion of a deuterium plasma drop.

For deuterium with initial radius of 1.5×10^{-2} cm and a W of 10^{10} W, t comes out equal 9×10^{-10} sec and ϵ to be 1.0 keV.

The results of some more complete calculations on deuterium and lithium drops are shown in Figs. 1–5. The results are in general agreement with the approximate formulas (20), (21), (26), and (27) even though the radii of these drops expands less than a factor of 2 before they become transparent. The approximations used in deriving these equations break down for drops which are so small that they become transparent before they expand appreciably. For very small drops one cannot increase the energy per particle by decreasing the drop size as predicted by (27). If the plasma is heated very fast so that expansion is unimportant, then we may estimate the temperature we can reach by again equating Kr_0 to 1. From Eq. (19) this gives

$$kT = 2.5 \times 10^{-6}[Z^{2/3}n_e^{4/3}(\ln \Lambda)^{2/3}r_0^{2/3}]/Z^{4/3}, \qquad (28)$$

kT in eV. For $Z = 1$, $n_e = 2.4 \times 10^{21}$, $\nu = 4.35 \times 10^{14}$, and $r_0 = 10^{-2}$ cm we obtain a temperature of 525 eV. While one could certainly reach somewhat higher temperatures than this since the plasma is still being heated, a large increase in temperature above this will be difficult since the absorption coefficient drops off rapidly and the expansion will soon become important.

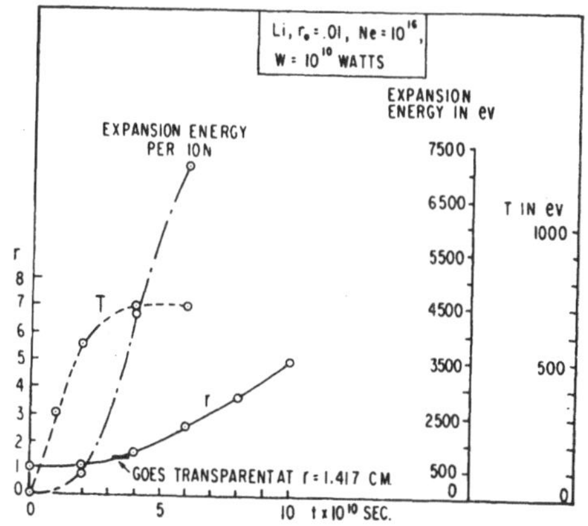

FIG. 3. Heating and expansion of a lithium plasma drop.

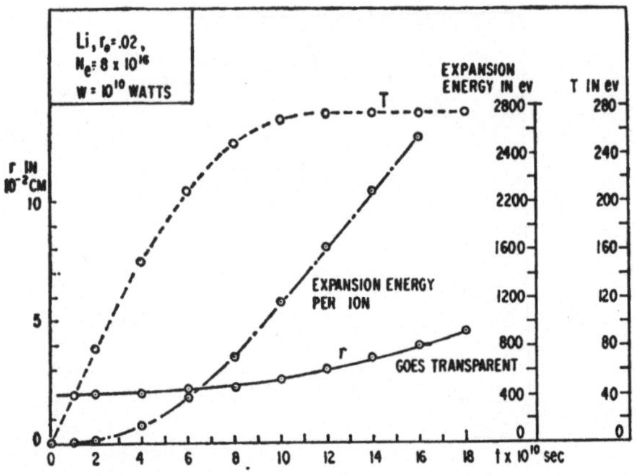

FIG. 4. Heating and expansion of a lithium plasma drop.

9. EXPANSION AFTER HEATING AND PLASMA TRAPPING

After the plasma became transparent it still expands rapidly and it cools as a result. If one assumes that no energy is added after transparency is reached, then Eqs. (15), (16), and (17) governing the expansion can again be solved and one finds

$$T = T_1(r_1/r)^2, \tag{29}$$

$$\dot{r}^2 = \dot{r}_1^2 + [3(N_e + N_i)/\bar{M}]kT_1[1 - r_1^2/r^2], \tag{30}$$

where r_1, T_1, and $\dot{r}_1$ are the radius, temperature, and radial velocity of the plasma at the time it becomes transparent.

We see that the plasma temperature drops rapidly with expansion and that by the time the radius has increased by a factor of 10 the temperature is down to a few eV. All the energy ends up as energy of radial expansion.

At this point one must check on whether or not the plasma recombines. At these densities the principle mechanism for recombination is due to three particle encounters, two electrons and one ion. A theory for such recombination for hydrogen has been worked out by Hinnov and Hirschberg.[7]

[7] E. Hinnov and J. Hirschberg, Phys. Rev. 125, 795 (1962).

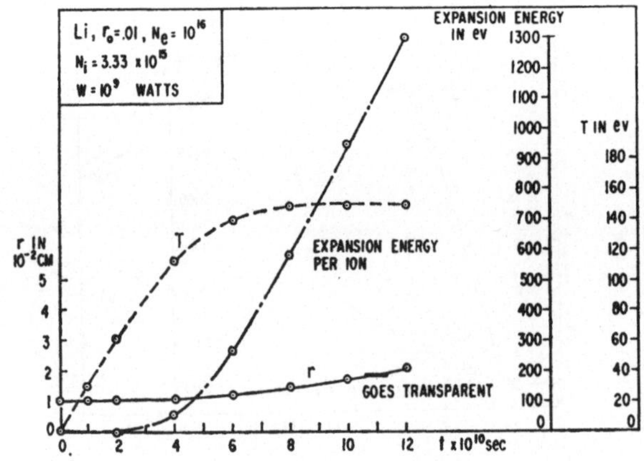

FIG. 5. Heating and expansion of a lithium plasma drop.

According to these authors the rate of recombination dn_e/dt is given by

$$dn_e/dt = 5.6 \times 10^{-27}(kT)^{-9/2}n_e^3, \qquad (31)$$

with kT in eV.

The characteristic recombination time is

$$\tau = 1.8 \times 10^{26}(kT)^{9/2}/n_e^2, \qquad (32)$$

where τ is in sec.

If we substitute (29) and (23) into this formula we obtain

$$\tau = 1.8 \times 10^{26}[(kT_1)^{9/2}/n_{e1}^2](r_1/r)^3, \qquad (33)$$

where 1 refers to the plasma state at the time it becomes transparent. This time should be compared to the time required for the plasma to double its radius

$$\tau = r/v_{ex}, \qquad (34)$$

where v_{ex} is the expansion velocity. If these times are equated we find for r,

$$r = 3.66 \times 10^6[(kT_1)^{9/8}/n_{e1}^{1/2}]r_1^{3/4}v_{ex}^{1/4}. \qquad (35)$$

If we take $kT_1 = 300$ eV, $n_{e1} = 5 \times 10^{20}$, $r_1 = 2.5 \times 10^{-2}$, $v_{ex} = 3 \times 10^7$ (expansion energy of 1 keV for deutrons) which roughly correspond to the values given in Fig. 2, then (35) gives for r, $r = 0.4$ cm. Thus the radius has increased by roughly a factor of 20, the density has decreased by a factor of 8000, and the temperature has fallen to $\frac{3}{4}$ eV.

For three-particle recombination the ionization energy is returned to the plasma and this tends to keep the temperature up and prevent further recombination. The drop in temperature should be almost arrested at this point. If it is assumed that the temperature remains constant with recombination just supplying the work done in expansion then we get

$$E_i \, dN = 4\pi r_2^3 n_2 kT_2(dr/r), \qquad (36)$$

$$\Delta N = \frac{4\pi r_2^3 n_2 kT_2}{E_i} \ln \frac{r}{r_2} = \frac{3NkT_2}{E_i} \ln \frac{r}{r_2}, \qquad (37)$$

where 2 refers to the state at the time recombination starts and E_i is the ionization energy for the atoms. We see from (37) that when kT_2 is much smaller than E_i, that even for relatively large r/r_2, $\Delta N/N$ is small. From this it would appear that we could neglect recombination. Of course cooling by radiation will enhance the recombination rate. Straight radiative recombination is negligible because of its small cross section. However, radiation by highly excited states produced by three-body recombination might be important. On the other hand recombination is reduced if the laser continues to operate during the expansion since the drop never goes completely transparent and even a small increase in temperature over that used above greatly reduces its rate. Finally if we attempt to catch the plasma in a magnetic field (see below) the currents set up in the plasma will probably keep the electrons sufficiently warm to prevent recombination. This problem needs to be investigated in much more detail.

If the organized motion of expansion can be randomized the plasma will again acquire a high temperature. One method of achieving this would be to allow two expanding plasma drops to collide with each other. Since the ions have most of the energy this would result in a plasma with a high ion temperature and a low electron temperature.

Many possibilities suggest themselves for trapping the plasma in a magnetic field. The plasma expansion is ultimately stopped due to the work required to push aside the magnetic field. Creating a cavity of volume V in the magnetic field of strength B requires an amount of energy W,

$$W = (B^2/8\pi)V. \qquad (38)$$

If B is in kG, V is in cm^3, and W is in J, this formula reads,

$$W = 3.98 \times 10^{-3}B^2V. \qquad (39)$$

For a field of 10 kG a volume of 2.5 cm^3 is required to absorb 1 J of energy. The maximum volume to which the plasma can expand is obtained by equating W to the plasma energy and we see for B's in the 10 kG range and W in the range of a few joules this volume is of the order of 10 cm^3.

When the plasma expands against the magnetic field surface currents are set up in it. The energy required to maintain these currents against the ohmic dissipation is supplied by the radial expansion. Thus the ordered expansion energy may again be converted to heat. Further such a plasma expanding against a magnetic field is subject to strong Rayleigh–Taylor instabilities. These may cause turbulence which effectively randomizes the organized motion. On the other hand the resultant plasma turbulence may result in plasma escaping across the field lines. Experiments with various fields configuration are required to determine how effectively the plasma can be trapped.

10. CONCLUSION

It appears that laser pulses with powers in the range of 10^{10} W lasting for a few nsec can, by vaporizing a small speck of material, produce plasmas with particle energies in the keV range. It should be possible to do many interesting experiments on such plasmas. Among the possible experiments which might be carried out are the following:

(1) The interaction of a rapidly expanding plasma drop with a magnetic field. Can the plasma be trapped or does trubulence result which takes the plasma across the magnetic field? If the plasma can be trapped one might employ this means to fill

thermonuclear machines. One might also mock up the interaction of supernovae with the interstellar magnetic field and possibly also solar flares and their interaction with the Earth's magnetic field.

(2) Investigation of collisionless shocks. If two such expanding plasma drops collide the conditions for establishing a collisionless shock are easily met. Another possibility is to embed the drop in a low-density plasma. As it expands it could act as a driver for an intense shock in this plasma.

(3) Recombination in the expanding drop might provide some information on recombination in dense plasmas. Such recombination might also provide a means for obtaining energetic neutral particles.

(4) One might use the drop to study the interaction of radiation with dense plasma as, for example, absorption, scattering, and harmonic generation.

ACKNOWLEDGMENTS

The author wishes to thank John L. Johnson for calling his attention to this problem and also for many interesting discussions of the problem. He is also indebted to Lyman Spitzer, Jr., for valuable discussions and encouragement.

This work was performed under the auspices of the U. S. Atomic Energy Commission under Contract AT(30-1)-1238.

SOME RESULTS OF THE SELF-SIMILARITY MODEL*

Heinrich Hora

Rensselaer Polytechnic Institute and

Institut für Plasmaphysik, Garching

ABSTRACT

The self-similarity model of expanding laser produced plasma is derived from the hydrodynamic equations. An analytical iterative solution taking into account the varying plasma diameter results in remarkably higher electron temperatures. Numerical iteration with the condition of total transfer of the radiation to the plasma in overdense states shows a good agreement with measurements of Engelhardt et al. at large spherical suspended targets and agreement with measurements of Sigel et al. at hydrogen foils. On the basis of detailed numerical calculations of Mulser et al., either heating by self focusing, or higher thermal conductivity than Spitzer's values are concluded due to collective effects at conditions of high density, if conditions of degeneration and Coulomb logarithms less than unity are present.

I. INTRODUCTION

The following considerations are devoted to some applications of the thermokinetic self-similarity expansion model to laser produced plasmas, as it was used by Basov and Krokhin[1], Dawson[2], Engelhardt[3], Hora[4], Haught and Polk[5], Papoular et al.[6], and Lubin et al.[7] The reasonable success of this model for interpreting many measurements[5,6,7,8] and the generalization of the theory for nonspherical plasmas[9] shows a high degree of reality reached by this model to describe the thermokinetic expansion properties of a large group of laser produced plasmas.

*Presented at the Workshop on "Laser Interaction and Related Plasma Phenomena", Rensselaer Polytechnic Institute, Hartford Graduate Center, June 9-13, 1969.

Reprinted from: *Laser Interaction and Related Plasma Phenomena,* Vol. 1, H. Schwarz and H. Hora (eds.), Plenum, New York, 1971, pp. 365–382.

The theory will not cover the nonlinear processes involved in the interaction of the laser radiation with plasmas[10], an effect which is connected very probably with the keV-ions first observed by Linlor[11]. Further problems are given regarding the number of theoretical models describing a thermokinetic shock wave mechanism produced by the interaction of laser radiation with matter[12 to 18].

As we have pointed out before[8], and as discussed quantitatively in much detail[18], the essential difference in both theories is the question of the time scale of the heating mechanism. The similarity expansion model presumes that the distribution of the energy absorbed at the front of incidence of the radiation is distributed fast enough to the whole volume of the irradiated material, if it is limited geometrically as a free sphere or a thin foil. The heating of the material is then homogeneous and the gas dynamic or thermokinetic process can be described by the self-similarity expansion model. If the laser pulse length is too short and the thermal conductivity too small, the heating will be an asymmetric shock wave process [12 to 18].

Somehow, the shock wave process and the similarity expansion can be considered as consecutive mechanisms. The self-similarity model ignores many details of the energy transfer from radiation to plasma and of the energy distribution within the material, while the shock wave process covers numerous properties of the initial processes by considering the equation of state of the primary material, the Saha equation, the thermal conductivity, different temperatures of electrons and ions, etc. As it was pointed out in detail by Mulser[15], a general numerical treatment of the case of plane geometry covers both the shock wave process and the following self-similarity expansion (see Fig. 10 of Mulser[15]). The result of a good applicability of the self-similarity process to certain experiments is no argument against the shock wave mechanism but an expression of the neglibility of the details of the starting mechanism which is determined by the shock wave process.

This paper discusses some experiments differing essentially from the usual cases[5 to 7] which can be explained by the self-similarity model. The problems in connection with the shock wave process are considered in section V, bringing up an indirect conclusion that high density deviations of the usual thermal conductivity may occur due to collective effects or others. Section II gives a rigorous derivation of the basic formulae of the self-similarity model from the hydrodynamic equations, demonstrating limitations and general properties. A specific application of the generalized formulae to an experimental case is derived in Sect. III, and Sect. IV discusses the comparison of the model with measurements of laser interaction with free falling aluminum balls[8] or thin hydrogen foils[19].

II. RELATIONS BETWEEN THE SELF-SIMILARITY MODEL AND THE HYDRODYNAMIC EQUATIONS

A detailed derivation of the self-similarity model from the hydrodynamic equations will demonstrate the various assumptions which are necessary for the validity of the model. The mathematical purity of the assumptions of Basov and Krochin[1] and Dawson [2] will be pointed out as well as mathematical details of the assumptions of Haught and Polk[5]. An analytical discussion of solutions of the hydrodynamic equations of Fabre and Vasseur[20] on the basis of the results of Jacquinot et al.[21] confirms the numerical results of Fader[22] for the self-similarity solutions.

The hydrodynamic equations for a thermokinetic model are the equation of continuity, the equation of motion, and the equation of conservation of energy to find time and space dependent solutions of the ion density n_i, the temperature T and the ion velocity v of the plasma. The equation of continuity is

$$\frac{\partial n_i}{\partial t} + \underset{\sim}{\nabla}\cdot(n_i\underset{\sim}{v}) = 0 \tag{1}$$

and the equation of motion with the pressure p, the mass m_e of the electron and m_i of the ion is

$$n_i(1 + Z\frac{m_e}{m_i})m_i\frac{d}{dt}\underset{\sim}{v} = -\underset{\sim}{\nabla}p \tag{2}$$

In the equation of conservation of energy with the ion charge Z and Boltzmann's constant K ,

$$\frac{\partial}{\partial t}\frac{n_i m_i}{2}(1 + Z\frac{m_e}{m_i})\underset{\sim}{v}^2 = -\frac{\partial}{\partial t}(1 + Z)n_i KT + w \tag{3}$$

where w is the power density of external energy input (from the laser light) or radiation loss, the change of temperature T in addition to w is only due to the change of kinetic energy, indicated by the index k:

$$\frac{\partial}{\partial t}\frac{n_i m_i}{2}(1+Z\frac{m_e}{m_i})\underset{\sim}{v}^2 = -KT\frac{\partial}{\partial t}(1+Z)n_i - (1+Z)n_i K\frac{\partial}{\partial t}T_k + w \tag{4}$$

In order to formulate the total change of the temperature

$$\frac{\partial T}{\partial t} = \frac{\partial T_k}{\partial t} + \frac{\partial T_{th}}{\partial t} \tag{5}$$

we have to take into account the change by thermal conduction (index th) which is

$$\frac{\partial T_{th}}{\partial t} = -\kappa\Delta T \tag{6}$$

using[23]

$$\kappa = 20 \left(\frac{2}{\pi}\right)^{3/2} \frac{(KT)^{5/2} K \delta_L}{m_e^{1/2} e^4 Z \ln \Lambda} \tag{7}$$

δ_L is Spitzer's correction to electron interaction against the case of the Lorentz-gas which increases from 0.225 to 1 for Z from 1 to ∞ . With respect to Eqs.(5)to(7) one gets finally the complete equation of energy conservation

$$\frac{\partial}{\partial t} \frac{n_i m_i}{2} (1 + Z \frac{m_e}{m_i}) \underset{\sim}{v}^2 = - \frac{\partial}{\partial t} (1 + Z) n_i KT - \kappa \Delta T + w \tag{8}$$

Equations (1), (2) and (8) are general hydrodynamic equations to solve n_i, v, and T in space and time at given initial conditions.

If a spherical symmetric plasma is considered with a spatial dependence only on the radial coordinate r with the radial component v_r of the velocity, we get the equation of continuity from Eq. (1)

$$\frac{\partial}{\partial t} n_i + \frac{\partial}{\partial t} n_i v_r + \frac{2n_i}{r} v_r \tag{9a}$$

Using the equation of state

$$p = n_i (1 + Z) KT \tag{9b}$$

in Eq. (2) we get the equation of motion for the spherical case

$$\frac{d}{dt} n_i m_i (1 + Z \frac{m_e}{m_i}) v_r = - \frac{\partial}{\partial r} n_i (1 + Z) KT \tag{9c}$$

and finally the equation of energy conservation from Eq.(8)

$$\frac{\partial}{\partial t} \frac{n_i m_i}{2} (1 + Z \frac{m_e}{m_i}) v_r^2 = - \frac{\partial}{\partial r} (1 + Z) n_i KT - \kappa \frac{1}{r^2} \frac{\partial}{\partial r} (r^2 \frac{\partial}{\partial t} T) + w \tag{9d}$$

We then have to solve the basic Eqs. (9) together with the initial conditions at a time $t = t_o T(r, t = t_o)$, $n_i(r, t = t_o)$ and $v_r(r, t = t_o)$, to find solutions T, N and $\underset{\sim}{v}$ as a function of r and t.

We shall now derive the formulae of the self-similarity model which were used by Dawson[2], starting from the general radially symmetric hydrodynamic Eqs.(9). We multiply the equation of motion (9c) with v_r

$$\frac{1}{2} \frac{d}{dt} n_i m_i (1 + Z \frac{m_e}{m_i}) v_r^2 = - v_r \frac{\partial}{\partial r} p \tag{10}$$

and integrate over the whole volume of the spherical plasma with the radius R of the plasma surface. This leads to

$$- \int_0^R v_r \left(\frac{\partial p}{\partial r}\right) 4\pi r^2 dr = \frac{1}{2}\frac{d}{dt}\int_0^R n_i m_i \left(1 + Z\frac{m_e}{m_i}\right) v_r^2 \, 4\pi r^2 dr \tag{11}$$

It is essential to point out that the procedure of integration in Eq. (11) and in the following equations gives a loss of information. Instead of details of $n_i(r,t)$ we can expect only average values. In this sense a still further restriction is made, when Eq. (11) is evaluated using the averaged values of the functions under the integrals.

If the pressure within the spherical plasma is assumed to be a constant value p from r = 0 till r = R − ε and decreases to p = 0 at r = R, then the left hand side of Eq. (11) can be written

$$-\lim_{\varepsilon\to 0} v_r \, 4\pi R^2 \int_{R-\varepsilon}^{R} \frac{\partial p}{\partial r} dr = - v_r \, 4\pi R^2 (P(R) - R(R - \varepsilon))$$

$$= 4\pi R^2 p v_r(R) \tag{12}$$

Constant p at constant T gives an averaged constant n_i in the inner of the sphere. Assuming additionally a linear velocity profile

$$v_r(r) = v_{ro}\frac{r}{R} \qquad (0<r<R) \tag{13}$$

which expresses the properties of a similarity expansion, then the right hand side of Eq. (11) leads to

$$\frac{1}{2}\frac{d}{dt}\int_0^R n_i m_i (1 + Z\frac{m_e}{m_i}) v_r^2 \, 4\pi r^2 dr = \frac{1}{2}\overline{M}\frac{d}{dt} v_r(R)^2 \tag{14}$$

with the abbreviation of an averaged mass $\overline{M} = \frac{3}{5} n_i m_i (1+Z\frac{m_e}{m_i})$ Combining Eqs. (14) and (12) we find

$$4\pi R^2 p \frac{dR}{dt} = \frac{1}{2}\overline{M}\frac{d}{dt} v_r(R)^2 = \frac{1}{2}\overline{M}\frac{d}{dt}\left(\frac{dR}{dt}\right)^2 \tag{15}$$

A further integration over the volume is of interest, when similar to the assumptions for Eq. (15) n_i is used constant in the averaged sense within the plasma sphere and a linear velocity profile Eq. (13) is used. In this case one gets from the left hand side of Eq. (9d)

$$\frac{\partial}{\partial t}\int_0^R \frac{n_i m_i}{2}\left(1+Z\frac{m_e}{m_i}\right) v_r^2 \, 4\pi r^2 dr = \frac{\partial}{\partial t}\frac{n_i m_i}{2}\left(1+Z\frac{m_e}{m_i}\right)\frac{v_o}{R^2} 4\pi\int_0^R r^4 dr \tag{16}$$

Using

$$\frac{3}{5} n_i m_i \left(1+Z\frac{m_e}{m_i}\right) v_o = \overline{M} v_o = \frac{2}{3} p \tag{17}$$

one finds for the integral of the left hand side of Eq. (9d)

$$\frac{\partial}{\partial t}\frac{2}{3}4\pi pR^3 \tag{18}$$

Integrating the right hand side of Eq. (9d) over the plasma volume and assuming a constant temperature $\left(\frac{\partial}{\partial r}T = 0\right)$, an equation of energy conservation with the total power W absorbed in the plasma after reduction of the radian loss is

$$4\pi pR^2 \frac{\partial}{\partial t}R = -\frac{3}{2}K\frac{dT}{dt}\left[\frac{4\pi}{3}R^3(1+Z)n_i\right] + W \tag{19}$$

Equations (9b), (15), and (19) are the same which Dawson derived from the base of phenomenological combination of gas kinetic laws. The essential points in the derivation from the hydrodynamic Eq. (9) was that the energy transfer from the radiation has to be assumed in such a way that the plasma temperature can be assumed spatially constant at each instant and that a linear velocity profile, Eq. (13), is valid. The properties of an averaged density n_i and an averaged mass M, and a steep pressure decrease at the surface, are consequences of the integration procedures; these are mathematically correct but they involve a loss of details of mostly unnecessary information.

The question is a little different than that of the preceding result, when details of the density profile $n_i(r)$ at varying times are to be evaluated. To reach a certain statement on the actual density profile, Haught and Polk[5] started from a linear decrease of the density $n_i(r)$ from the center of the plasma to the surface. Simultaneously, a linear velocity profile, as given by Eq. (13) should be conserved. We show that both linear profiles are not conservable in time. Substituting $n_i = n_{io}(1 - r/R)$ with the normalization $n_{io} \propto R(\varepsilon)^{-3}$ at spatial invariant T in the equation of motion (9c), we find

$$\frac{dv_r}{dt} = \frac{(1+Z)KT}{\left(1+Z\frac{m_e}{m_i}\right)m_i R(1-r/R)} \tag{20}$$

a non-linearly varying acceleration profile, even with a pole at r=0. If, at any time t_o, the velocity profile should be linear and the density profile at t_o and further on linear, the velocity profile must change its linearity at $t > t_o$ due to Eq. (20).

One way that a linear velocity profile (self-similarity motion) can be conserved at the condition of spatial independent temperatures, can be verified if the density profile is Gaussian [20,21]. Such a profile

$$n_i(r,t) = \frac{n_o}{\pi^{3/2}\ell^3} \exp(-r^2/\ell^2) \tag{21}$$

where the length ℓ is only a function of t, produces an acceleration in the equation of motion (9c)

$$\frac{\partial^2 r}{\partial t^2} = - \frac{KT(1+Z)}{m_i(1+Z\frac{m_e}{m_i})n_i} \frac{\partial}{\partial r} n_i = \frac{KT(1+Z)}{m_i(1+Z\frac{m_e}{m_i})} \frac{2r}{\ell^2} \tag{22}$$

If the starting velocity $v_r(r,t=0)$ is zero or linear with r, the acceleration[22] conserves this property. The bell-like density profile expands similarly. The limitation of the model with the Gauss-profile is given by the fact that the real plasma has a definite surface, while the Gauss-profile distributes the plasma to any distance. It is interesting that the Gauss-like density profile was observed by interferometry of plasma produced by lasers from thin foils[24], that a direct numerical calculation based on the hydrodynamic equations of a starting linear density profile assimilates after a certain time a Gaussian profile[22] and that the hydrodynamic numerical calculation for thin rectangular density profiles (foils of hydrogen) will be developed also into a kind of a Gaussian profile[15].

III. SELF-SIMILARITY MODEL FOR OVERDENSE PLASMA WITH VARYING RADIUS

An application of the self-similarity model, Eqs. (9b), (15), and (19), to interpret some experimental results[8,25] made it necessary to develop some generalizations with respect to the energy transfer from the radiation to the plasma which occurs even at times when the plasma interior is overdense (light frequency ω is less than the plasma frequency ω_p). A further step was to include the varying cross section of a spherical plasma during irradiation within constant laser focus due to the heating and expansion. It was reported[5] that this fact was included into a numerical application of the self-similarity model. Here we can give some analytical expressions, demonstrating immediately the physical properties. A numerical application is based on an iteration procedure with a definite numerical stabilization.

As evaluated by Dawson[2], the solution of Eqs. (9b), (15), and (19) for an energy input $W_l = W$ starting at a time t_o and remaining constant gives the solutions derived by Basov and Krokhin[1], for the time dependence of the laser heated sphere of the radius R and temperature T

$$R^2 = [R_o^2 + \frac{10}{9} G] \tag{23}$$

$$KT = \frac{W_1 t}{3N_t(1+Z)} \frac{2R_o^2 + \frac{5}{9} G}{R_o^2 + \frac{10}{9} G} \tag{24}$$

where R_o is the initial radius of the target before t_o and

$$G = \frac{W_1 t^3}{N_i m_i} \tag{25}$$

The total number of ions N_i was used

$$N_i = n_i(t_o) \frac{4\pi}{3} R_o^3 . \tag{26}$$

The model used[1,2] implies a Z constant in time which is quite reasonable, especially in the following discussion of aluminum targets with respect to the conduction electrons in the cold material.

A modification in the time variation of the input power W is made taking into account that the plasma sphere expands and changes its cross section of energy transfer. We assume a power density in the focus which is spatial constant and has a time dependence as a step function and constant for $t > t_o$. We use the further assumption that all energy incident within the cross section of the plasma is transferred to the plasma as long as the plasma is overdense $(\omega_p > \omega)$ with the ion density n_i

$$\omega_p^2 = \frac{4\pi e^2}{m_e} Z n_i \tag{27}$$

This assumption is based on the fact that plasmas produced from solids or at overdense gas breakdown by laser pulses larger than 2 nsec and with intensities higher than 10^{10} W/cm^2 have a negligible reflectivity [8,19,26,27]. The details of the energy transfer from the skin depth to the whole plasma are assumed to be fast enough, as it will be discussed later. The power input then has the formulation

$$W(t) = \frac{R(t)^2}{R_o^2} W_1 \tag{28}$$

Where W_1 is the constant power input of the starting cross sections. Using this time dependent W(t), we cannot solve the Eqs. (9b), (15) and (19) immediately. We use an iteration by calculating, as done in Eqs. (23) and (24), a first iteration $R_1(t)$, $T_1(t)$ using $W = W_1$ and a second iteration $R_2(t)$, $T_2(t)$ using $W = R_1(t)^2 W_1/R_o^2$ etc. The second iteration is then (avoiding the index 2)

$$R^2 = R_o^2 + \frac{10}{9} G(1 + \frac{1}{18R_o^2} G) \tag{29}$$

$$KT = \frac{W_1}{3N_i(1+Z)} \frac{2R_o^2 + 5G/3 + 80G^2/(81R_o^2) - 200G^3/(1458R_o^4)}{R_o^2 + (10G/9)(1 + G/18R_o^2)} \quad (30)$$

The difference of this solution as compared with the solutions Eqs. (23) and (24) is obvious, when we evaluate the time t_{TP} at which the plasma becomes transparent ($\omega_p = \omega$). In the case of ruby laser radiation ($\omega = 2.7\times10^{15}$ Hz) and an initial density $n_o = 6\times10^{22}\text{cm}^{-3}$, as is the case for solid hydrogen or solid aluminum, we find from Eq. (23) with Eq. (25) for such an R, for which $(R/R_o)^3 = n_o/n_{co}$ with the cut-off density, given by Eq. (27) at $n_i = n_{co}$ for $\omega_p = \omega$

$$t_{TP}^{(1)} = (7.78\ r_o^2\ N_i m_i/W_1)^{1/3} \quad (31)$$

For the solution with varying cross section, from Eq. (29) there follows in the same way

$$t_{TP} = (7.15\ r_o^2\ N_i m_i/W_1)^{1/3} \quad (32)$$

The higher energy input in this case makes the plasma transparent a little earlier.

A higher difference of both cases can be seen from the temperature reached by the plasma. The maximum temperature of the plasma is where a total energy input $W_1 t$ is so short that the expansion of the plasma is negligible[2,4]. In this case the temperature is[2]

$$T = T_{max} = 2/[3KN_i(1+Z)] \quad (33)$$

We consider the temperature T at the time $t = t_{TP}$ and find in the case of Eq. (24)

$$T^{(1)} = 0.32T_{max} \quad (34)$$

and in the case of varying cross section

$$T = 0.583T_{max} \quad (35)$$

The temperature is increased nearly by a factor of two although $t_{TP} < t_{TP}^{(1)}$.

A numerical program to solve R(t) and T(t) for a more general input power W in Eqs. (9b), (15) and (19) was based on the following assumptions:

Evaporation, ionization, and recombination were neglected. The focal region was approximated by a boxlike intensity profile constant in space. The intensity had a time dependence of variable forms (rectangular, triangular, symmetric or with steepened rise time), tailored[7] with triangular initial pulse and triangular main pulse, expressed by $W_1(t)$ given by the laser power within the area of the cross section of the plasma at $t=t_o$. The geometry of the cross section of the plasma with the laser radiation is given by a factor

$$f = \begin{cases} R^2/R_o^2 & \text{for} \quad R < R_F \\ R_F^2/R_o^2 & \text{for} \quad R \geq R_F \end{cases} \tag{36}$$

with the focus diameter R_F. The energy transfer due to the absorption given by the optical absorption constant K is approximated with respect to the spherical geometry by

$$g = \begin{cases} 1 & \text{for} \quad n_i \geq n_{co} \\ 1 - \dfrac{1 - (1 + 2KR)\exp(-2KR)}{2K^2R^2} & \text{for} \quad n_i < n_{co} \end{cases} \tag{37}$$

The power input into the plasma at negligible radiation losses is then

$$W(t) = W_1(t)\ g(t)\ f(t) \tag{38}$$

The time dependent functions R(t) and T(t) are solved from the following system of equations where the total mass of the plasma is $M = 4\pi R_o^3 n_o m_i/3$

$$R^2(t) = R_o^2 + \frac{20}{3M}\int_o^t d\tau \int_o^{\tau} d\tau' \int_o^{\tau'} d\tau'' W[R(\tau''), T(\tau''), \tag{39}$$

and

$$KT(t) = \frac{M}{5N_i(1+Z)}\left\{-\left(\frac{dR}{dt}\right)^2 + \frac{10}{3M}\int_o^t d\tau W[R(\tau), T(\tau)]\right\} \tag{40}$$

The numerical program used the absorption constant K from the collision produced dispersion relation of plasmas[28]. The solution was verified by iteration with the first step $R_1(t)$ T_1 (t) as described before, put into $W(R_1(t), T_1(t))$ and Eqs. (39) and (40) solved to find the second iteration $R_2(t)$, $T_2(t)$ and with these values a $W(R_2(t), T_2(t)$ was used to solve the third iteration $R_3(t)$, $T_3(t)$ etc. The iteration was performed until the R and T-values differed by less than 10^{-4} from the values of the former iteration. In the following examples the necessary number of iterations was six or less. A check of the numerical stability of this procedure was given by the result that the temperature T(t) decreased to zero at large t compared with the laser pulse length t_L, defining $W_1=0$ at $t> t_L$ in Eq. (38).

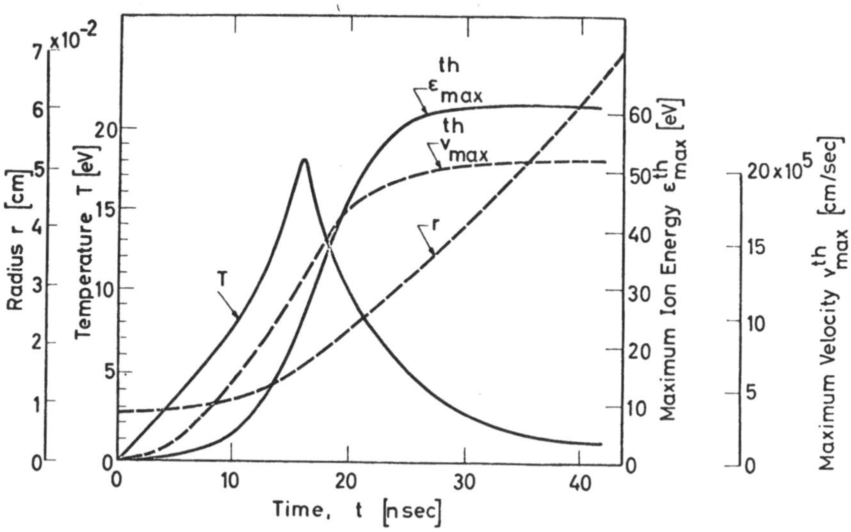

Fig. 1

Numerical calculation by iteration of Eqs. (39) and (40) to evaluate the time dependence of the plasma radius R, temperature T, maximum velocity v_{max}^{th} and ion energy ε_{max}^{th} of the plasma surface for an aluminum ball of 80 μ radius irradiated by a laser pulse of 4.3 Joule energy and a rectanglar pulse length of 16 nsec.

IV. COMPARISON WITH MEASUREMENTS

The results of the self-similarity model of the formulation described here will be compared with experiments. A first comparison is performed with experiments at Westinghouse[8] where single aluminum balls of 50 to 150 μ radius were irradiated by focused laser pulses of 2 to 5 Joule energy and 15 to 35 sec pulse length. A numerical solution of the time dependence of the ball radius R(t) and the temperature T(t) for a case measured[8] is shown in Fig. 1. The velocity of the plasma surface $v(t)_{max}^{th} = dR/dt$ is evaluated and also the maximum ion energy $\varepsilon_{max}^{th}(t)=m_i v^2/2$. Further, the amount of the total absorbed energy was evaluated, taking into account the geometry of the growing plasma within the laser focus and the absorption constant. The results were that the calculated absorption of energy was equal to the measured values within 15% and the dependence of the amount of the absorbed energy on the ball radius and the pulse length showed the same systematic variation as seen from the measurements.

The final maximum ion energy of the expanding plasma was measured from side-on framing camera pictures[4].

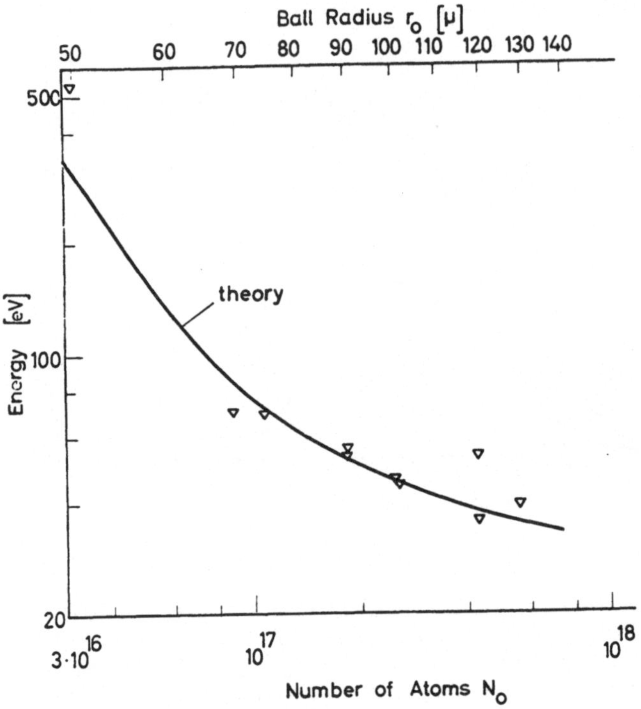

Fig. 2

Measured maximum ion energies of plasmas produced from aluminum balls of varying ball radius at irradiation of laser pulses of about 70 MW power and 30 nsec pulse length (▽) compared with theoretical values based on the self-similarity model (curve).

These values fit very well a theoretical slope with a triangular pulse shape, the parameters of which were determined by the measured pulse shape (Fig. 2). The index "th" in ε_{max}^{th} expresses the inner thermal part of the created plasma that was studied, while an outer part of plasma[8] had properties of a nonlinear surface mechanism, explaining Linlor's keV-ions[11] on the basis of a theory of a pondermotive interaction of the laser radiation with the plasma[10] if self-focusing produces power densities of 10^{14} W/cm^2. This is easily possible[29] in the case discussed.

The result of Eqs. (34) and (35) indicates a higher temperature of the plasma if the self-similarity model is applied in the way described. This increase of the temperature was observed by Thomson scattering experiments[8] in principle - besides the possibility that recombination mechanisms increase the electron temperature.

A further comparison of the self-similarity model with experiments is possible at measurements of the time t_{TP} at which a plasma produced from a solid foil of polythene[30] or solid hydrogen[19] becomes transparent. With respect to the results of the nonspherical case of the self-similarity model[9], the expansion properties along one axis of an elliptic plasma are independent of the properties of the other axes. Therefore, we can begin the procedure for solid hydrogen and ruby laser radiation from Eq. (31) to evaluate the transparency time where r_o is the thickness of the foil and

$$W_1 = W_o \, r_o^2/r_F^2 = P\pi r_o^2 \tag{41}$$

W_o is the input laser power at the front of the layer, where the laser beam has a radius r_F and P is the laser intensity. Using the density = 0.1 g/cm3 of solid hydrogen we find from Eq. (31)

$$t_{TP} = (7.78 \, \frac{4}{3} \, r_o^3 \, \rho/P)^{1/3} \tag{42}$$

In experiment[19] with W_o = 200 MW and $P = 2.4\text{x}10^{12}$ W/cm^2, Eq. (42) results in

$$t_{TP} = 3.5 \times 10^{-7} \, r_o \tag{43}$$

with a foil thickness r_o in cm, and for W_o = 500 MW

$$t_{TP} = 4.75 \times 10^{-7} \, r_o \; . \tag{44}$$

To compare these values with the measured transparency time[19] t_D we have to add to t_{TP} the time between the beginning of the laser pulse and the creation of the plasma, which is of about 5 nsec in agreement with analogous experiments. Eqs. (43), and (44) fit very well the measurements as shown in Fig. 3.

V. CONCLUSIONS

The self-similarity model is an analytically treatable method for describing symmetric expansion of not too strongly extended laser produced plasmas[1 to 9]. It implies a good enough thermal conduction to reach a uniform distribution of the energy which is transferred in many cases only from the side of illumination of the plasma where the energy is absorbed only within the inhomogeneous skin depth.

The validity of this model was confirmed by many measurements with targets of LiH of about 10 μ radius suspended free in vacuum in the focal center of a laser beam [5 to 7]. A special proof of the self-similarity

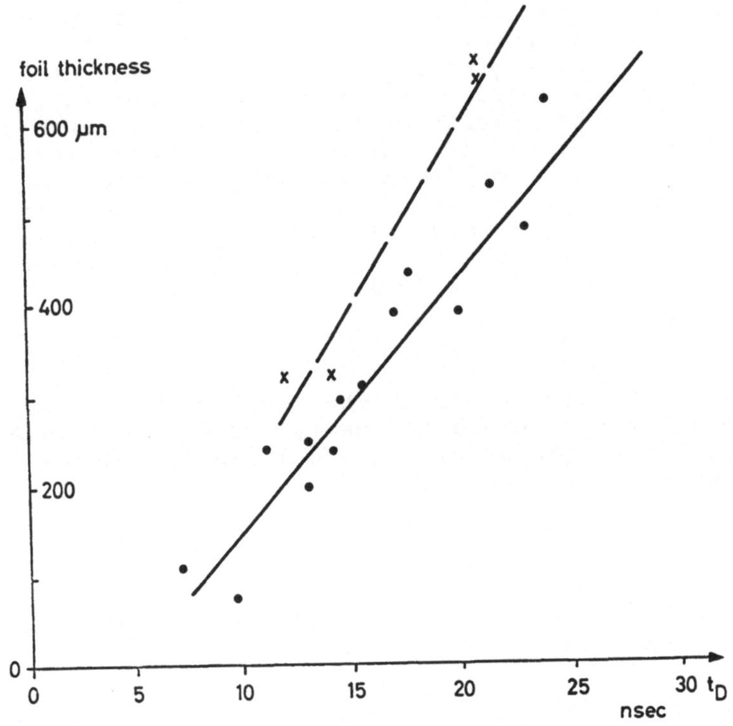

Fig. 3

Measured (o,x) values of the delay time of transparency t_P of solid hydrogen foils of given thickness compared with the calculated (lines) transparency time t_{TP} of the self-similarity model, when $t_P - t_{TP}$ is 5 nsec. The laser intensities were 200 MW (———) and 500 MW(– – – –).

expansion of the plasma was the measured spherical symmetry of the plasma[5] and the maximum energy of ions of some 100 eV in agreement with the thermokinetic processes. Larger targets, i. e. of 20 μ radius showed no spherical symmetry, and charge collecting measurements of the expanding plasma differentiated two groups of plasma[31]. Free targets of aluminum balls of 50 to 150 also showed two parts, one outer part with an asymmetrical nonlinear behaviour, and an inner spherical part. If the self-similarity model is applied in the way demonstrated here, the agreement with the measurements of the inner part is satisfactory with respect to

1) the measured amount of the total absorbed energy, if the energy absorption in the overdense state is assumed totally within the varying cross section of the laser beam and the plasma;

2) the final maximum ion energy of the thermal core;

3) the increased electron temperature according to Thomson scattering experiments.

The assumption of total energy absorption in the over-dense case allows a very good fit of calculated transparency times with measurements[19] of plasmas produced from solid hydrogen foils up to thicknesses of 600 μ.

A numerical calculation of the interaction process on the basis of the hydrodynamic equation with forces due to the thermodynamic states after thermalization of the radiation includes in a general way the details of the initial processes (shock waves, Saha equation, etc.) and the following expansion mechanism[15]. This following process is comparable with the self-similarity model. The agreement of the transparency times of the self-similarity model with measurements allow the conclusion that the initial processes are fast enough for an application of the self-similarity model. The details calculated for the initial process on the basis of the usual constants of optical absorption and thermal conductivity result in larger times of thermal exchange and larger transparency times[15]. So the indirect conclusion can be drawn that the constants of absorption and thermal conductivity based on the macroscopic theory[23] are not valid in plasmas of the cases considered. The problem is indeed obvious, because densities higher than 10^{23} cm^{-3} at temperatures of a few eV have been calculated, while it is known from the calculation of absorption constant[28] that fully ionized plasmas of 10^{23} cm^{-3} electron density degenerate below temperatures of 100 eV and have larger impact parameters than the Debye length (Coulomb logarithm less than unity) below 10 eV. If in this way an increased thermal conductivity could be concluded, it would be of interest to find a relation to the concluded decrease of the electric conductivity due to collisionless damping and instabilities in laser produced plasmas[32]. Another mechanism of a fast heating of the whole plasma in the hydrogen foils would be a self focusing process[29] which develops pipes of high energy transfer. A further point of view in this discussion was a consideration of the three-dimensional properties in the hydrogen layer[19].

ACKNOWLEDGEMENT

Parts of the presented results were derived during the author's work at the Westinghouse Research Center. Stimulating discussions with A. G. Engelhardt, A. V. Phelps, T. V. George, and J. L. Pack are gratefully acknowledged.

This work was performed as part of the agreement to conduct joint research in the field of plasma physics between the Institute für Plasmaphysik, Munich-Garching, and Euratom.

REFERENCES

1. N. G. Basov, O. N. Krokhin — Quantum Electronis, Proc. IIIrd Int. Congress, Paris (Feb.1963) Ed. by P. Grivet and N. Bloembergen, Dunod, Paris 1964, Vol. I, p. 1373.
2. J. M. Dawson — Phys. Fluids 7, 981 (1964).
3. A. G. Engelhardt — Bull. Amer.Phys. Soc. 9, 305 (1964).
4. H. Hora — Institut f. Plasmaphysik, Garching, Report IPP 6/23 (1964).
5. A. F. Haught, D. H. Polk — Phys. Fluids 9, 2047 (1966).
6. P. E. Faugeras, M. Mattioli, R. Papoular — AIAA Fluid and Plasma Dynamics Conf. Los Angeles, Calif. June 1968.
7. M. J. Lubin, H. S. Dunn, W. Friedman — Plasma Physic and Cont. Thermon. Fusion, Conf. Proc. Novosibirsk (Aug. 1968), Vienna 1969 p. 945.
8. A. G. Engelhardt, T. V. George, H. Hora, J. L. Pack — Phys. Fluids 13, No. 1 (1970).

   T. V. George, A. G. Engelhardt, J. L. Pack, H. Hora, G. Cox — Bull. Amer. Phys. Soc. 13, 1553 (1968).
9. J. Dawson, P. Kaw, B. Green — Phys. Fluids 12, 875 (1969).
10. H. Hora, D. Pfirsch, A. Schlüter — Z. Naturforschung 22A, 278 (1967).
11. W. I. Linlor — Appl. Phys. Letters 3, 210 (1963) Phys. Rev. Letters 12, 383 (1964)

    H. Opower, E. Burlefinger — Phys. Letters 16, 37 (1965)

    H. Opower, W. Kaiser, H. Puell, W. Heinicke — Z. Naturforschung 22A, 1392 (1967).

12.	Yu. V. Afanasyev O. N. Krokhin G. V. Sklizkov	IEEE J. Quantum Electronics QE2, 483 (1966)
13.	A. Caruso B. Bertotti P. Giupponi	Nuovo Cim. 45B, 176 (1966)
	A. Caruso R. Gratton	Plasma Physics 10, 867 (1968)
	A. Caruso A. deAngelis B. Gratton S. Martellucci	Phys. Letters 29A, 316 (1969)
14.	R. G. Rehm	Bull. Amer. Phys . Soc. 13,879 (1968)
15.	P. Mulser S. Witkowski	Phys. Letters 28A, 703 (1969)
	P. Mulser	Thesis TH Munich (1969) Institut f. Plasmaphysik, Garching, Report IPP 3/95 (1969)
16.	J. W. Shearer W. S. Barnes	Numerical Calculations of Plasma Heating by Means of Subnanosecond Laser Pulses; p.307 of these proceedings
17.	C. Fauquignon F. Floux	Phys. Fluids (to be published)
	F. Floux D. Cognard G. deGiovanni	Transactions, 9th International Conf. on Phenomena in Ionized Gases, Bukarest, Aug. 1969.
	J. L. Bobin Y. A. Durand Ph.P. Langer G. Tonon	J. Appl. Phys. 39, 4184 (1968)
18.	A. Caruso	Shock Wave Process, p.289 of these proceedings
19.	R. Sigel	Thesis TH Munich (1969) (to be published)
	R. Sigel K. Büchl P. Mulser S. Witkowski	Phys. Letters 26A, 498 (1968)
20.	E. Fabre P. Vasseur	J.de Physique 29, 123 (1968)

21. J. Jacquinot C. Leloup F. Waelbroeck	Rapp. C.E.A., No. 12.2617 (1964)
22. W. J. Fader	Phys. Fluids 11, 2200 (1968)
M. Mattioli	Association Euratom C.E.A. - Fontenay-Report EUR-CEA-FC-523 (1969)
23. L. Spitzer, Jr.	Physics of fully Ionized Gases Interscience New York 1956
24. J. Bruneteau E. Fabre H. Lamain P. Vasseur	Phys. Letters 22A, 37 (1967)
25. A. G. Engelhardt H. Hora T. V. George J. L. Pack	Bull. Amer. Phys. Soc. 13, 887 (1968)
26. R. W. Minck W. G. Rado	J. Appl. Phys. 37, 355 (1966)
27. N. G. Basov B. A. Boiko O. N. Krokhin O. G. Semenov G. V. Sklizkov	ZhTF 38, 1973 (1968)
28. H. Hora	Institut für Plasmaphysik, Garching, Report IPP 6/27 (1964)
H. Hora H. Müller	Institut für Plasmaphysik, Garching, Report IPP 3/81; (6/71 (1968))
29. H. Hora	Z. Physik 226, 156 (1969)
30. W. G. Griffin J. Schlüter	Phys. Letters 26A, 241 (1968)
31. T. Yamanaka N. Tsuchimori T. Sasaki Ch. Yamanaka	Technol. Report Osaka University 18, 155 (1968)
32. F. Schwirzke	Proc. IIIrd. Europ. Conf. Contr. Fusion, Utrecht, June 1969, Wolters-Publ., Groningen 1969
F. Schwirzke R. G. Tuckfield	Phys. Rev. Letters 22, 1284 (1969)
D. K. Bhadra	Phys. Fluids 11, 234 (1968)

Hydrogen Plasma Production by Giant Pulse Lasers *

PETER MULSER

Institut für Plasmaphysik, Garching bei München

The problem of interaction of an intense laser beam with solid hydrogen is theoretically investigated in a one-dimensional plane geometry. The time dependent distributions of density, temperature and velocity of the produced plasma as well as those of the solid are found by numerical solution of the hydrodynamic equations for various laser powers. The maximum temperature can be approximately expressed in closed form as a function of the laser intensity and time. The calculations allow for the influence of thermal conduction and viscosity. At laser intensities above 1.8×10^{11} W/cm^2 the plasma frequency rises above that of the laser in the transition sheath between the hot plasma and the cold solid. The problem of absorption and reflection of laser radiation in this region is investigated.

I. Introduction

Nowadays giant pulse lasers reach intensities of 10^{12} W/cm^2 and more if focused with lenses of suitable focal length to an area of 10^{-4} cm^2. Emission thereby lasts between about 2 and 50 nsec. The interaction of such radiation fields with matter greatly differs from the usual picture of low-intensity electromagnetic waves passing through a medium. Numerous experiments since 1963 have demonstrated that a sufficiently intense laser will make transparent dielectrics opaque and cause a hot, dense plasma cloud to form on the surface.

Near the ruby laser frequency ($\omega = 2.73 \times 10^{15}$ sec^{-1}) many insulators such as hydrogen appear completely transparent; with normal light sources their coefficients of absorption are almost zero. If, however, there are some free electrons in the insulator, these in the intense radiation field of a laser can gain so much energy that they quickly increase in number as a result of collisional ionization, and strong absorption occurs. CARUSO et al. [1], for example, have calculated that at a radiation intensity of 10^{11} W/cm^2 a 100 μ thick hydrogen foil becomes opaque in less than 10^{-10} sec if there are 10^6 free electrons per cm^3 present when irradiation begins, i. e. a vanishingly small number compared with the number of atoms, which for solid hydrogen is 4.5×10^{22} cm^{-3}. No matter where the first free electrons originate – whether from multiphoton absorption, from impurities in which they were loosely bound or from external charging – it was found experimentally time and again that with sufficient laser intensity a highly absorbent layer of free electrons forms in a very short time. Any study of subsequent processes will therefore start from a certain initial ionization.

In the field of fusion research one is mainly interested in the complete evaporation of solid or liquid drops of low atomic weight in a focused laser beam. The production, heating, and expansion of such a plasma was therefore studied theoretically in a spherical model on the assumption that the matter in the focus is fully ionized from the outset, and that the energy input is uniformly absorbed in the drop as long as its diameter does not exceed that of the focus as a result of expansion [2–4].

* Auszug aus der von der Fakultät für Maschinenwesen und Elektrotechnik der Technischen Hochschule München zur Erlangung des akademischen Grades eines Doktors der Naturwissenschaften genehmigten Dissertation über „Erzeugung von Wasserstoffplasma durch Riesenimpulslaser" des Dipl.-Phys. PETER MULSER. Tag der Promotion 31. 7. 1969.

Sonderdruckanforderungen an Dr. P. MULSER, Institut für Plasmaphysik, Experimentelle Plasmaphysik 3, *D-8046 Garching* bei München.

[1] A. CARUSO, B. BERTOTTI, and P. GIUPPONI, Nuovo Cim. 45 B, 176 [1966].
[2] J. M. DAWSON, Phys. Fluids 7, 981 [1964].
[3] A. F. HAUGHT and D. H. POLK, Phys. Fluids 9, 2047 [1966].
[4] W. J. FADER, Phys. Fluids 11, 2200 [1968].

Reprinted from: *Zeitschrift für Naturforschung*, **25A**, 282–295 (1970).

The assumption of uniform energy absorption greatly facilitates calculation of heating and expansion and affords useful information on the properties of the plasma after a lengthy period of time. It does not, however, provide a satisfactory answer to the physically interesting question in what way and how quickly the solid is converted to plasma at high laser intensities. A major advance in this direction was made by AFANASYEV et al [5], who investigated plasma production in plane geometry, and then by CARUSO et al. [6] These authors consider the case of a time constant light flux incident on the surface of a solid occupying a half-space and then make a dimensional analysis of the dynamic process. This allows the plasma parameters averaged in space to be determined in closed form as functions of the laser intensity, time, and absorption coefficient.

In order, however, to determine the distributions of the density, temperature, and velocity of the plasma produced, the local conservation laws of hydrodynamics have to be solved simultaneously with the equations of optics. Only thus is it possible to treat such questions as the influence of thermal conduction and viscosity, propagation and reflection of light in the high density region, recombination in the plasma, or the problem of ponderomotive forces [7].

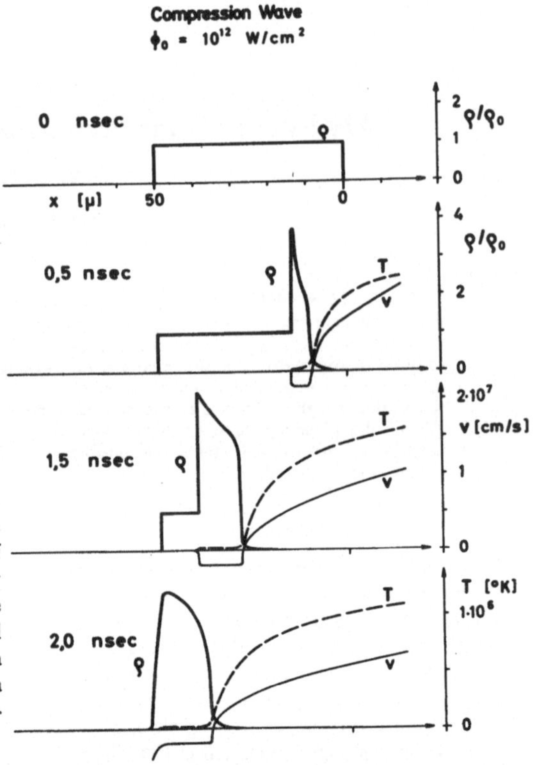

Fig. 1. Shock wave in solid hydrogen at various times for $\Phi_0 = 10^{12}$ W/cm². The shock velocity is 2.7×10^6 cm/sec. Undistrubed foil above. Irradiated from the right.

II. The Model

Basic physical information is ensured by making all calculations one-dimensional and using laser pulses of constant intensity. For this purpose we consider an infinitely extended foil of thickness D and initial density ϱ_0 on the one surface of which a laser of constant power density Φ_0 is incident starting at the time $t = 0$ (Fig. 1). A small number of free electrons should be present in the foil so that a noticeable fraction of the light can be absorbed from the outset and is converted into heat. This causes a change in the number of free electrons and hence in the light absorption; in addition a strong pressure gradient forms at the surface and matter escapes into the vacuum. Part of the laser light can now penetrate more deeply into the target, to the extent to which the expanding plasma becomes rarefied, and heats more matter.

This gasdynamic process can be adequately described in a one-fluid model taking thermal conductivity and viscosity into account. For this purpose all parameters of interest such as the density ϱ, velocity v, pressure p, energy per unit mass ε, and laser intensity Φ are expressed as functions of the Lagrange parameter a and the time t, where a is chosen such that x is identical with a at the time $t = 0$, i.e. a is the initial position of the volume element under consideration. The laws of conservation of mass, monentum, and energy (see, for example [8], Vol. I, p. 4 et seq. and p. 71)

[5] YU. V. AFANASYEV, O. N. KROKHIN, and G. V. SKLIZKOV, IEEE J. Quant. El. QE—2, 483 [1966].

[6] A. CARUSO and R. GRATTON, Plasma Phys. 10, 867 [1968].

[7] H. HORA, Phys. Fluids 12, 182 [1969].

[8] YA. B. ZELDOVICH and YU. P. RAIZER, Physics of Shock Waves and High-Temperature Hydrodynamic Phenomena, Acad. Press, New York 1966.

$$\varrho_0 = \varrho(\partial x/\partial a), \quad (1)$$

$$\varrho_0 \frac{\partial v}{\partial t} = \frac{\partial}{\partial a}\left\{-p + \frac{4}{3}\frac{\varrho}{\varrho_0}\mu\frac{\partial v}{\partial a}\right\}, \quad (2)$$

$$\varrho_0 \frac{\partial \varepsilon}{\partial t} = \frac{\partial \Phi}{\partial a} + \frac{\partial}{\partial a}\left(\frac{\varrho}{\varrho_0}\varkappa\frac{\partial T}{\partial a}\right) + \left\{-p + \frac{4}{3}\frac{\varrho}{\varrho_0}\mu\frac{\partial v}{\partial a}\right\}\frac{\partial v}{\partial a} \quad (3)$$

together with the relation for the velocity

$$v(a,t) = \partial x(a,t)/\partial t \quad (4)$$

then afford a complete description of the processes involved. The coefficients of viscosity and thermal conductivity are denoted by μ and $\varkappa$ in Eqs. (2) and (3). The three terms at the r.h.s. of energy equation represent consecutively energy input due to irradiation, thermal conduction, and mechanical work. In the energy range under consideration the radiation losses of the plasma are small and are not taken into account, cf. [2].

All of the calculations here apply to hydrogen. The number of free electrons is expressed in terms of the degree of ionization, which is defined as the ratio $\eta = n_e/(n_n + n_i) = n_e/n_H$, where n_n is the neutral particle density and n_H the total number of hydrogen nuclei per unit volume. Quasi-neutrality is thereby assumed, i. e. $n_e \approx n_i$. n_H, T and η are related by the Sahy equation (see Section V):

$$\frac{\eta}{1-\eta} = \left(\frac{2\pi m_e k}{h^2}\right)^{3/2} \frac{T^{3/2}\exp\{-U_i/kT\}}{n_H}. \quad (5)$$

Here the partition-function ratio for ions and hydrogen atoms $\Sigma^+/\Sigma^0 = 1/2$ was used. As an equation of state we take the ideal gas law for the plasma as well as for the high density region:

$$p = (1+\eta)\, n_H k T. \quad (6)$$

The internal energy per unit mass is then the sum of the thermal energy and ionization energy:

$$\varepsilon = \frac{1}{m_H}\left\{\frac{3}{2}(1+\eta)kT + U_i\,\eta\right\} \quad (7)$$

(m_H = mass of hydrogen atom, U_i = ionization energy per atom). In order to satisfy the energy balance, the ionization energy of the atomic hydrogen is increased by half the dissociation energy of H_2, i. e. $U_i = 13.6 + 2.24 = 15.84$ eV. The local light intensity $\Phi(x,t)$ should be related to the laser intensity Φ_0 by the absorption law

$$\Phi(x,t) = \Phi_0 \exp\{-\int_x^\infty \alpha(x,t)\,dx\}; \quad (8)$$

$\alpha(x,t)$ denotes the local absorption coefficient. (For α and the validity of Eq. (8) see Appendix I and Section IV.) According to [9], p. 462, the viscosity μ is given by the following expression:

$$\mu \approx \mu_i = 1.365 \frac{3\, m_i^{1/2} k^{5/2}}{4(2\pi)^{1/2} e^4} \frac{T^{5/2}}{\ln \Lambda} \text{ [cgs grad]}.$$

The coefficient of thermal conductivity according to [10] is

$$\varkappa = 20\left(\frac{2}{\pi}\right)^{3/2} \frac{k^{7/2}\,\varepsilon\,\delta_T}{m_e^{1/2} e^4} \frac{T^{5/2}}{\ln \Lambda} \quad \text{[cgs grad]}$$

with $\varepsilon = 0.4189$ and $\delta_T = 0.2252$. $\ln \Lambda$ is calculated in all transport parameters according to the following formulae [11]:

$$\Lambda = \Lambda_K = \frac{3\,k^{3/2}T^{3/2}}{2\,e^3 \pi^{1/2} n_e^{1/2}} \quad \text{for } T \leqq 4.2\times 10^5\,°\text{K},$$

$$\Lambda = \left(\frac{4.2\times 10^5}{T}\right)^{1/2} \Lambda_K \quad \text{for } T \geqq 4.2\times 10^5\,°\text{K}.$$

All calculations are made for ruby laser step pulses, i. e. the laser starts with the intensity Φ_0 at the time $t=0$, and this remains constant for the entire heating period. The density of solid hydrogen ϱ_0 is chosen as initial density, and for $t=0$ the temperature T and flow velocity v are everywhere set equal to zero; the thickness of the foil is assumed to be $D = 50\,\mu$. An initial degree of ionization $\eta = 7\times 10^{-3}$ was chosen. The numerical calculations were performed for intensities $\Phi_0 = 10^{11}$, 10^{12} and 10^{13} W/cm^2. The numerical integration method is outlined in Appendix II.

III. Numerical Results

The shock wave: As soon as the foil is exposed to laser radiation some of the light is absorbed and converted into internal energy, η increasing in accordance with the Saha equation (5). As a result, $50\,\mu$ foils quickly become opaque at laser intensities $\Phi_0 = 10^{11}$ W/cm^2, because with the rising number of free electrons the absorbing region reduces to a small layer at the surface and hot plasma streams off into space. At a power density $\Phi_0 = 10^{12}$ W/cm^2

[9] I. P. Shkarofsky, T. W. Jonston, and M. P. Bachynski, The Particle Kinetics, Addison-Wesley Publ. Co., Reading Mass. 1966.

[10] L. Spitzer and R. Härm, Phys. Rev. **89**, 977 [1953].

[11] L. Spitzer, Physics of Fully Ionized Gases, Intersc. Publishers, Inc., New York 1956.

a pressure of over 300 kbar builds up in less than 0.2 nsec in the region of strongest light absorption, and a shock wave with a velocity of 2.7×10^6 cm/sec is formed in the foil (Fig. 1; the laser is directed from the right). Behind the shock front the hydrogen, now compressed in a ratio of about 1 : 4, has attained the temperature 1.5×10^4 °K. After 2 nsec the compression wave reaches the free back surface of the 50 μ foil, a rarefraction wave runs back to the right, and the foil as a whole begins to move. The shock velocity depends, of course, on the laser power. At $\Phi_0 = 10^{11}$ W/cm² it is 1.1×10^6 cm/sec and at $\Phi_0 = 10^{13}$ W/cm it is 6.7×10^6 cm/sec. The compression ratio in all cases is about 4, as it should be for strong shocks in atomic gases ($\gamma = c_p/c_v = 5/3$). The shock wave in Fig. 1 has a somewhat different profile to that in [12] because the Saha equation was used here with the result that γ is no longer constant everywhere.

The hot plasma: On the laser side of the compressed foil the density drops sharply as a result of the strong temperature rise caused by heating. Fig. 2 shows the spatial distribution of plasma density, temperature, laser intensity, and velocity for $\Phi_0 = 10^{12}$ W/cm² at the times 2, 7, 12, 20, and 30 nsec. The left ends of these curves have to be imagined as joining on to the points in Fig. 1 where the density (normalized to the undisturbed solid) has the value 1. Because of the high electron density this is also the region of strongest absorption, as can be seen from the diagram for the laser intensity. All curves show similar behaviour here: The transition region is characterized by very high gradient of density, temperature and velocity which only decrease slightly at later times. As seen above the total conservation of momentum causes the foil to recede to the left. This is indicated in Fig. 2 by the pronounced displacement of the curves relative to one another. For the same reason more and more plasma streams to the left with time (negative velocities in the fourth diagram of Fig. 2). After about 55 nsec the foil is completely evaporated and transformed into plasma. Very similar diagrams are obtained for other power densities. Fig. 3 gives the temperature distribution at various times for $\Phi_0 = 10^{13}$ W/cm². The time development of the temperature maximum for three different power levels can be seen in Fig. 4 (solid curves).

[12] P. MULSER and S. WITKOWSKI, Phys. Letters **28** A, 703 [1969].

The quantity of plasma produced is shown in Fig. 5 as a function of time. The right ordinate gives the number N of atoms evaporated from an area of 10^{-4} cm². On the left ordinate the corresponding depth d of the evaporated layer is plotted. The foil is completely converted into plasma as soon as d reaches the value 50 μ. The increase of d per unit time is equivalent to a velocity much smaller than that of the shock wave. This velocity after 10 nsec, for example, is $\dot{d} = 1.04 \times 10^5$ cm/sec for $\Phi_0 = 10^{12}$ W/cm².

As a result of the large difference in density between plasma and compressed solid (see Fig. 1 and 2) it follows that the major part of the incident laser energy is transferred to the plasma. In fact only about 8% of energy go into the dense cold phase. During irradiation and as long as there is still a residue of cold matter present the ratio of the total thermal and kinetic energy is about 45% to 55%.

Similarity relations: In [5, 6] a dimensional analysis is made for the parameters of interest when heating infinitely thick foils. This is successful on the assumption that the heating process does not depend on the following parameters: initial density, ionization energy, thermal conductivity and viscosity. The following relation is then obtained for the quantity of plasma produced:

$$N \sim \Phi_0^{1/2}\, t^{3/4}. \qquad (9)$$

Although all the above mentioned parameters are allowed for in our computations it can be seen from Fig. 5 that the relation (9) is very well satisfied. The two bottom curves deviate so little that they could not be plotted separately. For $\Phi_0 = 10^{13}$ W/cm² the dependence can no longer be described exactly by (9), but the deviations can largely be ascribed to the finite foil thickness. The proportionality constant can only be obtained by numerical calculation, and it deviates from that determined in [5, 6] (in [5] a factor of about 2; cf. [13], Fig. 2). If it is assumed that the sets of curves in Fig. 2 are similar to one another at other time independent laser intensities, a relation for the maximum temperature T_{max} can be derived:

$$T_{max} \sim \Phi_0^{1/2}\, t^{1/4}. \qquad (10)$$

This relation is also well satisfied (Fig. 4). If we apply a laser intensity of $\Phi_0 = 10^{12}$ W/cm² and

[13] P. MULSER and S. WITKOWSKI, Phys. Letters **28** A, 151 [1968].

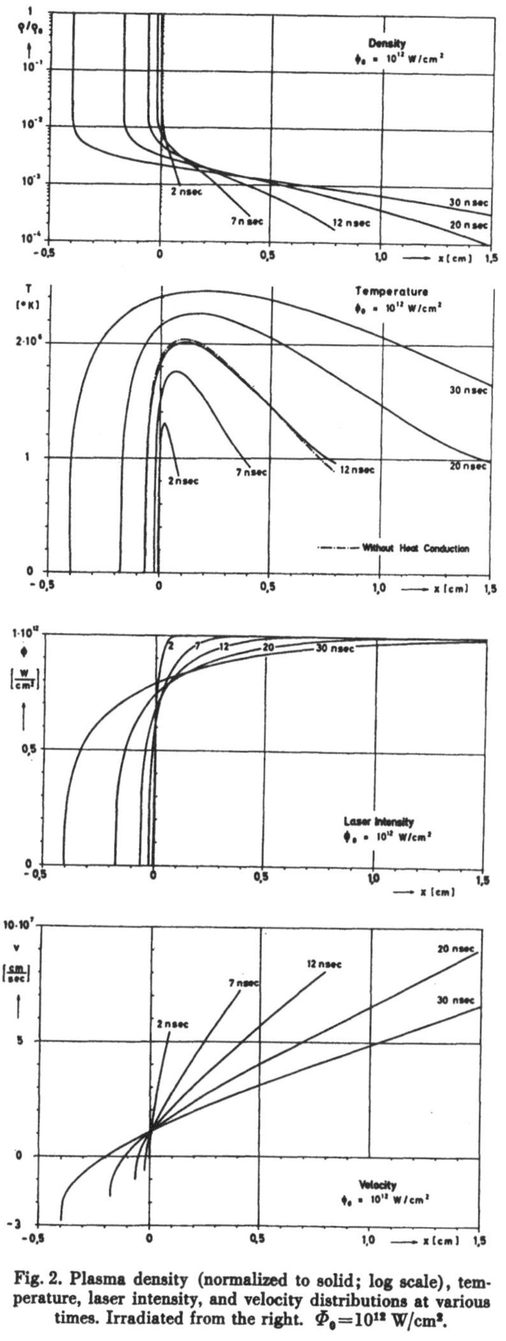

Fig. 2. Plasma density (normalized to solid; log scale), temperature, laser intensity, and velocity distributions at various times. Irradiated from the right. $\Phi_0=10^{12}$ W/cm².

choose the proportionality constant so that T_{max} from Eq. (10) coincides at 15 nsec with the value determined numerically without thermal conduction, these curves lie between the solid and dash-and-dot curves. At $\Phi_0=10^{11}$ W/cm² the influence of the ionization energy causes a small deviation: The true temperature (solid curve) lies below the curve calculated from the relation (10), where the ionization energy is disregarded. The largest deviations from (10) again occur at $\Phi_0=10^{13}$ W/cm² due to the finite thickness of the target.

Influence of thermal conductivity and viscosity: The numerical calculations with and without viscosity show deviations of a maximum of 4% only at $\Phi_0=10^{13}$ W/cm². The mean deviation is less than 1%. More important is the influence of thermal conductivity. For comparison the numerical results without thermal conduction are plotted (dash-and-dot curves) in Figs. 2, 3, 4, and 5: its influence is best pronounced at the highest intensity $\Phi_0=10^{13}$ W/cm² but does not exceed a few percent. At lower intensities it is still smaller and can be neglected.

IV. On the Reflection of the Laser Light

The use of the absorption coefficient in Appendix I in conjunction with the Saha equation leads to the high spatial gradients in the transition region (Fig. 2), where the normalized density increases from 2×10^{-2} to 1. The transition from given to complete ionization in the intensity range considered occurs between 0.6 and 1 μ, i.e. over about a vacuum wavelenght of the ruby laser. This causes appreciable changes of the refractive index (4a) (Appendix I) over a wavelenght. In addition, the plasma frequency ω_p exceeds the light frequency up to a factor of 3 in this region. The question arises how the light propagates in the transition region. Appreciable reflection is sure to be encountered here. As far as the author is aware, the problem has only been treated once before in this connection, by assuming the simplifying WKB condition[14]. For layers of such high optical inhomogeneity the wave structure of the light has to be allowed for in calculating reflection and absorption. This means in our case that in the transition region the wave Eq. (2a) would have to be solved simultaneously with the equations of motion (1), (2),

[14] J. Dawson, P. Kaw, and B. Green, Phys. Fluids **12**, 875 [1969].

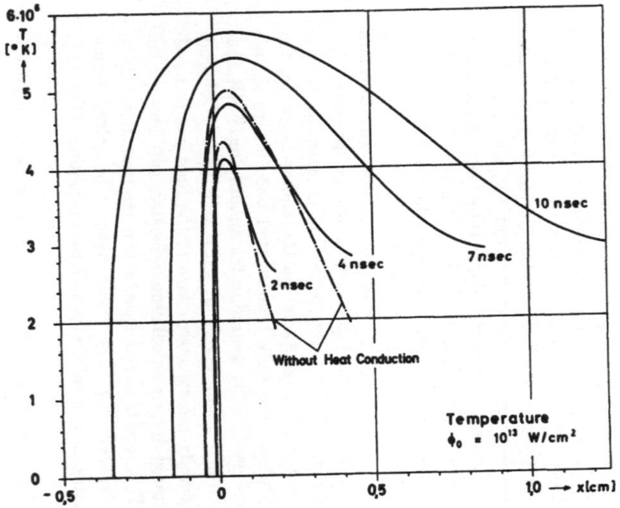

Fig. 3.

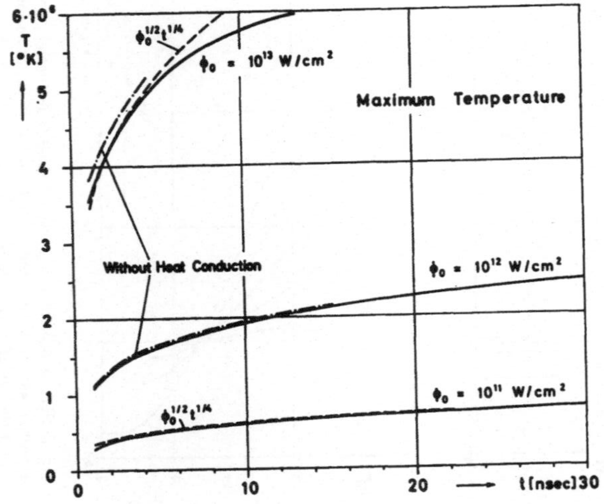

Fig. 4.

Fig. 3. 50 μ foil. Temperature distribution for power density $\Phi_0 = 10^{13}$ W/cm².

Fig. 4. Time development of the temperature maximum for $\Phi_0 = 10^{11}$, 10^{12}, and 10^{13} W/cm². $T_{max} \sim \Phi_0^{1/2} t^{1/4}$.

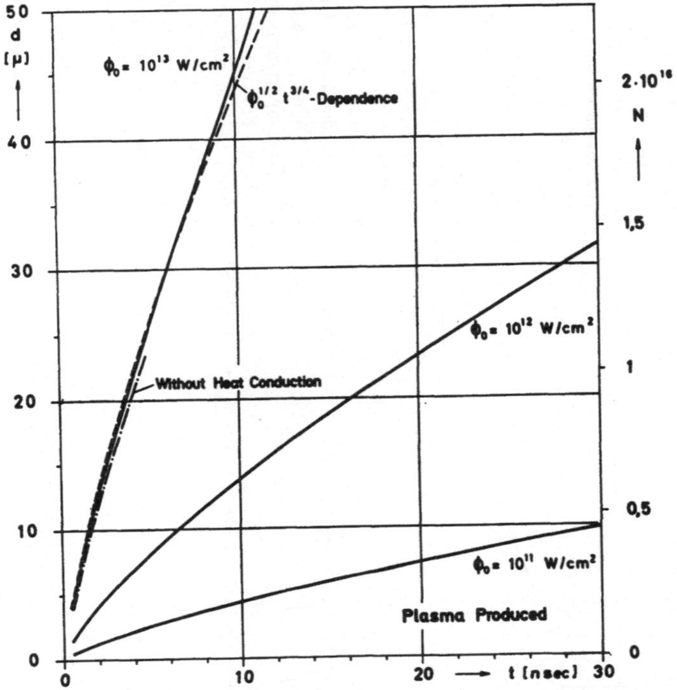

Fig. 5. Quantity of plasma produced as a function of time for various intensities. Left ordinate: thickness of evaporated layer; right ordinate: particle number per 10^{-4} cm². Relation $N \sim \Phi_0^{1/2} t^{3/4}$ is very well satisfied.

and (3). Since this is a cumbersome procedure, only time independent density profiles (fitted to the numerical results described above) for purely transverse electromagnetic waves are investigated in the following. In the numerical results of Section III the elctron density rises steeply to a maximum in the region of greatest absorption and then drops very quickly to almost zero towards the shock wave. The real part of the refractive index varies as follows: In the low density plasma $\mathrm{Re}(n^2)$ is only slightly below 1; with increasing electron density it tends to zero or may even drop below it; finally, in the shocked region it resumes a value of about 1 (the value in solid hydrogen is $n = 1.112$). We therefore consider an inhomogeneous layer of thickness d in which the particle density is determined by

$$n_{\mathrm{H}} = n_{\mathrm{H}}^0 \sin^2\left(\frac{\pi x}{2d}\right) \qquad -d \leqq x \leqq 0$$

and the temperature by

$$T = T_0 \cos^2\left(\frac{\pi x}{2d}\right) \qquad -d \leqq x \leqq 0 .$$

The number of free electrons is again governed by the Saha Eq. (5). The value 5×10^5 °K was chosen for T_0, while n should cover the range from 10^{21} to a few $10^{22}\,\mathrm{cm}^{-3}$. The local refractive index n is then determined by (4a) of Appendix I. On the left and right of the profile n should be $n = 1$.

The Maxwell equations yield the energy conservation law in the following way:

$$\mathfrak{j}\,\mathfrak{E} + \varepsilon_0 c^2 \nabla (\mathfrak{E} \times \mathfrak{B}) + \frac{\partial}{\partial t} \{\tfrac{1}{2}\,\varepsilon_0\,\mathfrak{E}^2 + \tfrac{1}{2}\,\varepsilon_0 c^2 \mathfrak{B}^2\} = 0 .$$

In real notation $\mathfrak{j}\,\mathfrak{E}$ thereby represents the work irreversibly performed by the electromagnetic field per unit time and volume. Since we set $E = E(x)\,e^{-i\omega t}$ and, accordingly, $\mathfrak{j}$ complex, this yields for the light absorption averaged over an oscillation period

$$\frac{\partial \Phi}{\partial x} = \langle \mathrm{Re}(j) \cdot \mathrm{Re}(E) \rangle = \tfrac{1}{2} \mathrm{Re}(\sigma) E E^{*}. \qquad (11)$$

E has to satisfy the stationary wave Eq. (2a), and σ is related to n by (3a). The simple absorption law $\partial\Phi/\partial x = \alpha\,\Phi$ following from Eq. (8) and taken as basis for the numerical calculation in Section III thus has to be replaced in strongly inhomogeneous media by the relation (11). In the case of weak optical inhomogeneity, of course, relation (11) leads back to the expression (8). In order to solve the wave equation, which can only be done numerically for the above profiles, one has to satisfy the condition that only a transmitted wave may occur behind the layer. The wave equation is therefore split into the equivalent first-order system:

$$(E^- + E^+)' = i\,k(E^- - E^+),$$
$$(E^- - E^+)' = i\,k\,n^2(E^- + E^+).$$

Where $E^+ + E^- = E$ (see [15], p. 11). Here, in front of the layer ($x \geqq 0$), E^- denotes the laser wave and $E^+(x)$ the wave reflected by the layer. On the side of the emerging wave ($x = -d$) we have to set $E^+ = 0$. The reflection and transmission coefficients R and Tr are calculated from

$$R = \left(\frac{E^+ E^{+*}}{E^- E^{-*}}\right)_{x=0} \quad \text{and} \quad Tr = \frac{(E^- E^{-*})_{x=-d}}{(E^- E^{-*})_{x=0}} .$$

The results for four profiles with different values of n_{H}^0 are shown in Fig. 6. The bottom curve in each case gives the variation of the real part of n^2 in a layer of thickness $d = \lambda$. As long as the electron density n_e remains everywhere below the value $2.34 \times 10^{21}\,\mathrm{cm}^{-3}$, the plasma frequency is lower than the ruby laser frequency ω, and $\mathrm{Re}(n^2)$ is positive everywhere. This is the case in the top two pictures; in the bottom two pictures the electron density reaches higher values inside the layer, and $\mathrm{Re}(n^2)$ becomes negative. At the boundaries $x = 0$ and $x = -d$ of the layer the electron density drops to zero and n^2 continuously attains the value 1. The curve denoted by $|E|$ gives the magnitude of the electric field $|E^- + E^+|$ for a monochromatic wave with amplitude $|E^-| = 5$ units that is incident from the right (dashed line in Fig. 6). The total coefficients of reflection R, transmission Tr, and absorption A of the profiles are given at the bottom of each picture. R remains below 25% as long as the plasma frequency is everywhere below the light frequency. In the four layers drawn there is superposition of the incident and reflected waves so that $|E|$ appears to be modulated and elevated in part, in the fourth picture by a factor of almost 2. The curve denoted by $A(x)$ represents for every point the energy absorbed in the layer up to this point; $\langle \mathrm{Re}(j) \cdot \mathrm{Re}(R) \rangle$ determines the slope of $A(x)$. Calculating the absorbed energy for the profiles in question according to the WKB approximation (8) still yields good agreement with Eq. (11) for the

[15] P. Mulser, IPP 3/85 [1969].

first picture, but the deviations grow with increasing reflection. Equation (8) would give the strongest absoprtion for the profile with the highest electron density, which is not at all the case using the exact Eq. (11): Because of the strong reflection in the latter case only a little light penetrates into the region of high electron density, and there is a decrease in absorption.

The transmission Tr, reflection R, and absorbed energy A for such inhomogeneous layers of thickness $d=0.5\,\lambda$, $1\,\lambda$, $1.5\,\lambda$, and $2\,\lambda$ are shown in Fig. 7. The curves result from the fact that for each profile thickness n_{H}° covers the whole range from 10^{21} to a few $10^{22}\,\mathrm{cm}^{-3}$. In each case we chose as abscissa the square of the maximum normalized plasma frequency $(\omega_p/\omega)^2$ since, with constant T_0 of course, $(\omega_p/\omega)^2$ is a function of n only. With increasing electron density the transmission quickly decreases, and the reflection rises strongly. The absorption behaviour already mentioned, viz. A rises to a maximum and drops towards higher electron densities, is particularly pronounced for $d=2\,\lambda$.

It emerges from the foregoing remarks that the absorption calculated according to Eq. (8) is close to the values determined from the correct Eq. (11) as long as reflection is not too large. According to Fig. 7 this is the case when ω_p does not exceed the laser frequency ω at any point of the profile or, according to Sections II and III, when the laser power Φ_0 does not exceed the value $1.8\times10^{11}\,\mathrm{W}^5\mathrm{cm}^2$. It can therefore be regarded as proved that already at relatively low intensities $\Phi_0 \geqq 2\times10^{11}\,\mathrm{W/cm}^2$ there is an overdense region $(\omega_p>\omega)$ in the plasma in the immediate vicinity of the cold phase. The limiting intensity is probably lower still because our calculations made no allowance for the reduction of the ionization energy, which at high densities can be quite large but is very difficult to state numerically (according to the Debye theory it should be about 8 eV); in addition, the Saha equation always gives too low values for the degree of ionization in the heating phase if the recombination time is not short enough (cf. Section V). As soon as we have $\omega_p>\omega$, of course, the wave equation has to be solved in order to obtain exact information on the electron densitites that actually occur.

At the point in the plasma where the WKB condition is no longer satisfied the original intensity Φ_0 in our intensity range has dropped to at least 1/4 owing to absorption. This immediately explains why only a small amount of reflection is observed in the experiment [16]; for even with total reflection in the overdense layer the remaining intensity would be reduced by at least 3/4 in the region of low electron

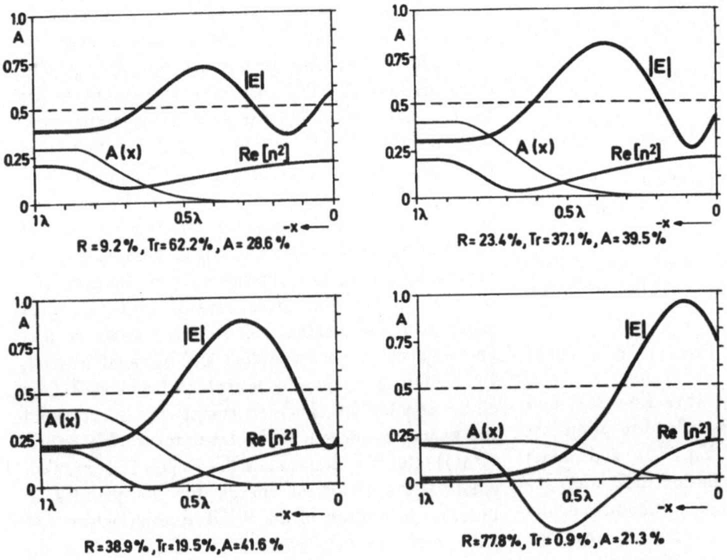

Fig. 6. Variation of the real part of n^2, variation of $|E|$ and normalized light absorption $A(x)$ for profiles of thickness $d=\lambda$ and various electron densities. R, Tr, A reflection, transmission and absorption coefficients. Irradiated from the right. Arbitrary units on right ordinate.

[16] R. Sigel, Thesis Technische Hochschule München (1969) or IPP 3/96 [1969].

density on its way back, i. e. 1/16 of the original intensity at most would be observed as reflected light. It is not until sub-nanosecond laser pulses of extremely high intensity are applied that considerable reflection of up to 45% [17] occurs because in this case the expanded plasma in front of the transition region is absent.

V. Discussion

The basic equations in Section II are subject to two esesntial conditions; firstly, that the plasma production and heating processes can be described by hydrodynamic theory and, secondly, that a thermal model with $T_i = T_e = T$ (T_e, T_i electron and ion temperature) may be used. The momentum equation (2) is valid if the pressure in the numerical results changes only slightly over a mean free path l_i; i. e. $l_i \frac{1}{p} \frac{\partial p}{\partial x} \ll 1$, or if, according to [9], p. 258 and 438 and with the aid of Eq. (1),

$$4.3 \times 10^{-18} \frac{T^2}{\ln \Lambda} \frac{1}{n_i T} \frac{\partial (n_i T)}{\partial a} \ll 1 \quad [\text{cgs grad}].$$

This inequality is in fact very well satisfied in all calculations. — The critical point for the application of a thermal model with the electron temperature equal to the ion temperature is in the transition region from the cold phase to the thin, hot plasma. Figures 3 and 5 show that the temperature here rises almost linearly to 10^6 °K and enters the dense phase between 0.5 and 1 nsec with the velocity $\dot{d} = 6.5 \times 10^5$ cm/sec. Since this increase in the calculation covers four space steps $\Delta a = 2 \times 10^{-5}$ cm (see Appendix II), we get the characteristic reference time $\tau_0 = 4\,\Delta a/\dot{d} = 1.2 \times 10^{-10}$ sec. At the point $T = 10^6$ °K the normalized particle density is $N = n_H/n_0 = 0.1$. The following relaxation times ($\ln \Lambda = 5$) are now calculated according to [9] for the values:

Thermalization time for ions:

$$\tau_{ii} \approx \frac{1}{\nu_{ii}} = 6 \times 10^{-12}\,\text{sec};$$

Thermalization time for electrons:

$$\tau_{ee} \approx \tau_{ii} \left(\frac{m_e}{m_i}\right)^{1/2} = 1.4 \times 10^{-13}\,\text{sec};$$

Equipartition time for $T_i = T_e$:

$$\tau_{ei} \approx \tau_{ee} \frac{m_i}{4\,m_e} = 6.5 \times 10^{-11}\,\text{sec}.$$

These times are appreciably shorter than τ_0, and so it seems certain that there is equilibrium in the transition region as well.

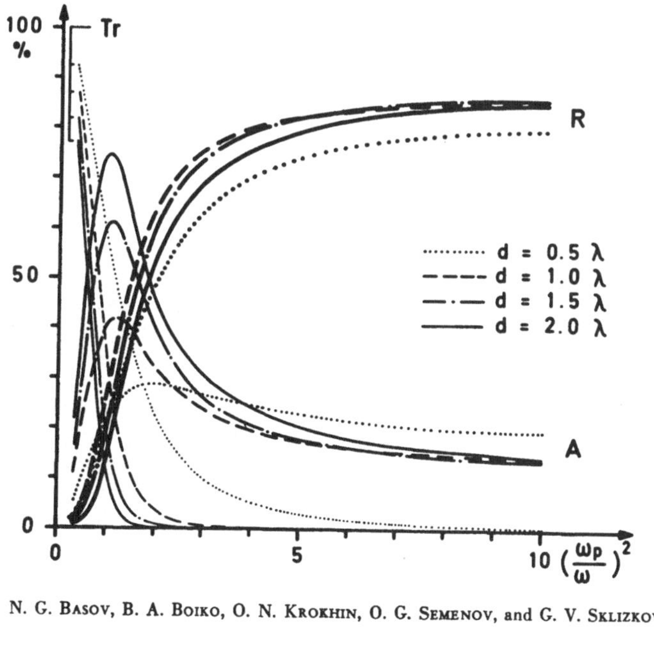

Fig. 7. Transmission *Tr*, reflection *R*, and absorption *A* as functions of the maximum normalized plasma frequency for layer thicknesses $d = 0.5$, 1, 1.5, and 2 λ.

[17] N. G. Basov, B. A. Boiko, O. N. Krokhin, O. G. Semenov, and G. V. Sklizkov, ZHTF 38, 1973 [1968].

Saha equilibrium is obtained when the ionization and recombination rates are sufficiently high. Estimates according to [8], Vol. I, p. 389 et seq. yield ample ionization rates for applying the Saha equation. It is more difficult to obtain concrete information on the recombination times. Should these in fact be very long (which is certainly the case for very low degrees of ionization well below 0.7%), the Saha equation yields a lower limit for the electron density in the narrow, not completely ionized part of the transition region. This, however, is completely sufficient for our purposes since our aim in this connection has been to show that a strongly absorbing layer has to form in a plasma produced from solid hydrogen and that there are layers in the transition region where the plasma becomes overdense ($\omega_p > \omega$). The value for the initial degree of ionization used in Section III, $\eta = 7 \times 10^{-3}$ is arbitrary to a certain extent. Test calculations showed, that the special choice of the value for the initial ionization only has an effect in the first few nanoseconds, provided it is not too small (e. g. 10^{-10}) since ionisation-recombination equilibrium is not even approximately obtained in such a case.

The transition region is characterized by high densities and strong gradients. There is some uncertainty here as regards the equation of state and transport parameters. The absorption coefficient (6a), for example, may assume extremely high values. Numerical test calculations with widely varying coefficients for the dense region ($2 \cdot 10^{-2} \leqq \varrho/\varrho_0 \leqq 1$) – more than a factor 10^3 for the absorption coefficient – did show differences in the density, temperature, and velocity distributions in the immediate vicinity of the solid (up to a factor of 2), but such variations have almost no influence ($\leqq 6\%$) on the properties of the thin plasma ($\varrho/\varrho_0 \leqq 10^{-2}$), and the characteristic behaviour described in Section III persists in all regions. The curves in Fig. 5 for the quantity of plasma produced proved to be particularly insensitive (deviations below 2%).

In these calculations the collision frequency in the absorption coefficient is assumed to be caused by purely thermal motion. This is justified as long as the oscillation energy represents only a small fraction of the thermal energy of the plasma, as is the case here. In the coefficient of thermal conductivity allowance was made only for Coulomb interaction since only very close to the solid the plasma is not completely ionized and there thermal conduction is almost insignificant because of the lower temperature. Taking the electron-neutral collisions into account reduces thermal conduction in this region even more.

The true equation of state for hydrogen in the cold compressed phase is not known. The experimental investigations in [18] show very high compressibility for liquid hydrogen, and the assumption of Lennard-Jones interaction between the molecules does not yield correct results. On the other hand, we are not interested here primarily in the shock wave, but only insofar as it can modify the calculations for the hot plasma. It may be stated with certainty that the true compressibility of the solid hydrogen is between zero (incompressible solid) and that of an ideal gas, but closer to the latter. In order to obtain some idea of the influence of the equation of state of the cold phase on the heating of the plasma, the case of an incompressible solid was also calculated. The differences in temperature and density after 14 nsec in the plasma produced are shown in Fig. 8. The maximum error in the density is 18%, but it is under 10% over wide ranges. The largest relative error in the temperature is found to be 7.2%. It should be pointed out, however, that Fig. 8 misrepresents the values as too high, this being due to the fact that the incompressible case was calculated only for an infinitely extended solid (no reaction present), while the solid curves refer to a 50 μ foil accelerated to the left. The special form of the equation of state in the solid phase has very little influence on the quantity of plasma produced. After 20 nsec the difference in the quantity of plasma for compressible and incompressible hydrogen is less than 1%.

Even if the electron densities calculated with the Saha equation are too low in the transition region, they are still sufficient for forming a very strongly absorbing layer which prevents uniform heating of the foils and results in a strong shock wave. The conditions when small pellets are irradiated are similar, provided the laser intensity rises fast enough. "Burn-through" from one side of a foil has been convincingly demonstrated [16]. This, however, does not contradict the experimental findings that the plasma obtained by evaporating a drop expands almost isotropically after a relatively short time.

[18] M. VAN THIEL and B. J. ALDER, Mol. Phys. 10, 427 [1966].

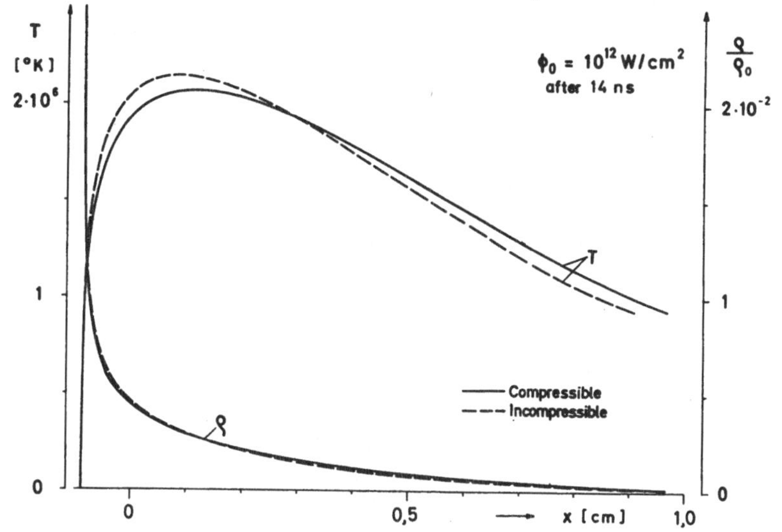

Fig. 8. Differences of temperature and density in the plasma for a compressible (ideal gas equation) and an incompressible solid.

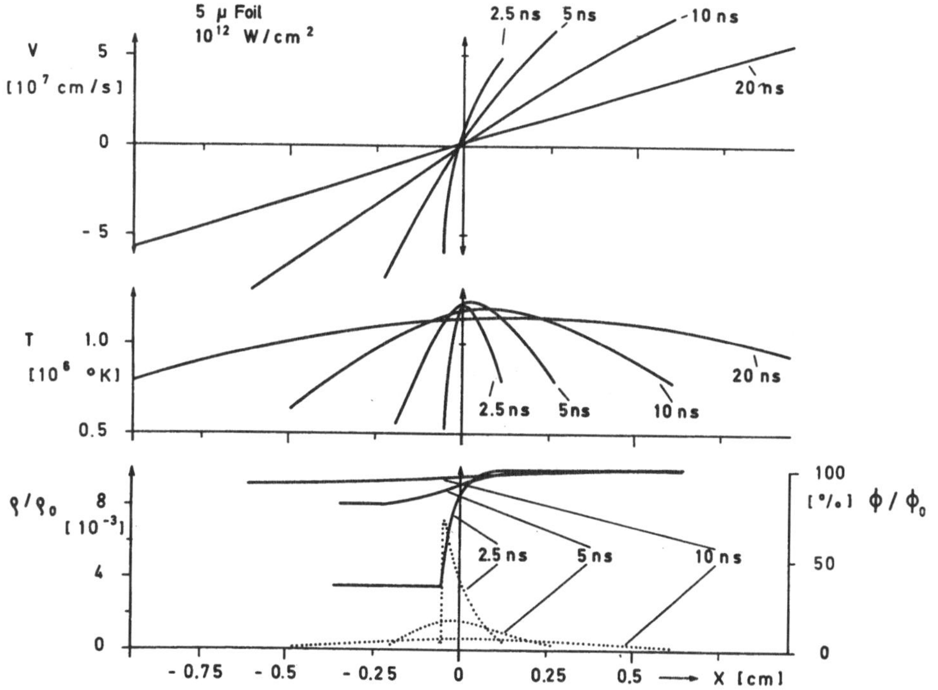

Fig. 9. Heating of a 5 μ thick foil and development of a symmetric plasm distribution. The solid curves below give the intensity, the dotted curve the density distribution. $\Phi_0 = 10^{12}$ W/cm². Irradiated from the right.

This can be seen by following up the heating of a thin foil beyond its burn-through time in our plane model, as in Fig. 9. One can readily trace the development to a symmetrical distribution of velocity, temperature and density.

All of the results in this paper refer to a one-dimensional plane model. With the laser intensities at present available such a case cannot be achieved experimentally. In experiments actually conducted the plasma is always more or less rotation symmetric since the high radiation densities required can only be obtained by focusing the laser light. The analogue calculations for such a geometry were felt, however, to be too cumbersome, especially since the heating process can readily be studied qualitatively in a one-dimensional model and the deviation estimated. In the experiments conducted hitherto the hot plasma can also expand to the sides, some of it already escaping from the radiation field of the laser after a short time. This reduces the temperatures attained theoretically in the one-dimensional case, which means that for a given incident energy the temperature has to be distributed over more particles with the result that a larger quantity of plasma is produced (cf. the comparison of theory and experiment in [16]).

Appendix I
Absorption Coefficient

In the Maxwell equations

$$c^2 \nabla \times \mathfrak{B} = \mathfrak{j}/\varepsilon_0 + \dot{\mathfrak{E}}, \quad \nabla \times \mathfrak{E} := -\dot{\mathfrak{B}}$$

the electric field strength is represented by that of a monochromatic light wave. Neglecting the polarization current one gets for the current density

$$j = n_e e \dot{y}$$

(y perpendicular to x). Furthermore, the equation of motion of free, not too fast electron is

$$\ddot{y} + \nu \dot{y} = (e/m) E \tag{1a}$$

where the Lorentz force is neglected and ν is the collision frequency. By substituting the solution of Eq. (1a) in the expression for j one obtains the following stationary wave equation for a plane electromagnetic wave $E = E(x, t)\, e^{-i\omega t}$ with amplitude varying slowly with time:

$$E'' + k^2 n^2(x, t)\, E = 0 \tag{2a}$$

and j itself is proportional to E: $j = \sigma \cdot E$, where the high frequency conductivity σ is related to the refractive index n by

$$(i\,\sigma/\varepsilon_0\,\omega) = n^2 - 1 \tag{3a}$$

($k = \omega/c$ is the wave vector in vacuum). The complex refractive index $n = n_R + i\, n_I$ is then expressed by means of the electron plasma frequency $\omega_p = (n_e e^2/\varepsilon_0 m_e)^{1/2}$ as follows:

$$n^2 = (n_R + i\, n_I)^2 = 1 - \left(\frac{\omega_p}{\omega}\right)^2 \frac{1}{1 + (\nu/\omega)^2} + i \frac{\nu}{\omega} \left(\frac{\omega_p}{\omega}\right)^2 \frac{1}{1 + (\nu/\omega)^2} . \tag{4a}$$

If $\left| \frac{1}{2 k^2 n^2} \left\{ \frac{n''}{n} - \frac{3}{2} \left(\frac{n'}{n} \right)^2 \right\} \right| \ll 1$, the solution of the wave equation is

$$E = \frac{E_0}{\sqrt{n}} \exp\{i k \int_x^\infty n\, dx\} = \frac{E_0 \exp\{-k \int_x^\infty n_I\, x\}}{\sqrt{n}} \exp\{i k \int_x^\infty n_R\, dx\}$$

(WKB approximation; see, for example [19], p. 181). As long as n_I is small relative to n_R, one obtains for the intensity using the relation for the group velocity $v_{gr} = c\, n_R$ (s. [19], p. 232)

$$\Phi(x,t) = \frac{\varepsilon_0 v_{gr}}{2} \frac{E_0 E_0^*}{n_R} \exp\{-2k \int_x^\infty n_I\, dx = \frac{\varepsilon_0 c}{2} E_0 E_0^* \exp\{-2k \int_x^\infty n_I dx\} = \Phi_0 \exp\{-\int_x^\infty \alpha\, dx\} \tag{5a}$$

with the absorption coefficient $\alpha = 2 k n_I$. Solving (4a) for n_I yields for α

$$\alpha = 2k \sqrt{\frac{1}{2} \left\{ \sqrt{\left[1 - \left(\frac{\omega_p}{\omega}\right)^2 \frac{1}{1+(\nu/\omega)^2}\right]^2 + \left[\frac{\nu}{\omega} \left(\frac{\omega_p}{\omega}\right)^2 \frac{1}{1+(\nu/\omega)^2}\right]^2} - \left[1 - \left(\frac{\omega_p}{\omega}\right)^2 \frac{1}{1+(\nu/\omega)^2}\right] \right\}} . \tag{6a}$$

for $\nu \ll \omega$ and $\omega_p < \omega$ (6a) reduces to the most comonly used formula

$$\alpha = \frac{\omega_p^2 \nu}{c^2 \omega^2 [1 - (\omega_p/\omega)^2]^{1/2}} . \tag{6a'}$$

The expression (6a) is somewhat more general than the formula (6a'), and so (6a) is always used in

[19] V. L. GINZBURG, The Propagation of Electromagnetic Waves in Plasmas, Perg. Press Ltd., Oxford 1964.

this paper. For the temperatures expected it holds that $h\nu \ll kT$. The induced emission must therefore be taken into account. The way in which α is derived means, however, that it is already contained in (6a) (cf., for example [8], p. 124).

The collision frequency ν has two components $\nu = \nu_{ei} + \nu_{en}$ corresponding to the electron-ion and electron-neutral collisions; ν_{ei} is taken from SPITZER [11], p. 83.

$$\nu_{ei} = \frac{\pi^{3/2} e^4 Z^2 n_H \eta}{2 m_e^{1/2} (2kT)^{3/2} \gamma_E} \ln \Lambda \qquad [\text{cgs grad}].$$

For ν_{en} we take

$$A(n_H/n_0)(1-\eta)$$

(n_0 particle density in solid hydrogen). According to [20] the mean value for A is $A = 4.5 \times 10^{14}\ \text{sec}^{-1}$, this being roughly in agreement with theoretically determined cross sections.

It should also be mentioned that Spitzer's formula for ν_{ei} in hydrogen is not very different from that used by Dawson. The difference in fact is

$$\frac{\nu_{ei}^{\text{Spritzer}}}{\nu_{ei}^{\text{Dawson}}} = \frac{16}{3\pi} \gamma_E = 0.988\,.$$

Appendix II
Numerical Integration of the Equations of Motion

For numerical integration of the equations of motion (1), (2), (3), and the velocity (4) the explicit difference scheme listed below proved suitable. The lattice points have the indices (l, m) l for time, m for space; the time step is Δt, the space step Δa. The term q denotes the artificial viscosity, for which the expressions in [21], p. 216 were used. The expression for the thermal current was rearranged as follows:

$$-\varkappa \frac{\varrho}{\varrho_0} \frac{\partial T}{\partial a} = -\varkappa' \frac{\varrho}{\varrho_0} T^{5/2} \frac{\partial T}{\partial a} = -\frac{2}{7} \frac{\varrho}{\varrho_0} \varkappa' \frac{\partial T^{7/2}}{\partial a}$$

Difference scheme

1) $$\varrho_0 \frac{v_m^{l+1/2} - v_m^{l-1/2}}{\Delta a} = \frac{1}{\Delta a} \Big\{ p_{m+1/2}^l - p_{m-1/2}^l + \frac{4}{3\,\Delta a} [\varrho_{m-1/2}^l \mu_{m-1/2}^l (v_{m-1}^{l-1/2} - v_m^{l-1/2}) - \varrho_{m+1/2}^l \cdot (v_m^{l-1/2} - v_{m+1}^{l-1/2})] \frac{1}{\varrho_0} \Big\} - \frac{\Delta q}{\Delta a}$$

2) $$v_m^{l+1/2} = \frac{x_m^{l+1} - x_m^l}{\Delta t},$$

3) $$\varrho_{m+1/2}^{l+1} = \frac{\varrho_0 \Delta a}{x_m^{l+1} - x_{m+1}^{l+1}},$$

4) $$\varrho_0 \frac{\varepsilon_{m+1/2}^{l+1} - \varepsilon_{m+1/2}^l}{\Delta t} = \frac{1}{\Delta a} \Big\{ \Phi_m^l - \Phi_{m+1}^l + (v_{m+1}^{l-1/2} - v_m^{l-1/2}) \Big[p_{m+1/2}^l + q_{m+1/2}^l - \frac{4}{3\varrho_0 \Delta a} \varrho_{m+1/2}^l \mu_{m+1}^l \cdot (v_{m+1}^{l-1/2} - v_m^{l-1/2}) \Big] + \frac{2}{7\varrho_0 \cdot 2\Delta a} [\varrho_{m-1/2}^l \varkappa'^l_{m-1/2} + \varrho_{m+1/2}^l \varkappa'^l_{m+1/2})(T^{7/2\,l}_{m-1/2} - T^{7/2\,l}_{m+1/2})] - \frac{2}{7\varrho_0 \Delta a} [(\varrho_{m+1/2}^l \varkappa'^l_{m+1/2} + \varrho_{m+3/2}^l \varkappa'^l_{m+3/2})(T^{7/2\,l}_{m+1/2} - T^{7/2\,l}_{m+3/2})] \Big\}.$$

The signs of the individual terms are due to the fact that the m-subscripts are counter to the a-direction. Solution of ε for the temperature T was done by the nested interval method.

Step sizes: For given $\Delta a = 2 \times 10^{-5}$ cm and laser intensities below $4 \times 10^{12}\ \text{W/cm}^2$ a time step $\Delta t = 10^{-12}$ sec was chosen; for higher Φ_0 the Δt was half as large. This is because the time step has to be made at least so small that the inequality $|\Delta v_m| \Delta t < |\Delta x_m|$ is satisfied at all lattice points. A further condition is imposed, on the ratio $\Delta t/\Delta a$ and on Δa itself. Owing to the finite space and time step sizes it is not possible to determine the velocity of the plasma front, i. e. the velocity of the plasma-vacuum interface; the calculation has to be truncated towards the vacuum at a finite thick-

[20] W. P. ALLIS and S. C. BROWN, Phys. Rev. 87, 419 [1952].

[21] R. D. RICHTMYER, Difference Methods for Initial-Value Problems, Intersc. Publ. Inc., New York 1957.

ness. This may result in errors propagating into the interior of the plasma. The difference scheme was therefore tried out on an adiabatic rarefaction wave for which the solution can also be given in closed form. Excellent agreement was obtained using the above step sizes. In calculating the thermal conduction it was ensured that no heat can flow into the vacuum by introducing an appropriate boundary condition. After every 500 time steps the conservation of energy and momentum was checked by integrating the appropriate quantities over space. The relative errors are below 0.1%.

This work was undertaken as part of the joint research programme between the Institut für Plasmaphysik and Euratom.

Acknowledgement

The author wishes to thank Prof. Dr. R. Wienecke for his support and encouragement. He is also indebted to Dr. S. Witkowski for numerous discussions.

OPTICAL CONSTANTS OF FULLY IONIZED HYDROGEN PLASMA FOR LASER RADIATION †

H. HORA, Hannelore WILHELM
Institut für Plasmaphysik
Garching near Munich
Federal Republic of Germany

ABSTRACT. The refractive index n and the absorption constant K are numerically evaluated for fully ionized hydrogen plasmas as functions of the electron temperature T and atomic density N for the case of ruby, neodymium-glass, and CO_2 laser radiation and for the second harmonics of each type of radiation on the basis of the two-fluid model of hydrodynamics using Spitzer's collision frequency. Limitations due to degeneration, too small Debye length, and non-linear effects are given. The absorption constant is compared with the values of the quantum theory of inverse bremsstrahlung without collective effects, and with the Dawson-Oberman theory.

1. INTRODUCTION

Since the first successful measurement of laser light scattering by plasmas [1] and the advent of plasma production by lasers for the purpose of controlled thermonuclear reactions [2], knowledge of the optical properties of plasmas at the frequencies of laser radiation has become increasingly important. An evaluation of the optical constants – the refractive index n and the absorption constant K – on the basis of the dispersion relation of the two-fluid model [3] has already been given [4] for radiation having the frequency of the ruby laser. These results were used in several cases of laser-produced plasma [5] and reached high interest in view of controlled thermonuclear fusion, after the first neutrons were observed in laser-produced plasma [6] and the existence of controlled thermonuclear fusion reactions was proved carefully with an efficiency of only two magnitudes less than the theta pinches [7].

In this paper we report the numerical calculations of the optical constants for the radiation of ruby, neodymium glass and CO_2 lasers and their second harmonics for fully ionized hydrogen. Therefore, all questions of recombination and line radiation at low temperatures are neglected. The absorption is based on the usual collision process. The stimulated emission processes are included, and the results are of a linear type. Non-linear absorption processes are not treated.

2. OPTICAL CONSTANTS

The complex refractive index $\tilde{n}$ for a plasma without external magnetic fields is given by the dispersion relation derived from the Maxwellian equations and the equations of the two-fluid model without pressure terms and neglecting relativistic motions of the plasma [3]:

$$\tilde{n}^2 = 1 - \frac{\omega_p^2}{\omega^2} \frac{1}{1 - i\frac{\nu}{\omega}} \tag{1}$$

where ω is the frequency of the light, and the plasma frequency ω_p is given by

$$\omega_p^2 = 4\pi N_e e^2/m_e \tag{2}$$

Here N_e is the density, e the charge and m_e the mass of the electron. The electron collision frequency ν introduced by Spitzer [8] is, in Gaussian units,

$$\nu = \frac{\omega_p^2 \pi^{3/2} m_e^{1/2} Z e^2 \ln \Lambda}{8\pi \gamma_E(Z)(2kT)^{3/2}} \tag{3}$$

where we have used the Boltzmann constant k, the atomic number Z of the fully ionized plasma, Spitzer's correction $\gamma_E(Z)$ [8], and the Coulomb logarithm with

$$\Lambda = \frac{3}{2Z e^3}\left(\frac{k^3 T^3}{\pi N_e}\right)^{1/2} \tag{4}$$

expressing the ratio between Debye length and impact parameter. With $[N_e]$ = cm^{-3} and [T] = eV we obtain

$$\nu = \frac{8.64 \times 10^{-7}}{\gamma_E(Z)} \frac{Z N_e}{T^{3/2}} \ln\left(1.55 \times 10^{10} \frac{T^{3/2}}{Z N_e}\right) \tag{5}$$

---

† This work was performed as part of the joint research program of EURATOM and the Institut für Plasmaphysik.

Reprinted from: *Nuclear Fusion*, **10**, 111–116 (1970).

which is slightly more general than before [4] in the case of ruby lasers and where the value [8] $\gamma_E(1) = 0.582$.

The real part of $\tilde{n}$, the refractive index n, is found from Eq. (1) to be

$$n = \sqrt{\tfrac{1}{2}(A+B)} \tag{6}$$

The imaginary part of $\tilde{n}$ is the absorption coefficient κ, which is related to the absorption constant K by

$$K = 2\frac{\omega}{c}\kappa = 2\frac{\omega}{c}\sqrt{\tfrac{1}{2}(A-B)} \tag{7}$$

where

$$A = B^2 + \left(\frac{\nu}{\omega}\frac{\omega_p^2}{\omega^2+\nu^2}\right)^2; \qquad B = 1 - \frac{\omega_p^2}{\omega^2+\nu^2}$$

The simple meaning of K can be seen in the special case of a homogeneous plasma from the attenuation of the intensity I of the electromagnetic radiation propagating along x

$$I = I_0 \exp(-Kx) \tag{8}$$

The following figures give a numerical evaluation of K and n for Z = 1 for the case of ruby, neodymium-glass, and CO_2 laser radiation (Figs 1 to 3) and for the second harmonics of each of these types of radiation (Figs 4 to 6). At temperatures exceeding 10 eV the figures demonstrate an extremely large change of the optical constants on varying the electron density at $\omega_p \cong \omega$. This very close neighbourhood of $\partial K/\partial N_e$ or $\partial n/\partial N_e$ to a pole induced difficulties of the numerical treatment, which could be avoided by a special procedure only.

3. LIMITATIONS OF VALIDITY

In Figs 1 to 6 the curves are fully drawn where the linear absorption constants can be derived from the dispersion relation based on the two-fluid model.

Limitations are given by the assumption of Boltzmann statistics and the condition that the Debye length must be larger than the impact parameter ($\Lambda \geq 1$). Boltzmann statistics is valid at temperatures T larger than ten times the Fermi temperature

$$T > 10\,\zeta_0 = 10\frac{h^3}{2m_e}\left(\frac{3n_e}{8\pi}\right)^{2/3} = 3.65 \times 10^{-14} n_e^{2/3} \tag{9}$$

using Planck's constant h and $[n_e] = cm^{-3}$. Figures 1 to 6 show dashed curves where the calculation is continued on the basis of Boltzmann statistics but inequality (9) is not fulfilled. All curves are ending at $\Lambda = 1$, which means from Eq. (4) for $\Lambda \geq 1$

$$n_e \leq \frac{9k^3T^3}{4Z^2e^2} = 2.42 \times 10^{20}\, T^3 \tag{10}$$

where $[T] = eV$ and $[n_e] = cm^{-3}$.

A further limitation of the optical constants of Figs 1 to 6 is given by non-linear effects. One type of those processes is the creation of a non-linear force due to the direct interaction of the radiation with the plasma [9] producing non-linear collisionless absorption of the radiation. This force exceeds the thermokinetic (gas-dynamic) forces if the increase of the electrodynamic momentum flux density due to collective effects over the vacuum value is larger than the gas-dynamic power density. This relation is approximated at the cut-off density ($\omega_p \lesssim \omega$) and sufficiently high temperatures (T > 10 eV) by the condition that the energy of the electrons of coherent oscillation exceeds the energy of random motion of temperature T.

The non-linear effect is important if the amplitude of the electric field of the radiation in the vacuum E exceeds

$$E^* = \sqrt{\frac{4m_e\omega^2}{e^2}kT} \tag{11}$$

This limitation, corresponding to a laser intensity I^*, is in the case of a ruby laser with $[T] = eV$

$$E^* = 1.263 \times 10^8 \sqrt{T}\,(V/cm); \quad I^* = 2.12 \times 10^{13}\, T\,(W/cm^2);$$

for neodymium-glass laser

$$E^* = 8.27 \times 10^7 \sqrt{T}\,(V/cm); \quad I^* = 9.06 \times 10^{12}\, T\,(W/cm^2)$$

and for CO_2 lasers

$$E^* = 8.27 \times 10^6 \sqrt{T}\,(V/cm; \quad I^* = 9.06 \times 10^{10}\, T\,(W/cm^2)$$

The same limitation (11) determines a non-linear decrease of the collision frequency [10], the beginning of a special mechanism of self-focusing [11], of a two-stream instability [12], and of a change of the absorption constant as it was evaluated, at least, for lower densities than the critical value ($\omega_p \ll \omega$) [13]. Another type of non-linear processes is the absorption by a decay into plasma waves [14].

4. COMPARISON WITH OTHER CALCULATIONS

The quantum-mechanical calculation of the inverse bremsstrahlung gives an absorption constant K_{Br} which can be compared with the two-fluid values K of Eq. (7) at densities much below the

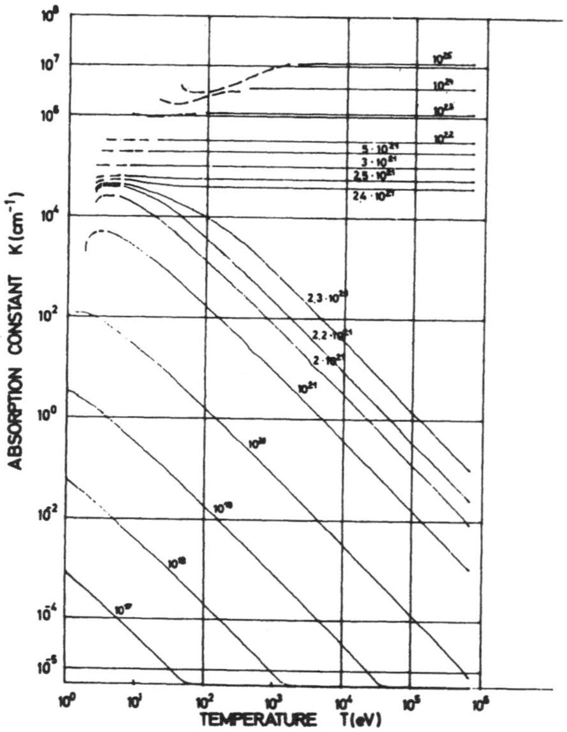

FIG. 1a. Absorption constant K for ruby laser radiation as a function of the electron temperature T and atomic densities N in cm^{-3} (denoted as curve parameters) for fully ionized hydrogen and isotopes.

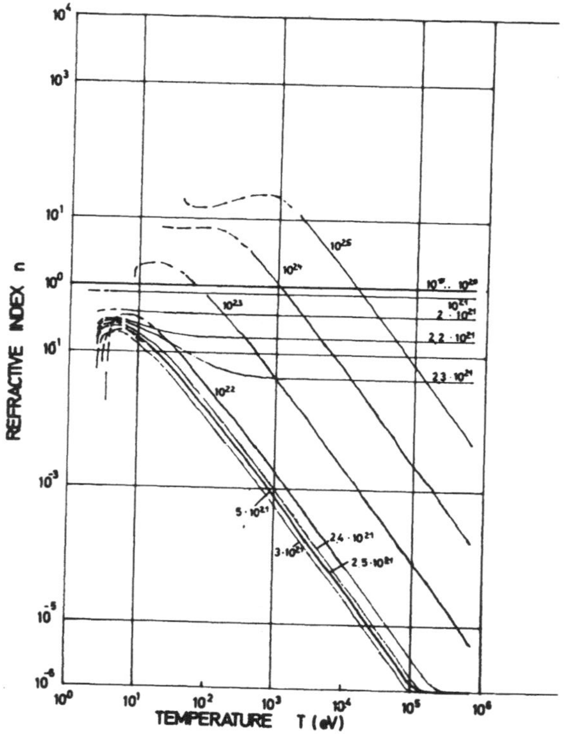

FIG. 1b. Refractive index n for ruby laser radiation as a function of the electron temperature T and atomic densities N in cm^{-3} (denoted as curve parameters) for fully ionized hydrogen and isotopes.

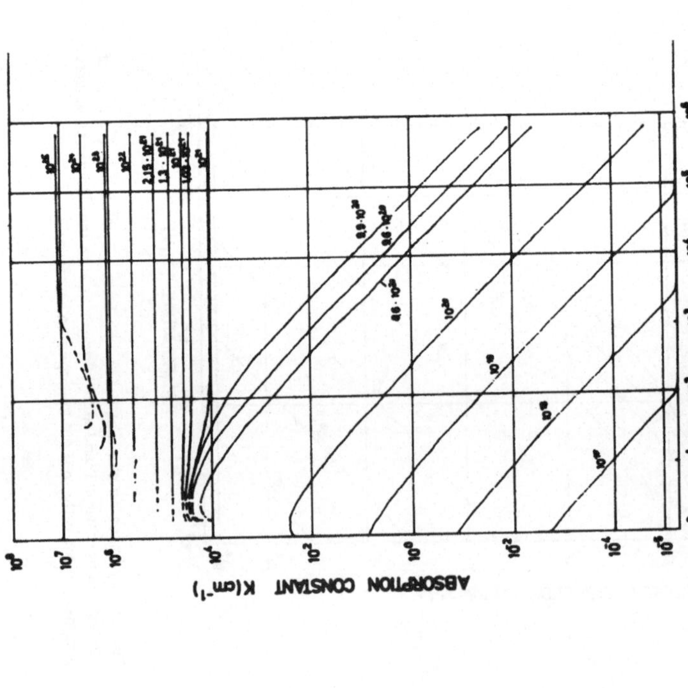

FIG. 2a. Absorption constant K for neodymium glass laser radiation as a function of the electron temperature T and atomic densities N in cm^{-3} (denoted as curve parameters) for fully ionized hydrogen and isotopes.

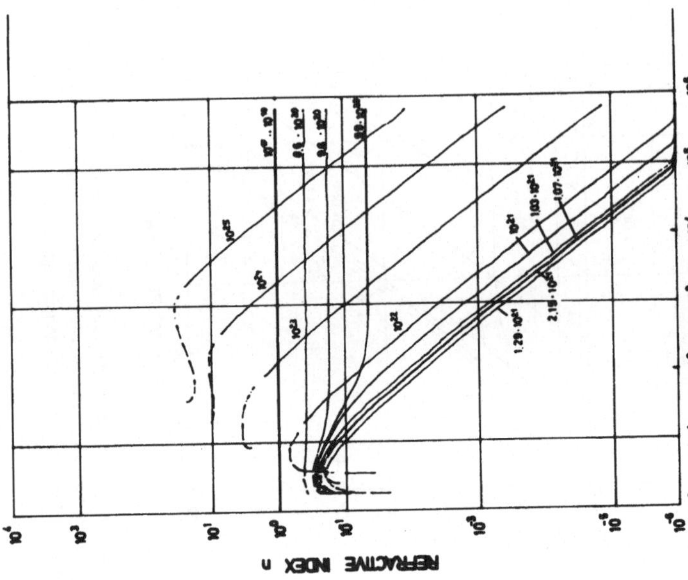

FIG. 2b. Refractive index n for neodymium glass laser radiation as a function of the electron temperature T and atomic densities N in cm^{-3} (denoted as curve parameters) for fully ionized hydrogen and isotopes.

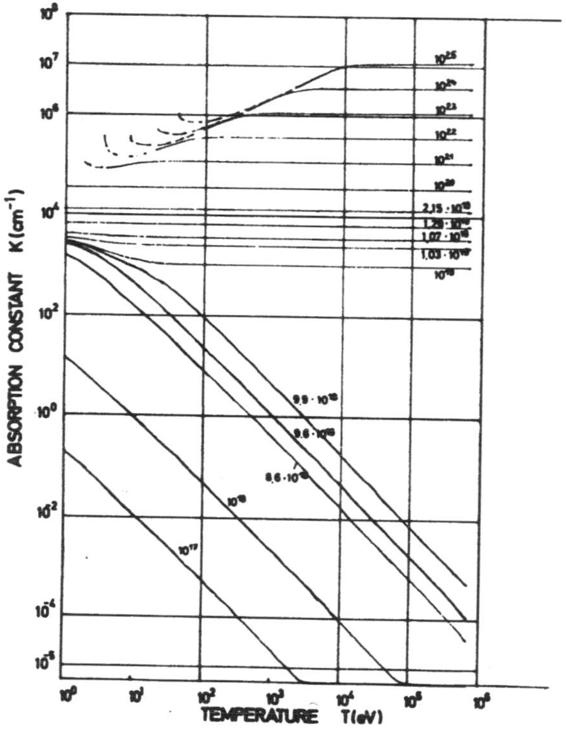

FIG. 3a. Absorption constant K for CO_2 -laser radiation as a function of the electron temperature T and atomic densities N in cm^{-3} (denoted as curve parameters) for fully ionized hydrogen and isotopes.

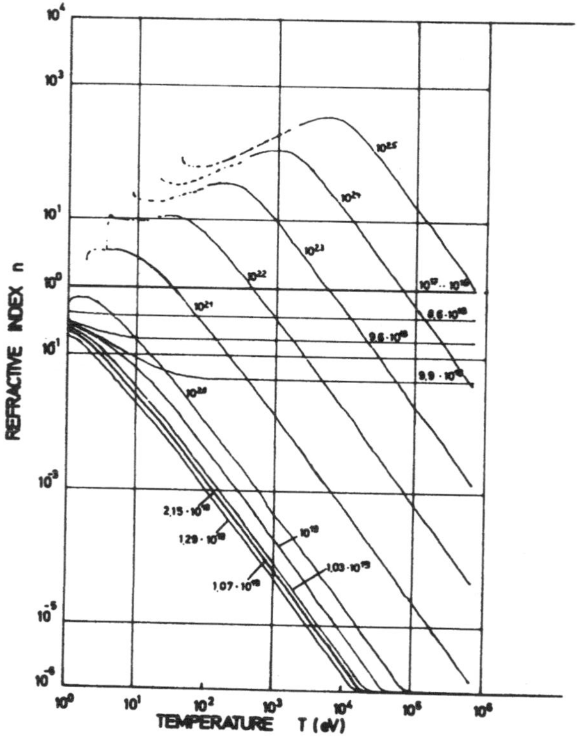

FIG. 3b. Refractive index n for CO_2 -laser radiation as a function of the electron temperature T and atomic densities N in cm^{-3} (denoted as curve parameters) for fully ionized hydrogen and isotopes.

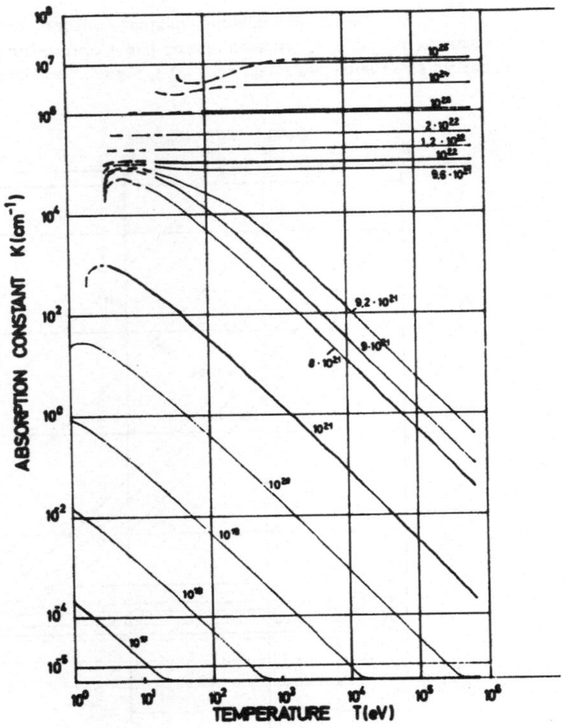

FIG. 4a. Absorption constant K for the second harmonics of ruby laser radiation as a function of the electron temperature T and atomic densities N in cm^{-3} (denoted as curve parameters) for fully ionized hydrogen and isotopes.

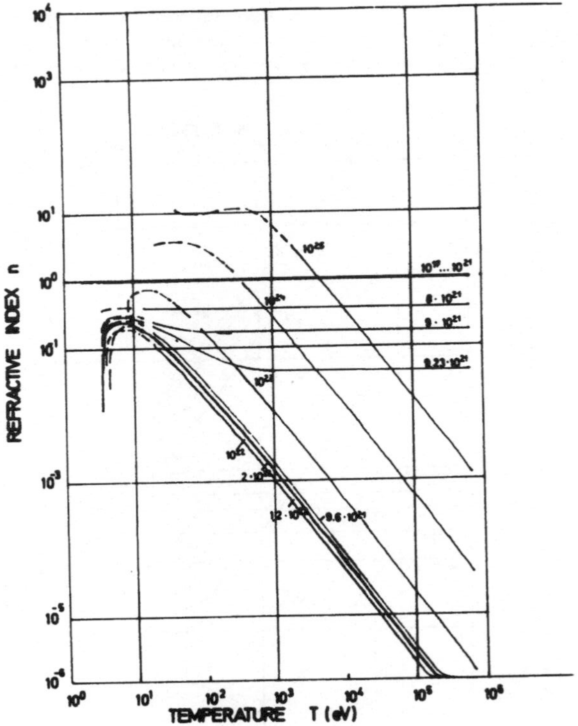

FIG. 4b. Refractive index n for the second harmonics of ruby laser radiation as a function of the electron temperature T and atomic densities N in cm^{-3} (denoted as curve parameters) for fully ionized hydrogen and isotopes.

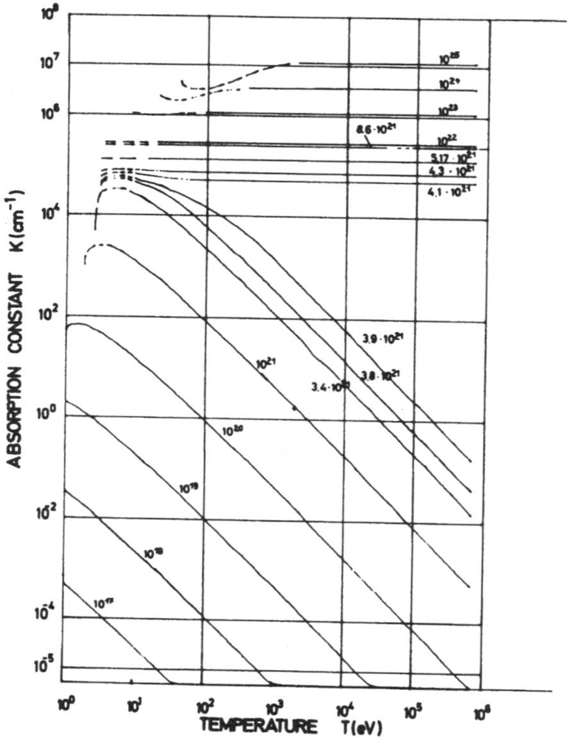

FIG. 5a. Absorption constant K for the second harmonics of neodymium glass laser radiation as a function of the electron temperature T and atomic densities N in cm^{-3} (denoted as curve parameters) for fully ionized hydrogen and isotopes.

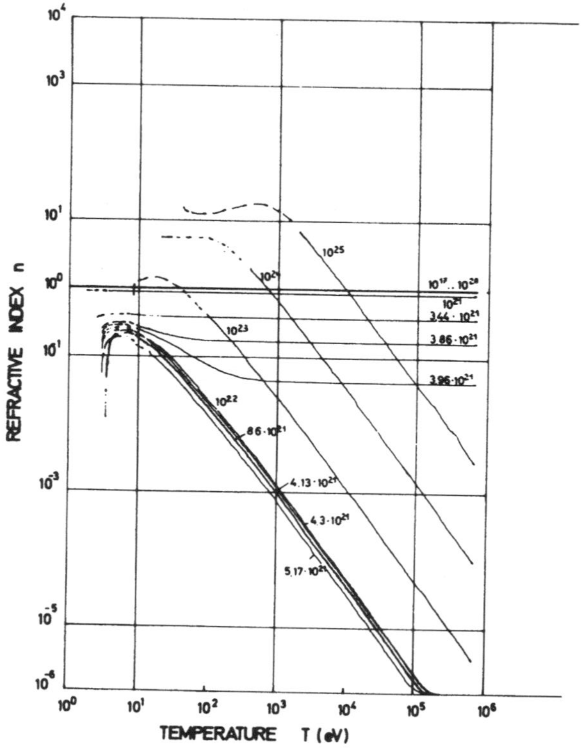

Fig. 5b. Refractive index n for the second harmonics of neodymium glass laser radiation as a function of the electron temperature T and atomic densities N in cm^{-3} (denoted as curve parameters) for fully ionized hydrogen and isotopes.

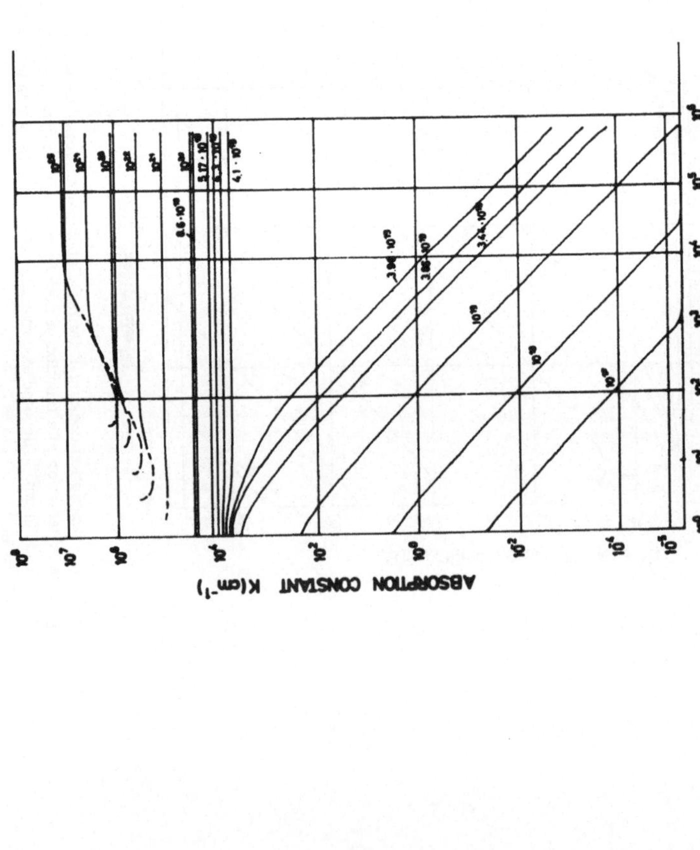

FIG. 6a. Absorption constant K for the second harmonics of CO_2 laser radiation as a function of the electron temperature T and atomic densities N in cm^{-3} (denoted as curve parameters) for fully ionized hydrogen and isotopes.

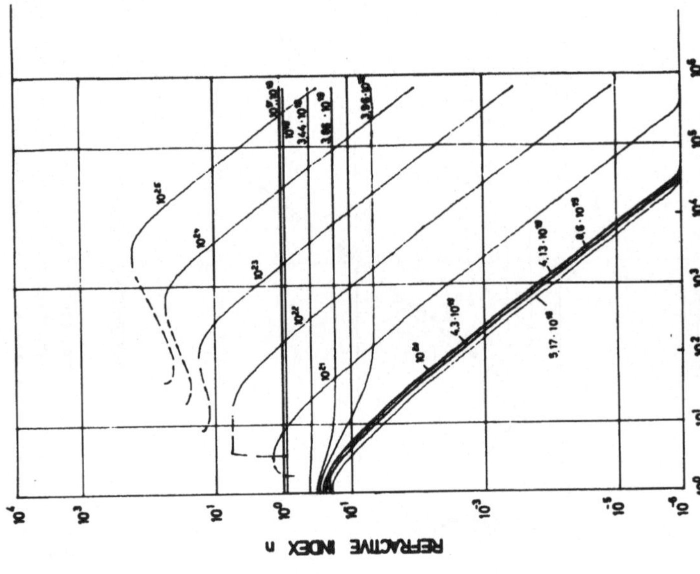

FIG. 6b. Refractive index n for the second harmonics of CO_2 laser radiation as a function of the electron temperature T and atomic densities N in cm^{-3} (denoted as curve parameters) for fully ionized hydrogen and isotopes.

cut-off density ($\omega_p \ll \omega$). These cases are the 45°-branches of the curves in Figs 1 to 6. Using a Gaunt factor g [15] with [T] = eV, $[N_e]$ = cm^{-3}

$$g = 1.2695(7.45 + \ln T - (1/3)\ln N_e) \qquad (12)$$

we find a ratio of the two absorption constants

$$\frac{K}{K_B} = 0.324\frac{\ln \Lambda}{\gamma_E(Z)\,g} = \frac{0.883}{\gamma_E(Z)}\left(1 - \frac{0.76 + \frac{2}{3}\ln Z}{7.45 + \ln T - \frac{1}{3}\ln N_e}\right) \qquad (13)$$

which is very close to unity. The quantum-mechanical calculation of K_p for $\omega_p \geq \omega$, including the collective effects, is still not available.

Dawson and Oberman [16] calculated the absorption constant K_{DO} on the basis of a microscopic theory. If $\omega_p \ll \omega$ we obtain

$$\frac{K}{K_{DO}} = \frac{1.18}{\gamma_E(Z)} \qquad (14)$$

and at $\omega_p \geq \omega$ we found a quite good agreement of K and K_{DO} which is possible by numerical calculation only [4]. Equation (14) demonstrates that the absorption constants of Dawson and Oberman are smaller than K by a factor of 2 at Z = 1 (hydrogen). The difference decreases at higher Z. This rough agreement proves, that the statically derived collision frequency ν is applicable even at such high frequences as that of light. The difference due to γ_E indicates that Spitzer's model presents some further details.

Calculations for Z = 1 to 4 and for the same cases as given in Figs 1 to 6 are available in report IPP3/83 on request from the authors.

ACKNOWLEDGEMENT

Very helpful discussions on this work with Dr. D. Pfirsch are gratefully acknowledged.

REFERENCES

[1] FÜNFER, B., KRONAST, KUNZE, H.J., Phys. Lett. 5 (1963) 125.
[2] BASOV, N.G., KROKHIN, in Quantum Electronics, (Proc. Conf. Paris, 1963) (BLOEMBERGEN, N., GRIVET, D., Eds) 2 Dunod, Paris (1964) 1373.
[3] SCHLÜTER, A., Z. Naturf. 5a (1950) 72.
[4] HORA, H., Institut für Plasmaphysik, Garching, Report 6/23 (1964); USAEC-Rep. NCR-TT 1193 (1965); HORA, H., Institut für Plasmaphysik, Garching, Report 6/27.
[5] KUNZE, H.J., Z. Naturf. 20A (1965) 801; MACE, P.N., Los Alamos Scientific Laboratories, Report 3369-UC-34, Phys. TID-4500 (1965); SALAT, A., Institut für Plasmaphysik, Garching, Report 6/49 (1966); OPOWER, H., PRESS, W., Z. Naturf. 21A (1966) 344; HAUGHT, A.F., POLK, D.H., Physics Fluids 9 (1966) 2047; ENGELHARDT, A.G., Westinghouse Research Report WERL-3472-5 (1967); TUCKFIELD, R.G., SCHWIRZKE, F., Plasma Physics 11 (1969) 11; KÜPPER, F.D., Euratom-FOM Rijnhuizen Rep. 68-44 (Febr. 1968); OPOWER, H., PUELL, H., HEINICKE, W., KAISER, W., Z. Naturf. 22A (1967) 1392; CAVALIERE, A., GIUPPONI, P., GRATTON, R., Phys. Lett. 25A (1967) 636; SUN, K.H., HICKS, J.M., EPSTEIN, L.M., SUCOV, E.W., J. appl. Phys. 38 (1967) 3402; BHADRA, D.K., Physics Fluids 11 (1968) 234.
[6] BASOV, N.G., KRUIKOV, P.G., ZAKHAROV, S.D., SENATSKY, Yu.V., TCHEKALIN, S.V., IEEE J. Quantum Electronics, QE-4, (1968) 864.
[7] FLOUX, F., COGNARD, D., BOBIN, J.-L., DELOBEAU, F., FAUQUIGNON, C., C.r. Acad. Sc. Paris 269B (1969) 697.
[8] SPITZER, L., Jr., Physics of Fully Ionized Gases, Interscience, New York (1956).
[9] HORA, H., PFIRSCH, D., SCHLÜTER, A., Z. Naturforsch. 22a (1967) 278; SCHLÜTER, A., Plasma Physics 10 (1968) 471; HORA, H., Physics Fluids 12 (1969) 182; Ann. Phys. 22 (1969) 402; Z. Physik 226 (1969) 156.
[10] KAW, P.K., SALAT, A.R., Physics Fluids 11 (1968) 2223; KIDDER, R.E., Paper presented at the AIAA Symposium, Los Angeles, Calif. (June 24, 1968).
[11] KAW, P., Appl. Phys. Lett. 15 (1969) 16.
[12] DAWSON, J., Reflection of Light, "Workshop on Laser Interaction and Related Plasma Phenomena", Hartford (June 1969), (SCHWARZ, H., HORA, H., Eds) Plenum Press, New York (1969).
[13] HUGHES, T.P., NICHOLSON-FLORENCE, M.B., J. Phys. A (Proc. Phys. Soc.) Ser. 2 1 (1968) 588; RAND, S., Phys. Rev. 136 (1964) B231.
[14] BORMATICI, M. CAVALIERE, A., ENGELMANN, F., Lett. Nuovo Cim. 1 (1969) 713; TSYTOVICH, V.N., SCHWARTSBURG, A.B., Soviet Phys. Tech. Phys. 12 (1967) 425.
[15] SMARD, S.F., WESTFOLD, K.C., Phil. Mag. 40 (1949) 831.
[16] DAWSON, J.M., OBERMAN, C., Physics Fluids 5 (1962) 517.

LASER-INDUCED INSTABILITIES AND ANOMALOUS ABSORPTION IN DENSE PLASMAS†

Donald F. DuBois

Los Alamos Scientific Laboratory

Los Alamos, New Mexico 87544

INTRODUCTION

In the last three years or so, the subject of radiation-induced parametric instabilities in plasmas has become an exceedingly active field of research. These instabilities have been observed in many experiments. High power transmitting antennas can excite these instabilities in the F-layer of the ionosphere where the resulting enhanced wave fluctuations are observed by radar backscatter techniques. Numerous laboratory experiments using Q-machines, DP devices and discharges and employing radiation sources in the radio-frequency to microwave regime have studied these instabilities. Many detailed theoretical predictions have been verified by these experiments, but at this early point there also remain many unanswered questions. Perhaps the most exciting but least understood application of these effects is toward the heating of laser produced plasmas for laser-induced fusion.

This is such a rapidly advancing subject that it is impossible to do it justice in a forty-five minute talk. I will therefore limit myself to the simplest cases which have received the most study. These instabilities were first seriously studied in 1965. (References 1, 2, and 5a.) At that time, however, available laser intensities were four orders of magnitude below the threshold intensities needed to excite these instabilities. Today available power densities are several orders of magnitude above this threshold.

†Research supported in part by A.F.O.S.R. contract #AFF 44 620-73-C-003. Presented at the Third Workshop on "Laser Interaction and Related Plasma Phenomena" held at Rensselaer Polytechnic Institute, Troy, New York, August 13-17, 1973.

Reprinted from: *Laser Interaction and Related Plasma Phenomena,* Vol. 3A, H. Schwarz and H. Hora (eds.), Plenum, New York, 1974, pp. 267–289.

The basic processes which interest us particularly for laser-fusion research are anomalous absorption and anomalous backscattering processes, all of which can be classed as parametric instabilities. These processes can be represented schematically as in Fig. 1. The anomalous absorption processes are three wave decay processes in which an intense electromagnetic wave is transformed by nonlinear interaction into two predominantly longitudinal plasma waves which ultimately give their energy to random particle motion

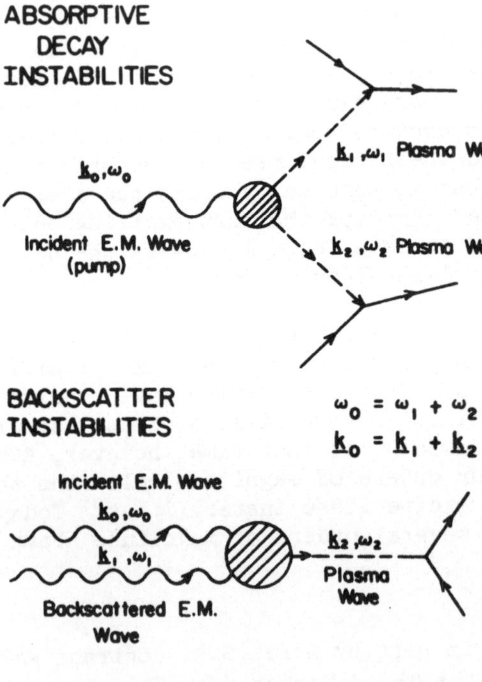

Figure 1. Schematic representation of three wave parametric instabilities. a) Absorptive decay instabilities b) Backscatter instability. (Decay of plasma waves is indicated by inverse Cerenkov effect into particle energy.)

via Landau or collisional damping. Thus electromagnetic energy is converted into plasma energy. In the case of an isotropic plasma ω_1 may correspond to a high frequency electron-plasma (Langmuir) waves $[\omega_1 = (\omega_{pe}^2 + 3k_1^2 v_e^2)^{\frac{1}{2}}]$ and ω_2 to an ion-acoustic wave $[\omega_2 \approx k_2(T_e/m_i)^{\frac{1}{2}}]$. We call this the electron-ion decay instability $(\omega_o \sim \omega_{pe})$.(1,2) If ω_1 and ω_2 both correspond to Langmuir waves, we call this the electron-electron decay instability $(\omega_o \sim 2\omega_{pe})$.(3) The absorptive parametric instabilities appeared at first sight to be just what was needed for plasma heating of laser-fusion pellets.(4) However, computer simulations, experiments, and some preliminary theory shows that these instabilities often produce very high energy tails on the electron velocity distribution which have detrimental preheating effects on laser-fusion schemes which rely on extreme compression of fuel pellets.

The backscatter instabilities in an isotropic plasma have two names: Stimulated Raman scattering (SRS)[5] if the excited plasma wave is a Langmuir wave, and stimulated Brillouin scattering (SBS)[6] if ω_2 corresponds to an ion-acoustic wave. In the inhomogeneous plasma encountered in laser-plasma interaction, these instabilities mostly backscatter the incident laser light. They do relatively little plasma heating and may prevent the radiation from reaching the region of density where the absorptive instabilities can act and where appreciable linear absorption by inverse bremsstrahlung can occur.

SUMMARY OF LINEAR THEORY

Next I will give a brief sketch of a phenomenological linearized theory of these instabilities which can all be treated from a common point of view. In a homogeneous, but not necessarily isotropic plasma, we can write the space-time Fourier transform of Maxwell's equations in the following from

$$\Lambda_{ij}(\underline{k},\omega)E_j(\underline{k},\omega) = \frac{4\pi i}{\omega} J_i(\underline{k},\omega) \tag{2.1}$$

$$\Lambda_{ij}(\underline{k},\omega) = \frac{k^2c^2}{\omega^2}\left[\delta_{ij} - \frac{k_ik_j}{|k|^2}\right] - \varepsilon_{ij}(\underline{k},\omega) \tag{2.2}$$

where $\varepsilon_{ij}(\underline{k},\omega)$ is the dielectric tensor which takes into account the linear induced currents in the plasma. The current $J_i^{NL}(k,\omega)$ arises from nonlinear induced currents which result from the beating of the intense pump wave

$$E^o{}_j(\underline{r},t) = \frac{E^o}{2} j\, e^{i(\underline{k}_o\cdot\underline{r} - \omega_o t)} + c.c. \tag{2.3}$$

with waves excited in the plasma. If the pump is sufficiently weak, we can write J^{NL} as a series and Eq. (2.1) becomes[1]

$$\Lambda_{ij}(\underline{k},\omega)E_j(\underline{k},\omega) = \chi^{(1)}_{ijk} E^o_j E_k(\underline{k}-\underline{k}_o, \omega-\omega_o) + \chi^{(1)}_{ijk} E^{o*}_j E_k(\underline{k}+\underline{k}_o, \omega+\omega_o) + \text{terms of form } \chi^{(n)}(E^o)^n E_k(\underline{k}-n\underline{k}_o, \omega-n\omega_o) \quad . \tag{2.4}$$

The coefficients such as $\chi^{(1)}_{ijk}$ (which depend on $\underline{k}$, $\underline{k}_o$, ω, ω_o) are called nonlinear susceptibility coefficients. This is the point where microscopic plasma physics comes in. Either from the kinetic equation (e.g., Vlasov eq.) or the fluid equations, we must compute the nonlinear response of the current to the electric fields in the plasma. I will not have time in this talk to discuss the details of these important but straightforward calculations.[1,3,7] Many cases have been considered by various workers. The χ_{ijk} (and higher order) coefficients have important symmetry properties which again I will not discuss here.

For the purposes of this talk, we will regard these coefficients as given phenomenological quantities. The advantage of this formulation, which essentially is that commonly used in nonlinear optics, is that all the parametric instabilities in a plasma can be viewed from a common point of view. On the other hand, the approach using the particle kinetic equations or fluid equations which emphasizes the particle variables rather than the associated wave variables seems to be favored by plasma physicists. In this approach it is harder to see the common features of all these instabilities.

Equation (2.4) as written has been linearized by neglecting terms such as $(2\pi)^{-4}\int d^3k'd\omega' \chi^{(1)}_{ijk} E_j(\underline{k}'\omega') E_k(\underline{k}-\underline{k}',\omega-\omega')$ which are of second order or higher in the "weak" Fourier coefficients. Such terms must be retained to study the nonlinear theory of the saturation level of these instabilities. In the linearized theory, we also assume that the pump parameters are fixed (the origin of the term parametric). Then the expansion on the right-hand side of Eq. (2.4) may contain arbitrarily high powers of $|E_o|$. More detailed considerations[3,7] show that only the leading terms of this expansion are required for a nearly collisionless plasma if

$$d_{oe} = \frac{eE_o}{m_e\omega_o^2} << \frac{\lambda}{2\pi} \quad . \tag{2.6}$$

---

[1] repeated indices are summed

That is, if the excursion amplitude of an electron in the pump field, d_{oe}, is much less than the wavelength of the plasma waves λ involved. In cases of large E_o, the coefficients of the beat amplitudes are often Bessel functions of argument (d_{oe}/λ).(3,7)

Although it is not an essential restriction in carrying out the linearized theory, I will assume for simplicity that the directed velocity $\underline{v}_o = (e\underline{E}_o/m\omega_o)$ of electrons in the oscillating laser field is much less than the thermal velocity $v_e = (k_B T_e/m_e)^{\frac{1}{2}}$. This guarantees that $|\underline{k} \cdot \underline{d}_{oe}| = (v_o/v_e)(k/k_{De})(\omega_{pe}/\omega_o) \ll 1$ provided $k \ll k_{De}$ -- the electron Debye wavenumber. For $k_B T_e = 1$ keV and a laser intensity $I_o = cE_o^2/4\pi$ of 10^{15}W/cm^2, the ratio is $v_o^2/v_e^2 \approx 3\times10^{-2}$.

Even in the linearized theory we have an infinite set of coupled equations for the various Fourier beat amplitudes. This set of equations can often be truncated by noting that waves of a given polarization can only be resonantly excited if their linear dispersion relation is satisfied. We introduce a complete set[1] of polarization vectors $\hat{e}_{i\alpha}(\alpha=1,2,3)$ which diagonalize the Maxwell operator $\Lambda_{ij}(\underline{k},\omega)$ in Eq. (2.1)

$$\Lambda_{ij}(\underline{k},\omega)\ \hat{e}_{j\alpha}(\underline{k},\omega) = \Lambda_\alpha(\underline{k},\omega)\ \hat{e}_{i\alpha}(\underline{k},\omega) \tag{2.7}$$

The (usually complex) solutions of $\Lambda_\alpha[\underline{k},\hat{\omega}_\alpha(\underline{k})] = 0$ for $\hat{\omega}_\alpha$ in terms of $\underline{k}$ give the linear dispersion relation for waves of polarization $\hat{e}_{i\alpha}$. In general these solutions are complex values $\hat{\omega}_\alpha = \omega_\alpha - i\lambda_\alpha$ for fixed $\underline{k}$. In other words, $\Lambda_\alpha^{-1}(k,\omega)$ is the linear response function for the harmonic oscillator representing the mode with polarization $\hat{e}_\alpha(k,\omega_\alpha)$ and wave vector $\underline{k}$. For the example of an homogeneous isotropic plasma, we have three mutually perpendicular polarization vectors $\hat{e}_{i\alpha}(\alpha=1,2,3)$ where $\hat{e}_{i1} = \hat{k}_i$ with $\Lambda_1(k,\omega) = -\epsilon_L(k,\omega)$ where ϵ_L is the usual longitudinal dielectric function. For $\alpha = 2,3$, we have two mutually perpendicular transverse polarization vectors and $\Lambda_\alpha(k,\omega) = (k^2c^2/\omega^2) - \epsilon_T(k,\omega)$ where ϵ_T is the transverse dielectric function. We represent the coefficients $E_i(\underline{k}\omega)$ as a superposition of these polarization vectors with coefficients $\mathcal{E}_\alpha(\underline{k}\omega)$. Then, barring certain degeneracies, it is often the case that if the frequency and wavevector matching condition

$$\begin{aligned} \omega_o(\underline{k}_o) &= \omega_1(\underline{k}_1) + \omega_2(\underline{k}_2) \\ \underline{k}_o &= \underline{k}_1 + \underline{k}_2 \end{aligned} \tag{2.8}$$

is satisfied for two modes, α=1 and α=2, that only these two modes can be simultaneously resonantly excited by the parametric inter-

---

[1]Actually, a biorthogonal set since Λ_{ij} is non-hermitian.

action and the Eqs. (2.4) reduce to a 2×2 set:

$$\Lambda_1(\underline{k},\omega)\,\mathcal{E}_1(\underline{k},\omega) = \chi^{(1)}_{102}\,\mathcal{E}_o\,\mathcal{E}_2(\underline{k}-\underline{k}_o,\omega-\omega_o)$$

$$\Lambda_2(\underline{k}-\underline{k}_o,\omega-\omega_o)\,\mathcal{E}_2(\underline{k}-\underline{k}_o,\omega-\omega_o) = \chi^{(1)}_{201}\,\mathcal{E}^*_o\,\mathcal{E}_1(\underline{k},\underline{\omega}) \qquad (2.9)$$

where $\chi^{(1)}_{102}$ is the projection of $\chi^{(1)}_{ijk}$ on the polarizations of the three waves involved

$$\chi^{(1)}_{102} = \hat{e}_{i1}\,\chi^{(1)}_{ijk}\,\hat{e}_{jo}\,\hat{e}_{k2} \qquad (2.10)$$

where $\hat{e}_{jo}$ is the polarization vector of the pump. We see here that the beating of one Fourier component with the pump acts as a source or driver for the other Fourier component. This "bootstrapping" or regenerative effect is the source of the instability. The effect is strong only if the two driven harmonic oscillators are <u>simultaneously</u> near resonance, a condition which leads to Eq. (2.8).

The dispersion relation for the coupled wave problem is obtained by setting the determinant of Eqs. (2.9) to zero. If we solve this for complex ω for fixed values of $\underline{k}$, we find that there are temporally growing solutions corresponding to instability provided the pump amplitude is greater than a threshold value. If we define a quantity Γ of dimensions sec^{-1}

$$\Gamma^2(\underline{k}) = \frac{\chi_{102}\chi_{201}|\mathcal{E}_o|^2}{\left.\dfrac{\partial \mathrm{Re}\Lambda_1}{\partial\omega}\right|_{\omega=\omega_1}\left.\dfrac{\partial \mathrm{Re}\Lambda_2}{\partial\omega}\right|_{\omega_o-\omega=\omega_2}} \qquad (2.11)$$

the threshold condition can be written as[1,7]

$$\Gamma^2(\underline{k}) > \Gamma^2_m = \gamma_1\gamma_2 \qquad (2.12)$$

the threshold is determined by the uncoupled wave damping decrements. These relations hold for all three-wave decay-type parametric instabilities including those in a magneto-plasma provided only that the uncoupled waves are sharply defined as modes, i.e., $\gamma_1/\omega_1 \ll 1$, $\gamma_2/\omega_2 \ll 1$. The quantity Γ is the growth rate of the coupled modes for perfect frequency matching, $\Delta\omega = \omega_o - \omega_1 - \omega_2 = 0$, in the <u>absence</u> of damping ($\gamma_1 = \gamma_2 = 0$).

Near threshold, the growth rate depends critically on $\gamma_1\gamma_2$

and $\Delta\omega$:

$$\gamma_g(\varepsilon_o, k) = \frac{\gamma_1\gamma_2}{\gamma_1+\gamma_2}\left\{\frac{\Gamma^2(\overline{k})}{\gamma_1\gamma_2}\frac{1}{1+\left(\frac{\Delta\omega}{\gamma_1+\gamma_2}\right)^2} - 1\right\} \tag{2.13}$$

Notice that for $\Delta\omega=0$, $\gamma_g=0$ for $\Gamma^2=\Gamma_m^2$. For $\Delta\omega\neq 0$, a higher value of Γ^2 is required to produce instability ($\gamma_g > 0$).

In this near-threshold limit, γ_g behaves linearly with $|\mathcal{E}_o|^2$. In the electron-ion decay instability and in SBS when the pump is sufficiently high above threshold so that the growth rates becomes comparable in magnitude to the ion acoustic frequency, the dispersion relation predicts[9] that the growth rate then is proportional to $|\mathcal{E}_o|^{2/3}$.

If the solution of the dispersion relation is used to relate ω to $\underline{k}$ in Eqs. (2.9), we can solve for the amplitude ratio $\mathcal{E}_1(\underline{k},\omega)/\mathcal{E}_2(\underline{k}-\underline{k}_o, \omega-\omega_o)$. The result can be summarized in the so called Manley-Rowe relations which simply express the fact that for a three wave decay process for every pump photon absorbed, a plasmon appears in the modes 1 and 2. Letting n_o, n_1, and n_2 represent the quantum occupation numbers (or more strictly, the change in these quantities), we have

$$n_o = n_1 = n_2 \tag{2.14}$$

or since the energy in the mode α is related to n_α by $W_\alpha = n_\alpha h\omega_\alpha$, we have

$$\frac{W_o}{\omega_o} = \frac{W_1}{\omega_2} = \frac{W_2}{\omega_3} \tag{2.15}$$

Thus the energies in the coupled modes are, in the ratios of their frequencies, the greatest energy in the highest frequency mode. If the SBS and SRS instabilities are strongly excited, this means that most of the absorbed pump energy will appear in the scattered light wave and relatively little will be absorbed in the lower frequency plasma waves. In Table I, the properties are listed of the four decay-type parametric instabilities which can be excited in a non-magnetized plasma by an intense electromagnetic wave pump. The geometry of the $\underline{k}$ vectors, which satisfy the matching condition, relative to the pump polarization is shown, the 3 frequency dispersion relations involved, the form of Γ^2 for each instability. Numerical values of the pump intensity $I_o = E_o^2c/8\pi$ in Watts/cm^2 at threshold for the case of an Nd-glass laser with $n_c = 10^{21}$ cm^{-3} at the critical density ($\omega_o=\omega_{pe}$) and $T_e = 1$ keV and for the case of a CO_2 laser with $n_c = 10^{19}$ cm^{-3}. For the two absorptive decay

TABLE I

GEOMETRY	FREQUENCIES $\omega_o = \omega_1 + \omega_2$	Γ^2	THRESHOLDS W/cm^2 Nd	CO$_2$
e-i Decay 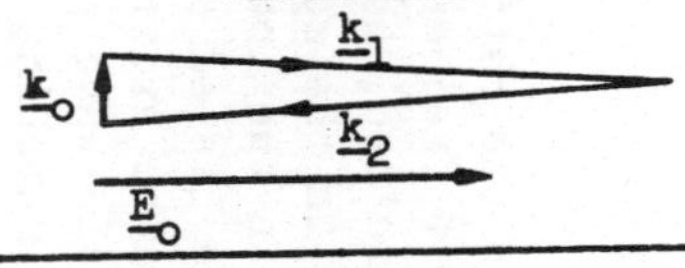	$\omega_o = \omega_{em}(k_o)$ $\omega_1 = \omega_L(k_1)$ $\omega_2 = \omega_a(k_2)$	$\frac{\Lambda_o^2}{16} (\hat{\underline{k}} \cdot \hat{\underline{e}}_o)^2 \omega_1 \omega_2$ $(\Lambda_o \equiv \frac{E_o^2}{4\pi n T_e}$	Nd 10^{13} 10^{11} $(T_e >> T_i)$	10^{10}
e-e Decay 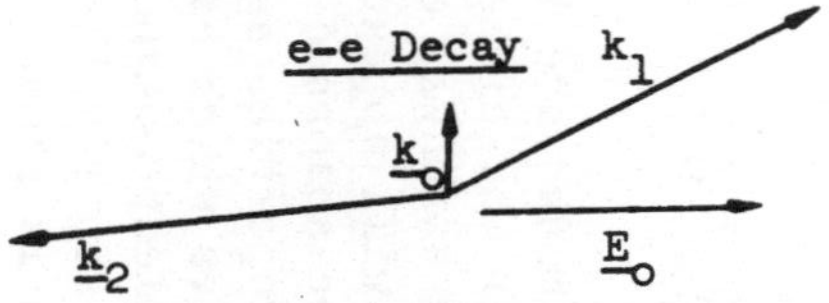	$\omega_o = \omega_{em}(k_o)$ $\omega_1 = \omega_L(k_1)$ $\omega_2 = \omega_L(k_2)$	$\frac{\Lambda_o^2}{64} \frac{k_o^2}{k_D^2} (\hat{\underline{k}} \cdot \hat{\underline{k}}_o)^2 (\hat{\underline{k}} \cdot \hat{\underline{e}}_o)^2 \omega_1 \omega_2$	$\sim 10^{13}$	10^9
SBS 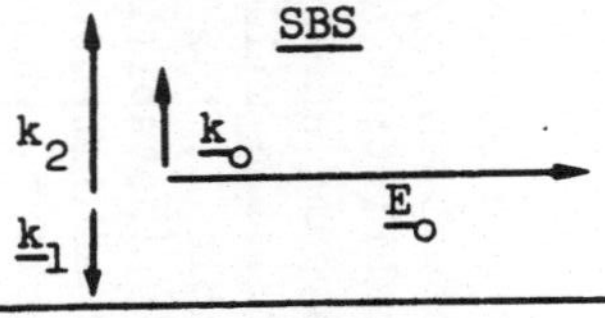	$\omega_o = \omega_{em}(k_o)$ $\omega_1 = \omega_{em}(k_1)$ $\omega_2 = \omega_a(k_2)$	$\frac{\Lambda_o^2}{16} (\hat{\underline{e}}_1 \cdot \hat{\underline{e}}_o)^2 \frac{\omega_{pe}^4}{\omega_o^4} \omega_1 \omega_2$	$\sim 10^{13}$	10^{10}
SRS 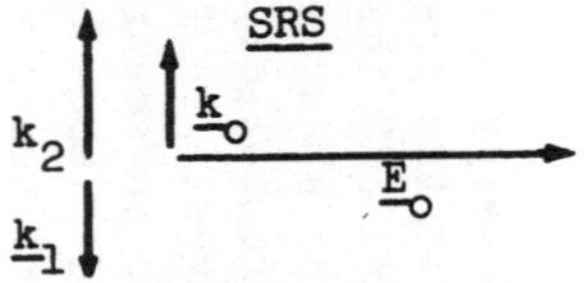	$\omega_o = \omega_{em}(k_o)$ $\omega_1 = \omega_{em}(k_1)$ $\omega_2 = \omega_L(k_2)$	$\frac{\Lambda_o^2}{16} \frac{k_2^2}{k_D^2} (\hat{\underline{e}}_1 \cdot \hat{\underline{e}}_o)^2 \frac{\omega_{pe}^2}{\omega_o^2 \omega_1^2} \omega_1 \omega_2$	$\sim 10^{13}$	10^9

Electromagnetic $\omega_{em}(k) = (\omega_{pe}^2 + k^2c^2)^{1/2}$; Langmuir $\omega_L(k) = (\omega_{pe}^2 + 3k^2v_e^2)^{1/2}$; Ion Acoustic $\omega_a = k(\gamma T_e/m_i)^2$

instabilities, only the cases $k_o << k_1, k_2$ are shown and for SRS and SBS only the geometry for backscatter is shown.

There are several cases of parametric instabilities in which one mode is actually not well defined or is a quasi-mode. This is the case for the e-i decay instability[1] or SBS for $T_e = T_i$ for which the ion acoustic mode is not well-defined. In this case, the process we are discussing may be more correctly described as induced scattering (or conversion) from ions off transverse waves into Langmuir waves or transverse waves (see Fig. 2).

In the case in which mode 2 is a longitudinal wave, this growth rate is given by[1]

$$\gamma_g(E_o,k) \simeq \frac{\chi_{102}\chi_{201}|\mathcal{E}_o|^2}{\left.\frac{\partial \mathrm{Re}\Lambda_1}{\partial\omega}\right|_{\omega_1}} \left| \mathrm{Im}\, \frac{1}{\varepsilon(k-k_o,\omega_1-\omega_o)} \right| - \gamma_1 \qquad (2.16)$$

(This assumes that the real parts coupled mode frequencies are not shifted far from their uncoupled values.) The condition for maximum growth, i.e., maximum $\mathrm{Im}\,\varepsilon^{-1}(k-k_o,\omega_1-\omega_o)$ provides a matching condition $\omega_1(k) - \omega_o = -1.7\,(k/k_{De})\sqrt{m_e/m_i}\,\omega_p$ which can be obtained from tables of the plasma dispersion function.

The simple two-mode coupling scheme fails in several important cases. In these cases, the "anti-stokes" mode with Fourier component $\omega_o+\omega$, $\underline{k}_o+\underline{k}$ can also be resonantly excited in addition to the Fourier components at $\omega,\underline{k}$ and $\omega_o-\omega$, $\underline{k}_o-\underline{k}$ already considered. In this problem, we have 3 coupled equations to solve and thus a somewhat more complicated dispersion relation.[8,9] Two important cases arise:

1) All three modes resonantly coupled: This can occur in the electron-ion decay instability if $2\omega_2 << \gamma_1$ and for forward Brillouin and Raman scattering. The instability threshold is generally higher if the third mode can be resonantly coupled.[8]

2) Purely growing mode.[9] In this case, the stokes and anti-stokes modes $\omega_o-\omega$ and $\omega_o+\omega$ are resonantly coupled but the mode ω is shifted far from a resonance. For the case $k_o=0$, i.e., a homogeneous pump field, this leads to a root of the dispersion relation which is unstable but has a zero real frequency, i.e., a purely growing instability with wavevector $\underline{k}$. The Fourier components at $\omega_o\pm\omega$ then oscillate at exactly the pump frequency ω_o and grow in time with the growth rate $\mathrm{Im}\omega$. When $k_o\neq0$, this mode with wavevector $\underline{k}$ retains a small real frequency proportional to k_o. The

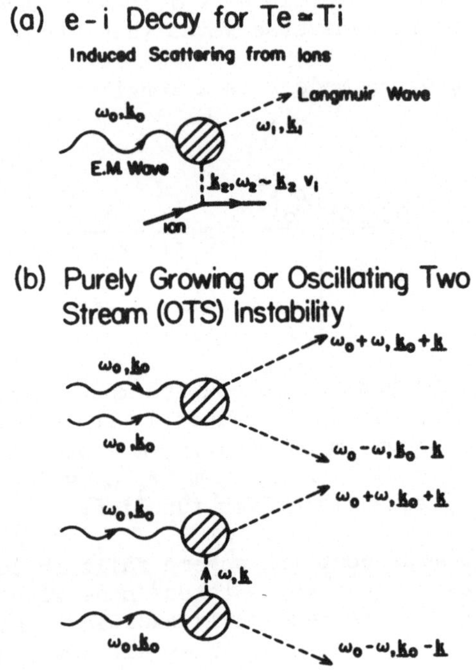

Figure 2. a) Decay instability for $T_e = T_i$ which is best represented as induced scattering or conversion of E.M. waves into Langmuir waves from ions. b) Schematic representation of mode coupling for the purely growing instability which involves an off resonant ion wave (ω,$\underline{k}$).

threshold and growth rate for this instability is nearly equal to that of the electron-ion decay instability if the electron (T_e) and ion temperatures (T_i) are nearly equal. It has a higher threshold if $T_e >> T_i$. This instability excites electrostatic waves with $k > k_c$. It appears that the quasi-linear decay of shorter wavelength waves excited by the purely-growing instability are responsible for the high energy electron tails observed in the computer simulations.(10a) In the forward scattering geometry of SRS and SBS, i.e., when the scattered light wave is in the same direction as the laser wave, the "anti-stokes" mode $\underline{k} + \underline{k}_o, \omega+\omega_o$ can likewise be resonantly excited. The resulting zero frequency or low frequency instability goes by the name of the self-focusing instability(10b) or the modulational instability(10c) depending on the relative orientation of $\underline{k}$ and $\underline{k}_o$. A schematic diagram of the wave coupling involved in these instabilities is shown in Fig. 2.

PARAMETRIC INSTABILITIES IN INHOMOGENEOUS PLASMAS

All actual experiments are performed in bounded or inhomogeneous plasmas. In these situations, it is necessary to understand the spatial behavior of the parametric instabilities. Consider first a plane-layered plasma whose properties vary only in the z direction. We also restrict attention to the case in which the pump electromagnetic wave propagates with $\underline{k}_o$ in the z direction, polarized in the transverse direction. (Thus, we avoid the important problems of tunneling and longitudinal wave resonance associated with oblique incidence.)

Consider for a moment a <u>homogeneous</u> plasma. The dispersion relation arising from Eqs. (2.9) can also be solved for complex $\underline{k}$ for fixed real values of ω. Then for given values of k_x, k_y, we can find complex solutions for k_z corresponding to spatially growing (or convective) modes provided

$$\gamma_1\gamma_2 = \Gamma_m^2 << \Gamma^2(k) << \frac{|v_1 v_2|}{4}\left(\frac{\gamma_1}{v_1} - \frac{\gamma_2}{v_2}\right)^2 \qquad (3.1)$$

where $\underline{k} = (k_x, k_y, \mathrm{Re}k_z)$ and v_1, v_2 are the group velocity components of the uncoupled waves along the z (or pump) direction. These quantities generally depend on $(k_x, k_y, \mathrm{Re}k_z)$. In this regime, a wave packet of unstable modes will grow as it moves in space but will not grow in time at a fixed point in space.

However, one can show by several methods of analysis[11,12] that for the backward wave case $v_1 v_2 < 0$, the dispersion relation leads to absolute temporal instability (growth in time at a fixed spatial point) provided

$$\Gamma^2 > \Gamma_T^2 = \frac{|v_1 v_2|}{4}\left(\frac{\gamma_1}{v_1} - \frac{\gamma_2}{v_2}\right)^2. \tag{3.2}$$

This threshold is generally higher than that (Γ_m) for spatial growth except for the $\omega_o = 2\omega_{pe}$ decay instability for which $\gamma_1/v_1 = -\gamma_2/v_2$ so that $\Gamma_T = \Gamma_m$. For SRS and SBS there is an intermediate pump region of spatial growth where $\Gamma_m < \Gamma < \Gamma_T$. In the absolute instability regime for SBS and SRS, there is spatial growth as well as temporal growth; the peak of the back scattered pulse moves with a velocity $[v_1+v_2]/2$ and grows. This spatial growth may be more significant than the temporal growth for short laser pulses.[13] In this case, the absolute regime may not differ significantly from the convective regime.

In an inhomogeneous plasma, the $\underline{k}$ matching condition $\underline{k}_o = \underline{k}_1+\underline{k}_2$ can only be satisfied locally since k is not a constant over the ray path of the wave. The region over which matching can be approximately satisfied is called the interaction region. For the convective instabilities, the waves will grow exponentially in space only in this limited region and thus the gradients limit the size to which the excited waves can grow. Of course, nonlinear effects may still be a limiting factor if this spatial amplification is sufficiently large. We impose the rather arbitrary criterion for threshold in this case that the wave (power) grows by a factor e^A where, say, A is 5 in the interaction region. The stronger the gradient the smaller the interaction region and so the pump amplitude must be stronger to achieve the same total spatial amplification over the interaction region. This type of reasoning leads to the criterion[14,12]

$$A < \frac{2\pi\Gamma^2}{\kappa' \, v_1 v_2} \tag{3.3}$$

with $\kappa' = d(\Delta k)/dz$ where $\Delta k(z) = k_{zo}(z)-k_{z_1}(z)-k_{z_2}(z)$ is the local mismatch of the z components of the wavevectors calculated in the WKB approximation for the three waves; k' is inversely proportional to the scale height, H, of the density gradient. In Table II are shown the threshold conditions dependent on the density gradient for our 4 instabilities. The factor in parenthesis multiplying $\gamma_1\gamma_2$ in these formulas is the enhancement factor of the threshold due to gradient effects. The rough order of magnitude of this factor is given for the same laser-plasma parameters as used in Table I.

Note that the electron-ion decay instability threshold is the least strongly effected by the gradient since in the geometry chosen, the excited electrostatic waves propagate normal to the density gradient and thus "stay" in the interaction region except for diffraction spreading of the waves along the z direction.(15a) SBS and SRS have their thresholds raised more strongly by the gradient because the backscattered waves rapidly propagate out of the interaction region. The enhancement factor for SBS is dominated by the temperature gradient if $k^2c^2/\omega_p^2 \gg 1$ since the acoustic mode frequency depends only on temperature. In the plane-layered geometry considered here, the sidescattering SRS or SBS process in which the scattered light wave propagates normal to the gradient also has a lower inhomogeneity threshold.(15b,c) The plasma wave involved propagates at 45° to the gradient direction. Side-scattering which has a higher homogeneous threshold and lower growth rate may dominate backscatter for sufficiently strong gradients provided the laser beam width is large enough to permit appreciable growth in the transverse direction.(15c) The electron-electron decay instability usually behaves an an absolute instability since $\underline{k}_1 = -\underline{k}_2$ if $k_1 \gg k_o$ so that $\gamma_1/v_1 = -\gamma_2/v_2$. The propagation velocity of this coupled mode ($v_o = (\bar{v}_1+\bar{v}_2)/2$) is zero and we find only temporal growth in the interval Δz surrounding the interaction region.(12) In this interval, the wavegrowth may saturate for large times due to inhomogeneity effects(13) or may be limited by nonlinear effects.

Recent work indicates an important sensitivity of the backward wave, inhomogeneous problem to boundary conditions. If a linear mismatch $k(z) = \kappa'(z)$ is assumed and also that the right and left going waves decay at $z = \pm\infty$, respectively, it can be shown(13) that no absolute instability exists in an inhomogeneous plasma, i.e., a pulse will begin to grow at the absolute instability growth rate but will ultimately saturate and die away as the pulse spreads out of the interaction region. However, if the homogeneous plasma is bounded (as it always is), an absolute instability which grows exponentially in the asymptotic time limit can be excited even when the amplitudes of the oppositely traveling waves are taken to be zero at the entrance boundaries.(12)

NONLINEAR THEORY

The nonlinear theory of the parametric instabilities is, of course, the most poorly understood area of interest. In this regard, computer simulations have been very helpful in indicating important nonlinear effects, especially in cases of very strong pumping. I will concentrate on some recent results on the saturation of the electron-ion decay instability which is the case that has received the most attention from the "analytical theorists."

The question of most interest is "how much of the electromagnetic energy of the pump can be converted into heating of

TABLE II

STABILITY	CONVECTIVE GROWTH THRESHOLD	ENHANCEMENT FACTOR
e-i DECAY	$\Gamma^2 > \gamma_1\gamma_2(1 + \frac{A}{2}\frac{\omega_{p_e}}{\gamma_c}\frac{1}{Hk_1})$ (H is scale height of density gradient)	H = .01-.1 cm 1 - 10
SRS	$\Gamma^2 > \gamma_1\gamma_2(1 + \frac{A}{4\pi}\frac{\omega_o^2}{\omega_1\omega_2}\frac{k_1 c^2}{\gamma_c^2 H})$	$10^4 - 10^5$
e-e DECAY	NONE	
SBS	$\Gamma^2 > \gamma_1\gamma_2(1 + \frac{A}{2\pi}\left[\frac{\omega_2}{\gamma_2}\frac{\omega_o}{\gamma_c}\frac{1}{k_o H} + \frac{k_o^2 c^2}{\omega_p^2}\frac{H}{H_T}\right])$ (H_T is scale height of temperature gradient)	$10 - 10^2$
PURELY GROWING (OTS)	NONE	

electrons and ions?" In this process, we have electromagnetic energy of the pump transformed into energy associated with coherent electrostatic plasma waves which is ultimately converted into random particle energy. Since the rate of energy loss by the pump by the parametric process is exactly equal to the energy gained by the plasma waves, we can define a nonlinear conductivity (or sometimes misleadingly identified with an effective collision frequency, ν_{eff}).[16]

$$\sigma^{NL}(E_o)\, E_o^2 \equiv \frac{\nu_{eff}}{4\pi} E_o^2 = \int \frac{d^3k}{(2\pi)^3} \left[\gamma_g(E_o,k) - \gamma_g(o,k)\right] \frac{\langle |E_k|^2 \rangle}{4\pi} \tag{4.1}$$

Here the left hand side is the effective rate of pump energy depletion and the right hand side is the rate of growth of the energy of the excited electrostatic waves. (Note -- we must subtract the growth rate for $E_o = 0$, a negative contribution corresponding to damping, since we want only that part of the growth rate due to the pump.) Alternatively, the nonlinear conductivity can be viewed as arising from the interaction of the pump wave with the enhanced density fluctuation levels arising from the instability.[4] The well-known Dawson-Oberman formula predicts an increased σ^{NL} arising from the enhanced ion density fluctuation level. Not so well-known is the increased σ^{NL} due to enhanced electron-density fluctuations which is usually a much weaker effect but may dominate in a case to be discussed below. It is clear that to compute the wave intensity $I(\vec{k}) = \langle |E_k|^2 \rangle$, we need a saturated or nonlinear theory of the instability since E_k grows without bound as $\exp[\gamma_g(E_o,k)t]$ in the linear theory, and no finite steady state wave energy would result.

In the linearized theory discussed above for fixed plasma parameters, only a finite region of $\underline{k}$ space is unstable, i.e., corresponds to $\gamma_g(E_o,k) > 0$. This is the lense shaped region shown in Fig. 3a which is centered at k_m for which the frequency matching condition $\Delta\omega(k_m) = \omega_o - \omega_L(k_m) - \omega_a(k_m) = 0$ is exactly satisfied. The unstable region around k_m increases in size as ε_o increases. For the cases $k_m \lesssim .2k_{De}$, the primary saturation mechanism of the electron-ion decay instability for equal electron and ion temperatures has been shown[7, 16, 17] to be the process of induced scattering of Langmuir waves from ions. This procers is diagramed schematically in Fig. 3b. Note that this is a ladder of successive induced scattering processes beginning with the primary interaction which is the induced conversion of a transverse e.m. wave off ions into a Langmuir wave which in turn acts as a pump to scatter into still another Langmuir wave, etc. The system is stabilized by scattering waves out of the unstable region of $\underline{k}$ space into a stable region in which they are collisionally or Landau damped. Recent detailed numerical solutions[18] of the nonlinear equations describing this process have produced a saturated wave spectrum concentrated in umbrella shaped regions as shown schematically in Fig. 4a. These regions spread toward values of k lower than k_m as shown schematically in Fig. 4b which shows the angle averaged intensity $\bar{I}_k = \frac{1}{2}\int_{-1}^{1} d\mu\, I(k,\mu)$ where μ is the cosine of the angle between $\underline{k}$ and $\underline{\vec{E}}_o$. Early analytical work[16,17] which neglected the fine structure of the spectrum resulted in a curve such as the dotted curves in Fig. 4b. The spectrum gets higher and spreads to lower k as E_o increases resulting in a total enhanced electrostatic wave energy proportional to E_o^4, a result also found in the detailed numerical results. The peaks in the fine structure of the spectrum are separated by multiples of roughly $\Delta k = 2\omega_a(k_m)/V_g(k_m) = (2/3)\, k_D \sqrt{m_e/m_i}$ where $\omega_a \simeq k_m c_s$ is an ion acoustic frequency, and $V_g(k_m)$ is the group velocity of the Langmuir waves. This saturated wave spectrum can be used in Eq. (4.1) to compute a nonlinear conductivity (or effective collision frequency)[19]

$$\sigma^{NL}(E_o) = \eta\, \frac{\omega_{pe} E_o^2}{16\pi n T_e} \tag{4.2}$$

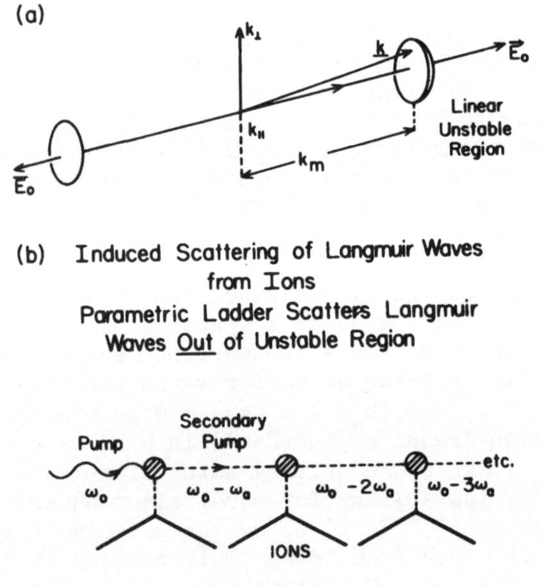

Figure 3. a) Linearly unstable regions of k space for e-i decay instability are the lens shaped regions at $k \simeq k_m$ centered about the direction of the pump $\vec{E}_o$. b) Schematic representation of the saturation process of induced scattering of Langmuir waves from ions.

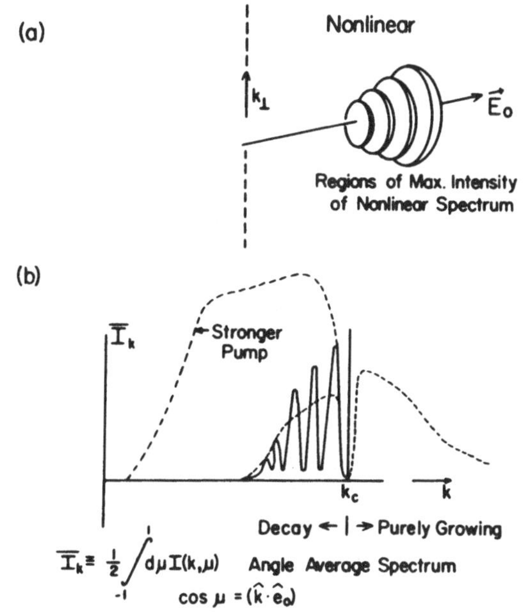

Figure 4. a) Regions of maximum intensity of the saturated e-i decay instability spectrum are shown to the right.

b) Angle averaged spectrum of Langmuir waves. Solid lines represent structure found in detailed numerical solutions. Dashed lines show results of approximate analytical theory. Spectrum in purely growing region is purely schematic.

where η varies between 1/3 and 4/3 depending on the details of the nonlinear theory used.[19] As E_o increases, the spectrum described above spreads to $k = 0$, and for higher values of E_o, the total energy goes as E_o^2 (instead of E_o^4) with the result that σ^{NL} asymptotically approaches a constant value independent of E_o. Application of the Manely-Rowe relations to the electron-ion decay instability shows that the energy in the low frequency mode (ion acoustic waves) is much less than the energy in Langmuir waves. Thus it can be shown[16] that this enhanced conductivity arises from enhanced electron-density fluctuations rather than ion fluctuations.

Because of the flow of energy from pump into electrostatic waves, the amplitude of the pump inside the plasma is less than that predicted by linear theory for the same boundary conditions. To compute the steady state pump amplitude self-consistently inside an inhomogeneous plasma, it is necessary to solve the wave equation

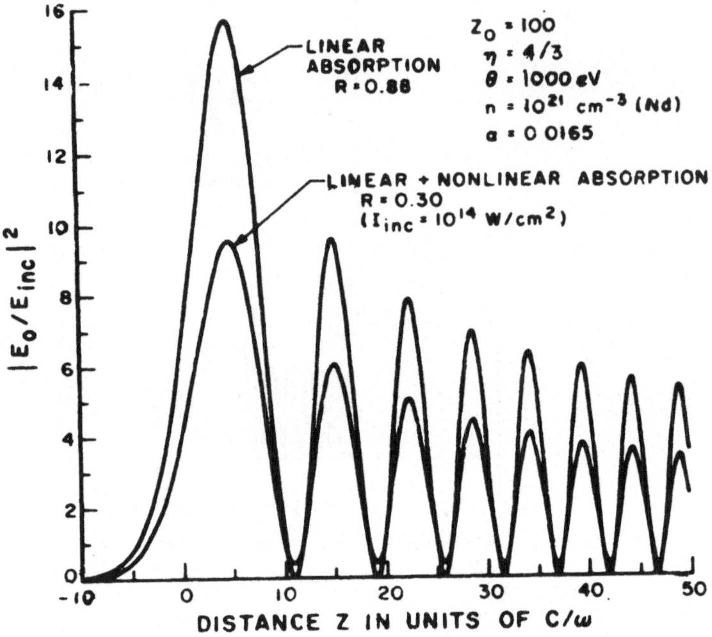

16
14
12
10
8
6
4
2
0
$|E_0/E_{inc}|^2$
LINEAR ABSORPTION R = 0.88
LINEAR + NONLINEAR ABSORPTION R = 0.30 (I_{inc} = 10^{14} W/cm²)
Z_0 = 100
η = 4/3
θ = 1000 eV
n = 10^{21} cm^{-3} (Nd)
a = 0.0165
-10
0
10
20
30
40
50
DISTANCE Z IN UNITS OF C/ω

(a)

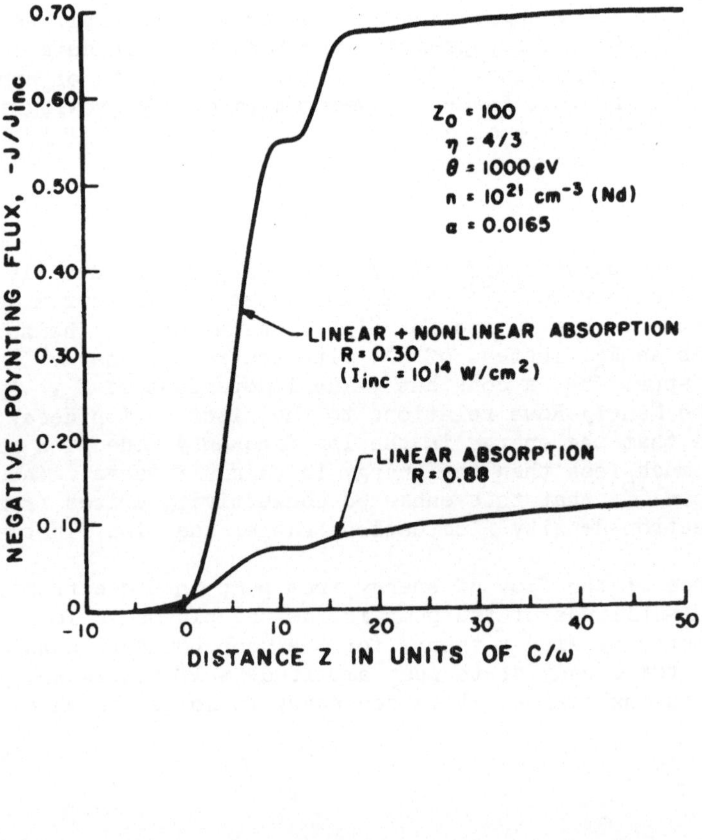

0.70
0.60
0.50
0.40
0.30
0.20
0.10
0
NEGATIVE POYNTING FLUX, $-J/J_{inc}$
Z_0 = 100
η = 4/3
θ = 1000 eV
n = 10^{21} cm^{-3} (Nd)
a = 0.0165
LINEAR + NONLINEAR ABSORPTION R = 0.30 (I_{inc} = 10^{14} W/cm²)
LINEAR ABSORPTION R = 0.88
-10
0
10
20
30
40
50
DISTANCE Z IN UNITS OF C/ω

(b)

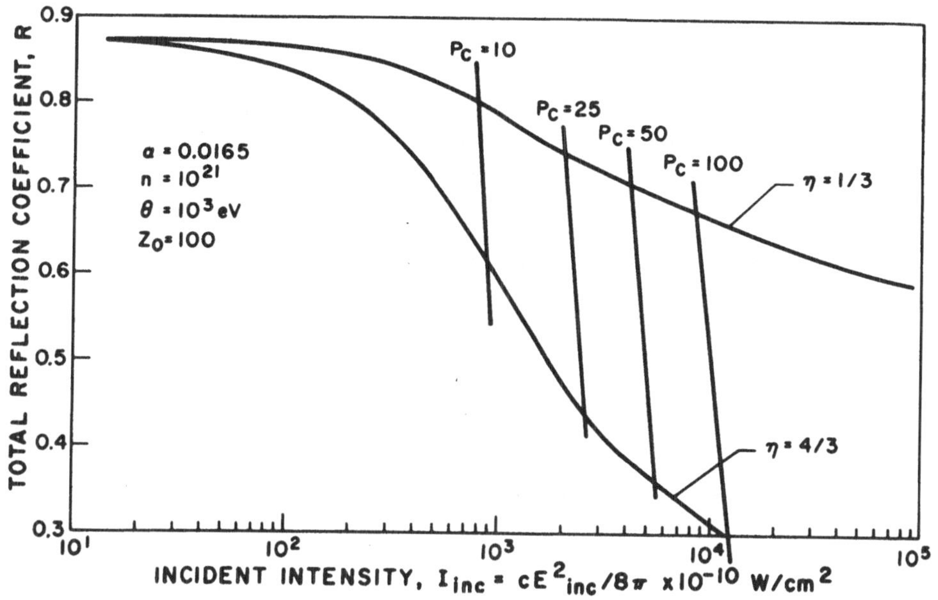

(c)

Figure 5. a) Radiation amplitude squared (in units of incident amplitude) versus distance (in units of c/ω). The incident intensity is 10^{14}W/cm^2. The plasma parameters correspond to a Nd-glass laser (λ = 1.06μm) incident on a deuterium plasma ($\alpha = \sqrt{m_e/m_i}$ = 0.0165). b) Negative Poynting Flux (in units of incident flux) versus distance for same parameters as above. c) Total reflection coefficient versus incident laser intensity from an overdense plasma for parameters above. Upper and lower curves correspond to values 1/3 and 4/3 respectively for the scaling parameter η. Vertical lines labelled by values of $P_c = E_o^2/E_{th}^2$ the ratio of local pump at the largest maxima in Fig. 5a to threshold power.

$$\nabla^2 E_o(\underline{r}) + \frac{\omega_o^2}{c^2}\left(1 - \frac{\omega_p^2(r)}{\omega_o^2}\left[1 - \frac{i\nu_{ei}}{\omega_o} - \frac{i\sigma^{NL}(E_o,r)}{\omega_o}\right]\right)E_o(r) = 0 \tag{4.3}$$

Here the term proportional $\sigma^{NL}(E_o,r)$ represents (the time derivative of) the currents associated with the excited electrostatic waves which react back on the pump field and the term involving ν_{ei} is the usual collisional absorption. In an inhomogeneous plasma, $\sigma^{NL}(E_o,r)$ can be obtained from a homogeneous calculation by replacing the plasma parameters by their local values if these do not vary appreciably over the distance of an electron mean free path. In Fig. 5, typical numerical results are shown for a plane-layered plasma whose density varies linearly in the z direction.[19] The nonlinear conductivity discussed above was used in these calculations, with some additional modifications to take into account Landau damping which limits the instability to a narrow region in z near the reflection point $\omega_o = \omega_{pe}$ in which the matched k_m values of the excited Langmuir waves are much smaller than $k_{De}(z)$. In Fig. 5a, the ratio of local wave intensity squared to incident laser intensity is plotted versus distance from the reflection point z = 0 for which $\omega_p(o) = \omega_o$ for parameters typical of Nd-glass laser interactions. Here z_o is the density scale height H in units of free space wavelengths c/ω. The upper curve includes only linear collisional absorption while the lower curve includes the nonlinear absorption for an incident intensity of 10^{14}W/cm^2. Note that the standing wave pattern and swelling behavior persist for rather strong nonlinear absorptions. In Fig. 5b, the net Poynting flux of the pump wave versus distance is plotted for the same two cases; the regions of largest slope indicating the regions of maximum absorption of energy. In Fig. 5c, the total reflection coefficient is plotted as a function of incident intensity for typical Nd-glass laser parameters.

This is, of course, a very idealized theory since it neglects the effects of strong gradients which themselves may saturate the instability and effect of the backscatter instabilities which may reduce the electromagnetic energy which reaches the absorption region near $\omega_o = \omega_p$. In this regard, one should consider E_{inc} to be the net intensity incident on the absorption region, subtracting away the backscattered intensity. This calculation is included here only as an example of a self-consistent theory. The problem of oblique incidence with E_o in the plane of incidence is of great interest. Here the incident electromagnetic wave can directly resonantly excite a longitudinal wave via the density gradient. This wave in turn can drive a parametric instability. It can be shown[20] that a maximum absorption of 50% can arise from this effect. In a sufficiently weak gradient, the normal incidence case can lead to more absorption but in stronger gradients, oblique incidence absorption will dominate.

The nonlinear theory on which the calculations above are based has a number of important limitations. First, if $T_e \approx T_i$ as assumed, the purely growing instability discussed above has a threshold nearly equal to that of the decay instability. As mentioned above, this instability excites a zero frequency mode and a high frequency electrostatic wave with frequency exactly equal to ω_o and $k > k_c = \sqrt{(\omega_o^2 - \omega_p^2)/3V_e^2}$. This high frequency mode can decay by the induced scattering process into an electrostatic wave in the decay instability region $k < k_c$. This decay provides another parametric ladder process which tends to stabilize the purely growing instability. The spectrum for $k < k_c$ looks like Fig. 5b except that the peaks are displaced to lie between those in Fig. 5b. When $\omega_o > \omega_{pe}$, this process can increase the energy in the decay region ($k < k_c$) and thereby increase the nonlinear conductivity. This effect can be incorporated into the factor η in Eq. (4.2). In addition, computer simulations and numerical solution of coupled fluid equations show that the wave energy remaining in the region $k > k_c$ spreads toward larger k as P increases.[10] These high k waves also contribute to σ_{NL} and provide low phase velocity waves to accelerate electrons to produce the high energy tails observed in the simulations and in experiments.[10,21]

If $T_e \gg T_i$, the ion waves become weakly damped and the threshold for the decay instability is lower than that for the purely growing instability by a factor of $\sqrt{m_e/m_i}$. In this case, it is possible to excite only the decay instability. The nonlinear theory is complicated by the necessity of including ion wave nonlinearities. Some attempts[22a,b] have been made to do this. There are indications[22] that the usual weak-turbulence theories for the ion waves are not adequate to describe the saturation of the ion waves and that an orbit-perturbation approach may be necessary. However, under most conditions it is found that the spectrum, and more certainly, the total wave energy at saturation is unchanged from the result found for $T_e = T_i$, i.e., the inclusion of ion wave nonlinearities has little effect. Even in the case of Langmuir waves, the orbit-perturbation approach may provide the strongest nonlinearities for wavenumbers $k \gtrsim .2k_D$ where Landau damping effects begin to become important.[23] In this theory, the parametrically excited waves are locally absorbed in the unstable region of $\underline{k}$ space by a sort of resonance broadened Landau damping in which the resonance width increases with the wave energy. The saturated spectrum is therefore roughly confined to the unstable region of the linear theory. The total wave energy predicted by this theory[23] is in qualitative accord with computer simulations[24] but the wave spectrum is not broadened in k as in the simulations.[10] The induced scattering saturation mechanism discussed above also does not lead to results completely in accord with simulations.[25] This may be due to the inability of the simulation codes to treat close range particle-particle collisions accurately, or to a failure of the random phase approximation made in weak turbulence theory or the inapplicability of this approximation to the simulations.

A recent investigation of the weak turbulence saturation of the electron-electron decay instability ($\omega_o \approx 2\omega_{pe}$) has shown that induced scattering from ions is again the strongest nonlinear effect. Preliminary results[26] indicat that the total saturation wave energy in this case is proportional to $(k_o/k_m)^4\ (m_i/m_e)E^4$ where k_o is the pump wave number and k_m the matched wavenumber $(\omega_o = 2\sqrt{\omega_p^2+3k_m^2v_e^2})$. This saturation level is generally much lower than that obtained for the electron-ion decay instability. Similar considerations can be applied to saturation of SRS.

Computer simulations[5e, 10, 24] have been extremely important in demonstrating the dominant nonlinear effects for cases of strong pumping $v_o/v_e \gtrsim .1$. Here such highly nonlinear phenomena as particle trapping and wave breaking become important. In the backscatter instabilities, pump depletion is seen to be an important saturation mechanism.

Among the important remaining theoretical problems is the need to study regimes in which the analytical weak turbulence theory and the computer simulations both apply so that we can better understand the limitations of both techniques.

ACKNOWLEDGMENTS

I wish to thank Dr. M.V. Goldman and Dr. B. Bezzerides for their comments concerning this review.

REFERENCES

1. D. F. DuBois and M. V. Goldman, Phys. Rev. Letts., 14, 544 (1956) and Phys. Rev., 164, 207 (1967).
2. V. P. Silin, Sov. Phys., -JETP- 21, 1127 (1965).
3. M. V. Goldman, Annals of Phys. (N.Y.) 38, 95 (1966).
4. P. K. Kaw and J. M. Dawson, Phys. Fluids, 12, 2586 (1969).
5. a) M. V. Goldman and D. F. DuBois, Phys. Fluids, 8, 1404 (1965); b) M. V. Goldman, Proceedings of the 2nd Orsay Summer Institute of Plasma Physics-Nonlinear Effects in Plasmas. Edited by C. G. Kalman and M. Feix, Gordon Breach and Co. (1969); c) N. Bloembergen and Y. R. Shen, Phys. Rev., 141, 298 (1966); d) C. G. Comisar, ibid, p. 200; e) D. W. Forslund, J. M. Kindel and E. L. Lindman, Phys. Rev. Letts., 30, 739 (1973).
6. L. M. Gorbunov, Sov. Phys. JETP, 28, 1220 (1969); L. M. Gorbunov and V. P. Silin, Tech. Phys., 14, 1 (1969); Forslund, et al., ref. 7.
7. D. F. DuBois, in Statistical Physics of Charged Particle Systems, edited by R. Kubo and T. Kihara (Syokabo, Tokyo and Benjamin, New York, 1968).
8. D. F. DuBois and M. V. Goldman, Phys. Rev. Letts., 19, 1105 (1967).
9. K. Nishikawa, J. Phys. Soc. Japan, 24, 916 1152 (1968).

10. a) J. J. Thomson, R. J. Faehl and W. L. Kruer, Phys. Rev. Letts. 31, 918 (1973); b) P. Kaw, G. Schmidt, and T. Wilcox, Phys. Fluids 16,1522 (1973); c) J. Drake, P. Kaw, Y. C. Lee, G. Schmidt, C. S. Liu, and M. N. Rosenbluth, UCLA Report (to be published).
11. E. S. Cassedy and C. R. Evans, Journ. Appl. Phys. 43, 4453 (1972).
12. D. F. DuBois and E. A. Williams, U. of Colorado Report #1005 (1973) and D. F. DuBois, D. W. Forslund and E. A. Williams (to be published).
13. R. B. White, C. S. Liu, and M. N. Rosenbluth, Phys. Rev. Letts. 31, 520 (1973) and J. M. Kindel (private communication).
14. N. M. Rosenbluth, Phys. Rev. Letts., 29, 565 (1972).
15. a) F. W. Perkins and J. Flick, Phys. Fluids, 14, 2012 (1971); b) C. S. Liu, R. B. White and N. M. Rosenbluth, Phys. Rev. Letts. 31, 697 (1973); c) M. A. Mostrom, D. R. Nicholson, A. N. Kaufman (to be published).
16. D. F. DuBois and M. V. Goldman, Phys. Rev. Letts. 28, 481 (1972).
17. E. Valeo, F. Perkins, and C. Oberman, Phys. Rev. Letts., 28, 340 (1972).
18. J. A. Fejer and Y. Y. Kuo, Phys. Rev. Letts. 29, 1667 (1972); F. Perkins and E. Valeo (to be published).
19. D. F. DuBois, M. V. Goldman, and D. McKinnis, Phys. Fluids 16, 2257 (1974).
20. a) V. L. Ginzburg, "Propagation of Electromagnetic Waves in Plasmas," Pergamon Press, 1970, 2nd Edition; b) J. P. Freidberg, R. W. Mitchell, R. L. Morse, and L. I. Rudsinski, Phys. Rev. Letts., 28, 795 (1972); c) R. B. White, C. S. Liu, M. N. Rosenbluth (to be published).
21. J. Katz, J. Weinstock, W. L. Kruer, J. S. DeGroot, UCRL Report-74334, to be published.
22. a) V. V. Pustovalov and V. P. Silin, Zh. E.T.F., 59, 2215 (1970); b) D. F. DuBois and M. V. Goldman, Vull. Am. Phys. Soc., 17, 994 (1972).
23. B. Bezzerides and J. Weinstock, Phys. Rev. Letts., 28, 994 (1972).
24. W. L. Kruer, P. Kaw, J. M. Dawson, and C. Oberman, Phys. Rev. Letts., 24, 987 (1970).
25. B. B. Godfrey, C. E. Rhoades, and K. A. Taggart, Phys. Fluids (to be published).
26. D. F. DuBois and M. V. Goldman (to be published).

Nonlinear Confining and Deconfining Forces Associated with the Interaction of Laser Radiation with Plasma

HEINRICH HORA*
Westinghouse Research Laboratories, Pittsburgh, Pennsylvania

The nonlinear interaction of an intense light wave with an inhomogeneous plasma produces a macroscopic motion. A rigorous treatment of this interaction based on the ponderomotive force description leads to a general equation of motion. In a plasma with collisions and high electron density, the resulting force density can be interpreted as an expression of the radiation pressure. Below special densities there results a nonlinear collisionless force whose direction is toward decreasing density, and which produces a deconfinement. The magnitude of this force has a polarization dependence only in the third-order terms of the spatial dependence of the density. The total deconfining recoil momentum transferred to the inhomogeneous transition layer is evaluated. The theory of the nonlinear collisionless acceleration is used to explain the experimentally observed properties of the fast part of plasmas produced by lasers from isolated single small aluminum balls and thick solid targets.

I. INTRODUCTION

The interaction of laser radiation with plasma has been discussed from several points of view. There have been investigations of ion and molecular scattering of light,[1] of light mixing,[2] and of self-focusing or defocusing.[3] With a view to estimating possible applications to controlled thermonuclear fusion a number of investigations[4] have considered plasma heating as the result of laser light absorption by means of the usual collision processes.[5] In addition to these studies there remains to be treated the important case of the macroscopic motion of a plasma due to the interaction of the electromagnetic field.

The macroscopic motion is relevant to the observation of a substantial acceleration of plasma up to kiloelectron volt ion energies.[6] Surprisingly, the acceleration was directed against the laser light. A previous description[7] of a nonlinear collisionless interaction of the laser light with the inhomogeneous transition region of a plasma showed that the interaction had the same sign as that observed,[6] and that the measured high values of ion energies could be reproduced when simplifying assumptions were used. The results of the theory of this article are consistent with more detailed recent observations of plasmas produced from isolated single small aluminum balls.[8]

The first steps in the deduction of the theory are just as difficult as those in the similar situation of considering the microwave interaction with plasmas.[9] If a ponderomotive force description based on the momentum-flux-density tensor is used, the resulting forces can be either positive or negative, depending on whether the description starts from Maxwell's or Lorentz's theory of the continuum. This problem in determining the correct sense of the ponderomotive force is not new.[10] A partial resolution of

---

* Now with Institut für Plasmaphysik, Garching, Germany, and Rensselaer Polytechnic Institute—Hartford Graduate Center, East Windsor Hill, Connecticut.

[1] See for example, E. E. Salpeter, Phys. Rev. **122**, 1663 (1961); E. Fünfer, B. Kronast, and H. J. Kunze, Phys. Letters **5**, 125 (1963); T. V. George, L. Goldstein, L. Slama, and M. Yokoyama, Phys. Rev. **137**, A369 (1965); and H. Röhr, Z. Physik **200**, 295 (1968).

[2] N. M. Kroll, A. Ron, and N. Rostoker, Phys. Rev. Letters **13**, 83 (1964); A. Salat, Z. Naturforsch. **20a**, 689 (1965).

[3] S. A. Akhmanov, D. P. Krindach, A. P. Sukhorkukov, and R. V. Khokhlov, Zh. Eksp. Teor. Fiz. Pis. Red. **6**, 509 (1967) [JETP Letters **6**, 38 (1967)].

[4] N. G. Basov and O. N. Krokhin, Zh. Eksp. Teor. Fiz. **46**, 171 (1964) [Sov. Phys.—JETP **19**, 123 (1964)]; A. G. Engelhardt, Bull. Am. Phys. Soc. **9**, 305 (1964); J. M. Dawson, Phys. Fluids **7**, 981 (1964).

[5] J. M. Dawson and C. Oberman, Phys. Fluids **5**, 517 (1962).

[6] W. I. Linlor, Appl. Phys. Letters **3**, 210 (1963); Phys. Rev. Letters **12**, 383 (1964); H. Opower and E. Burlefinger, Phys. Letters **16**, 37 (1965); H. Opower and W. Press, Z. Naturforsch. **21a**, 344 (1966); J. C. Bryner and G. H. Sichling, Bull. Am. Phys. Soc. **12**, 1161 (1967).

[7] H. Hora, D. Pfirsch, and A. Schlüter, Z. Naturforsch. **22a**, 278 (1967).

[8] A. G. Engelhardt, H. Hora, T. V. George, and J. L. Pack, Phys. Fluids (to be published).

[9] H. Motz and C. J. H. Watson, in *Advances in Electronics and Electron Physics*, L. Marton, Ed. (Academic Press Inc., New York, 1967), Vol. 23, p. 153.

[10] W. Weizel, *Lehrbuch Theoretische Physik* (Springer-Verlag, Berlin, 1949), Vol. I, p. 360.

Reprinted from: *The Physics of Fluids*, **12**, 182--191 (1969).

the dilemma has been found[7] by using the two-fluid model to describe the macroscopic plasma behavior. The right sign of the force was determined and a first description by the momentum flux density was sketched. The two-fluid and momentum-flux models are in agreement for perpendicular incidence of light on a plane inhomogeneous plasma. In the case of oblique incidence, however, the two-fluid model shows insufficiencies, as will be shown here.

The general deduction of the theory given here is based on the Lorentz theory, as it is treated by Landau and Lifshitz.[11] The equation of the force density of the plasma is then given by the ponderomotive force. This force is intrinsically nonlinear, for a force quantity is combined with products of at least two field strengths, each of which are force quantities.

Further, the question of an inhomogeneous plasma with collisions will be discussed. It will be found that at high electron densities a confining force occurs in the same direction as the light producing the radiation pressure. By evaluating the total momentum transferred by the collisionless deconfining acceleration of the plasma, one finds the region of the electron density for which the deconfining process (acceleration of plasma against the laser light) is more dominant than the thermal absorption process.

II. PONDEROMOTIVE FORCE

The force density $\mathbf{f}$ of a plasma in the presence of electric and magnetic fields $\mathbf{E}$ and $\mathbf{H}$ is given by the ponderomotive force description[11]

$$\mathbf{f} = \nabla \cdot \boldsymbol{\sigma} - \frac{\partial}{\partial t} \frac{\mathbf{E} \times \mathbf{H}}{4\pi c}, \qquad (1)$$

where t is time and c is the velocity of light. This relation is also defined to be the equation of motion. In Eq. (1) the three-dimensional stress tensor $\boldsymbol{\sigma}$ has the following components:

$$\sigma_{\phi\chi} = -p_{\phi\chi} - \left[\frac{\mathbf{E}^2}{8\pi}\left(\tilde{n}^2 - \rho \frac{\partial \tilde{n}^2}{\partial \rho}\right) + \frac{\mathbf{H}^2}{8\pi}\right]\delta_{\phi\chi} + \tilde{n}^2 \frac{E_\phi E_\chi}{4\pi} + \frac{H_\phi H_\chi}{4\pi}. \qquad (2)$$

$p_{\phi\chi}$ is the ϕ, χth component of the pressure tensor, $\delta_{\phi\chi}$ is the Kronecker delta, the magnetic permeability is unity, and $\tilde{n}$ is the complex index of refraction; i.e.,

$$\tilde{n}^2 = 1 - \frac{\omega_p^2}{\omega^2 + \nu^2}\left(1 + i\frac{\nu}{\omega}\right). \qquad (3)$$

In Eq. (3) ω is the radian frequency of the electromagnetic field, and ν is the total collision frequency of the plasma electrons given[12] by

$$\nu = \frac{\omega_p^2 \pi^{\frac{3}{2}} m_e^{\frac{1}{2}} Z e^2 \ln \Lambda}{8\pi \gamma_E(Z)(2\mathcal{K}T)^{\frac{3}{2}}}. \qquad (4)$$

Here m_e is the electron mass, e the electronic charge, n_e the electron density, $\ln \Lambda$ the Coulomb logarithm, $\mathcal{K}$ the Boltzmann constant, and T the temperature; γ_E, which is a function of the ionic charge Z is Spitzer's correction due to electron–electron collisions. The electron plasma frequency ω_p is defined by the relation

$$\omega_p^2 = 4\pi e^2 n_e / m_e. \qquad (5)$$

The density ρ in Eq. (2) can be replaced by the well-known relation[11]

$$\rho \frac{\partial \tilde{n}^2}{\partial \rho} = \tilde{n}^2 - 1, \qquad (6)$$

which is derived using the theory of continuous media [see Eq. (15.14) of Landau–Lifschitz]. Thus, Eq. (2) can be modified to take the form

$$\sigma_{\phi\chi} = -p_{\phi\chi} - (\mathbf{E}^2 + \mathbf{H}^2)\,\delta_{\phi\chi}/8\pi + \tilde{n}^2 E_\phi E_\chi/4\pi + H_\phi H_\chi/4\pi. \qquad (7)$$

Equation (7) can also be written as

$$\sigma_{\phi\chi} = -p_{\phi\chi} + U_{\phi\chi} - \frac{\omega_p^2 E_\phi E_\chi}{4\pi(\omega^2 + \nu^2)}\left(1 + i\frac{\nu}{\omega}\right), \qquad (8)$$

where the Maxwellian stress tensor $\mathbf{U}$ has the form

$$4\pi(U_{\phi\chi}) = \begin{pmatrix} \frac{1}{2}(E_x^2 - E_y^2 - E_z^2 + H_x^2 - H_y^2 - H_z^2) & E_xE_y + H_xH_y & E_xE_z + H_xH_z \\ E_xE_y + H_xH_y & \frac{1}{2}(-E_x^2 + E_y^2 - E_z^2 - H_x^2 + H_y^2 - H_z^2) & E_yE_z + H_yH_z \\ E_xE_z + H_xH_z & E_yE_z + H_yH_z & \frac{1}{2}(-E_x^2 - E_y^2 + E_z^2 - H_x^2 - H_y^2 + H_z^2) \end{pmatrix}. \qquad (9)$$

[11] L. D. Landau and E. M. Lifshitz, *Electrodynamics of Continuous Media* (Pergamon Press, Ltd., Oxford, 1966), p. 242.

[12] L. Spitzer, Jr., *Physics of Fully Ionized Gases* (Interscience Publishers, Inc., New York, 1950).

Consequently, the force density in the plasma takes the form,

$$\mathbf{f} = -\nabla\cdot\mathsf{p} + \nabla\cdot\left[\mathsf{U} - \frac{\omega_p^2}{4\pi(\omega^2+\nu^2)}\left(1+i\frac{\nu}{\omega}\right)\mathbf{EE}\right] - \frac{\partial}{\partial t}\frac{\mathbf{E}\times\mathbf{H}}{4\pi c}. \qquad (10)$$

With the aid of Maxwell's equations, Eq. (10) can be rewritten as

$$\mathbf{f} = -\nabla\cdot\mathsf{p} + \frac{1}{c}\mathbf{j}\times\mathbf{H} + \frac{1}{4\pi}\mathbf{E}\nabla\cdot\mathbf{E} - \frac{1}{4\pi}\frac{\omega_p^2}{\omega^2+\nu^2}\left(1+i\frac{\nu}{\omega}\right)\mathbf{E}\nabla\cdot\mathbf{E} - \frac{1}{4\pi}\frac{\omega_p^2}{\omega^2+\nu^2}\left(1+i\frac{\nu}{\omega}\right)\mathbf{E}\cdot\nabla\mathbf{E} - \frac{1}{4\pi}\mathbf{EE}\cdot\nabla\frac{\omega_p^2}{\omega^2+\nu^2}\left(1+i\frac{\nu}{\omega}\right), \qquad (11)$$

where the current density $\mathbf{j}$ was introduced by the second of Maxwell's equations. Equations (10) and (11) are the most general expressions for the force density (i.e., the equation of motion) of a plasma in an electromagnetic field. The procedure for getting real quantities will be discussed when the formulas are applied.

If the equation of motion is deduced not from the ponderomotive force description but from the macroscopic theory of the two-fluid model,[13] we find

$$\mathbf{f} = -\nabla\cdot\mathsf{p} + \frac{1}{c}\mathbf{j}\times\mathbf{H} + \mathbf{j}\cdot\nabla\frac{1}{\omega^2}\frac{\partial\mathbf{E}}{\partial t}, \qquad (12)$$

when the gravitational forces are neglected. The first two terms of this equation are identical with those in Eq. (11). The third term in Eq. (12) is identical for $\nu = 0$ (collisionless plasma) with the second from last one in Eq. (11) because of the relation

$$\mathbf{j}\cdot\nabla\frac{1}{\omega^2}\frac{\partial\mathbf{E}}{\partial t} = -\frac{1}{4\pi}\frac{\omega_p^2}{\omega^2}\mathbf{E}\cdot\nabla\mathbf{E}, \qquad (13)$$

which is valid when the fast oscillation properties are neglected. The remaining three terms in Eq. (11) do not appear in the case of the two-fluid model. In fact the last three terms in Eq. (11) are less important in the usual cases of the macroscopic plasma theory. From the general point of view it should be emphasized that while the term (13) is reproduced by Schlüter's model,[13] it is not verified by the deduction[12] of the macroscopic equations from the microscopic theory. The third term on the right-hand side of Eq. (11) is due to space charges, which were neglected by Schlüter[13] because of the assumed quasineutrality. A more general equation of motion is used in the well-known theory of radio frequency acceleration of plasmas [see Eq. (12) on page 271 of Ref. 9] but the last term in (11) of this paper is usually neglected. Thus, the microwave theory takes into account the space charge term because an inhomogeneous electromagnetic wave with a longitudinal component produces periodically oscillating space charges. But, as will be shown, only an equation of motion with all terms present in Eqs. (10) or (11) is sufficient to evaluate the nonlinear acceleration effects in laser produced plasmas.

It should be mentioned that the complete equation of motion, (10) or (11), can be deduced from the basic equations of the continuum mechanics in the relativistic formulation.[14] In the present treatment the description using the ponderomotive force is preferred, because the nonlinear interaction of light on plasma has the character of a radiation pressure (confining or deconfining as the case may be).

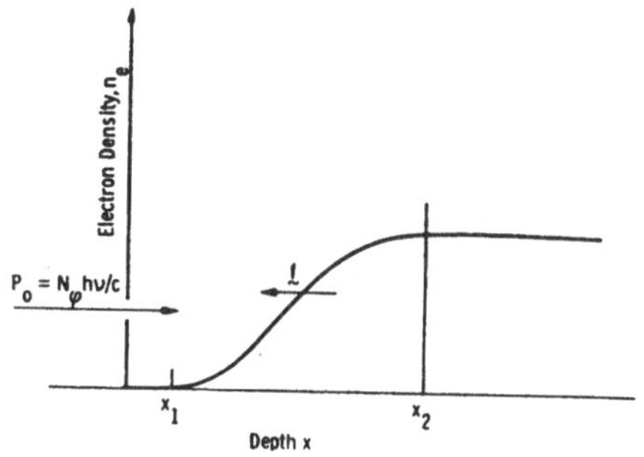

FIG. 1. Electron density n_e as a function of the depth x of a plane inhomogeneous plasma layer between vacuum ($x < x_1$) and a homogeneous plasma ($x > x_2$). The arrows show the momentum P_0 of incident N_φ photons and the resulting force density f in the inhomogeneous layer.

III. PERPENDICULAR INCIDENCE OF RADIATION

The force density as given by Eqs. (10) or (11) will now be evaluated for the case of an inhomogeneous plasma, located between vacuum and the homogeneous interior of a plasma. It is assumed, that the plasma is one dimensional (i.e., in Fig. 1 the inhomogeneity is between x_1 and x_2), and the light is perpendicularly incident and penetrates in the positive x direction. With these conditions one has

$$\frac{\partial}{\partial y} = \frac{\partial}{\partial z} = 0, \quad E_x = E_z = 0, \text{ and } H_x = H_y = 0, \qquad (14)$$

when the electromagnetic wave is linearly polarized with the direction of oscillation parallel to y. It is

[13] A. Schlüter, Z. Naturforsch. 5a, 72 (1950).

[14] D. Pfirsch (private communication).

convenient to split the force density (10) into a thermokinetic part $\mathbf{f}_{th} = -\nabla\cdot\mathbf{P}$ and a nonlinear part $\mathbf{f}_{NL}$, viz.,

$$\mathbf{f} = \mathbf{f}_{th} + \mathbf{f}_{NL}. \tag{15}$$

From Eqs. (10) and (14) it follows that for the case of perpendicular incidence

$$\mathbf{f}_{NL} = -\frac{\mathbf{i}_1}{8\pi}\frac{\partial}{\partial x}(E_y^2 + H_z^2) - \frac{\partial}{\partial t}\frac{E_y H_z}{4\pi c}\,\mathbf{i}_1. \tag{16}$$

The unit vectors are $\mathbf{i}_1$, $\mathbf{i}_2$, and $\mathbf{i}_3$ in the directions of the x, y, and z axes.

In the following, average values of the force density during one period of the light wave will be considered. One can presume that the switch-on process of the light wave is very slow in comparison with the frequency of the laser light, so that the neglect of the time average of the Poynting term, the last term in Eq. (16), is possible, i.e.,

$$\overline{\mathbf{f}_{NL}} = -\frac{\mathbf{i}_1}{8\pi}\frac{\partial}{\partial x}(\overline{E_z^2} + \overline{H_z^2}). \tag{17}$$

Equation (17) which gives the mean force density will be applied to the case of a slow spatial variation of the refractive index in order to fulfill the WKB condition,

$$\theta = \frac{c}{2\omega}\frac{1}{|\tilde{n}|^2}\frac{\partial\,|\tilde{n}|}{\partial x} \ll 1. \tag{18}$$

The complex electric and magnetic field strengths are then

$$\mathbf{E} = \mathbf{i}_2\frac{E_v}{(\tilde{n})^{\frac{1}{2}}}\exp(iF)\exp\left(\mp\frac{k(x)}{2}x\right), \tag{19}$$

and

$$\mathbf{H} = \mathbf{i}_3 E_v(\tilde{n})^{\frac{1}{2}}\exp(iF)\exp\left(\mp\frac{k(x)}{2}x\right), \tag{20}$$

where

$$F = \omega\left(t \mp \int^x \frac{\mathrm{Re}\,[\tilde{n}(\xi)]}{c}\,d\xi\right)$$

$$\text{and}\quad k(x) = \frac{1}{x}\frac{\omega}{c}\int^x \mathrm{Im}\,[\tilde{n}(\xi)]\,d\xi. \tag{21}$$

In Eq. (21) the magnitudes of the real and imaginary part of the index of refraction $\tilde{n}$ are used. As shown in Fig. 1, the upper sign denotes a wave propagating into the $+x$ direction, whereas the lower sign denotes propagation in the $-x$ direction. E_v is the electric field amplitude in vacuum. Terms with derivatives of $\tilde{n}$ in space are neglected within this section, for these only give terms of second and higher order in the time-averaged expressions to follow.

By selecting only the real parts of the expressions for the nonlinear terms in Eq. (10), one finds from Eq. (19)

$$E_y^2 = E_v^2\left[\left(\mathrm{Re}\,\frac{1}{(\tilde{n})^{\frac{1}{2}}}\right)^2\cos^2 F + \left(\mathrm{Im}\,\frac{1}{(\tilde{n})^{\frac{1}{2}}}\right)^2\sin^2 F + \tfrac{1}{2}\,\mathrm{Re}\left(\frac{1}{(\tilde{n})^{\frac{1}{2}}}\right)\mathrm{Im}\left(\frac{1}{(\tilde{n})^{\frac{1}{2}}}\right)\sin^2 F\right]\exp\,[\mp k(x)x]. \tag{22}$$

This leads to the time-averaged value

$$\overline{E_y^2} = \tfrac{1}{2}E_v^2\frac{1}{|\tilde{n}|}\exp(\mp kx). \tag{23}$$

In the same way from Eq. (20) one gets

$$\overline{H_z^2} = \pm\frac{E_v^2}{2}|\tilde{n}|\exp(\mp kx). \tag{24}$$

The averaged nonlinear force density (17) is then

$$\overline{\mathbf{f}_{NL}} = -\mathbf{i}_1\frac{E_v^2}{16\pi}\frac{\partial}{\partial x}\left(\frac{1}{|\tilde{n}|} + |\tilde{n}|\right)\exp(\mp kx) \tag{25a}$$

$$= \mathbf{i}_1\frac{E_v^2}{16\pi}\frac{1 - |\tilde{n}|^2}{|\tilde{n}|^2}\exp(\mp kx)\frac{\partial}{\partial x}|\tilde{n}| \pm \mathbf{i}_1\frac{E_v^2}{16\pi}\frac{1 + |\tilde{n}|^2}{|\tilde{n}|}\frac{2\omega}{c}\,\mathrm{Im}\,(\tilde{n})\exp(\mp kx). \tag{25b}$$

In the collisionless case with $k = 0$ and $\tilde{n}$ taken from Eq. (3), the second term of Eq. (25b) vanishes and one obtains the previous result[7]

$$\overline{\mathbf{f}_{NL}} = \mathbf{i}_1\frac{E_v^2}{16\pi}\frac{\omega_p^2}{\omega^2\tilde{n}^2}\frac{\partial\tilde{n}}{\partial x}. \tag{26}$$

This expression indicates a deconfining collisionless acceleration because the direction of the force density is towards decreasing plasma densities, and is independent also of the direction of the incident laser light. Since Eq. (26) was derived for the collisionless case, the force density can rise to infinitely high values as $\tilde{n} \to 0_+$. In the case of collisions, a limitation on the magnitude of the nonlinear force is determined by the minimum of $|\tilde{n}|$ in Eq. (25). When the electron temperature increases, this minimum of $|\tilde{n}|$ does not approach zero as rapidly as the minimum of the real part of $\tilde{n}$ which is the usual index of refraction. From Eq. (3) one finds

$$\min(|\tilde{n}|) = \left(\frac{\nu}{\omega}\right)^{\frac{1}{2}} \quad \text{at} \quad \omega_p = \omega$$

$$\text{and} \tag{27}$$

$$\min\,[\mathrm{Re}\,(|\tilde{n}|)] = \frac{\nu}{\omega} \quad \text{at} \quad \omega_p = 2^{\frac{1}{2}}\omega, \quad \text{if} \quad \nu \ll \omega.$$

Here, the dependence of ν on the temperature T is given by Eq. (4).

Another result of Eq. (25b), is that the second term on the right-hand side indicates an acceleration of the plasma in the same direction as the laser light, i.e., a confinement is produced when the geometry of Fig. 1 is considered. This acceleration can be interpreted as a collision produced radiation pressure of the light within the inhomogeneous plasma in analogy with the usual radiation pressure of homogeneous media. When the direction of wave propagation in Fig. 1 is changed from $+\mathbf{i}_1$ to $-\mathbf{i}_1$, the sign of the second term on the right-hand side of Eq. (25) also changes. The radiation pressure is always in the direction of the light.

We now ask for what conditions the collisionless nonlinear deconfining force is larger than the usual radiation pressure. This is the case, if the right-hand side of Eq. (25) is negative; i.e., if

$$\frac{1-|\tilde{n}|^2}{|\tilde{n}|}\left|\frac{\partial}{\partial x}|\tilde{n}|\right| = \xi(1+|\tilde{n}|^2)\frac{2\omega}{c}\,\mathrm{Im}\,(\tilde{n}), \quad \text{for} \quad \xi \geq 1. \qquad (28)$$

High force densities are possible only for very small $|\tilde{n}|$. This restricts our consideration to ω_p near ω. With Eq. (18) one can then rewrite Eq. (28) in the form

$$1 = \frac{\xi}{\theta} D(n_e, T) \qquad (29)$$

where

$$D = \frac{1}{2^{1/2}}\frac{\nu}{\omega}\frac{\omega_p^2}{\omega^2+\nu^2-\omega_p^2}, \quad \text{assuming} \quad \nu \ll \omega. \qquad (30)$$

In the case of laser produced plasmas, n_e and T^{-1} will decrease with time during the early stages of the absorption of radiation by the plasma. D is then monotonically decreasing with respect to Eqs. (4) and (5). Consequently, ξ increases with time, and the nonlinear force density (25) changes from a confining case of radiation pressure to a deconfining acceleration. The details of this process cannot be followed without special assumptions about the individual plasma described. We now try to determine the region, where the WKB approximation is applicable and where the radiation pressure is smaller than 20% of the nonlinear collisionless deconfining force. In order to insure applicability of the WKB approximation [see Eq. (18)] it is reasonable to select $\theta < 0.25$. The second condition dictates that $\xi > 5$. These two constraints on θ and ξ determine an upper limit to n_e or ω_p which we denote by ω_p^*, i.e.,

$$\omega_p^2 \leq \omega_p^{*2} = \frac{-0.1 \pm [0.01 + 0.4\omega^2(1/\omega a T^{3/2} - 0.1/a^2T^3)]^{1/2}}{2(1/\omega a T^{3/2} - 0.1/a^2T^3)}. \qquad (31)$$

Here the quantity a is defined by [cf. Eq. (4)]

$$a = \frac{\omega_p^2}{\nu T^{3/2}} = \frac{8\pi\gamma_E(Z)(2K)^{3/2}}{\pi^{1/2} m_e^{1/2} Z e^2 \ln \Lambda}. \qquad (32)$$

In Fig. 2 the variation of ω_p^*/ω is given as a function of the temperature T for Z equal to 1, and the Coulomb logarithm $\ln \Lambda$ equal to 5 since we are only discussing cases of high density. Further, Spitzer's correction $\gamma_E(Z)$ which is due to electron–electron collisions is set equal to unity. In Eq. (31) only the upper sign of the square root is possible. The expression in the brackets under the square root sign of Eq. (31) is positive for

$$T > \left(0.1\,\frac{\omega}{a}\right)^{2/3} = 0.69 \text{ eV}. \qquad (33)$$

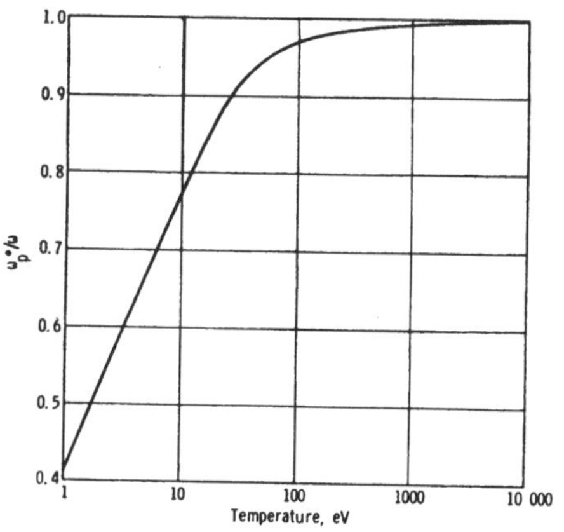

FIG. 2. Plasma frequency ω_p^* which determines the electron densities below which the radiation pressure can be neglected compared to the deconfining acceleration.

IV. OBLIQUE INCIDENCE OF RADIATION

A. The WKB Solution

Now to be considered for a collisionless plasma is the case of a plane electromagnetic wave obliquely incident at an angle α_0 in the x-y plane which is the plane of incidence. The wave propagates from vacuum into a stratified inhomogeneous layer between x_1 and x_2 having a density profile similar to that described in Fig. 1. As a condition for neglecting the collision produced radiation pressure as compared with the collisionless nonlinear acceleration, one can adopt the same criterion that has been found for perpendicular incidence [see Eq. (31) and Fig. 2]. As in Sec. III we restrict our consideration to a reflectionless propagation of the waves.

In this case the first- and second-order WKB conditions with $\nu = 0$ in Eq. (3) are

$$\theta(\alpha) = \frac{1}{2}\frac{c}{\omega}\frac{1}{\tilde{n}^2}\frac{1}{\cos^2\alpha}\left|\frac{d(\tilde{n}\cos\alpha)}{dx}\right| \ll 1, \qquad (34)$$

and

$$\psi(\alpha) = \frac{1}{2}\frac{c^2}{\omega^2}\frac{1}{\tilde{n}^3}\frac{1}{\cos^2\alpha}\left|\frac{d^2(\tilde{n}\cos\alpha)}{dx^2}\right| \ll 1, \qquad (35)$$

where the angle α of the propagation direction of the wave in the plasma follows Snell's law,

$$\tilde{n}^2(x)\sin^2\alpha(x) = \sin^2\alpha_0. \tag{36}$$

First to be resolved is the question whether it is possible using the WKB conditions (34) and (35) to write the general linearly polarized obliquely incident plane wave as a sum of two parts—one with oscillating E vector parallel to the plane of incidence (subscript ||) and the other with its E vector perpendicular to this plane (subscript $\perp$). It is well known that this separation is not possible[15] in the most general case. In order to obtain the electric vector in the perpendicular case, one has to solve the equation

$$\left(\frac{\partial^2}{\partial x^2}+\frac{\partial^2}{\partial y^2}+\frac{\omega^2}{c^2}\tilde{n}^2(x)\right)\mathbf{E}_\perp = 0. \tag{37}$$

The WKB solution gives

$$\mathbf{E}_\perp = \mathbf{i}_3\frac{E_v(\cos\alpha_0)^{1/2}}{[\tilde{n}^2(x)-\sin^2\alpha_0]^{1/4}}\cos G, \tag{38}$$

where

$$G = \pm\frac{\omega}{c}\cdot\left(\int^x[\tilde{n}^2(\xi)-\sin^2\alpha_0]^{1/2}d\xi+\tilde{n}(x)y\sin\alpha(x)\right)+\omega t. \tag{39}$$

The upper sign is to be chosen for an electromagnetic wave whose propagation vector has a component towards the plasma. The lower sign is for the opposite direction. From Eq. (38) one gets from the Maxwellian equations

$$\mathbf{H}_\perp = \pm\frac{E_v(\tilde{n}\cos\alpha_0)^{1/2}}{(\cos\alpha)^{1/2}}(\mathbf{i}_1\sin\alpha-\mathbf{i}_2\cos\alpha)\cos G - \mathbf{i}_2\frac{c}{\omega}\frac{E_v(\cos\alpha_0)^{1/2}}{2\tilde{n}^{3/2}(x)\cos^{5/2}\alpha(x)}\frac{d\tilde{n}}{dx}\sin G. \tag{40}$$

In the case of $E_\parallel$, the electric vector is no longer perpendicular to the direction of propagation as it was for the general case, and the electric field is no longer divergenceless. The components of the electric field strength $\mathbf{E}$ are given by[15]

$$\left(\frac{\partial^2}{\partial x^2}+\frac{\partial^2}{\partial y^2}+\frac{\omega^2}{c^2}\tilde{n}^2(x)\right)E_{\parallel x} - \frac{\partial}{\partial x}\nabla\cdot\mathbf{E}_\parallel = 0, \tag{41}$$

and

$$\left(\frac{\partial^2}{\partial x^2}+\frac{\partial^2}{\partial y^2}+\frac{\omega^2}{c^2}\tilde{n}^2(x)\right)E_{\parallel y} - \frac{\partial}{\partial y}\nabla\cdot\mathbf{E}_\parallel = 0. \tag{42}$$

The last terms in Eqs. (41) and (42) couple all components of the electric field, so that a representation as the sum of two differently polarized components is not possible. In the case of the WKB condition, however, the last terms in Eqs. (41) and (42) are negligible, as will now be shown. Using the relation

$$\tilde{n}(x)\nabla\cdot\mathbf{E}_\parallel = -2E_{\parallel x}\frac{d\tilde{n}}{dx}, \tag{43}$$

which can be derived from Eq. (19.19) of Ginzburg,[15] for (41) and (42) one gets

$$\left(\frac{\partial^2}{\partial x^2}+\frac{\partial^2}{\partial y^2}+\frac{\omega^2}{c^2}\tilde{n}^2(x)\right)E_{\parallel x} + 2\frac{\partial}{\partial x}\left(E_{\parallel x}\frac{d\ln\tilde{n}(x)}{dx}\right) = 0, \tag{44}$$

and

$$\left(\frac{\partial^2}{\partial x^2}+\frac{\partial^2}{\partial y^2}+\frac{\omega^2}{c^2}\tilde{n}^2(x)\right)E_{\parallel y} + 2\frac{\partial E_{\parallel x}}{\partial y}\frac{d\ln\tilde{n}(x)}{dx} = 0. \tag{45}$$

Comparison with the WKB conditions (34) and (35) shows that the time-averaged last term in Eq. (44) is always $(\theta + 2\psi)$ times the $\tilde{n}^2(x)$ term in the first bracket of Eq. (44). Because θ and ψ from Eqs. (34) and (35) are very much smaller than 1, neglecting the last term in Eq. (44) is possible. Neglecting the last term in Eq. (45) in comparison with the $\tilde{n}^2(x)$ term in the first brackets in Eq. (45) is possible in the same way when

$$2tg\alpha(x)\sin\alpha(x) \le 1,\quad \text{i.e., when } \alpha \lesssim 40°. \tag{46}$$

So, with the conditions given by Eqs. (34), (35), and (46), one can separate the two cases of polarization and can solve Eqs. (44) and (45) separately without the coupling terms to get

$$E_\parallel = \frac{E_v(\cos\alpha_0)^{1/2}}{[\tilde{n}(x)\cos\alpha(x)]^{1/2}}\cdot[-\mathbf{i}_1\sin\alpha(x)+\mathbf{i}_2\cos\alpha(x)]\cos G. \tag{47}$$

Substitution of Eq. (47) into Maxwell's equations gives

$$\mathbf{H}_\parallel = \pm\mathbf{i}_3\frac{E_v(\tilde{n}\cos\alpha_0)^{1/2}}{(\cos\alpha)^{1/2}}\cos G - \mathbf{i}_3\frac{c}{\omega}\frac{E_v(\cos\alpha_0)^{1/2}(\frac{1}{2}-\sin^2\alpha)}{\tilde{n}^{3/2}\cos^{3/2}\alpha}\frac{d\tilde{n}}{dx}\sin G. \tag{48}$$

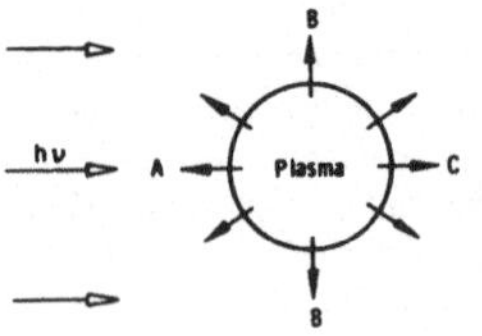

FIG. 3. Schematic demonstration of the nonlinear collisionless forces (full arrows) produced by a plane electromagnetic wave (hollow arrows) in the inhomogeneous transition region of a spherical plasma when obliquely incident plane waves are considered.

B. Calculation of the Force Density

For the geometry of Fig. 3, the time-averaged values of the following quadratic forms are zero;

[15] V. L. Ginzburg, *The Propagation of Electromagnetic Waves in Plasmas* (Pergamon Press, Ltd., Oxford, 1964), p. 205.

viz.,

$$\frac{\partial}{\partial y}\overline{\mathbf{EE}} = \frac{\partial}{\partial z}\overline{\mathbf{EE}} = \frac{\partial}{\partial y}\overline{\mathbf{HH}} = \frac{\partial}{\partial z}\overline{\mathbf{HH}} = 0. \quad (49)$$

The time-average force density (equation of motion) from Eq. (10) for a collisionless plasma ($\nu = 0$) is then

$$\begin{aligned}\bar{\mathbf{f}} = \frac{1}{8\pi}\mathbf{i}_1\frac{\partial}{\partial x}[(2\tilde{n}^2 - 1)(\overline{E_{\parallel x}^2} - \overline{E_{\parallel y}^2})\cos\beta \\ - \overline{E_{\perp x}^2}\sin^2\beta + (H_{\perp x}^2 - H_{\perp y}^2)\sin^2\beta - H_{\parallel x}^2\cos^2\beta] \\ + \frac{1}{4\pi}\mathbf{i}_2\frac{\partial}{\partial x}(\overline{\tilde{n}^2E_{\parallel x}E_{\parallel y}}\cos^2\beta + H_{\perp x}H_{\perp y}\sin^2\beta) \\ + \frac{1}{8\pi}\mathbf{i}_3\frac{\partial}{\partial x}\left(\frac{\tilde{n}^2}{2}\overline{E_{\parallel x}E_{\perp x}} + \overline{H_{\perp x}H_{\parallel x}}\right)\sin 2\beta, \quad (50)\end{aligned}$$

where β is the angle between the electrical vector $\mathbf{E}$ and the plane of incidence. The $\mathbf{i}_3$ component shows a coupling between parallel and perpendicular polarization. To evaluate this component, one finds from (38), (40), (47), and (48),

$$\begin{aligned}\tilde{n}^2E_{\parallel x}E_{\perp x} &= -\frac{\tilde{n}E_v^2\cos\alpha_0}{\cos\alpha}\sin\alpha\sin^2 G; \\ H_{\perp x}H_{\parallel x} &= \frac{\tilde{n}E_v^2\cos\alpha_0}{\cos\alpha}\sin\alpha\cos^2 G.\end{aligned} \quad (51)$$

As a result, the time-averaged bracket in the $\mathbf{i}_3$ component of Eq. (50) vanishes identically.

Because of the last result, one can construct an expression for the case of a general β by a simple addition of the two expressions valid for each polarization. For $E_\perp(\beta = \frac{1}{2}\pi)$ the expression of the time-averaged perpendicular component $\mathbf{f}_\perp$ of the force density is

$$\overline{\mathbf{f}_\perp} = \frac{1}{8\pi}\mathbf{i}_1\frac{\partial}{\partial x}(-\overline{E_{\perp x}^2} + \overline{H_{\perp x}^2} - \overline{H_{\perp y}^2}) + \frac{1}{4\pi}\mathbf{i}_2\frac{\partial}{\partial x}(\overline{H_{\perp x}H_{\perp y}}). \quad (52)$$

From (40) one finds

$$\begin{aligned}H_{\perp x}H_{\perp y} &= -\tilde{n}E_v^2\cos\alpha_0\sin\alpha\cos^2 G \\ &\mp \frac{c}{4\omega}\frac{E_v^2\cos\alpha_0\sin\alpha}{\tilde{n}^4\cos^3\alpha}\frac{d\tilde{n}}{dx}\sin 2G. \quad (53)\end{aligned}$$

From Eq. (39) the time-averaged value of Eq. (53) is

$$\overline{H_{\perp x}H_{\perp y}} = -\tfrac{1}{2}E_v^2\cos\alpha_0\sin\alpha_0 = \text{const}, \quad (54)$$

and as a result, the $\mathbf{i}_2$ component of $\overline{\mathbf{f}_\perp}$ vanishes in (52). From (38) and (40) one finds

$$-\overline{E_{\perp x}^2} + \overline{H_{\perp x}^2} - \overline{H_{\perp y}^2} = -E_v^2\cos\alpha_0 \cdot \frac{\cos^2\alpha_0 + \tilde{n}^2\cos^2\alpha}{2\tilde{n}\cos\alpha} + \frac{A}{2}. \quad (55)$$

No second-order term was found but only a third-order term given by

$$A = \frac{c^2}{4\omega^2}\frac{1}{\tilde{n}^3\cos^5\alpha}\left(\frac{d\tilde{n}}{dx}\right)^2. \quad (56)$$

The force density from (52) is then

$$\overline{\mathbf{f}_\perp} = \mathbf{i}_1\frac{E_v^2\cos\alpha_0}{16\pi}\left(\frac{1}{\cos^3\alpha}\frac{\omega_p^2}{\omega^2}\frac{1}{\tilde{n}^2}\frac{d\tilde{n}}{dx} - \frac{d}{dx}A\right). \quad (57)$$

In the case of $\mathbf{E}_\parallel(\beta = 0)$ one finds in (51)

$$\overline{\mathbf{f}_\parallel} = \frac{\mathbf{i}_1}{8\pi}\frac{\partial}{\partial x}\left[\overline{E_{\parallel x}^2}\left(1 - 2\frac{\omega_p^2}{\omega^2}\right) - \overline{E_{\parallel y}^2} + \overline{H_{\parallel x}^2}\right] + \frac{\mathbf{i}_2}{4\pi}\frac{\partial}{\partial x}\overline{\tilde{n}^2E_{\parallel x}E_{\parallel y}}. \quad (58)$$

The last term vanishes because of the constancy in space of

$$\overline{\tilde{n}^2E_{\parallel x}E_{\parallel y}} = \frac{E_v^2}{2}\cos\alpha_0\sin\alpha_0. \quad (59)$$

If one uses the equation of motion (12) of the two-fluid model[13] or that of the microwave interaction,[9] the $\mathbf{i}_2$ component of the force density (58) does not vanish. Only the general formulations (10) or (11) give the right results. Thus, the importance of the last three terms of Eq. (11) is demonstrated.

Using Eqs. (56) and (3) with $\nu = 0$ one finds

$$\begin{aligned}\overline{(2\tilde{n}^2 - 1)E_{\parallel x}^2 - E_{\parallel y}^2 - H_{\parallel x}^2} &= -E_v^2\cos\alpha_0 \\ &+ \left(\frac{\cos^2\alpha_0 + \tilde{n}^2\cos^2\alpha}{2\tilde{n}\cos\alpha} + \frac{A}{2}\cos^2\alpha(1 - 2\sin^2\alpha)^2\right),\end{aligned} \quad (60)$$

and finally from Eqs. (58) and (60)

$$\begin{aligned}\overline{\mathbf{f}_\parallel} = \mathbf{i}_1\frac{E_v^2\cos\alpha_0}{16\pi} &\cdot\left(\frac{1}{\cos^3\alpha}\frac{\omega_p^2}{\omega^2}\frac{1}{\tilde{n}^2}\frac{d\tilde{n}}{dx} - \frac{dA}{dx}\cos^2\alpha(1 - 2\sin^2\alpha)^2\right. \\ &+ 2A\frac{d\tilde{n}}{dx}\frac{\sin^3\alpha}{\sin\alpha_0}(2\sin^2\alpha - 1) \\ &\left.\cdot(3\sin 2\alpha - 2\cos^4\alpha - 1)\right). \quad (61)\end{aligned}$$

A comparison with the collisionless case of perpendicular incidence, Eq. (26), shows in the first order from Eqs. (57) and (61) that

$$\overline{\mathbf{f}(\alpha)} = \overline{\mathbf{f}(0)}\frac{\cos\alpha_0}{\cos^3\alpha}. \quad (62)$$

This expression has already been obtained by Schlüter using a specialized treatment.[16] In addition the force density of this analysis is only in the negative x direction, i.e., toward decreasing electron density, up to the third order in the spatial variation of the index of refraction in all cases of polarization, and is independent of the propagation direction of the light. The magnitude of the force density, however, is weakly dependent on the polarization

[16] A. Schlüter, Plasma Phys. 10, 471 (1968).

of the light in the third-order terms in $\tilde{n}$. The consideration of these third-order terms is also justified in the case of the WKB approximation, because it can be shown[17] that deviations of the WKB solution from the exact solution are exponentially small and are given by exp $(-\theta)$ where θ is much smaller than unity according to Eq. (18).

The validity of the force density of Eq. (50) is restricted to angles $\alpha \lesssim 40°$ [see Eq. (46)]. For a parallel light wave interacting with a spherical plasma, if one can approximate each part of the inhomogeneous layer of the plasma by a plane geometry, then one would expect a collisionless nonlinear acceleration of the plasma as indicated by the full arrows at A and C of Fig. 3. The acceleration which is not parallel with respect to the incident light will be larger by $\cos \alpha_0/\cos^3 \alpha$ in comparison with perpendicular incidence. At the points B of the plasma, this treatment loses validity because of the already discussed coupling of the differential equations of the field components. But the acceleration can be expected to be approximately as indicated by the arrows at B.

V. MOMENTUM TRANSFER TO THE INHOMOGENEOUS TRANSITION LAYER

The application of an equation of motion similar to Eq. (25) to solve the dynamical interaction process is extremely complicated due not only to the dependence of the refractive index $\tilde{n}$ on space and time, but also because of the attenuation of E_v by this process of collisionless absorption. Even if one considers the nonlinear collisionless case when the condition of Eq. (31) obtains, no significant simplification occurs. Consequently, the discussion of integrated expressions is of greater interest.

In the treatments of the previous sections, plane waves of constant amplitude were used. If a giant laser pulse a few nanoseconds wide between the times t_1 and t_2 is considered, the time variation of the amplitude E_v of the electric field strength in the vacuum is slow enough and does not change the applicability of all conditions assumed in this analysis. Also the picosecond substructure of laser spikes[18] does not change these assumptions. The total energy ϵ_L of the laser pulse is then

$$\epsilon_L = c \int_K dy\, dz \int_{t_1}^{t_2} \frac{E_v^2(t, y, z)}{8\pi}\, dt. \qquad (63)$$

Here the integration is performed across the entire cross section K (coordinates y and z) where the light interacts with the plasma. The momentum of all photons in the vacuum is then

$$P_0 = \frac{\epsilon_L}{c}. \qquad (64)$$

Under the condition (31) which gives the main interaction of the light beam with the plasma transition layer by the collisionless nonlinear process at sufficiently high temperatures T at the surface, the force given in Eq. (26) produces a total momentum P_{inh} transferred to a plasma between x_1 and x_2 in the direction of $\mathbf{f}$ shown in Fig. 1 given by

$$P_{\text{inh}} = \int_K dy\, dz \int_{x_1}^{x_2} dx \int_{t_1}^{t_2} \overline{\mathbf{f}_{\text{NL}}}\, dt. \qquad (65)$$

Using Eq. (26) and the procedure which led to this equation under collisionless conditions, one finds

$$P_{\text{inh}} = -\int_K dy\, dz \int_{t_1}^{t_2} dt \cdot \int_{x_1(t)}^{x_2(t)} dx \left[\frac{\partial}{\partial x} \frac{E_v^2(y, z, t)}{16\pi} \left(\frac{1}{|\tilde{n}|} + |\tilde{n}|\right)\right] \qquad (66a)$$

$$= -\frac{P_0}{2\,|\tilde{n}_2|}(1 - |\tilde{n}_2|)^2. \qquad (66b)$$

$\tilde{n}_2$ equals $\tilde{n}(x_2)$ by definition. Equation (66b) expresses the momentum of the accelerated inhomo-

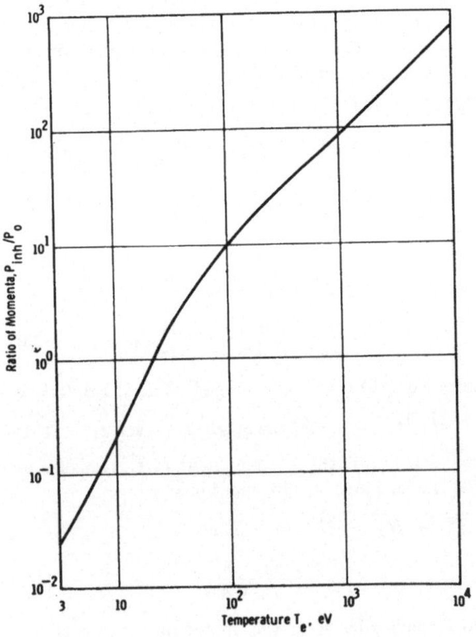

FIG. 4. Ratio of P_{inh} to P_0 as a function of electron temperature T_e. P_{inh} is the momentum transferred to the inhomogeneous plasma in the direction of decreasing density. P_0 is the momentum of the incident laser radiation in the vacuum.

[17] H. L. Berk, D. L. Book, and D. Pfirsch, J. Math. Phys. 8, 1611 (1967).

[18] S. L. Shapiro, M. A. Duguay, and L. B. Kreuzer, Appl. Phys. Letters 12, 36 (1968).

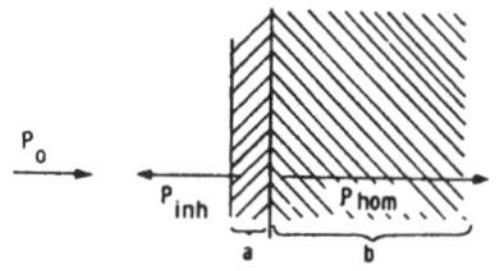

FIG. 5. Momentum transfer to an inhomogeneous plasma *a* from a laser beam. P_{hom} is the momentum of photons and plasma in the homogeneous part *b*.

geneous transition layer in terms of the momentum P_0 of the laser pulse in vacuum. To get a high momentum one needs low values of $|\tilde{n}_2|$. The magnitude of the momentum is limited by the extent to which the radiation pressure can be neglected, i.e., by how high T_e is in Eq. (31). In Fig. 4 the momentum P_{inh} is given for densities determined by ω_p^* [see Eq. (31) and Fig. 2]. We had pointed out before,[7] that the nonlinear collisionless interaction can transfer to the plasma a momentum much larger than the magnitude of the momentum given by all photons, because then the momentum at the front of the plasma is compensated by that at the back of a collisionless plasma. Here, we have found to what magnitude the momentum imparted by the nonlinear collisionless acceleration can increase for a plasma with collisions.

In Fig. 5 we are considering the total momentum P_{hom} transfered to an absorbing homogeneous plasma *b* and the photons in it, when the vacuum momentum of the photons is P_0 and the momentum transferred to the inhomogeneous layer *a* by the collisionless nonlinear process is P_{inh} as given by Eq. (66). The conservation of momentum gives

$$P_0 = P_{hom} - P_{inh}. \tag{67}$$

When all photons are absorbed in the homogeneous plasma, then the momentum transferred to *b* is

$$P_{hom} = P_0 + P_{inh} = \frac{P_0}{2}\left(\frac{1}{|\tilde{n}|} + |\tilde{n}|\right). \tag{68a}$$

In terms of P_0, we can write

$$\frac{P_{hom}}{P_0} = 1 + \frac{P_{inh}}{P_0}. \tag{68b}$$

For example, when the temperature T_e of the homogeneous plasma is 1 keV, then the momentum transferred to the plasma by the nonlinear mechanism can be 100 times the momentum of the incident laser light in vacuum. At higher temperatures this momentum transfer increases nearly linearly [see Eq. (68b) and Fig. 4]. This result may explain the measurement by Gregg and Thomas.[19] The momentum measured by these authors can be identified with P_{hom}. Neglecting a small variation with the laser pulse length, one finds for the dependence on the input laser energy ϵ_L

$$P_{hom} \propto \epsilon_L^{1-\zeta}, \tag{69}$$

where the number ζ is about 0.2 to 0.3 for different materials. This gives

$$P_{hom}^2 \propto \epsilon_L^{2-2\zeta} = \epsilon_L^{1.5}. \tag{70}$$

Thus, the measured recoil indicates an essentially nonlinear process.

The observation[19] of P_{hom} being 10^3 to 10^4 times P_0 could be explained by surface electron temperatures in excess of 10^4 eV (see Fig. 4). Since the primary light–plasma interaction takes place at the surface, such high values of T_e may not be too unrealistic. The ion temperatures can be much lower during this process. Consequently, the explanation given here may be interpreted as the basic nonlinear process leading to a shock wave, probably in the same sense as an explanation by Kidder has been mentioned[19] or in the sense of blast waves.[20] A purely thermodynamic expansion is less probable as an explanation for the measurements[19] because a thermodynamic process would lead to a relation $P_{hom}^2 \propto \epsilon_L$ instead of the nonlinear relation (70).

VI. CONCLUSIONS AND DISCUSSION

This treatment has verified the existence of substantial macroscopic acceleration of an inhomogeneous plasma by interaction with laser light. This nonlinear interaction results from a rigorous application of the ponderomotive force description based on Lorentz's theory. The deduced equation of motion is more general than that of the two-fluid model of the plasma and that used in the theory of microwave interaction with plasma.

The necessity of using all nonlinear terms of the general equation of motion is demonstrated, when the nonlinear collisionless acceleration of an inhomogeneous plasma at oblique incidence of light is calculated. As might be expected, the forces are only in the direction of lower plasma densities and tangential forces vanish only with the general equation. This result has been verified up to the third order in the spatial variation of the electron density. Deconfinement results because the acceleration is always in the direction of decreasing electron density. The magnitude of the acceleration varies for different cases of polarization only in terms of the third order in the spatial variation of the electron density.

In a plasma with collisions, the nonlinear acceleration is in the direction of the laser light, and can be interpreted as a radiation pressure when the electron density is equal to or higher than that which gives a plasma frequency equal to the light frequency. A temperature-dependent upper limit of the electron density has been evaluated for electron densities below which the radiation pressure term in the

[19] D. W. Gregg and S. J. Thomas, J. Appl. Phys. **37**, 2787 (1966).

[20] E. Panarella and P. Savic, Can. J. Phys. **46**, 183 (1968).

force density can be neglected, and only the nonlinear collisionless deconfining acceleration remains.

Integration of this force density in space and time gives the deconfining recoil momentum imparted to the inhomogeneous plasma in terms of the momentum of the photons in vacuum. Under certain reasonable conditions the transferred momentum can be 100 times or more the vacuum photon momentum. The high recoil momentum measured by Gregg and Thomas[19] may be explained by extreme heating of the outermost plasma front to initiate the nonlinear mechanism. Consequently, the observations of Gregg and Thomas[19] could be the result of a nonlinear interaction.

In addition, the theory given here can explain some of the observations in plasmas produced by lasers from single small aluminum balls.[8] While an inner part of plasma containing nearly all the plasma shows the properties of a thermal expansion, behavior of the outer part can be attributed to a surface mechanism. An averaged nonlinear absorption constant based on the recoil momentum shows that nonlinear acceleration at the surface of inhomogeneous aluminum plasmas[8] overwhelms the thermal absorption processes for electron temperatures above 130 eV.[8] The nonlinearity of the expansion is verified by a superlinear dependence of the maximum ion energy on the laser intensity, as was the case for measurements from a thick target of tantalum.[21]

ACKNOWLEDGMENTS

This work began while the author was with the Institut für Plasmaphysik at Garching-Munich and was completed at the Westinghouse Research Laboratories, Pittsburgh, Pennsylvania. Many extensive and illuminating discussions with D. Pfirsch (Munich), A. Schlüter (Garching), J. M. Dawson (Princeton, N.J.), and A. G. Engelhardt (Westinghouse Research Laboratories) are gratefully acknowledged.

The work at Garching was done under the auspices of a contract with Euratom. The work at Westinghouse Research was supported in part by the United States Atomic Energy Commission.

[21] S. Namba, D. H. Kim, T. Itoh, T. Arai, and H. Schwarz, in *Record of the IEEE 9th Annual Symposium on Electron, Ion and Laser Beam Technology, Berkeley, 9-11 May 1967*, R. F. W. Pease, Ed. (San Francisco Press, Inc., San Francisco, Calif.), p. 86.

Ponderomotive Force on Laser-Produced Plasmas

JOHN D. LINDL AND PREDHIMAN K. KAW
Plasma Physics Laboratory, Princeton University, Princeton, New Jersey 08540

The deconfining ponderomotive force exerted by an intense electromagnetic wave on a linearly inhomogeneous plasma layer has been analytically investigated. Since the ponderomotive force maximizes near the region where the dielectric constant $\epsilon' \to 0$, an exact solution of the wave equation is required for a correct estimate of the maximum force. For the case of oblique incidence with the electric vector in the plane of incidence, the solution of the exact wave equation leads to an interesting resonance effect which predicts a force higher than that obtained by earlier workers using a WKB procedure. Physically, this enhanced ponderomotive force arises because of the large nonuniform longitudinal fields that are generated in the region $\epsilon' \approx 0$ for this geometry.

I. INTRODUCTION

Hora, Pfirsch, and Schlüter,[1] Schlüter,[2] and Hora[3] analytically investigated the nonlinear interaction of an intense laser beam with an inhomogeneous plasma layer, and have shown that such an interaction leads to a deconfining (i.e., directed away from regions of high particle density) ponderomotive force on the layer. Physically, this force arises because of the general tendency of any nonuniform dielectric fluid to move toward regions of higher dielectric constant in the presence of an electric field.[4] The analyses in Refs. 1 and 3 (the details of Ref. 2 are not known) are based on the use of the WKB approximation for laser wave propagation in the inhomogeneous plasma layer. Explicit expressions have been derived for the ponderomotive force, and it is concluded that this force will have its maximum value for the critical layer in which the laser frequency ω matches the local plasma frequency ω_p. Unfortunately, in this critical layer the WKB approximation is least valid; the estimates of the maximum ponderomotive force made in the above papers are therefore of a rather limited range of applicability. Further, some interesting qualitative effects for the case of oblique incidence have been missed because of the use of WKB approximation.

A more rigorous way of computing the magnitude of the maximum ponderomotive force is to solve the wave equation exactly in the critical layer. One obvious limitation of this procedure is that exact solutions can be obtained only for some special density profiles.[5] However, for most of the situations (especially if the critical layer is far removed from the flat top of the density profile) it is a good approximation to consider the density variation in the critical layer to be linear. This then is the approach that we have adopted in the present paper.

Three cases have been examined in detail:

1. Normal incidence on a plane stratified layer.
2. Oblique incidence with the
   (a) electric vector normal to the plane of incidence, and
   (b) electric vector in the plane of incidence.

In cases 1 and 2(a), the exact analysis extends the range of validity of Hora's treatment and sets an accurate upper bound on the level of the force. The effect of the superposition of incident and reflected waves which leads to a standing wave pattern and a "ponderomotive bunching" effect has also been investigated, in contrast to earlier work. In case 2(b), an interesting resonance effect, not found by the WKB approximation, leads to a deconfining force which is much greater than that of the first two cases.

Momentum transfer to the plasma layer in the three cases is also evaluated.

II. PONDEROMOTIVE FORCE

As derived by Hora,[3] the expressions for the ponderomotive force components are in terms of the electric and magnetic field vectors. Most of the present work will, therefore, be concerned with an evaluation of the electric and magnetic field vectors, by solving the exact wave equation.

Reprinted from: *The Physics of Fluids*, **14**, 371--377 (1971).

Starting from Maxwell equations, one can derive the general wave equation for the propagation of electric and magnetic fields in an inhomogeneous plasma; these are

$$\nabla^2\mathbf{E} - \nabla(\nabla\cdot\mathbf{E}) + \frac{\omega^2}{c^2}\epsilon'(\omega, \nu)\mathbf{E} = 0 \tag{1}$$

and

$$\nabla^2\mathbf{B} + \frac{1}{\epsilon'(\omega, \nu)}[\nabla\epsilon'(\omega, \nu)\times(\nabla\times\mathbf{B})] + \frac{\omega^2}{c^2}\epsilon'(\omega, \nu)\mathbf{B} = 0, \tag{2}$$

where the symbols have their usual meanings and an exp $(-i\omega t)$ time dependence for the fields has been assumed. The complex dielectric permittivity for a cold plasma may be written

$$\epsilon' = 1 - \frac{\omega_p^2}{\omega^2 + \nu^2}\left(1 - \frac{i\nu}{\omega}\right), \tag{3}$$

where ν is a phenomenological collision frequency. Equations (1) and (2) are investigated in the neighborhood of the region where $\epsilon' \sim 0$; in this region, ω_p^2 is assumed to depend linearly on z. Thus, for $\nu/\omega \ll 1$

$$\epsilon' = -az + is, \tag{4}$$

where $a \approx [(1/n)(dn/dz)]$ is a measure of the density gradient and $s \equiv (\nu/\omega)$. It has been assumed here that the imaginary part of ϵ' is independent of the position coordinate and equal to its value at $\omega \simeq \omega_p$. This is a good approximation since $(\nu/\omega) \ll 1$ and the variation of Re ϵ' (which pases through a zero) with z is much more important than that of Im ϵ' (which approaches a constant). In general, only one of Eqs. (1) and (2) need be solved.

A. Normal Incidence

Since the dielectric constant is a function of z alone, we can call the medium plane layered. Consider an electromagnetic wave propagating along the z direction, i.e., normal to the plane stratifications. Equation (1) for this case assumes the form

$$\frac{\partial^2\mathbf{E}}{\partial z^2} + \frac{\omega^2}{c^2}(-az + is)\mathbf{E} = 0. \tag{5}$$

Without loss of generality, one can choose the x axis along the direction of the electric vector. Introduce

$$\zeta = \left(\frac{\omega}{ca}\right)^{2/3}(-az + is) = \rho^{2/3}(-az + is) \equiv \rho^{2/3}\xi. \tag{6}$$

Equation (5) then reduces to

$$\frac{d^2E}{d\zeta^2} + \zeta E = 0. \tag{7}$$

The solutions to Eq. (7) are the well-known Airy functions with complex arguments, which may be written in terms of Bessel functions of order $\frac{1}{3}$; thus

$$E = 3Aa_i(-\zeta) = A\zeta^{1/2}[J_{1/3}(\tfrac{2}{3}\zeta^{3/2}) + J_{-1/3}(\tfrac{2}{3}\zeta^{3/2})] \quad \mathrm{Re}\,\zeta > 0 \tag{8a}$$

$$= A(-\zeta)^{1/2}\{I_{-1/3}[\tfrac{2}{3}(-\zeta)^{3/2}] + I_{1/3}[\tfrac{2}{3}(-\zeta)^{3/2}]\} \quad \mathrm{Re}\,\zeta < 0. \tag{8b}$$

This type of solution has been discussed by Ginzburg.[5] It represents a standing electromagnetic wave in the region Re $\zeta > 0$ (arising because of a superposition of the incident and reflected waves) and an exponentially decaying evanescent wave in the region Re $\zeta < 0$. The constant A is to be determined from the boundary conditions. It is assumed that the plasma boundary is far from the reflection region or the critical layer, so that on the boundary one has $|\zeta| \gg 1$. One can then use the asymptotic form for the electric field at the plasma boundary; viz.,

$$\mathbf{E} = \frac{3A}{\pi^{1/2}}\zeta^{-1/4}\cos\left(\frac{2}{3}\zeta^{3/2} - \frac{\pi}{4}\right) \tag{9a}$$

$$= \frac{3A}{2\pi^{1/2}}\left(\frac{\omega}{ca}\right)^{-1/6}\epsilon^{-1/4}\left[\exp\left(\frac{-i\omega}{c}\int_z^{+is/a}\epsilon^{1/2}\,dz + i\frac{\pi}{4}\right) + \exp\left(+\frac{i\omega}{c}\int_z^{+is/a}\epsilon^{1/2}\,dz - i\frac{\pi}{4}\right)\right], \tag{9b}$$

where Eqs. (4) and (6) have been used. In Eq. (9b) the first term represents a forward-propagating wave [remember that an exp $(-i\omega t)$ time dependence has been assumed] and the second a backward propagating or a reflected wave. Now, if the vacuum–plasma boundary is assumed sufficiently diffuse, then there will be negligible reflection from the boundary itself and one can equate the forward propagating part of Eq. (9b) to the incident electric vector $\mathbf{E}_v$ at the boundary. This gives

$$A = \tfrac{2}{3}\pi^{1/2}\left(\frac{\omega}{ca}\right)^{1/6}E_v\exp\left(+\frac{i\omega}{c}\int_{-1/a}^{+is/a}\epsilon^{1/2}\,dz - \frac{i\pi}{4}\right)$$

$$\simeq \tfrac{2}{3}\pi^{1/2}\rho^{1/6}E_v\exp(-\rho s)\exp\left[-i\left(\tfrac{2}{3}\rho - \frac{\pi}{4}\right)\right], \tag{10}$$

where $s = (\nu/\omega)$ has been assumed to be much less than unity.

For normal incidence, the ponderomotive force is given by[3]

$$\mathbf{F}_0 = -\frac{\mathbf{e}_z}{8\pi}\frac{\partial}{\partial z}(\bar{E}_x^2 + \bar{H}_y^2). \tag{11}$$

One can relate H_y to E_x for our special geometry by using Maxwell's equations; thus,

$$H_y = +i\rho^{-1/3}\left(\frac{dE_x}{d\zeta}\right).$$

Multiplying Eq. (8a) for E_x and the corresponding

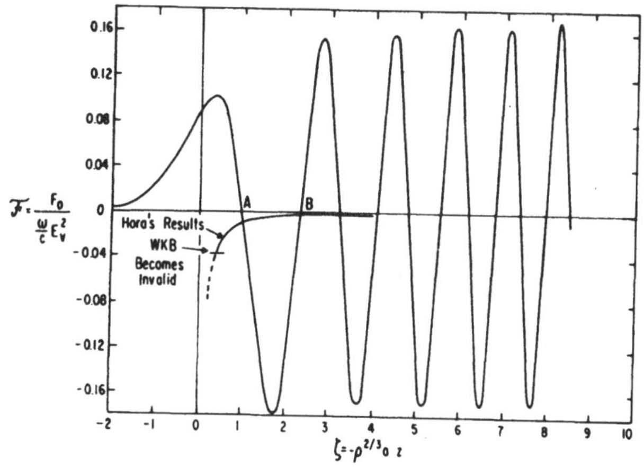

FIG. 1. Ponderomotive force for normal incidence as a function of distance from plane of reflection.

expression for H_y by exp $(-i\omega t)$, taking the real parts, squaring and adding them, and finally averaging over several periods of the laser frequency, one obtains from Eq. (11)

$$\mathbf{F}_0 = \mathbf{e}_z \frac{\omega}{c} E_v^2 \exp(-2\rho s)\{(a_{iR}a'_{iR} + a_{iI}a'_{iI}) + \rho^{-2/3}[a'_{iR}(\zeta_R a_{iR} - \zeta_I a_{iI}) + a'_{iI}(\zeta_I a_{iR} + \zeta_R a_{iI})]\}, \tag{12}$$

where a_{iR}, etc., are the Airy functions, the subscripts R and I denote the real and imaginary parts, and the prime denotes differentiation with respect to the argument. Hora's expression

$$\mathbf{F}_0 = \mathbf{e}_z \frac{E_v^2}{16\pi} \frac{\omega_p^2}{\omega^2 \epsilon} \frac{d}{dx} (\epsilon^{1/2}), \tag{13}$$

could be recovered, if only the first term (viz., the forward-propagating part) of the asymptotic form (9b) for the electric vector and the corresponding expression for the magnetic vector had been used. Thus, the interesting ponderomotive "bunching" effect, arising because of the standing wave pattern on the left of the reflection region, also is not included in Hora's treatment.[3]

Figure 1 is a plot of

$$F = \frac{F_0}{[(\omega/c)E_v^2]}$$

versus ζ for a collisionless plasma. It shows the typical force profile in the neighborhood of the reflection plane. Note that a positive force acts *in* the direction of the incident beam and acts as a confining force whereas a negative force does the reverse. In Fig. 1, the force to the left of A (viz., the first zero) acts on the main body of the cold plasma and pushes it inward. The force to the right of A is, however, oscillatory in space and leads to ponderomotive bunching. This may give rise to "striations" in the plasma. If one averages the oscillatory force to the right of A, one gets the net outward directed force on the plasma layer. Note that the largest contribution to the average force will come from the unbalanced force in the region AB; the rest of the maxima and minima nearly average out to zero.

In Fig. 1, the force profile due to Hora's expression [Eq. (13)] has also been plotted for comparison. A more appropriate comparison is the one between Eq. (12) and the force obtained using a standing wave solution obtained from the sum of incident and reflected WKB solutions. Such a comparison shows that the WKB approximation gives the correct *order of magnitude* of the force to the right of A. However, the absolute magnitudes can be off by more than a 100%. Furthermore, one also incorrectly predicts the phases of the fields (and hence the position of A, for example). Finally, to the left of A, the WKB solution breaks down and predicts an infinite force at $\zeta = 0$. Figure 1, on the other hand, clearly shows that the maximum ponderomitive force at any point in the plasma is of the order of a few tenths of $E_v^2(\omega/c)$ and is determined by the free space wavelength.

Table I illustrates the variation of the normalized ponderomotive force F at the first negative maximum, with the level of collisions. The force is seen to decrease as $\zeta_I = (\omega/ca)^{2/3}(\nu/\omega)$ increases. The values of ζ_I chosen are typical of laser produced plasmas; thus $\zeta_I = 1$ corresponds to a 100 μ radius plasma with a density of 10^{21} cm^{-3} and a temperature of few hundred volts.

TABLE I. Variation of the normalized ponderomotive force $F = F_s/E_s^2(\omega/c)$ at the first maximum with the collision frequency parameter $\zeta_I = (\omega/ca)^{2/3}(\nu/\omega)$ for the normal incidence.

ζ_I	0	0.2	0.4	0.6	0.8
F	0.176	0.320×10^{-2}	0.554×10^{-4}	0.863×10^{-6}	0.930×10^{-8}

B. Oblique Incidence

Consider a plane stratified layer of the type described earlier with ϵ' a function of z only. Assume (without loss of generality) that the incident k vector lies in the y-z plane; all of the field quantities are then independent of x. If one writes the general wave equation (1) in its component form, one finds that the equation for E_x decouples from those of E_y and E_z (which are coupled to each other). This circumstance arises only because ϵ' is chosen independent of x, and physically it means that in this case one can decompose an electromagnetic wave with arbitrary polarization into two waves: one with the electric vector in the plane of incidence and the other with electric vector normal to the plane of incidence. One is thus justified in discussing these two waves separately.

1. E Vector Normal to the Plane of Incidence

In this case $E_y = E_z = 0$. Let $E_x = F(z) \exp[\pm ik(z)\alpha(z)y]$, where $k(z) = (\omega/c)[\epsilon'(z)]^{1/2}$, $\alpha(z) = \sin\theta(z)$, and θ is the angle between the direction of propagation and the z direction. Substituting in the x component of Eq. (1) and demanding that the resultant equation be valid for all y, one obtains

$$\frac{d}{dz}[k(z)\alpha(z)] = 0 \quad \text{or} \quad k(z)\alpha(z) = \text{const} = k_0\alpha_0 \qquad (14a)$$

and

$$\frac{d^2F}{dz^2} = -\frac{\omega^2}{c^2}\epsilon'(1-\alpha^2)F = -\frac{\omega^2}{c^2}(\epsilon' - \sin^2\theta_0)F, \qquad (14b)$$

where $\alpha_0 = \sin\theta_0$ is the sine of the angle of incidence on the layer and $k_0 = (\omega/c)$ is the free-space wavenumber. Introducing $\epsilon' = -az + is$, $\zeta = (\omega/ca)^{2/3}(\epsilon' - \alpha_0^2)$, one gets

$$\frac{d^2F}{d\zeta^2} + \zeta F = 0.$$

This equation is identical to the one obtained for normal incidence; thus, the results obtained in the last section are valid with a redefinition of ζ. For this case the ponderomotive force is given by[3]

$$\mathbf{F}_{\alpha\perp} = -\frac{\mathbf{e}_z}{8\pi}\frac{\partial}{\partial z}(E_x^2 - H_z^2 + H_y^2) \qquad (15)$$

$$= -\frac{\mathbf{e}_z}{8\pi}\frac{\partial}{\partial z}[(1-\alpha_0^2)E_x^2 + H_y^2].$$

One can thus derive an expression for the ponderomotive force; this force always turns out to be less than that for normal incidence.

2. E Vector in the Plane of Incidence

In this case the electromagnetic wave exhibits a very interesting behavior in the neighborhood of the region $\epsilon' \approx 0$. The wave is reflected at a plane $z < z_0$, where z_0 is determined by $\epsilon(z_0) = \sin^2\theta_0$. Beyond this point, the wave begins to exhibit the usual evanescent character. However, as the point $\epsilon' \approx 0$ is approached, there is a large increase in the magnitude of the electric vector.[5] Thus, huge gradients in the electromagnetic fields are developed in the neighborhood of $\epsilon' \approx 0$, and one can find a very large ponderomotive force.

To examine this problem in some detail, it is convenient[5] to start from Eq. (2) for the magnetic vector. For the present special case, $\mathbf{H}$ has only one nonzero component, H_x; furthermore, E_y and E_z can readily be related to H_x by the equations

$$E_y = -\frac{ic}{\omega\epsilon'}\frac{\partial H_x}{\partial z} \qquad (16a)$$

and

$$E_z = \frac{ic}{\omega\epsilon'}\frac{\partial H_x}{\partial y}. \qquad (16b)$$

Using $H_x = G(z)\exp[\pm ik(z)\alpha(z)y]$ in the x component of Eq. (2) and proceeding as in Sec. IIB1, one obtains

$$k(z)\epsilon(z) = k_0\alpha_0 = (\omega/c)\sin\theta_0$$

and

$$\frac{d^2G}{dz^2} - \frac{1}{\epsilon'}\frac{d\epsilon'}{dz}\frac{dG}{dz} + \frac{\omega^2}{c^2}(\epsilon' - \sin^2\theta_0)G = 0.$$

Using Eq. (4) and introducing $\xi = -az + is$, $\rho = (\omega/ca)$, and $U(\xi) = \xi^{-1/2}G(\xi)$, this equation can be put in the form

$$\frac{d^2U}{d\xi^2} - \left(\rho^2(\xi + \alpha_0^2) + \frac{3}{4\xi^2}\right)U = 0. \qquad (17)$$

An approximate solution to Eq. (17), valid everywhere except near $\xi = 0$ and which represents a standing wave to the left of $\xi = \alpha_0^2$, can be written

$$U = \left(\frac{\pi\Delta}{2}\right)^{1/2}\exp\left(-\frac{i\pi}{12}\right)\left(\frac{S}{S'}\right)^{1/2}H_{1/3}^{(1)}(iS), \qquad (18)$$

where

$$S = \rho \int_{-\alpha_0^2}^{\xi} (\xi + \alpha_0^2)^{1/2}\, d\xi = \tfrac{2}{3}\rho(\xi + \alpha_0^2)^{3/2},$$

$$S' = \left(\frac{dS}{d\xi}\right) \quad \text{and} \quad \Delta = \rho(1 - \alpha^2)^{1/2}$$

$$\cdot \exp\,[-2\rho s - \tfrac{4}{3}i\rho(1 - \alpha^2)^{3/2}],$$

and the $H_{1/3}^{(1)}$ denotes the Hankel function of the first kind of order $\frac{1}{3}$. The constant in Eq. (18) is chosen so that the amplitude of the incident wave equals unity at the boundary of the inhomogeneous layer. For large ξ, the asymptotic form of the H_x obtained by using Eq. (18), agrees with the expression derived by Hora,[3] using the WKB approximation.

Another approximate solution to Eq. (17), valid everywhere to the right of $\xi = -\alpha_0^2$, is

$$W = A(\eta/\eta')H_1^{(1)}(i\eta), \tag{19}$$

where

$$\eta = \rho \int_0^{\xi} (\xi + \alpha_0^2)^{1/2}\, d\xi = \tfrac{2}{3}\rho[(\xi + \alpha_0^2)^{3/2} - \alpha_0^3],$$

$$\eta' = \frac{d\eta}{d\xi},$$

and A is an arbitrary constant. The two solutions, (18) and (19), should be close to each other everywhere except in the neighborhood of $\xi = 0$ [where (18) is not valid] and $\xi = -\alpha_0^2$ [where (19) is not valid]. Matching the solutions in the region $-\alpha_0^2 < \xi < 0$, one can evaluate A. Thus, the solution to Eq. (17) in the neighborhood of $\xi \approx 0$ is given by

$$U(\xi) = \xi^{-1/2}G(\xi)$$

$$= \left(\frac{\pi\Delta}{2}\right)^{1/2} \exp\left(\frac{i\pi}{4} - \tfrac{2}{3}\rho\alpha_0^3\right)\left(\frac{\eta}{\eta'}\right)^{1/2} H_1^{(1)}(i\eta). \tag{20}$$

If ξ is very small, we can use $\eta \simeq \rho\alpha_0\xi \ll 1$ and expand

$$H_1^{(1)}(i\eta) \simeq -\frac{2}{\pi\eta} - \frac{\eta}{\pi}\ln(\eta).$$

Using this series expansion in Eq. (20), and also using Eqs. (16a) and (16b), we find the following expressions for the electric and magnetic field vectors in the neighborhood of ξ small:

$$H_x(\xi) = -\left(\frac{2\cos\theta_0}{\pi\rho\sin^2\theta_0}\right)^{1/2} \exp(-\phi + i\psi)$$

$$= -G(0)\exp\left(i\frac{\omega}{c}\alpha_0 y\right), \tag{21a}$$

$$E_y(\xi) = -i\left(\frac{2\rho}{\pi}\cos\theta_0\sin^2\theta_0\right)^{1/2} \ln(\rho\sin\theta_0\xi)$$

$$\cdot\exp(-\phi + i\psi), \tag{21b}$$

$$E_z(\xi) = -\frac{1}{\xi}\left(\frac{2\cos\theta_0}{\pi\rho}\right)^{1/2} \exp(-\phi + i\psi)$$

$$= -\frac{\alpha_0 G(0)}{\xi}\exp\left(i\frac{\omega}{c}\alpha_0 y\right), \tag{21c}$$

where

$$G(0) = \left(\frac{2\cos\theta_0}{\pi\rho\sin^2\theta_0}\right)^{1/2}$$

$$\cdot\exp\left(-\rho s - \tfrac{2}{3}\rho\alpha_0^3 + \frac{i\pi}{4} - \tfrac{2}{3}i\rho\cos^3\theta_0\right),$$

$$\phi = \tfrac{2}{3}\rho\alpha_0^3 + \rho s,$$

$$\psi = \frac{\pi}{4} - \tfrac{2}{3}\rho\cos^3\theta_0 + \frac{\omega}{c}\sin\theta_0 y.$$

Equations (20) and (21) are valid only for relatively large angles of incidence because it was assumed that the region $\xi = 0$ is well separated from the region $\xi = -\alpha_0^2$. For arbitrary angles of incidence the problem is a bit more complicated; however, such a problem has been solved by Denisov[6] and his results apply here.

Note from Eqs. (21a)–(21c) that for ξ small H_x is nearly a constant and E_y and E_z become large. As a matter of fact, E_z is the largest field component near $\xi \to 0$ and, therefore, it will dominate the ponderomotive force. The magnitude of E_z is determined by the collision frequency (which determines the minimum value of ξ) and the function $\alpha_0\,|G(0)|$. For arbitrary angles of incidence, this latter function has been studied by Denisov.[6] It is found that for $\theta_0 = 0$, $\alpha_0\,|G(0)| = 0$; as θ_0 increases, this function increases and reaches a maximum; for still higher values of θ_0, it decreases. One may express this dependence on θ_0 as

$$\alpha_0\,|G(0)| = \frac{\exp(-\rho s)}{(2\pi\rho)^{1/2}}\,\Phi(\rho^{1/3}\alpha_0),$$

where Φ is a function which attains its maximum value of $(1.2)(2\pi\rho)^{1/2}$ for $\alpha_0\rho^{1/3} \simeq 0.8$ (for a plot of Φ vs θ_0, see Denisov[6]). For $\rho = 10^3$, this leads to an optimum angle of incidence of 4° to 5°. Note that $\rho = 10^3$ corresponds to an inhomogeneous layer 10^3 wavelengths thick since $\rho = \omega/ca = d/\lambda$, where d is the thickness of the layer and λ is the free space wavelength. Such a value of ρ is typical of laser-produced plasmas. We may finally write

$$E_z = \frac{\exp(-\rho s)}{(2\pi\rho)^{1/2}}\,\frac{\Phi(\alpha_0\rho^{1/3})}{\xi}\exp\left(i\frac{\omega}{c}\alpha_0 y\right). \tag{22}$$

For oblique incidence with the electric vector in the plane of incidence, the following expression for the ponderomotive force derived by Hora,[3] can be used:

$$F_{\alpha 1} = \frac{\mathbf{e}_y}{4\pi}\frac{\partial}{\partial z}(\epsilon'\bar{E}_z\bar{E}_y) + \frac{\mathbf{e}_z}{8\pi}\frac{\partial}{\partial z}[(2\epsilon' - 1)\bar{E}_z^2 - \bar{E}_y^2 - \bar{H}_x^2].$$

The term for the force in the $\mathbf{e}_y$ direction time-averages to zero leaving a force in the $\mathbf{e}_z$ direction only. To the lowest order, the force can be obtained by retaining only the E_z^2 term in the square bracket. Multiplying Eq. (22) by exp $(-i\omega t)$, taking the real part, squaring and averaging over several periods of the laser frequency, one obtains

$$F_{z_1} = \mathbf{e}_z \frac{\Phi^2 \exp(-2\rho s)}{16\pi^2 \rho} \frac{za^2}{(a^2z^2 + s^2)^2}. \quad (23)$$

This force has its maximum value for $z = \pm(s/3^{1/2}a)$, when it is equal to

$$F_{z_1m} = \mathbf{e}_z \frac{3^{3/2}}{256\pi^2} \frac{\omega}{c} \Phi^2 \frac{\exp(-2\rho s)}{\zeta_I^3} E_{v_1}^2, \quad (24)$$

where $\zeta_I = (\omega/ca)^{2/3}(\nu/\omega)$, as before, and E_v^2 is introduced on the right side to take care of the earlier normalization to unity.

A comparison of Eq. (24) with Eq. (12) (which is appropriate for normal incidence) shows that the oblique incidence force becomes higher when $\rho^{2/3}(\nu/\omega) = k_I < 0.30$ (for a typical ruby laser produced plasma with $\rho \simeq 10^3$, this corresponds to a temperature of about 1 keV) and becomes increasingly larger for decreasing (ν/ω), i.e., increasing temperature. This increase does not continue indefinitely. As the collision frequency becomes smaller, other dissipative processes assume an important role in the expression for ϵ'. As an example, if one retains the coupling of electromagnetic waves to plasma waves due to finite temperature and plasma inhomogeneity in the interaction region $\epsilon' \simeq 0$, ν should be approximately replaced by[5,6]

$$\nu_{eff} \simeq \nu + \omega\left(\frac{\kappa T a^2}{m\omega^2}\right)^{1/3}. \quad (25)$$

Equation (25) indicates that the absorption due to plasma wave coupling becomes comparable to the collisional absorption (for the same typical laser plasma parameters as above) when $\zeta_I \sim 0.1$. This value of ζ_I therefore determines the maximum ponderomotive force for oblique incidence; it turns out to be about 50 times the maximum obtained for normal incidence [Eq. (12)]. The above estimate should be treated with caution since it is based on Eq. (25), which is at best a crude approximation.

III. DISCUSSION

The authors have investigated the nonlinear interaction of an intense electromagnetic wave with a linearly inhomogeneous plasma layer and have estimated the magnitude of the ponderomotive force on the plasma. It is concluded that for normal incidence, the maximum ponderomotive force is determined by the free-space wavelength of the wave. An interesting new effect explored in this work is the ponderomotive bunching effect, arising because of the standing wave pattern on one side of the reflection plane. This ponderomotive bunching can lead to the breakup of the plasma layer into striations.

For oblique incidence with the electric vector normal to the plane of incidence, the force is always less than that for normal incidence. The most interesting case, however, is that of oblique incidence with the electric vector in the plane of incidence; in this case the maximum force can be considerably higher than in the first two cases and is only limited by the magnitude of the imaginary part of the dielectric constant ϵ' (e.g., collisions, finite temperature effects, etc.). Furthermore, in this case there is an interesting optimization of the maximum force with the angle of incidence; for typical laser-produced plasmas, the optimum angle seems to be around 4–5°. Qualitatively, the enhanced force arises because of the excitation of huge longitudinal fields in the region $\epsilon' \approx 0$. This physical picture also explains the optimization of the force with the angle of incidence. For, in the case of normal incidence, the wave stays transverse all the way to the reflection point and so there is no such enhancement. On the other hand, if the angle of incidence is too large, the region $\epsilon' \approx 0$ is far away from the reflection point $\epsilon' \approx -\alpha_0^2$; in this case, therefore, the wave amplitude decays considerably before reaching the resonant regions, and hence strong longitudinal fields cannot be excited to give a large force.

The treatment in this paper has been based on the use of cold plasma expressions for the dielectric constant. Thus, interesting finite temperature effects[7] have either been ignored or very crudely taken into account. Furthermore, since the longitudinal fields in the resonant regions may be quite large, nonlinear effects[8] may also become important. It should be mentioned here that these finite temperature and nonlinear effects may lead to a considerably enhanced absorption of the laser energy before any of the effects discussed in this paper can be produced. A detailed analysis should, therefore, take all of these effects into account.

It is of interest to obtain the net amount of momentum transfer to the inhomogeneous plasma layer. This can readily be obtained by using the analysis of Hora.[3] Thus,

$$P_{inh} = \int dy\, dx \int_{t_1}^{t_2} dt \int_{z_1}^{z_2} dz\, F.$$

For oblique incidence with the E vector in the plane of incidence, we have

$$P_{inh}'' \approx -\tfrac{3}{2} P_0 \frac{\exp(-2\rho s)}{\rho s^2}, \quad (26)$$

where

$$P_0 = \int dy\, dx \int_{t_1}^{t_2} dt \frac{E_0^2(x, y, t)}{8\pi}$$

is the incident momentum of the electromagnetic wave pulse. Note that for sufficiently small $s = (\nu/\omega)$, P_{inh}'' can be much greater than P_0; thus, the inhomogeneous plasma layer can acquire a huge directed momentum. The direction of the momentum is that of the density gradient, and is independent of the direction of incident electromagnetic wave. This momentum transfer to the plasma layer is balanced by momentum transfer in the opposite direction to the rest of the plasma. Thus,

$$P_{rest} = P_0 - P_{inh}.$$

In a similar manner, one can also estimate the momentum transfer for the case of E vector perpendicular to the plane of incidence. It is usually smaller than the above momentum transfer.

Steinhauer and Ahlstrom[9] have made a comparison of the magnitudes of thermokinetic and electromagnetic forces for a typical laser produced plasma. Using the expression for ponderomotive force derived by Hora[3] they conclude that thermokinetic effects dominate everywhere except in a thin region in the neighborhood of the critical layer for reflection. Their estimate of the thermokinetic force, however, contains a questionable averaging procedure which somewhat artificially makes the thermokinetic force depend only weakly on the electron–ion collision time. A correct comparison of the thermokinetic and ponderomotive forces actually involves the solution of equations of motion for the plasma, under the combined influence of the two forces; this is left for future investigation.

It is also of interest to note that for a small plasma blob isotropically illuminated by the laser beam, the ponderomotive force expansion, if important, will be anisotropic, i.e., one will see plumes of plasma coming out at small angles to the incident direction—since for them, the ponderomotive force is a maximum. The thermokinetic expansion, on the other hand, should be isotropic. This may, therefore, be one way of experimentally distinguishing between the two effects.

ACKNOWLEDGMENTS

Constructive criticism by J. M. Dawson and the referee is gratefully acknowledged.

This work was performed under the auspices of the United States Atomic Energy Commission, Contract AT(30-1)-1238.

[1] H. Hora, D. Pfirsch, and A. Schlüter, Z. Naturforsch. 22a, 278 (1967).
[2] A. Schlüter, Plasma Phys. 10, 471 (1968).
[3] H. Hora, Phys. Fluids 12, 182 (1969).
[4] W. Panofsky and M. Phillips, *Classical Electricity and Magnetism* (Addison-Wesley, Reading, Mass., 1962), p. 107.
[5] V. L. Ginzburg, *The Propagation of Electromagnetic Waves in Plasmas* (Addison-Wesley, Reading, Mass., 1969), pp. 193–198 and 213–228.
[6] N. G. Denisov, Zh. Eksp. Teor. Fiz. 31, 609 (1956) [Sov. Phys. JETP 4, 544 (1957)].
[7] J. Dawson, P. Kaw, and B. Green, Phys. Fluids 12, 875 (1969).
[8] P. Kaw and J. M. Dawson, Phys. Fluids 12, 2586 (1969).
[9] L. Steinhauer and H. G. Ahlstrom, Phys. Fluids 13, 1103 (1970).

NONLINEAR FORCES IN LASER-PRODUCED PLASMAS+

Heinrich Hora

Max-Planck-Institut für Plasmaphysik
Euratom Association, Garching, Germany
and Rensselaer Polytechnic Institute, Hartford, Conn., U.S.A.

ABSTRACT

The very recent observation of a correlation of the increase of reflectivity from laser produced plasmas with the onset of nuclear fusion reactions indicates a nonlinear mechanism. Numerical calculations of Shearer, Kidder and Zink of the plasma dynamics on interaction with light that allow for the nonlinear force, described before, on the basis of the ponderomotive interaction for collisionless dispersion effects indicated the increase of reflectivity due to this force. A review on this force is given and an exact numerical example of 10^{15} W/cm^2 laser intensity (Nd glass) in a plasma with a temperature of 10^3 eV and with a WBK-like density profile shows the predominance of the nonlinear force over the thermokinetic force. More detailed calculations of the threshold intensities I for the predominance of this force for a net plasma acceleration around $I^* = 10^{14}$ W/cm^2 are reported. These include the calculation of Steinhauer and Ahlstrom, where this threshold was given even for lower intensities but

---

+ Presented at the Second Workshop on "Laser Interaction and Related Plasma Phenomena" at Rensselaer Polytechnic Institute, Hartford Graduate Center, August 30 - September 3, 1971.

Reprinted from: *Laser Interaction and Related Plasma Phenomena,* Vol. 2, H. Schwarz and H. Hora (eds.), Plenum, New York, 1972, pp. 341–365.

so that temperatures exceeding 1o keV are then needed. The non-WBK-like density profile as treated analytically by Lindl and Kaw, resulted in the same net acceleration of the layer as in the WBK case, and the acceleration towards the nodes of the created standing wave was again very similar to the WBK case treated before by the author. For oblique incidence, where a Brillouin turbulence can occur, Lindl and Kaw found the possibility of a polarization dependent increase of the nonlinear force due to a Ginzburg-Denisov coupling of electromagnetic and electrostatic waves. Numerical calculations for non-WBK cases are reviewed, showing the predominance of the net acceleration and the increase of reflectivity. The forces in standing waves can exceed the thermokinetic forces even at intensities appreciably below I^*, as Mulser found from a dynamic numerical calculation. This fact allows the assumption that this mechanism may be one reason for the correlation of reflectivity and neutron production.

INTRODUCTION AND SYNOPSIS

Laser produced fusion plasmas show many very unexpected properties compared with the plasmas generated with lower laser powers than about 10^{14} W/cm^2, in which no fusion neutrons are produced. While the optical reflectivity of plasmas produced by laser intensities between 10^9 and 10^{13} W/cm^2 is remarkably low (only a few per cent although the radiation interacts with an overdense plasma)[1-4], a reflectivity of more than 3o % was reported from the first neutron producing plasma created by picosecond pulses[5]. Furthermore, it was observed that in the fusion experiment with nanosecond pulses of Floux et al.[6] the reflectivity was also always remarkably high when neutrons were produced, otherwise the reflectivity was small[7]. The same increase of reflectivity was correlated with other measurements of fusion neutrons, and a further correlation with the measurement of a temperature[8] of about 2 keV in addition to the usual temperature of about o.5 keV was found. The same increase of the reflectivity with the neutron production was also observed by Swain et al.[9], where the high reflectivity was observed only when a laser pulse impinged a few nanoseconds before the main pulse.

The increase of the reflectivity with increasing intensity could be seen from simplified numerical calculations using a rectangular laser pulse[7,8], and linear absorption processes only, but the experiment of Swaine et al. demonstrated that the preheating of the plasma and its creation of a slightly inhomogeneous plasma surface should not increase the reflectivity on the basis of the linear theory. It is to be underlined that a few months before Floux et al.[7] measured the correlation between reflectivity and neutron production for the first time, J. Shearer[10] et al. found from their numerical model WAZER that a remarkable increase of the reflectivity results if at high laser intensities the nonlinear force[11-13] is included in the calculations. At present it is too early to say whether this result completely explains the observed correlation, but it seems to be a first obvious indication of the action of nonlinear processes in high intensity laser produced plasmas. Indeed, there may be a number of other nonlinear processes involved, such as the nonlinear change of the collision frequency[13-15], the relativistic correction of the plasma frequency[16], coupling of the transverse waves with longitudinal waves[17], self-focusing with a low-power threshold[12,18] or with a high intensity threshold[19], two stream instabilities[20], or anomalous absorption at the cut-off density[21]. Nearly all of these processes have their threshold at about 10^{14} W/cm^2 and it would be puzzling if plasmas irradiated at higher intensities than 10^{14} W/cm^2 should be described only in terms of linear thermokinetic processes.

This contribution presents a review of the results for the theory of the nonlinear force, following the detailed description of the basic theory at the preceding Workshop[12]. From the outset it is to be pointed out that this type of nonlinear force we are treating is essentially connected with the strong changes of the absolute value of the complex refractive index $\tilde{\eta}$ insofar as the actual electrical field strength of the laser light $\underset{\sim}{E}$ is given by the field $\underset{\sim}{E}_v$ in vacuum as

$$\underset{\sim}{E} = \frac{\underset{\sim}{E}_v}{\sqrt{|\tilde{\eta}|}} \tag{1}$$

or somehow modified by absorption or standing wave processes. Only these variations of the fields can create forces of the kind $\nabla(E^2/|\tilde{\eta}|)$ which give a net acceleration of the whole plasma layer against the laser light and a recoil to the plasma interior by an increased radiation pressure (increased possibly by factors of a hundred and more). Other types of nonlinear forces such as the ponderomotive, electrostrictive or magnetostrictive type have been well known for a long time, e.g. radiation pressure or plasma confinement by standing waves[22] or similar configurations[23]. With the exception of Ref.[12] there is no case known where the variation of the refractive index was included generally. All that is important for the net acceleration and is mainly responsible for the Brillouin turbulences (see Ref.[12] p. 4o8), the increase of reflectivity[1o] and the polarization dependent increase of the net acceleration oblique incidence[24]. The standing wave process modified by the dispersion effects also seems to be important, as will be shown.

After sketching the derivation and the basic equations of the nonlinear force, the progress achieved in determining the range of validity is reported, including the paper of Steinhauer and Ahlstrom[25], where the predominance of the nonlinear force over the thermokinetic force is proved for laser intensities even below I^* if the electron temperature of the plasma exceeds 1o^{4} eV; the same treatment can be used to prove the predominance for $I > I^*$ at electron temperatures $T >$ 3o eV. The properties of the force are demonstrated by the special example of a plasma of constant temperature where the density profile satisfies the WBK condition. The analytic treatment of Lindl and Kaw for a collisionless plasma not fulfilling the WBK condition proves the agreement of the calculated nonlinear force with the value of the WBK case where the latter is valid, and beyond this range the nonlinear force is even larger than the formal value of the WBK case. At oblique incidence a resonance-like increase of the force is derived.[26] Finally, the numerical results of Mulser and Green are reported for non-WBK-like profiles of a collisional plasma, the nonlinear force being compared with the themokinetic force. Even at an intensity of neodymium glass lasers only a little higher than the threshold (I = 5x1o^{14} W/cm^{2}) and at low plasma temperature (T$_e$ 1oo eV), the net nonlinear forces are 2 to 3 times as high as the thermokinetic forces. The forces due to standing waves are 12 times as high as the thermokinetic forces. All these cases were treated for the very special condition of static density profiles. The more interesting case of dynamics was treated for the first time by Shearer, Kidder, and Zink[1o], some results of whose will be reported at the end of the paper. It is to be mentioned that the following short contribution of P. Mulser[27] reports on some further steps of a dynamic treatment.

BASIC CONCEPT OF THE NONLINEAR FORCE

The nonlinear force is given by direct interaction of the electromagnetic radiation with a plasma, thus causing the plasma to accelerate without any conversion of the radiation energy into thermal energy. There are different possible methods of derivation, always leading to the same expression for the force. The theory is based on the two-fluid model of plasma physics[11] or a combination of the single particle motion of the electrons in the plasma whose electric and magnetic fields $\underset{\sim}{E}$ and $\underset{\sim}{H}$ are determined by the collective properties of the plasma[12]. The most general derivation based on the nonrelativistic Lorentzian theory starts from the general equation of the ponderomotive force density[28]

$$\underset{\sim}{f} = \nabla \cdot \underset{\approx}{\sigma} - \frac{\partial}{\partial t} \frac{\underset{\sim}{E} \times \underset{\sim}{H}}{4\pi c} . \tag{2}$$

By this definition of Landau and Lifshitz both forces in the plasma are ponderomotive forces, the force generated by the gas dynamic pressure tensor $\underset{\approx}{p}$ and any force given by the electromagnetic field strengths $\underset{\sim}{E}$ and $\underset{\sim}{H}$, determining the stress tensor $\underset{\approx}{\sigma}$ in Eq.(2). The tensor components are

$$\sigma_{ij} = -p_{ij} - \left[\frac{E^2}{8\pi}\left(\tilde{\eta}^2 - \rho\frac{\partial\tilde{\eta}^2}{\partial\rho}\right) + \frac{H^2}{8\pi}\right]\delta_{ij} + \tilde{\eta}^2 \frac{E_i E_j}{4\pi} + \frac{H_i H_j}{4\pi}$$

where ρ denotes the density, and the complex refractive index $\tilde{\eta}$ of electromagnetic waves with a frequency ω is

$$\tilde{\eta}^2 = 1 - \frac{\omega_p^2}{\omega^2 + \nu^2}\left(1 + i\frac{\nu}{\omega}\right) . \tag{3}$$

Here the plasma frequency ω_p

$$\omega_p^2 = 4\pi e^2 n_e / m_e \tag{4}$$

is given by the charge e, the density n_e and the mass m_e of the electrons. The relativistic change of the electron mass[16] at neodymium glass laser intensities exceeding 10^{16} W/cm^2 will be neglected here because these intensities are about 100 times as high as these considered here. The collision frequency ν in Eq.(3) is given by the theory of the inverse bremsstrahlung[29] or by the Dawson-Oberman theory[30], or derived[31] from Spitzer's d.c. plasma resistivity[32], nearly the same values always being obtained. The collision frequency of the collision theory is

$$\nu = \frac{\omega_p^2 \pi^{3/2} m_e^{1/2} Z e^2 \ln\Lambda}{8\pi \gamma_E(Z)(2kT)^{3/2}} \tag{5}$$

with the Boltzmann constant k, the ion charge Z, with Spitzer's correction factor γ_E (being o.2 ... 1) and the Coulomb logarithm with Λ

$$\Lambda = \frac{3}{2Ze^3}\left(\frac{k^3T^3}{\pi n_e}\right)^{1/2} \tag{6}$$

The electron temperature T in Eq.(5) is such that the electron motion determines the dispersion relation, which means that the temperature is given by the random thermal motion according to a temperature T_{th} and by the coherent motion of the electrons in the electromagnetic field, given by the actual oscillation energy of the electrons ε_e^{osc}

$$T = T_{th} + \varepsilon_e^{osc}/k \tag{7}$$

In this way the nonlinear change of the collision frequency at high light intensities is included.

With an appropriate interpretation of the density factor in Eq.(3), Eq.(2) can be written

$$\underset{\sim}{f} = -\underset{\sim}{\nabla}\cdot\underset{\approx}{p} + \nabla\cdot\left[\underset{\approx}{U} - (\tilde{\eta}^2-1)\underset{\sim}{E}\underset{\sim}{E} - \frac{\partial}{\partial t}\frac{\underset{\sim}{E}\times\underset{\sim}{H}}{4\pi c}\right. \tag{8}$$

where the gas dynamic part is called the thermokinetic force

$$\underset{\sim}{f}_{th} = -\underset{\sim}{\nabla}\cdot\underset{\approx}{p} \tag{9}$$

and therefore

$$\underset{\sim}{f}_{NL} = \underset{\sim}{f} - \underset{\sim}{f}_{th} \tag{1o}$$

is essentially nonlinear owing to the quadratic forms of the electromagnetic field components. In the following the Poynting term (last term in Eq.(2)) can be neglected because of the much slower change of the radiation intensities compared with the laser periods. Averaging Eq.(8) over the time of one laser period neglects all effects occurring during these short times and results in the interesting long-term forces.

Some essential properties of the nonlinear force can be seen from the special case of perpendicular incidence of the laser radiation along the x-coordinate with linear polarization of the E-vector in the y-direction onto a stratified plasma. The time averaged nonlinear force is from Eqs.(8) and (1o)

$$\overline{\underset{\sim}{f}_{NL}} = \frac{i_x}{8\pi}\frac{\partial}{\partial x}\left(\underset{\sim}{E}_y^2 + \underset{\sim}{H}_8^2\right) \tag{11}$$

This force is well known from standing waves in a thin plasma[22], where the particles are driven towards the nodes. If then the refractive index is always very close to unity, no spatially averaged net motion of the plasma results. Any net motion can be expected only by dispersion effects, when the absolute value of the refractive index differs appreciably from unity, which only occurs in the neighborhood of the cut-off density.

In the following sections the nonlinear force is evaluated for general density profiles, of which the quantities E_y and H_z are especially calculated. To gain a better insight into the mechanism of the force, we first discuss the case of the WBK approximation which forces also agree with those derived from general non-WBK cases[24].

The WBK solution for a reflection-free penetrating wave with an amplitude E_v in vacuum is

$$\underset{\sim}{E} = \underset{\sim}{i}_y \frac{E_v}{\tilde{\eta}^{1/2}} \exp(iF) \exp(\mp\frac{K(x)}{2} x); \quad \underset{\sim}{H} = \underset{\sim}{i}_3 |\underset{\sim}{E}| \tilde{\eta} \tag{12}$$

where

$$F = \omega\left(t \mp \int^{x} \frac{\mathrm{Re}(\tilde{\eta}(\xi))}{c} d\xi\right) \text{ and } K(x) = \frac{1}{x}\frac{\omega}{c}\int^{x} \mathrm{Im}(\tilde{\eta}(\xi))\, d\xi . \tag{12a}$$

The condition of the WBK approximation is

$$\theta = \frac{c}{2\omega} \frac{1}{|\tilde{\eta}|^2} \frac{\partial|\tilde{\eta}|}{\partial x} << 1 . \tag{13}$$

Using Eq.(12) we find from Eq.(11)

$$\overline{\underset{\sim}{f}}_{NL} = \underset{\sim}{i}_x \frac{E_v^2}{16\pi} \frac{1-|\tilde{\eta}|^2}{|\tilde{\eta}|} \exp(-Kx) \frac{\partial}{\partial x}|\tilde{\eta}| + \underset{\sim}{i}_x \frac{E_v^2}{16} \frac{1-|\tilde{\eta}|^2}{|\tilde{\eta}|} \frac{2\omega}{c} \mathrm{Im}(\tilde{\eta}) \exp(-Kx) \tag{14}$$

At densities below the cut-off density, only the first term of the right-hand side dominates, which is the essential part of the nonlinear force, as can be seen if Eq.(13) is written for a collisionless plasma ($\nu = 0$, $K = 0$; $\mathrm{Im}(\tilde{\eta}) = 0$) using Eqs.(3) and (4)

$$\overline{\underset{\sim}{f}}_{NL} = \underset{\sim}{i}_x \frac{E_v^2}{16\pi} \frac{\omega_p^2}{\omega^2 \tilde{\eta}^2} \frac{\partial \tilde{\eta}}{\partial x} , \tag{15}$$

while at densities exceeding the cut-off density n_{eco}

(given by Eq.(4) if $\omega_p = \omega$) the second term of the right-hand side of Eq.(14) dominates. This term describes the usual radiation pressure, but increased by the refractive index variations.

NUMERICAL EXAMPLE OF THE WBK CASE

In Fig. 1 we report on the numerical calculation of the conditions and force densities of a special static case where a neodymium glass laser pulse with an intensity of 10^{15} W/cm^2 is perpendicularly incident on a plasma with local constant temperature of 10^3 eV and with a density profile such that the radiation can penetrate reflection-free according to the WBK condition (13). The term "static case" means that the evaluated light intensities, pressures and force densities are valid only during a time where the density profile can be considered constant. If this profile were an initial condition, the further dynamic development would obviously change the profile and the assumed local constance of the temperature.

Firstly, the electron density n_e was calculated on the basis of a value $\theta = 0.1$ in Eq.(13) with a monotonic increase along the depth x. This was performed by starting from a density difference Δn_e and calculating $|\tilde{\eta}|$ and $\Delta|\tilde{\eta}|$ from Eqs.(3), (4) and (5) using in Eq.(7) $T = T_{th}$ and then Δx from Eq.(13). The approximation $T = T_{th}$ in Eq.(7) causes a lower bound of the final values of the forces because the implication of the electron oscillation causes higher temperatures, hence lower values $|\eta|$, and then stronger force densities. The result in Fig. 2 is a very slow increase of $n_e(x)$ around the cut-off density, which for neodymium glass laser radiation is around 10^{21} cm^{-3}. The second step was to calculate the intensity

$$I = E^2 = \frac{E_v^2}{|\eta|} \exp(-K(x)x) \qquad (16)$$

following the numerical integration according to Eq.(12a). The strong increase of $(E/E_v)^2$ with depth up to values beyond 20 proves that the change of the $|\eta|$ values occurs much earlier than the decrease of I is influenced by the integral absorption constant K(x). Therefore, the maximum values of $|\eta|^{-1}$ are higher than 20. This results in an increase of the effective wavelength $\lambda = \lambda_0/|\eta|$ compared to the vacuum wavelength λ_0 by more than 20 as well. From this point of view, the nearly pathologically flat density profile from 25 to 65 μm (Fig. 1) is only over about four effective wavelengths.

The next step was to calculate the thermokinetic force density $f_{th} = -\nabla p = -kT\nabla n_e$, which is given by the pointed curve in Fig. 1. The nonlinear force f_{NL} was negative up to a depth of 52 μm and is drawn positive in Fig. 1. First we have to remark that the nonlinear force is larger than the thermokinetic force for a depth larger than 13 μm and can be 9 times as large as the maximum of the thermokinetic force. To use one scale in the logarithmic diagram we drew the total force $f = f_{NL} + f_{th}$ and its components with negative sign, meaning an acceleration of the plasma in the negative x-direction. For depths larger than 53 μm the nonlinear force f_{NL} is positive (directed in the +x-direction), and because its absolute value is very high the total force f is positive up to a depth of 68 μm, where the nonlinear force decreases strongly because of the absorption of the radiation. The depth from a little more than 53 μm to 68 μm is nearly the skin depth, which is so large because the effective wavelength is at some depth larger than 2o vacuum wavelengths. We continued to increase n_e monotonically for higher depth in an arbitrary manner. A detailed dynamic calculation may also result in a decreasing n_e within the highly shocked material below the radiation zone, but this question is irrelevant at this point.

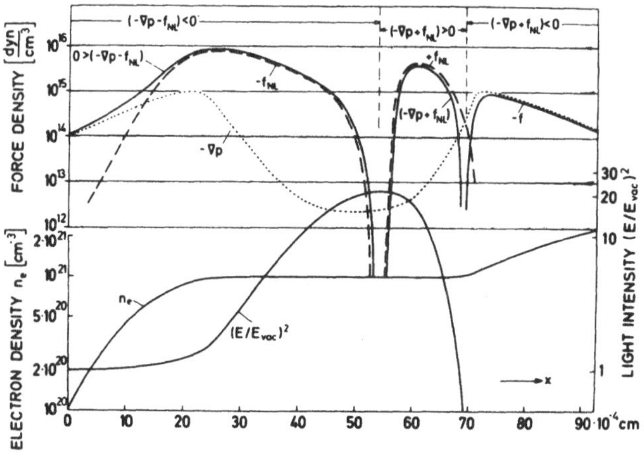

Fig. 1

Numerical calculation of a WBK-like density profile $n_e(x)$ of a plasma of temperature T = 1 keV. The radiation of $1o^{15}$ W/cm^2, incident from the left-hand side, produces an amplitude E of the electric field of the penetrating light wave compared with the amplitude E_v in vacuum before incidence. The resulting thermokinetic force f_{th} and the nonlinear force f_{NL} are calculated on the basis of the linear collision frequency.

RANGE OF PREDOMINANCE OF THE NONLINEAR FORCE

The example of the preceeding section demonstrated the predominance of the nonlinear force over the thermokinetic force at intensities of neodymium glass lasers of 10^{15} W/cm^2. This agrees with our former calculations[12], where the threshold of predominance was found to be near 10^{14} W/cm^2. The evaluation of this threshold was given in more detail by an iteration method[13] and was discussed by Steinhauer and Ahlstrom[25]. We review these treatments in three steps, namely a simplified consideration, the iteration procedure and the estimation of Steinhauer and Ahlstrom[25].

Simplified Consideration

For a plasma where the absorption is small up to the cut-off density, as shown in the preceding section, the oscillation energy of the electrons

$$\varepsilon_e^{osc} = \frac{\varepsilon_{ev}^{osc}}{|\tilde{\eta}|} \qquad (17)$$

is given by its value in vacuum, but increased by the changes of the refractive index $\tilde{\eta}$. The maximum value of

$$\left.\frac{1}{|\tilde{\eta}|}\right|_{max} = \frac{T^{3/4}}{a} \qquad (18)$$

has already been derived[12,13], where

$$a = \left[\frac{\omega_p^2 \pi^{3/2} m_e^{1/2} Z e^2 \ln\Lambda}{8\pi\omega\gamma_E (Z)(2k)^{3/2}}\right]^{1/2} (eV)^{3/4} \qquad (19)$$

has values around 2...4 $(eV)^{3/4}$ for neodymium glass lasers at temperatures around $10^2 \ldots 10^4$ eV. A simplified integration of the equation of motion was possible to evaluate the final energy ε_i^{trans} of the translational motion of the ions with a charge Z after the nonlinear acceleration where the details of the special density profile cancelled.

$$\varepsilon_i^{trans} = Z \left.\varepsilon_e^{osc}\right|_{max} = Z \varepsilon_{ev}^{osc} \cdot \frac{T^{3/4}}{a} \qquad (20)$$

The problem is to use the electron temperature T of Eq.(7). The temperature where the random electron energy is equal to the energy of coherent motion

$$T_{th} = \varepsilon_{ev}^{osc} \cdot \frac{T_{th}^{3/4}}{k\ a} \qquad (21)$$

leads with (at $\omega \approx \omega_p$)

$$\varepsilon_{ev}^{osc} = \frac{E_v^{1*2}}{16\pi n_{eco}} \frac{\omega_p^2}{\omega^2} \sim I^* \qquad (22)$$

to

$$E_v^{1*2} = 16\pi n_{eco}\ ak\ T_{th}^{1/4} \qquad (23)$$

In this way we find from Eq.(2o)

$$\varepsilon_i^{trans} = \begin{cases} Z \dfrac{E_v^2}{16\pi n_{eco}} \cdot \dfrac{T_{th}^{3/4}}{a}, & \text{if } \varepsilon_e^{osc} \ll T_{th} \\ Z \dfrac{E_v^8}{(16\pi a n_{eco} k^{3/4})^4}, & \text{if } \varepsilon_e^{osc} \gg T_{th} \end{cases} \qquad (24)$$

which shows resonance-like increase of the ion energy due to the nonlinear force at intensities a little higher than I^*. The damping mechanism can be explained qualitatively from the fact that at the start of this resonance the absorption due to the nonlinear force also acts and will result in an asymptotic limiting value of the ion energy ε_i^{trans}. The evaluation of Eq.(23) gives the threshold values with [T] = eV

$$E_v^{1*} = \begin{Bmatrix} 2.83\text{x}10^8 \\ 1.67\text{x}10^8 \end{Bmatrix} \cdot T_{th}^{1/8}\ \text{V/cm}; \quad I^{1*} = \begin{Bmatrix} 2.08\text{x}10^{14} \\ 7.5\ \text{x}10^{13} \end{Bmatrix} T_{th}^{1/4}\ \text{W/cm}^2$$

$$\text{for} \begin{Bmatrix} \text{ruby} \\ \text{Nd glass} \end{Bmatrix} \qquad (25)$$

This first approximation is indicated by the index 1 of the threshold values.

Iteration Method with WBK Approximation

The most direct way to compare the nonlinear force with the thermokinetic force is to go back to Eq.(8) and compare the gasdynamic pressure with the momentum flux density of the electromagnetic field (after subtraction of its vacuum value) before performing the spatial differentiation. With the WBK approximation we have the following relation, with the general threshold indicated by an asterix:

$$n_e(1-1/Z)kT_{th} = \frac{E_v^{*2}}{16}\left[\left(\frac{1}{|\tilde{\eta}|} + |\tilde{\eta}|\right)\exp_o - 1\right] . \qquad (26)$$

The exponential function given by the damping is to be determined separately and will be included as an open parameter between 1 and $1/e^2$. The fact that $|\tilde{\eta}|$ is to be used in the form of Eq.(17) with a temperature given by Eq.(7) causes the solution of Eq.(26) by an iteration only. For a set of E_v values with $\exp_o$ as a parameter we solved as a first step Eq.(26) for $|\tilde{\eta}| = a/T_{th}^{3/4}$ reaching $T_{th}^{(1)}$. With this value we calculated by iteration higher approximations (n+1)(n=1,2...)

$$n_e(1+1/Z)kT_{th}^{(n+1)} = \frac{E_v^{*2}}{16\pi}\left[\left(\frac{\left(T_{th}^{(n)} + \frac{E_v^{*2}\,T_{th}^{(n)3/4}}{16\pi a n_{eco} k}\right)^{3/4}}{a} + \frac{a}{\left(T_{th}^{(n)} + \frac{E_v^{*2}\,T_{th}^{(n)3/4}}{16\pi a n_{eco} k}\right)^{3/4}}\right)\exp_o - 1\right] \qquad (27)$$

When $T_{th}^{(n+1)} - T_{th}^{n}$ was less than o.5 eV the iteration was finished. The results are shown in Figs. 2 and 3. We see that the first approximation I^{1*} is very close to the case of $\exp_o$ = 1 to 1/2, reproducing again the thresholds near 10^{14} W/cm^2 for neodymium glass lasers.

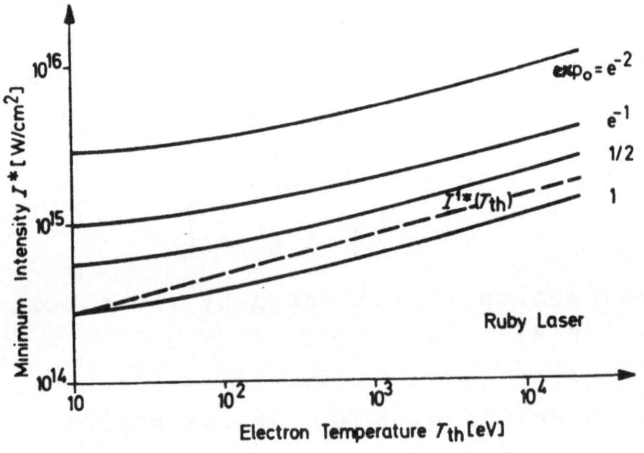

Fig. 2

Minimum intensity I^* for ruby laser to create larger nonlinear deconfining forces f_{NL} than thermokinetic forces $f_{th} = -\nabla n_e kT_{th}$. The undefined exponential function $\exp_o$ of collision induced attenuation is used as a parameter.

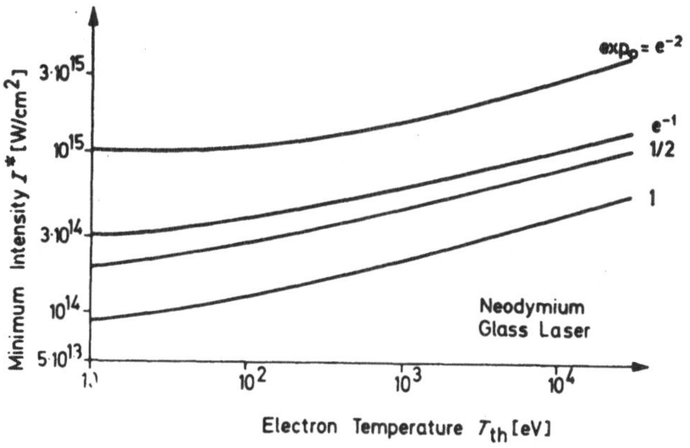

Fig. 3

Minimum intensity I^* for neodymium glass laser to create larger nonlinear deconfining forces f_{NL} than thermokinetic forces $f_{th} = -\nabla n_e kT_{th}$. The undefined exponential $\exp_o$ of collision induced attenuation is used as a parameter.

Steinhauer and Ahlstrom's Method

Steinhauer and Ahlstrom[25] compared the nonlinear force for the cases where no nonlinear changes of the collision frequency[12] take place, i.e. for neodymium glass laser intensities below 10^{14} W/cm^2. For a one-dimensional model, the thermokinetic pressure p due to the thermalization of the radiation is given by

$$\frac{dp}{db} = -\frac{2}{3}\nabla\, \overline{\tilde{E} \times \tilde{H}} \tag{28}$$

Using approximations for the time of an acoustic wave to travel one absorption length and using averaging procedures which were recognized to include somewhat artificially the dependence of the thermokinetic force on the electron-ion collision time (as was pointed out by Lindl and Kaw[24]), the ratio

$$\frac{|\tilde{f}_{th}|}{|\tilde{f}_{NL}|} \geq \frac{1.14\times 10^5}{T^{5/4}}\, Z^{1/2} \quad ; \quad [T] = \text{eV} \tag{29}$$

was given for the neodymium glass lasers. This means that the nonlinear force can exceed the thermokinetic force at low laser intensities only at temperatures exceeding 10^4 eV.

The temperature T in Eq.(29) is the quantity describing the energy of electrons with respect to the refractive index. Therefore, for high light intensities we have to use the value from Eqs.(7) and (17)

$$T = T_{th} + \frac{e\varrho\xi c}{k|\tilde{\eta}|} \qquad (3o)$$

Using the relation (18) with $T = T_{th}$ from Eq.(23) we get for $E \gg E^*$ a predominance of the second term on the right-hand side of Eq.(3o). With such a T the relation (29) reads[13]

$$\frac{|\mathbf{f}_{th}|}{|\mathbf{f}_{NL}|} \geq \left(\frac{7.2\text{x}1o^8}{E_v}\right)^8 Z^{1/2}; \; [E_v] = V/cm \qquad (31)$$

This means that at intensities a little higher than I^*, a resonance-like increase of the predominance of the nonlinear force can be expected. The assumptions for Eq.(31) did not include such types of absorption which cause the nonlinear collisionless acceleration of the plasma. This mechanism will again cause a damping of the resonance.

Though the condition of Steinhauer and Ahlstrom[25] was not derived in the most general way it reconfirms the predominance of the nonlinear force for intensities exceeding I^* of about $1o^{14}$ W/cm^2 and additionally proves a predominance at lower intensities as well if the temperature exceeds 1o keV. This result agrees also with the special numerical calculation of the preceding section, where at intensities of $1o^{15}$ W/cm^2 the nonlinear force was more than 8 times as high as the thermokinetic force and also agrees with the following numerical results of non-WBK-like profiles of plasmas with collisions.

ANALYTICAL NON-WBK CASE OF LINDL AND KAW

An analytical treatment of the nonlinear force was performed by Lindl and Kaw[24] where a special restriction to an analytically well known case[33] was necessary, with a linear density profile at the surface of the plasma. The solutions of the Maxwellian equations for E_y and H_z for evaluating the nonlinear force in the case of a one-dimensional geometry, Eq.(11), can begin by solving the wave equation

$$\frac{\partial^2 E_y}{\partial x^2} + \frac{\omega^2}{c^2}\tilde{\eta}^2 E_y = 0 \qquad (32)$$

where the complex refractive index $\tilde{\eta}$ (Eq.(3)) has the following spatial variation

$$\eta^2 = -ax + is \tag{33}$$

where

$$a \approx \frac{1}{|\tilde{\eta}|}\frac{\partial|\tilde{\eta}|}{\partial z} = \frac{2\theta\omega}{c} \quad ; \quad s = \frac{\nu}{\omega} \ll 1 \tag{34}$$

the definition of θ being taken from Eq.(13). The collision frequency ν is assumed to be very small and the density profile n(x) follows from Eq.(33) and (34)

$$\frac{4\pi e^2 n(x)}{m} = \omega_p^2(x) = \omega^2 \cdot \frac{x}{x_1} \tag{35}$$

The layer has a depth x_1 and is distributed from x = 0 to x. At $x = x_1$ is the reflection point where a collisionless plasma creates total reflection of the incident wave. The solution of the wave equation (32) is given using the substitution

$$\zeta = \left(\frac{\omega}{ca}\right)^{2/3} \cdot (-ax + is) \equiv \varrho^{2/3}\xi \quad ; \quad \varrho = \left(\frac{\omega}{ca}\right)^{2/3} \tag{36}$$

where the Eq.(32) reduces to

$$\frac{d^2E_y}{d\zeta^2} + \zeta E = 0 \quad , \tag{37}$$

which results in the well known Airy function a_i. In expressions of the Bessel functions of the order 1/3 we find

$$E_y = 3Aa_i(-\zeta) = \begin{cases} A\zeta^{1/2}\left[J_{1/3}\left(\frac{2}{3}\zeta^{3/2}\right) + J_{-1/3}\left(\frac{2}{3}\zeta^{3/2}\right)\right], & \text{if } \mathrm{Re}\,\zeta > 0 \\ A(-\zeta)^{1/2}\left[I_{-1/3}\left(\frac{2}{3}(-\zeta)^{3/2}\right) + I_{1/3}\left(\frac{2}{3}(-\zeta)^{3/2}\right)\right], & \text{if } \mathrm{Re}\,\zeta < 0 \end{cases} \tag{38}$$

with

$$A = \frac{2}{3}\pi^{1/2}\left(\frac{\omega}{ca}\right)^{1/6} E_v \exp\left(+\frac{i\omega}{c}\int_{-1/a}^{+is/a} \tilde{\eta}^{1/4}\,dz - \frac{i\pi}{4}\right) \tag{39}$$

The solution of the magnetic field strength is derived from the Maxwell equations

$$H_z = +i\varrho^{-1/3}\left(\frac{dE_y}{d\varrho}\right) \tag{40}$$

From Eq.(38) it can be seen[24] that for $x \geq 0$ there exists a totally reflected (standing) wave (exactly verified only in a collisionless plasma $\nu = 0$), as is to be expected, and for $x < 0$ a strongly damped wave occurs. The (time averaged) nonlinear force results simply by substituting Eqs.(38) and (40) in Eq.(11)

$$\bar{f}_{NL} = i_x \frac{\omega}{c} E_v^2 \exp(-2\rho s)\Big\{(a_{iR} a'_{iR} + a_{iI} a'_{iI})$$

$$+ \rho^{-2/3}\left[a'_{iR}(\zeta_R a_{iR} - \zeta_I a_{iI}) + a'_{iI}(\zeta_I a_{iR} + \zeta_R a_{iI})\right]\Big\} \quad (41)$$

where the real parts and the imaginary parts of the Airy functions are denoted by the indices R and I. The prime denotes differentiation with respect to the arguments.

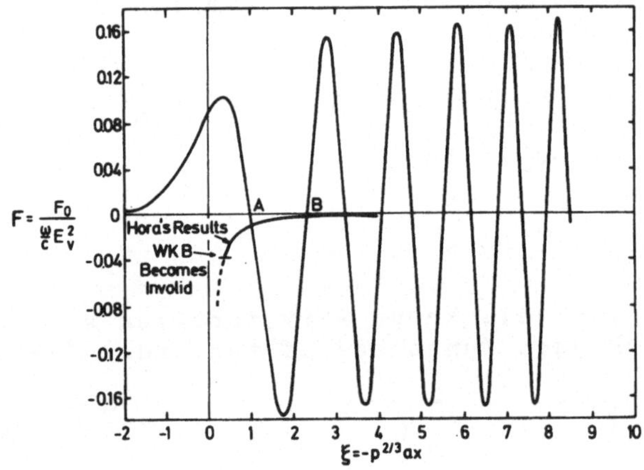

Fig. 4

Nonlinear force for a collisionless plasma of linear density profile at light incident from the right-hand side. The resulting standing wave creates the locally oscillating force, whose net acceleration is given by the double value of the monotonic force close to the abscissa. Comparison of the WBK case[12] with the rigorous case of the linear profile is given as derived by Lindl and Kaw[24].

Figure 4 reports on the numerical result of the evaluation of f_{NL} from Eq.(41) (oscillating curve). Very similar to our result of a standing wave with the WBK approximation[12] we find for the general case of Fig. 4 that the nonlinear force oscillates according to the acceleration of the plasma towards the nodes of the standing wave. In the neighborhood of the turning point ($x=x_1$; $\zeta=0$) the effective wavelength increases and the maxima and minima of the force decrease monotonically. The spatially averaged net force acting on the whole surface region of the plasma is double the monotonic curve close to the abscissa. This superposition of the net nonlinear force was also observed in the WBK case[12].

If Lindl and Kaw[24] use the incident wave only, they find the nonlinear force

$$\bar{\underset{\sim}{f}}_{NL} = \underset{\sim}{i}_x \frac{E_v^2}{16\pi} \frac{\omega_p^2}{\omega^2 \tilde{\eta}^2} \frac{d\tilde{\eta}}{dx} \qquad (\frac{\nu}{\omega} \ll 1) \qquad (42)$$

in agreement with our expression Eq.(15). A more detailed evaluation of Eq.(41) of a collisionless plasma ($\nu = 0$) was given by Kaw[34]. The force at the deepest maximum of Fig. 4 is then

$$f_{NL}^{Kaw} = \frac{0.1}{36} \frac{E_v^2}{x_1} (\frac{\omega x_1}{c}) \left[1 + (\frac{c}{\omega x_1})^{2/3} \cdot 0.6\right] \qquad (43)$$

This can be compared with the force in the WBK case, which by substituting Eq.(13) in Eq.(15) or (42)

$$f_{NL}^{WBK} = \frac{E_v^2}{16\pi} \frac{2\omega}{c} \theta \qquad (44)$$

because $\omega_p \approx \omega$. The ratio of Eqs.(43) and (44) is

$$\frac{f_{NL}^{Kaw}}{f_{NL}^{WBK}} = \frac{0.1}{36} \frac{16\pi}{2\theta} \left[1 + (\frac{c}{\omega x_1})^{2/3} \, 0.6\right] \qquad (45)$$

It is seen that at $(c/\omega x_1) \ll 1$ both forces are equal if the WBK parameter θ is 0.07. We note that for values $(c/\omega x_1) \gg 1$, in which case the WBK solution will become invalid, the rigorous solution of Kaw can even become very much larger than the formal (but invalid) value of the WBK solution. While the WBK solution will remain finite, in principle, as can be seen by combining Eqs.(13) and (15), Kaw's case could formally go to infinity but many other plasma physical conditions will then become critical.

Lindl and Kaw[24] also evaluated the solutions of the wave equations for the case of oblique incidence of the radiation on a plasma with linear increase of the electron density. The expressions for the ponderomotive force were taken from our former derivation of the case of oblique incidence[11]. While it was found for field strength derived from the WBK approximation that there was no strong difference of the plasma acceleration for the different cases of incidence, the linear density profile can cause an increase of the nonlinear force by a factor of 50 as high as for perpendicular incidence, if the electric vector of the light oscillates parallel to the plane of incidence. The reason is the influence of a certain function Φ, which is the result of the solution of the wave equation for this case, given by Denisov[35]. In this case a strong coupling of the electromagnetic waves and the longitudinal electrostatic waves occurs owing to finite temperature and plasma inhomogeneity in the interaction region, causing $|\tilde{\eta}| \ll 1$. There the collision frequency ν should be replaced by[33,35]

$$\nu_{eff} \simeq \nu + \omega\left(\frac{kTa^2}{m\,\omega^2}\right)^{1/3} \tag{46}$$

which, however, is a crude approximation. This strong increase of the nonlinear force at an optimum angle of incidence around 4 - 5° should be one of the possibilities to distinguish the nonlinear force from the thermokinetic processes.

The increased momentum P_{inh} transferred to the inhomogeneous layer was derived by Lindl and Kaw[24] on the basis of our relation[11,12]

$$P_{inh} = \int dy \; dz \int_{t_1}^{t_2} dt \int_{x_1}^{x_2} dx \; \underset{\sim}{f}_{NL} \tag{47}$$

where for oblique incidence with the $\underset{\sim}{E}$ vector in the plane of incidence

$$P_{inh} \approx -\frac{3}{2} P_o \frac{\exp(-2\rho s)}{\rho s^2} \tag{48}$$

where P_o is the momentum of the incident electromagnetic wave pulse at the times between t_1 and t_2

$$P_o = \int dy \; dz \int_{t_1}^{t_2} dt \; \frac{E_v^2(y,z,t)}{8\pi} \tag{49}$$

The quantity $s = (\nu/\omega)$ can become very small, especially at high electron temperatures or - as we have to point out - at high laser intensities following Eq.(7) at $I > I^*$. This results in another way in a very remarkable increase of the usual radiation pressure compared with the case treated before[36].

Lindl and Kaw[24] mentioned that the involved finite temperature and nonlinear effects may lead to enhanced absorption of the laser energy[20,21]. This may influence the nonlinear force positively or negatively.

NUMERICAL CASES

The numerical calculation of the field strengths $\tilde{E}$ and $\tilde{H}$ for a given density profile of a plasma permits numerical calculation of the nonlinear force according to Eq.(8) and (10) or, for perpendicular incidence of the radiation, Eq.(11). At the beginning only static solutions could be treated in the same sense as the analytic treatment of the preceding section was a static case where the force was calculated for a special density profile and no attention was paid to the temporal development of the density profile and of the forces, as a dynamic treatment takes into account.

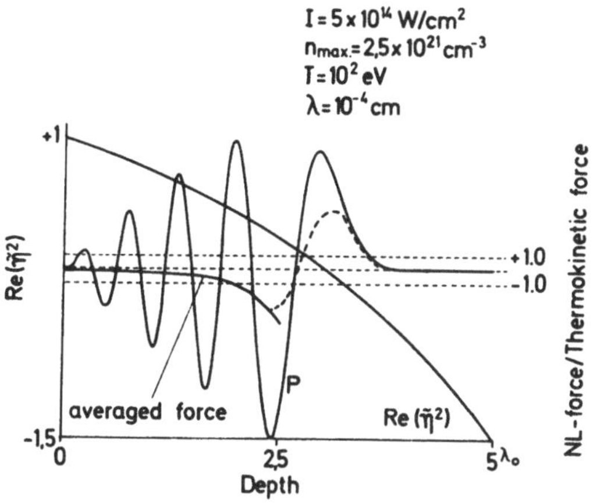

Fig. 5

Numerical calculation of the solutions of the wave equations for a density profile whose real part of the refractive index is parabolic, given by Mulser and Green[26]. The resulting standing wave at a plasma temperature T = 100 eV for neodymium glass lasers of 5×10^{14} W/cm^2 incident intensity (from the left-hand side) creates nonlinear forces whose value is given as ratio to the actual thermokinetic force. The net force is given as a spatially averaged value. It exceeds the thermokinetic force by a factor higher than 2.5.

Figure 5 demonstrates the result of Mulser and Green[26], where a parabolic profile of the electron density n(x), given by Eq.(3) and (4) is used

$$R_e(\tilde{\eta}^2) = 1 - \frac{4\pi e^2}{m(\omega^2+\nu^2)} \cdot n(x) \qquad (5o)$$

this value changing within the depth of five vacuum wavelengths from 1 (vacuum) to -1.5. The plasma is assumed to have a constant temperature of $1o^2$ eV and the intensity of the incident radiation is $5x1o^{14}$ W/cm^2 of neodymium glass lasers. The oscillating curve of Fig. 5 gives the ratio of the nonlinear force to the thermokinetic force, each calculated for the actual conditions. The oscillation is due to the fact that a depth of the cut-off density at less than 3 vacuum wavelengths below the vacuum can obviously cause nearly total reflection of the wave only. As can be seen from Fig. 5, the force may be nearly twelve times the gasdynamic force. Besides the forces driving the plasma towards the nodes of the standing wave, there is a net force accelerating the whole plasma layer, which is calculated by averaging over the spatial oscillation period Δ x

$$\bar{\underset{\sim}{f}}(x + \frac{1}{2}\Delta x) = \frac{1}{\Delta x}\int_x^{x+\Delta x} f(\xi)d\xi . \qquad (51)$$

We find that this averaged nonlinear force also exceeds the thermokinetic force by more than a factor of two. Averaging of the kind in Eq.(51) cannot be performed for $Re(\tilde{\eta}^2) < 1$, where only the aperiodic decrease of the nonlinear force results owing to absorption within the skin depth. Applying the Δx of the last period before this case, we get approximately the dashed curve in Fig. 5.

The larger net nonlinear force compared with the thermokinetic force is not unexpected, as can be seen from Fig. 3, where the threshold for the nonlinear force at 1oo eV is around $I^* = 3x1o^{14}$ W/cm^2 at reasonable exponential factors e^{-1}, which is only a little less than the calculated case[36] with $I = 5 \times 1o^{14}$ W/cm^2.

We also report a preliminary result of a dynamic numerical calculation of Shearer, Kidder and Zink[1o], where numerical simulation of the interaction of plasma with laser pulses of some picoseconds is treated. Figure 6 (top) shows the time dependence of a 15o psec neodymium glass laser pulse of $1o^{16}$ W/cm^2 intensity interacting with a plasma with a density profile of moderate slope over 15o vacuum wavelenghts crossing the cut-off

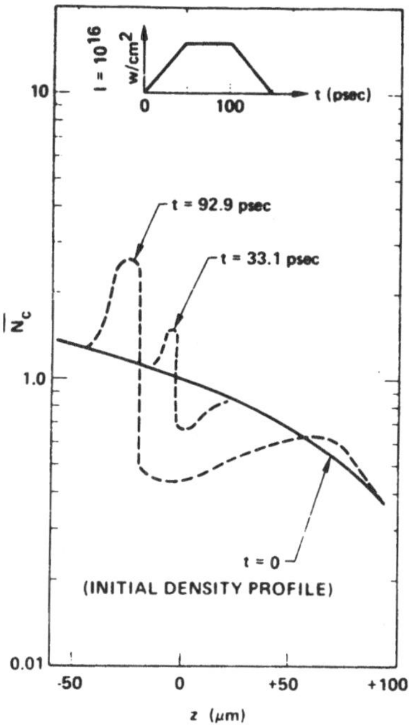

Fig. 6

Density profile at different times at a laser intensity described in the upper part of the figure in a dynamic computer program WAZER as given by Shearer[10,38]. The low density maximum is caused by the nonlinear force.

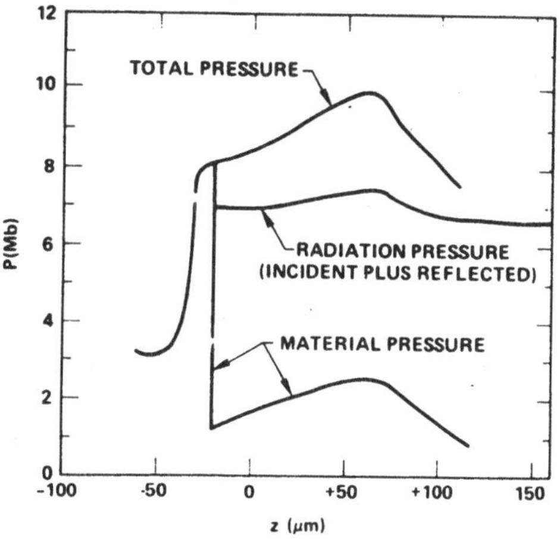

Fig. 7

Pressure vs distance at t =9o.9 psec for the WAZER computer run[38], see Fig. 6.

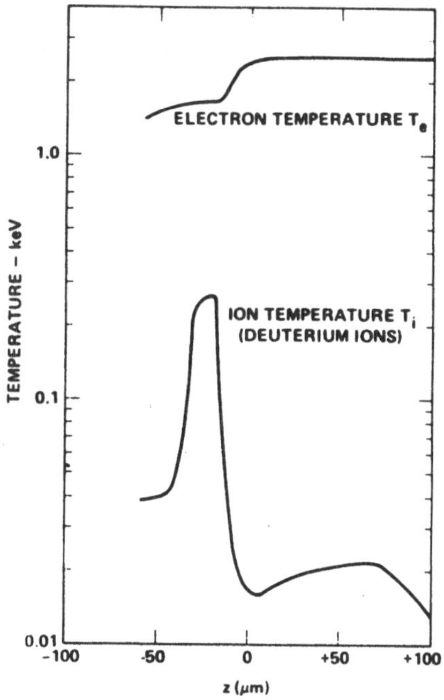

Fig. 8

Electron and ion temperature vs distance at t = 92.9 psec for WAZER computer run[38] of Fig. 6.

density at z=0. Figure 7 describes the pressure profile at the time of 9o.9 psec, where the material pressure p is distinguished from the electromagnetic energy-momentum-flux density (radiation pressure). The resulting temperature profiles at the time 92.9 psec (Fig. 8) demonstrates a state far away from thermal equilibrium between electrons and ions.

The result with respect to the nonlinear force was that the laser pulse is too short to generate a pronounced net motion of the plasma owing to this force. But it was established - and one can be sure that this is not due to numerical instabilities, which were certainly eliminated before the earlier steps of the program[37] - that the density profile was very strange at the times 33.1 psec and 92.9 psec (Fig. 6), showing strong maxima and minima. It is well known that the electron density can be monotonic or only have one maximum if only the thermokinetic force is taken into account. The strong bending of the profile causes an increase of the reflectivity of the laser light.

A further dynamic numerical treatment was performed by Mulser, the details of which will be reported separately[27]. It should merely be noted that the nonlinear force at the low intensity of $1o^{14}$ W/cm^2 for neodymium glass lasers creates in a plasma no strong net acceleration, as expectable from the higher threshold I^*, but maxima in the standing waves which are forty times higher than the thermokinetic forces.

CONCLUDING REMARKS

The nonlinear force causing a net acceleration of the plasma has a threshold around $1o^{14}$ W/cm^2 or at temperatures of $1o^4$ eV at lower intensities for ruby and neodymium glass lasers, as could be demonstrated analytically and by numerical examples of a WBK case (Fig.1) and of a non-WBK-like hyperbolic density profile (Fig.5). At oblique incidence and parallel orientization of the $\underset{\sim}{E}$-vector to the plane of incidence, a strong increase of the nonlinear force is possible owing to coupling of transversal and longitudinal waves in the plasma. It is still an open question how the nonlinear increase of the absorption at the cut-off density (Kaw and Dawson[2o]) influences the nonlinear force. It should be pointed out that positive changes of $|\tilde{\eta}|$ also cause strong gradients $\nabla|\tilde{\eta}|$ necessary for the nonlinear force.

The forces in standing waves, first treated under the aspects of changes of the refractive index under the conditions of the WBK case[12], more generally treated by Lindl and Kaw[24] and resulting in general numerical treatments[10,26,27], cause strong thermalization of the radiation and changes of the density profile. Therefore, the plasma is heated strongly and shows an increased reflectivity.

It is suggested that the standing-wave forces - which become effective even below the critical intensities I^* - are the dominating mechanism for increased heating (and neutron production) if the maxima of the nonlinear force exceed the thermokinetic force because of the anticipating acceleration of plasma fronts at wave modes (Brillouin resonances). Under these strongly disturbed conditions, the reflectivity is increased synchroneously.

REFERENCES

1. R.W. Minck and W.G. Rado, J. Appl. Phys. 37, 355 (1966).
2. A.G. Engelhardt, T.V. George, H. Hora and J.L. Pack, Phys. Fluids 13, 212 (1970).
3. R. Sigel, Z. Naturforsch. 25a, 488 (1970).
4. H. Salzmann, K. Eidmann and R. Sigel, Verhndl. Dtsch. Phys. Ges. (VI) 6, 407 (1971).
5. N.G. Basov, P.G. Kriukov, S.D. Zakharov, Yu.V. Senatsky and S.V. Tchekalin, IEEE J. Quantum Electronics QE-4, 864 (1968).
6. F. Floux, Laser Interaction and Related Plasma Phenomena (H. Schwarz and H. Hora, Eds.) Plenum 1971.
7. F. Floux, J.F. Benard, D. Cognard and A. Saleres, Second Workshop Laser Interaction and Related Plasma Phenomena, these Proceedings, p. 409.
8. K. Büchl, K. Eidmann, P. Mulser, H. Salzmann, R. Sigel and S. Witkowski, Paper CN 28-D-11, 4th Conf. on Plasma Physics and Controlled Thermonuclear Fusion Research, Madison, Wisc., June 1971.
9. S.W. Mead, R.E. Kidder and J.E. Swain, Report UCRL-73356, August 17, 1971, IEEE J. Quantum Electronics (submitted).
10. J.W. Shearer, R.E. Kidder and J.W. Zink, Bull. Am. Phys. Soc. 15, 1483 (1970).
11. H. Hora, D. Pfirsch and A. Schlüter, Z. Naturforsch. 22a, 278 (1967); A. Schlüter, Plasma Physics 10, 471 (1968).

12. H. Hora, Phys. Fluids 12, 182 (1969); H. Hora, Laser Interaction and Related Plasma Phenomena (H. Schwarz and H. Hora Eds.) Plenum, New York 1971, p. 383.
13. H. Hora, Opto-Electronics 2, 2o1 (197o).
14. S. Rand, Phys. Rev. 136,B 231 (1964).
15. T.B. Hughes and M.B. Nicholson-Florence, J. Phys. A (2) 588 (1968).
16. R.Kidder, paper presented at the Varenna Summer School (July 1969), UCRL-Preprint 71775 (1969).
17. A. Caruso, A. de Angelis, G. Gatti, R. Gratton and S. Martellucci, Phys. Lett. 33A, 29 (197o).
18. H. Hora, Z. Physik 226, 156 (1969).
19. P.K. Kaw, Appl. Phys. Lett. 15, 16 (1969).
2o. P.K. Kaw and J.M. Dawson, Phys. Fluids 12, 2586 (1969).
21. W.L. Kruer and J.M. Dawson, Phys. Fluids 14, 1oo3 (1971); Second Workshop Laser Interaction and Related Plasma Phenomena, these Proceedings, p. 317.
22. see for example H. Motz and C.J.H. Watson, in Advances in Electronics and Electron Physics (L. Marton, Ed.) Academic Press, New York, Vol. 23, p. 153.
23. A.V. Gorbunov and M.A. Miller, ZhETF 34, 242; 751 (1958) (Sov. Phys. JETP 7, 168; 515 (1958)).
24. J. Lindl and P. Kaw, Phys. Fluids 14, 371 (1971).
25. L.C. Steinhauer and H.G. Ahlstrom, Phys. Fluids 13, 11o3 (197o).
26. B. Green and P. Mulser, Verhandl. Dtsch. Phys. Ges. (IV) 6, 4o5 (1971).
27. P. Mulser, Second Workshop Laser Interaction and Related Plasma Phenomena, these Proceedings, p. 381.
28. L.D. Landau and E.M. Lifshitz, Electrodynamic of Continuous Media (Pergamon Press, Oxford, 1966) p. 242.
29. see e.g. C.W. Allen, Astrophysical Quantities, Athlon Press, London 1955.
3o. J.M. Dawson and C. Oberman, Phys. Fluids 5, 517 (1962).
31. H. Hora and H. Wilhelm, Nuclear Fusion 1o, 111 (197o).
32. L. Spitzer,Jr., Physics of Fully Ionized Gases, Interscience, New York (1956).
33. V.L. Ginzburg, The Propagation of Electromagnetic Waves in Plasmas (Addison-Wesley, Reading, Mass. 1969), pp. 193-198 and 213-228.
34. P. Kaw, (private communication, July 1969).
35. N.G. Denisov, ZhETF 31, 6o9 (1956); Sov. Phys. JETP 4, 544 (1957).

36. H. Hora, Ann. Physik (7) 22, 4o2 (1969).
37. J.W. Shearer and W.S. Barnes, Laser Interaction and Related Plasma Phenomena (H. Schwarz and H. Hora Eds.) Plenum, New York 1971, p. 3o7.
38. J.W. Shearer, Report Livermore Rad. Lab. UCID-15745 (Dec. 7, 197o).

PHYSICAL MECHANISMS FOR LASER-PLASMA PARAMETRIC INSTABILITIES*

Francis F. Chen†

University of California, Los Angeles

Los Angeles, California 90024

During the past year, there has been a profusion of theoretical results concerning parametric instabilities, anomalous backscattering, and other nonlinear interactions expected to occur at the large laser intensities needed to achieve fusion. It is often difficult for the experimentalist to gain an intuitive grasp of the important physical mechanisms involved. Fortunately, most of these phenomena are generalized parametric instabilities which can be explained, both qualitatively and quantitatively, by simple physical descriptions involving the ponderomotive force.

THE PONDEROMOTIVE FORCE

Originally invoked[1] in connection with laser interactions to explain fast-ion production, the ponderomotive force is nothing more than radiation pressure in a plasma. It is most readily understood from the single-particle point of view. Following Schmidt[2], we write the equation of motion for a single electron moving in a wave:

$$m \frac{d\underline{v}}{dt} = -e[\underline{E}(\underline{r}) + \frac{1}{c}\, \underline{v}\times\underline{B}(\underline{r})] \quad , \qquad (1)$$

*Presented at the Third Workshop on "Laser Interaction and Related Plasma Phenomena" held at Rensselaer Polytechnic Institute, Troy, New York, August 13-17, 1973.

†This work was supported by the U.S. Atomic Energy Commision, Contract AT(04-3)-34, Project 157, Mod. 6, Task II.

Reprinted from: *Laser Interaction and Related Plasma Phenomena,* Vol. 3A, H. Schwarz and H. Hora (eds.), Plenum, New York, 1974, pp. 291–313.

where

$$\underline{E}(\underline{r}) = \underline{E}_s(\underline{r})\cos\omega_0 t \tag{2}$$

and

$$\underline{B}(\underline{r}) = -\frac{c}{\omega_0}\ \nabla\times\underline{E}_s\sin\omega_0 t \quad \text{(Maxwell's equation)}\ . \tag{3}$$

The $\underline{v}\times\underline{B}$ term in Eq. (1) is v_0/c times smaller than the $\underline{E}$ term (v_0 will be defined below), and to lowest order the electron oscillates in the cirection of $\underline{E}$:

$$\underline{v}^{(1)} = -\frac{e}{m\omega_0}\underline{E}_s\sin\omega_0 t \quad . \tag{4}$$

We define the quiver velocity v_0 to be the peak value of $v^{(1)}$ in a plane-polarized wave ($E_0 = |E_s \max|$):

$$v_0 \equiv \frac{eE_0}{m\omega_0} \tag{5}$$

When the quantity v_0/c, which characterizes the strength of non-linear interactions, is small but finite, we can use $\underline{v}^{(1)}$ to solve Eq. (1) to the next order. We must also expand $\underline{E}(r)$ about the initial position $\underline{r}_0$, for Eq. (1) holds in the frame moving with the electron. Thus

$$\underline{E}(\underline{r}) \approx E(r_0) + \delta\underline{r}\cdot\underline{\nabla}\ \underline{E} \quad , \tag{6}$$

where $\delta\underline{r}$ is given by the integral of Eq. (4):

$$\delta\underline{r} = \frac{e}{m\omega_0^2}\underline{E}_s\cos\omega_0 t \quad . \tag{7}$$

Thus

$$m\frac{d\underline{v}^{(2)}}{dt} = -e[\delta\underline{r}\cdot\underline{\nabla}\ \underline{E}(\underline{r}_0) + \frac{1}{c}\underline{v}^{(1)}\times\underline{B}(\underline{r}_0)] \ . \tag{8}$$

Using Eqs. (7), (4), and (3) and averaging over time, we find the secular force acting on the electron:

$$\underline{f}_{NL} \equiv m < \frac{d\underline{v}^{(2)}}{dt} > = -\frac{e^2}{m\omega_0^2}\frac{1}{2}(\underline{E}_s\cdot\underline{\nabla}\ \underline{E}_s + \underline{E}_s\times\underline{\nabla}\times\underline{E}_s) \quad . \tag{9}$$

The $\underline{E}\times\nabla\times\underline{E}$ term gives rise to a drift of the electrons in the direction of the wave. This is caused by the wave magnetic field, which distorts the linear orbit of the electron into a figure-8 path, as described by Hora[1]. This term drives the backscattering instabilities. The $\underline{E}\cdot\underline{\nabla}\ \underline{E}$ term is due to the finite excursion of the oscillating electron, which brings it into regions of different field intensity. This term is equivalent to the $\underline{v}\cdot\underline{\nabla}\ \underline{v}$ convection term in the fluid equation. The electrostatic parametric instabilities are driven by this term.

The ponderomotive force F_{NL} is the force per unit volume, $n_e \underline{f}_{NL}$:

$$\underline{F}_{NL} = -\frac{\omega_p^2}{\omega_0^2}\underline{\nabla}\left(\frac{E_s^2}{16\pi}\right) = -\frac{\omega_p^2}{\omega_0^2}\underline{\nabla}\frac{\langle E^2\rangle}{8\pi}, \qquad (10)$$

where $\omega_p^2 = 4\pi n_e e^2/m$, and the average is over a wavelength.

Although the corresponding force on the ions is m/M times smaller, the force $\underline{F}_{NL}$ on the electrons is transmitted to the ions by self-consistent fields, provided (a) there are density gradients, and (b) $\underline{F}_{NL}$ varies in time more slowly than the ion plasma frequency ω_{pi}. As the electrons respond to $\underline{F}_{NL}$, they generate a charge-separation electric field $\underline{E}_1$ such that $\underline{F}_{NL} - n_e e\underline{E}_1 = 0$. The ions feel a force $n_i e\underline{E}_1$, and the total force density in the quasineutral plasma is therefore $\underline{F}_{NL}$.

Eq. (10) can also be derived[3] from the macroscopic viewpoint, considering the plasma to be a dielectric with $\varepsilon = 1-\omega_p^2/\omega^2$, but the physics is then more obscure. One can consider $\underline{F}_{NL}$ to be a radiation pressure

$$\underline{F}_{NL} = -\underline{\nabla}\frac{\langle E^2\rangle}{8\pi} - \underline{\nabla}\frac{\langle B^2\rangle}{8\pi}, \qquad (11)$$

but this expression does not contain the plasma density explicitly. To see how the density enters, consider first the case of a vacuum. If the wave were infinite, the field strength would be uniform, and $\underline{F}_{NL}$ would be zero. If the wave were reflected from a perfect mirror, $\underline{F}_{NL}$ would still vanish, because there is no preferred direction for it. Fig. 1 shows that in the standing wave pattern $\nabla\langle E^2\rangle/8\pi$ is always equal and opposite to $\nabla\langle B^2\rangle/8\pi$. On the other hand, if the wave were reflected from the cutoff layer of an

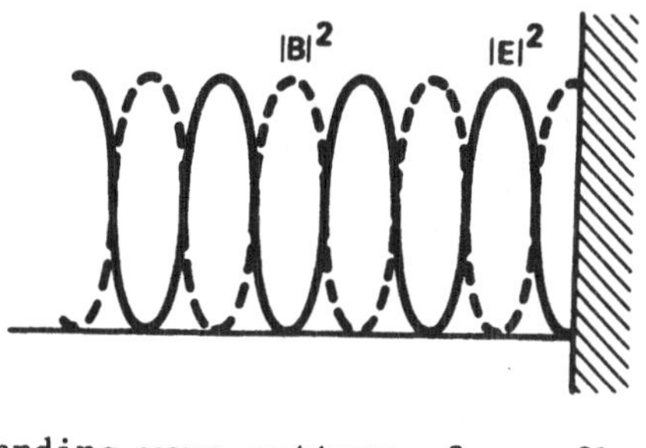

Fig. 1. Standing wave pattern of a reflected light wave in a vacuum.

inhomogeneous plasma, one would get the usual Airy-function pattern[4] of Fig. 2. As is well known, there is now a ponderomotive force,

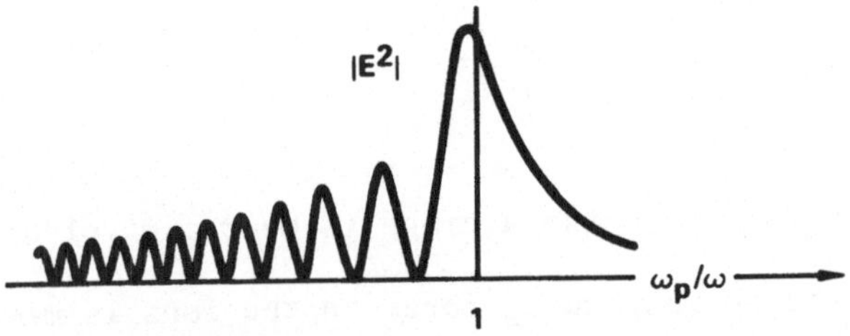

Fig. 2. Standing wave pattern of a light wave traveling along a plasma density gradient.

which reverses sign in each striation. This is because the oscillation of the plasma electrons causes $\underline{E}$ and $\underline{B}$ to have different amplitudes, so that $\nabla\langle E^2\rangle/8\pi$ and $\nabla\langle B^2\rangle/8\pi$ no longer cancel. The difference is proportional to the density. Since $ck/\omega = \varepsilon^{1/2}$ in a dielectric, Eq. (3) yields $\langle B^2\rangle = -\varepsilon\langle E^2\rangle$. Inserting this in Eq. (11), one has

$$\underline{F}_{NL} = (\varepsilon-1)\nabla \frac{\langle E^2\rangle}{8\pi} = -\frac{\omega_p^2}{\omega_0^2}\nabla\frac{\langle E^2\rangle}{8\pi} , \tag{12}$$

as before. The macroscopic approach is so circuitous and opaque that derivations based on the Maxwell stress tensor are often erroneous.

ELECTROSTATIC PARAMETRIC INSTABILITIES

The parametric decay of an electromagnetic wave with $\omega_0 \simeq \omega_p$ into electrostatic perturbations has been extensively studied in recent years because of the applications to both ionospheric heating and laser-pellet heating. There are two processes, depending on the sign of $\omega_0-\omega_e$, where ω_e is the Bohm-Gross frequency, given by

$$\omega_e^2 = \omega_p^2 + 3k^2v_e^2 . \tag{13}$$

For $\omega_0<\omega_e$, the oscillating two-stream (OTS) instability can occur, with $\mathrm{Re}\,\omega = 0$. For $\omega_0>\omega_e$, the parametric decay into an electron wave ω_e and an ion acoustic wave ω_i can occur. The reason for the difference is made clear by the ponderomotive force approach.

The Oscillating Two-Stream Instability

In the OTS instability, ω_0 is less than ω_e, and density ripples in the direction of $\underline{E}_0$ grow without propagating (Fig. 3). Let a

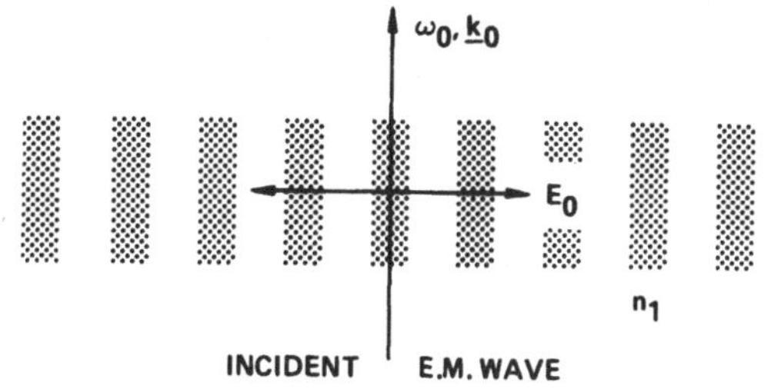

Fig. 3. Geometry of the OTS instability.

quasineutral density perturbation n_1 have $k \gg k_0$, so that k_0 can be assumed to be zero and $\underline{E}_0$ to be uniform. The motion of electrons in the direction of $-\underline{E}_0$ will give rise to a space charge oscillating at the frequency ω_0, as shown in Fig. 4 at an instant when $\underline{E}_0$ is in the +x direction. This space charge creates a field $\underline{E}_1$

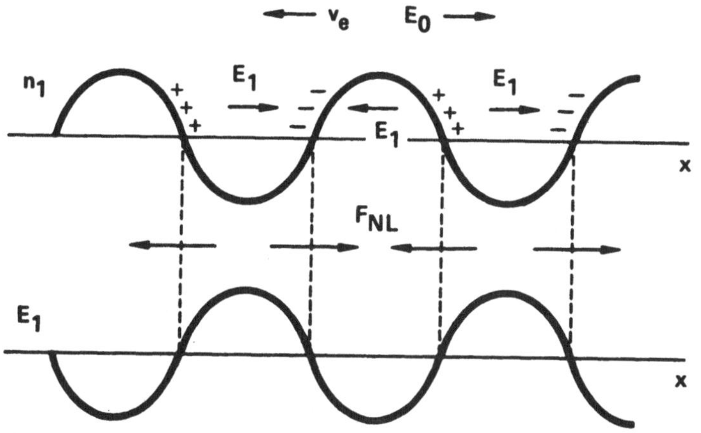

Fig. 4. Physical mechanism of the OTS instability.

with the phase shown in Fig. 4. For uniform $\underline{E}_0$, the ponderomotive force is given by

$$8\pi(\omega^2/\omega_p^2)\underline{F}_{NL} = -\nabla(\underline{E}_0+\underline{E}_1)^2 \simeq -2E_0\cdot\nabla E_1 = -2E_0\partial\underline{E}_1/\partial x \quad , \qquad (14)$$

and hence has the phase shown in the figure. Since the phase of $\underline{F}_{NL}$ is such as to move plasma into the regions of positive n_1, the density perturbation grows. Although these phase relationships hold for any $\omega_0<\omega_e$, excitation is easiest when ω_0 is near the natural oscillation frequency ω_e of the electron fluid. The electrons move in a standing wave, but the ion motion does not follow a natural mode of the ion fluid. By symmetry, the ion perturbation does not propagate in either direction, and this is an absolute instability with $\text{Re}\omega = 0$. It is possible[5] to recover from this ponderomotive force treatment the exact threshold and growth rates obtained by standard techniques[6]. A well-known theoretical result[6] is that the threshold for this instability depends on the damping rate γ_e of the electron waves, but not on the damping rate γ_i of ion waves. This is because the ions do not oscillate and suffer friction against another fluid but simply move slowly at the growth rate γ. By contrast, the electrons oscillate rapidly, and their motion is damped by collisions with ions and neutrals. Note that the $\underline{E}\times\nabla\times\underline{E}$ part of F_{NL} does not enter in this geometry, since it gives a drift along planes of constant density.

The Parametric Decay Instability

If ω_0 is larger than ω_e, the OTS mechanism does not work; instead, the incident wave decays into an electron wave ω_e and an ion acoustic wave $\omega_i = kc_s$. Consider an oscillator x with natural frequency ω_e driven by a force $F\cos\omega_0 t$:

$$\ddot{x} + \omega_e^2 x = F\cos\omega_0 t \ , \qquad x = \frac{F\cos\omega_0 t}{\omega_e^2 - \omega_0^2} \ . \tag{15}$$

If $\omega_0<\omega_e$, the displacement is in the same direction as the force, and the mechanism of Fig. 4 obtains. If $\omega_0>\omega_e$, the displacement has the opposite sign, and electrons would move in the $+\underline{E}_0$ direction. As Fig. 5 shows, the resulting F_{NL} then acts so as to destroy a density perturbation n_1, which would then decay rather than grow. However, if the density ripple of wavenumber k were an ion wave traveling with velocity c_s, the phase relation can be reversed. In a frame moving with the ion wave, the density perturbation would be at rest, as in the OTS case. In the limit $k\lambda_D \ll 1$, the plasma is quasineutral, and the electron fluid follows the motion of the ion fluid very closely. However, in the

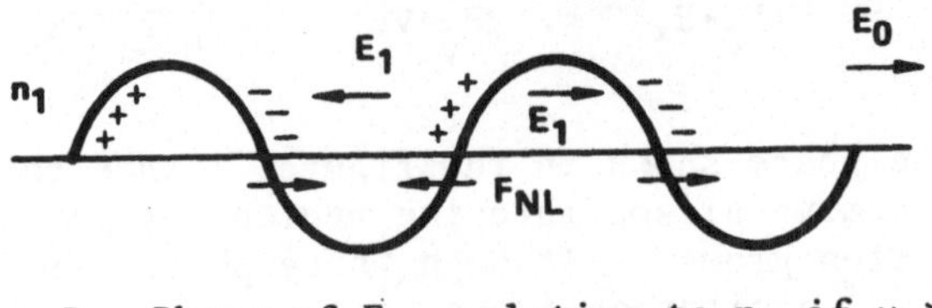

Fig. 5. Phase of F_{NL} relative to n_1 if $\omega_0>\omega_e$.

moving frame the electron fluid feels the pump field E_0 not at the frequency ω_0, but at the Doppler-shifted frequency $\omega_0' = \omega_0 - kc_s$. If $\omega_i = kc_s$ is sufficiently large, so that $\omega_0' < \omega_e$, the mechanism of the OTS instability can again operate, and the ion wave grows. Note that the critical condition $\omega_0 - kc_s = \omega_e$ is just the frequency-matching relation $\omega_0 = \omega_e + \omega_i$ for the decay process. Fig. 6A shows the usual parallelogram construction for ω and k matching in the parametric decay instability.

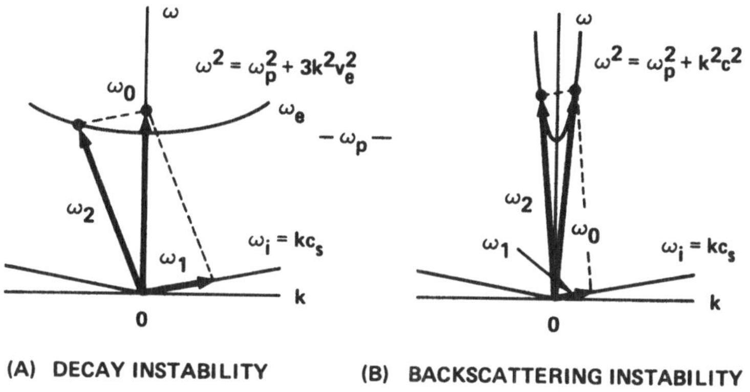

Fig. 6. Comparison of the ω and k matching conditions for electrostatic and electromagnetic parametric instabilities. In each case, ω_0 is the incident electromagnetic wave, and it decays into an ion wave ω_1 and another wave ω_2. In (A), ω_2 is an electron Bohm-Gross wave. In (B), ω_2 is a backscattered electromagnetic wave (Brillouin scattering). In Raman scattering, the ion wave is replaced by a Bohm-Gross wave.

BACKSCATTERING INSTABILITIES

Fig. 5B shows that backscattering is a parametric decay with one of the products being an electromagnetic wave. Since both ω_0 and ω_2 can be much larger than ω_p, backscattering can occur in an underdense plasma. For simplicity, we shall assume $\omega_0 >> \omega_p$ in describing the physical mechanism. Let ω_1 be electrostatic and ω_2 electromagnetic. The Stokes and anti-Stokes lines are given by $\omega_2 = \omega_0 \pm \omega_1$. In backscattering, the Stokes process (minus sign) is important, and the matching conditions are $\omega_0 = \omega_1 + \omega_2$, $\underline{k}_0 = \underline{k}_1 + \underline{k}_2$. The most likely process is for $\underline{k}_2$ to be opposite to $\underline{k}_0$, as shown in Fig. 7. Since $|k_2| \simeq |k_0|$ for $\omega_0 >> \omega_p$, we have $\underline{k}_2 \simeq -\underline{k}_0$ and, therefore, $\underline{k}_1 \simeq 2\underline{k}_0$.

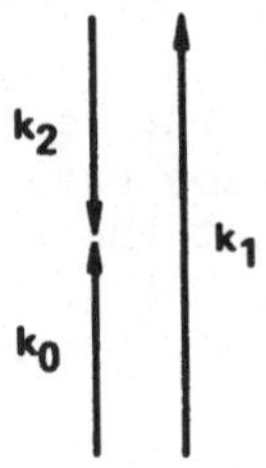

Fig. 7.

Let the incident wave be

$$\mathbf{E}_0 = E_0\hat{x}\cos(\omega_0 t - k_0 y).$$

The quiver velocity

$$\mathbf{v}_0 = -\frac{eE_0}{m\omega_0}\sin(\omega_0 t - k_0 y)\hat{x} \tag{16}$$

then leads $\mathbf{E}_0$ by 90°. We look for a ponderomotive force $\mathbf{F}_{NL}$ at the relatively low frequency ω_1 of the electrostatic wave; this frequency results from the beating of the incident wave ω_0 with the backscattered wave ω_2. The relevant cross terms in Eq. (8) then give

$$\mathbf{F}_{NL} = -\frac{n_0 e^2}{m\omega_0^{\ 2}}\langle \mathbf{E}_0\cdot\nabla\,\mathbf{E}_2 + \mathbf{E}_2\cdot\nabla\,\mathbf{E}_1\rangle - \frac{en_0}{c}\langle \mathbf{v}_0\times\mathbf{B}_2 + \mathbf{v}_2\times\mathbf{B}_0\rangle . \tag{17}$$

The $\mathbf{E}\cdot\nabla\,\mathbf{E}$ terms vanish by geometry in straight backscattering, and $\mathbf{F}_{NL}$ is due entirely to the $\mathbf{v}\times\mathbf{B}$ terms. In Fig. 8A the two electromagnetic waves are shown at an instant of time when the maxima of

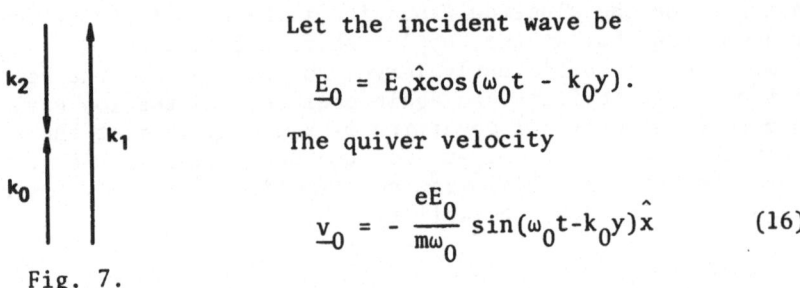

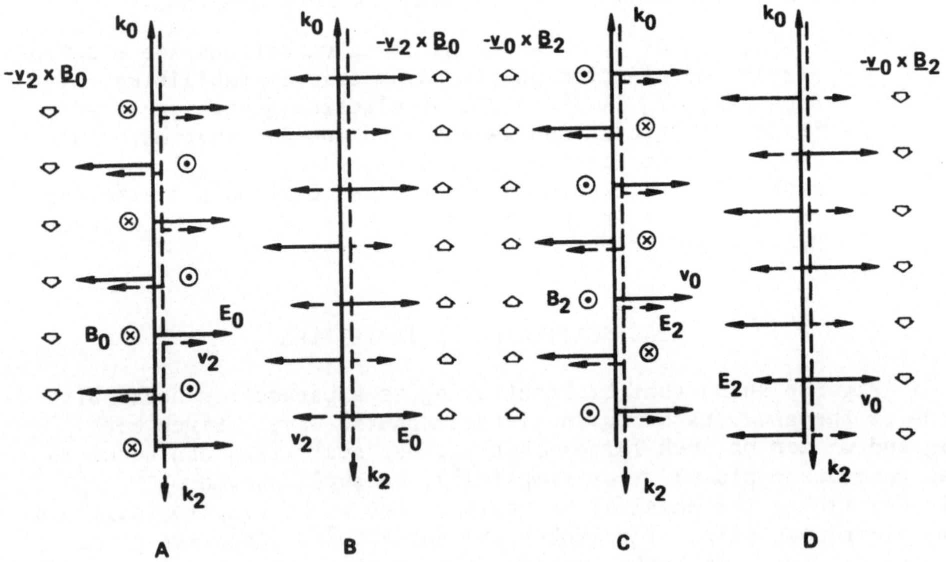

Fig. 8. Phase relationships in backscattering. The incident wave is indicated by solid lines; the backscattered wave by dashed lines. They heavy arrows indicate the ponderomotive force.

$\underline{E}_0$ and the quiver velocity $\underline{v}_2$ of the scattered wave are coincident. Taking $\underline{B}_0$ to be in the direction of $\underline{k}_0\times\underline{E}_0$, one finds that the part of $\underline{F}_{NL}$ due to $-\underline{v}_2\times\underline{B}_0$ is downward. As the waves pass, this term oscillates between 0 and its maximum value and hence has a finite average (over the ω_0 time scale) which depends on position. Fig. 8B shows the situation 1/4 of a period later. $\underline{E}_0$ has shifted upward and $\underline{v}_2$ downward; hence, $-\underline{v}_2\times\underline{B}_0$ is upward at the locations shown. Making use of the fact that $\underline{v}$ leads $\underline{E}$ by 90°, one obtains Figs. 8C and 8D for the phases of the $-\underline{v}_0\times\underline{B}_2$ term at the times corresponding to Figs. 8A and 8B, respectively, It is seen that the two terms in F_{NL} add in phase in such a way as to cause electrons to bunch up and form a density ripple with half the wavelength of the electromagnetic waves ($k_1 = 2k_0$). Thus, a density perturbation n_1 results from the nonlinear interference of the incident and backscattered waves, as depicted in Fig. 9.

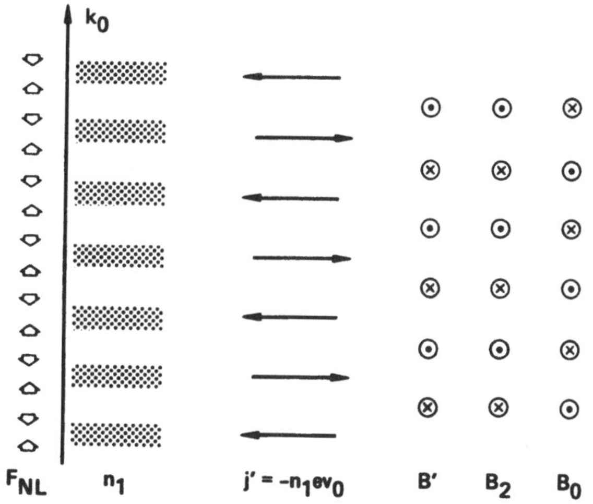

Fig. 9. Physical mechanism of backscattering instabilities.

It remains to show that n_1 causes the wave ω_2 to grow at the expense of the wave ω_0, closing the feedback loop for the instability. Since $\underline{v}_0$ is perpendicular to ∇n_1, we do not have the charge separation mechanism of the OTS instability. Instead, the effect is a spatial modulation of the oscillating current $-nev_0$ caused by $\underline{v}_0$. Let us neglect the motion of the density ripples and consider a phase 1/8 of a period after that of Fig. 8C. The maxima of $\underline{v_0}$ and n_1 are then aligned, and the current modulation $\underline{j}' = -n_1 e \underline{v_0}$ is as shown in Fig. 9. This current generates a magnetic field $\underline{B}'$ with the phase shown. At this instant of time, examination of Fig. 8 shows that the phases of $\underline{B_2}$ and $\underline{B_0}$ are as given in Fig. 9. It is seen that $\underline{B}'$ is in phase with $\underline{B_2}$ and out of phase with $\underline{B_0}$, as needed for excitation of $\underline{B_2}$. The ponderomotive force approach can be used[7] to derive the exact results for these instabilities.

The density perturbation, as yet unspecified, may or may not be a natural mode of the plasma. We distinguish several cases:

1) If $\omega_1 = \omega_e$ (an electron plasma wave), the process is called stimulated Raman scattering (SRS).

2) If $\omega_1 = kc_s$ (an ion acoustic wave), the process is called stimulated Brillouin scattering (SBS).

3) If ω_1 is not a normal mode, the density perturbation can still exist if E_0 is large enough to maintain it against diffusion. This has been called resistive quasimode scattering.

4) If $\omega_1 = k_1 v_e$ or $k_1 v_i$, where v_e and v_i are thermal velocities, there can be interaction with resonant particles to give instability. This has been called induced Compton scattering or nonlinear Landau growth, a process that can be important when the natural modes are inhibited, say, by Landau damping.

Homogeneous Plasmas

The thresholds and growth rates for SBS and SRS in a homogeneous plasma have been computed by a number of authors[7-11] and are listed here for future reference. Note that v_0 is the peak

	Threshold	Growth Rate	
SBS	$\frac{v_0^2}{v_e^2} = \frac{8\gamma_i \nu_{ei}}{\omega_i \omega_0}$	$\gamma_0 \simeq \frac{1}{2}\frac{v_0}{c}(\omega_0/\omega_i)^{1/2}\omega_{pi}$	(18)
SRS	$\frac{v_0^2}{c^2} = \frac{2\omega_p^2}{\omega_0^2}\frac{\gamma_e}{\omega_p}\frac{\nu_{ei}}{\omega_0}$	$\gamma_0 \simeq \frac{1}{2}\frac{v_0}{c}(\omega_0\omega_p)^{1/2}$	(19)

quiver velocity in a plane-polarized wave. Here $v_e^2 = KT_e/m$, γ_i and γ_e are the ion and electron wave damping rates, and $(\omega_p^2/\omega_0^2)(\nu_{ei}/2)$ is the damping rate of the electromagnetic waves. The growth rates are for the intermediate-intensity regime (see below), where $\gamma_0 \propto E_0$. Note that γ_0 depends very weakly on density -- hence the possibility of anomalous reflection from the outer layers of a plasma. The collisional thresholds above are easily exceeded but are not relevant to actual experiments.

Finite Homogeneous Plasmas

If the interaction region is limited in size either by the length of the plasma or the depth of focus of the laser beam, it is not sufficient to exceed the threshold; the growth rate must be large enough to overcome convection of wave energy out of the region. This problem has been treated by Pesme et al.[11], whose results may be summarized as follows. If ℓ is the length of the interaction region, γ_0 is the homogeneous-plasma growth rate given above, and V_1 and V_2 are the group velocities of the product waves, an absolute instability is possible if $V_1V_2 < 0$ and

$$\frac{\gamma_0\ell}{|V_1V_2|^{1/2}} > \frac{\pi}{2} \quad \text{or} \quad \frac{\gamma_0\ell}{(cc_s)^{1/2}} > \frac{\pi}{2} \quad \text{for SBS} \quad . \tag{20}$$

This result holds for weak damping such that $\gamma_0 >> \gamma_T$, where

$$\gamma_T \equiv (\gamma_1\gamma_2)^{1/2} = (\gamma_i\gamma_{ei}/2)^{1/2}(\omega_p/\omega_0) \quad \text{for SBS} \quad . \tag{21}$$

γ_T is just the threshold given in Eqs. (18) and (19).
We shall discuss specifically the Brillouin case, which has a higher nonlinear saturation level than Raman scattering. In SBS, ion Landau damping for reasonable ratios T_e/T_i causes γ_i/ω_i to be non-negligible. In such a case, the growth length is greatly increased by the damping, and Eq. (20) is valid only if γ_0 further satisfies $\gamma_0 >> \gamma_c$, where

$$\gamma_c \equiv \frac{1}{2}|V_1V_2|^{1/2}\left|\frac{\Gamma_1}{V_1} - \frac{\Gamma_2}{V_2}\right| \simeq \frac{1}{2}\gamma_i(c/c_s)^{1/2} \quad . \tag{22}$$

In most cases γ_0 lies in the range $\gamma_T < \gamma_0 < \gamma_c$, when no absolute instability is possible, but the waves can grow in space regardless of the sign of V_1V_2. The condition for a large number of e-foldings in ℓ is

$$\gamma_0{}^2\ell/c\gamma_i >> 1 \quad . \tag{23}$$

This condition is considerably more stringent than Eq. (20) and greatly reduces the large disparity between the SRS and SBS thresholds given in Eqs. (18) and (19).

These results can be understood physically as follows. In

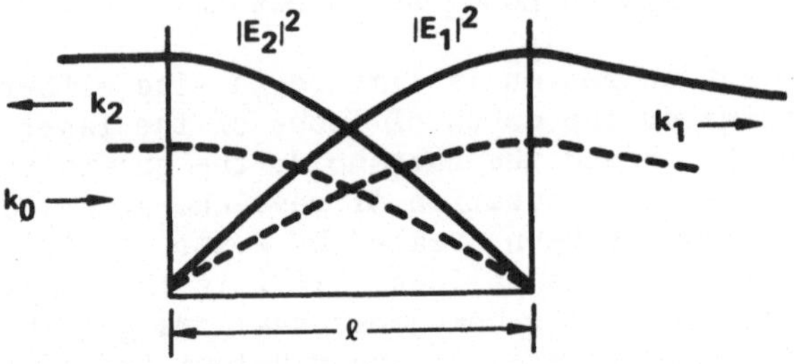

Fig. 10. Spatial variation of wave intensities in backscattering in a finite interaction region.

Fig. 10, we have shown schematically the behavior of the wave intensities in the weak-damping case (absolute instability). The pump wave is from the left; the ion wave grows from thermal noise from the left; and the backscattered wave grows from essentially zero amplitude from the right. To avoid questions of reflection from plasma boundaries, we assume the interaction region is limited by the focal depth alone, and the waves propagate with weak damping outside the region. The growth of each wave depends on the amplitude of the other one when it exits from the region. The light wave e-folds $\gamma_0\ell/c$ times in the length ℓ, while the ion wave e-folds $\gamma_0\ell/c_s$ times, c and c_s being the respective group velocities. Since each wave bootstraps the other, the net e-folding of the normal mode is the product of these two:

$$\frac{\gamma_0\ell}{c}\cdot\frac{\gamma_0\ell}{c_s} = \frac{\gamma_0{}^2\ell^2}{cc_s} > 1 \ . \tag{24}$$

This is essentially Eq. (20). Thus it is the geometric mean of the group velocities which determines the convection.

In the intermediate damping case, γ_e [Eq. (22)] is determined by the ion wave damping, the light wave damping term being negligible. The damping length is c_s/γ_i. The growth length is $(cc_s)^{1/2}/\gamma_0$, since we have seen that $(cc_s)^{1/2}$ is the effective group velocity of the normal mode. For the damping length to exceed the growth length, we require

$$\frac{c_s}{\gamma_i} > \frac{(cc_s)^{1/2}}{\gamma_0}\ , \quad \gamma_0 > \gamma_i(c/c_s)^{1/2} = 2\gamma_c \tag{25}$$

This is the meaning of γ_c. If $\gamma_0 < \gamma_c$, the ion wave loses energy by damping more rapidly than by convection out of the growth region. In the length ℓ, the ion wave exponentiates $\gamma_0\ell/c_s$ times while being damped $\gamma_i\ell/c_s$ times. The net exponentiation is then γ_0/γ_i. Meanwhile, the light wave e-folds $\gamma_0\ell/c$ times. The product of the two is

$$\frac{\gamma_0}{\gamma_i} \cdot \frac{\gamma_0\ell}{c} = \frac{\gamma_0^2\ell}{c\gamma_i} \quad . \tag{26}$$

Eq. (23) is just the condition that this be >>1.

Inhomogeneous Plasmas

If the plasma is infinite but inhomogeneous, a different effect occurs. The plasma wave has a different wavelength in each part of the plasma, so that the matching conditions $\omega_0 = \omega_1+\omega_2$, $\underline{k_0} = \underline{k_1}+\underline{k_2}$ can be satisfied only over a finite region. As a wave travels away from the point of perfect matching, it grows until the phase mismatch is so large that the proper phasing for growth is lost. The wave equation has a turning point there, and the wave propagates without growth from there on. The number of e-foldings is given by Eq. (24), with ℓ replaced by the distance ℓ_t between turning points. A turning point occurs when the phase mismatch $\int\Delta k dx \simeq \Delta k \ell_t$ is of order 1, where $\Delta k \simeq k'\ell_t$, $k' \equiv dk/dx$. Thus $\ell_t^2 \simeq 1/k'$, and Eq. (24) gives the approximate condition $\gamma_0^2/cc_s k' >> 1$ for appreciable growth. The exact number of e-foldings in E_2 is[12]

$$\pi\gamma_0^2/cc_s k' >> 1 \quad . \tag{27}$$

In SRS, k' is determined by the density gradient length L_n, since the Bohm-Gross dispersion relation depends sensitively on ω_p. In SBS, it is the temperature gradient length L_T which is important in the ion wave dispersion relation. Using these relations and the values of γ_0 from Eqs. (18) and (19), one obtains the following "inhomogeneous thresholds" from Eq. (27):

SBS $$\frac{v_0^2}{v_e^2} > \frac{\omega_0^2}{\omega_p^2}\frac{8}{k_0 L_T} \tag{28}$$

SRS $$\frac{v_0^2}{c^2} > \frac{2}{k_0 L_n} \quad . \tag{29}$$

Note that there is no density dependence in Eq. (29). This is because the density dependences in γ_0 and k' exactly cancel. Thus,

SRS can occur in a very underdense plasma. On the other hand, both Eq. (23) and Eq. (28) for SBS show that the threshold power varies as n_0^{-1} (when Landau damping dominates).

Nonlinear Behavior

Analysis of the nonlinear regime[9] shows that the growth rate (both temporal and spatial) varies with pump amplitude in the manner shown in Fig. 11. Region B is the region of linear growth rate described by Eqs. (18) and (19). Since the electrostatic wave, say ω_i, is driven by the ponderomotive force, the ion wave equation is of the form

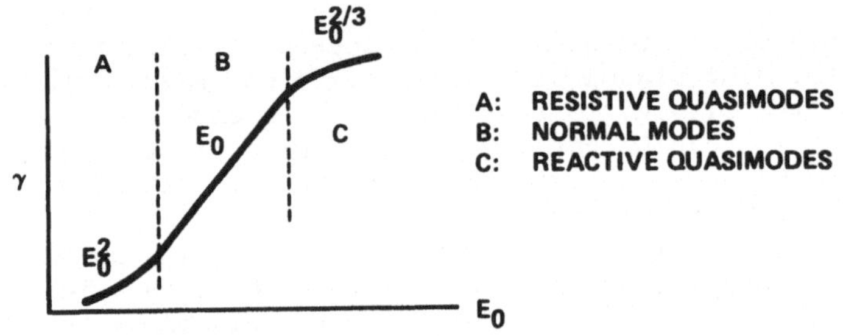

Fig. 11. Behavior of the growth rate of parametric instabilities with pump amplitude.

$$\omega^2 - \omega_i^2 + 2i\gamma_i\omega \propto F_{NL} \propto E_0^2 \quad . \tag{30}$$

In region B, the damping term is negligible, and the quadratic equation yields $\gamma \propto E_0$. In region A, only broad, damped responses rather than normal modes are excited. In this case, $2i\gamma_i\omega$ term dominates in Eq. (30), and γ is proportional to E_0^2. In region C, the ponderomotive force is so strong that the right-hand side of Eq. (30) determines the frequency ω rather than the natural frequency ω_i. The $E_0^{2/3}$ dependence of γ is a result of optimizing the phase shift $\omega-\omega_i$, a process for which we have not found a simple physical explanation. The result is that a factor ω appears in the denominator of the right-hand side of Eq. (30), making the equation a cubic in ω, so that $\gamma \propto E_0^{2/3}$. This behavior at large amplitudes is quite general[6].

Nonlinear saturation levels in collisionless plasmas with $\omega_0 \simeq 10\omega_p$ have been investigated by numerical simulations in one dimension[9,13]. Although SRS grows faster than SBS, the former is easily saturated so that no more than 50% of the incident energy is reflected. Possible saturation mechanisms for SRS are electron trapping, electron heating followed by increased Landau damping, mode-coupling followed by nonlinear Landau damping, and modulational instabilities. The last mechanism[8] is the refraction of plasma waves away from regions of high density because the index of refraction for them is less than unity. The waves then pile up in regions of low density, and the ponderomotive force causes the density to decrease further there. The large resulting density perturbations stabilize the backscattering by the inhomogeneity effect. Ion waves are not subject to these saturation mechanisms, and consequently SBS is expected to grow to large amplitudes. In numerical simulations, there are relaxation oscillations with reflected power reaching 99%. Given enough time, however, the light pressure eventually causes the incident wave to bore through any finite thickness of plasma in spite of SBS instabilities. At the point where $\omega_0 = 2\omega_p$, a particularly strong interaction occurs which causes a rift in the plasma. This has a simple physical explanation: SRS occurring at this point generates a scattered wave $\omega_2 = \omega_p$. This wave cannot propagate out of the plasma and is trapped.

SIDESCATTERING

Consider now the orientation of wave vectors as the angle θ between $\underline{k_1}$ and $\underline{k_0}$ is changed, as in Fig. 12. Since $|k_0|$ and $|k_2|$ are fixed at $\simeq \omega_0/c$ for $\omega_0 \gg \omega_p$, the locus of the tip of the $\underline{k_1}$ vector lies on a circle. It is clear from the geometry that $\underline{k_0}$ and $-\underline{k_2}$ are at an angle 2θ.

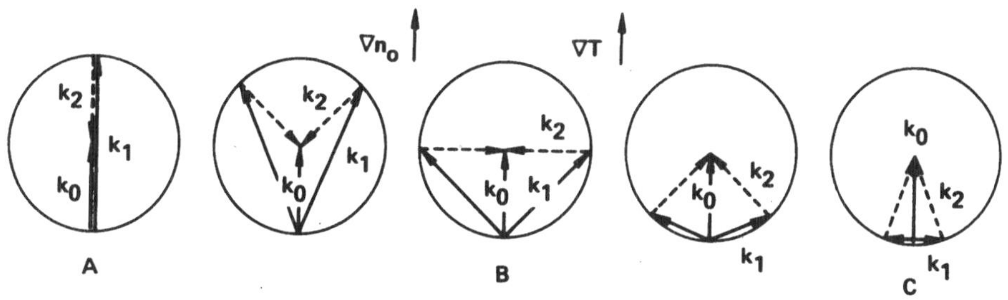

Fig. 12. Vector relations for (A) backscattering, (B) sidescattering, and (C) forward scattering. The incident, electrostatic, and scattered electromagnetic waves are $\underline{k_0}$, $\underline{k_1}$, and $\underline{k_2}$ respectively; and $\underline{k_0} = \underline{k_1} + \underline{k_2}$.

Case A in Fig. 12 is backscattering. Case C is forward scattering; this is a weak interaction because $\underline{k_1}$, and hence the electrostatic field, is small. Case B, sidescattering, is important if $\underline{k_0}$ is parallel to the plasma gradients. The wave $\underline{k_2}$ then propagates at right angles to the gradients and does not suffer the phase mismatch due to plasma inhomogeneity. For SRS, Mostrom et al.[14] find that the growth rate in an inhomogeneous plasma is limited only by the beam radius $\underline{a}$ (cm) and the damping rate γ_e. They have also calculated statistically the required number G of e-foldings from the thermal level. For CO_2, the threshold intensities are:

$$\text{Sidescattering:} \quad I_s > 4.5\, G_s\, \gamma_1/\omega_p a \times 10^{12}\ \text{W/cm}^2 \tag{31}$$

$$\text{Backscattering:} \quad I_b > 0.7\, G_b/L_n \times 10^{12}\ \text{W/cm}^2 \tag{32}$$

where $G_s \simeq 35$ and $G_b \simeq 30$. Eq. (32) agrees with Eq. (29) for $G_b = 2\pi$. Thus, depending on the ratio a/L_n, sidescattering may be as important as backscattering if the plasma is inhomogeneous.

The physical mechanism of sidescattering is illustrated in Fig. 13.

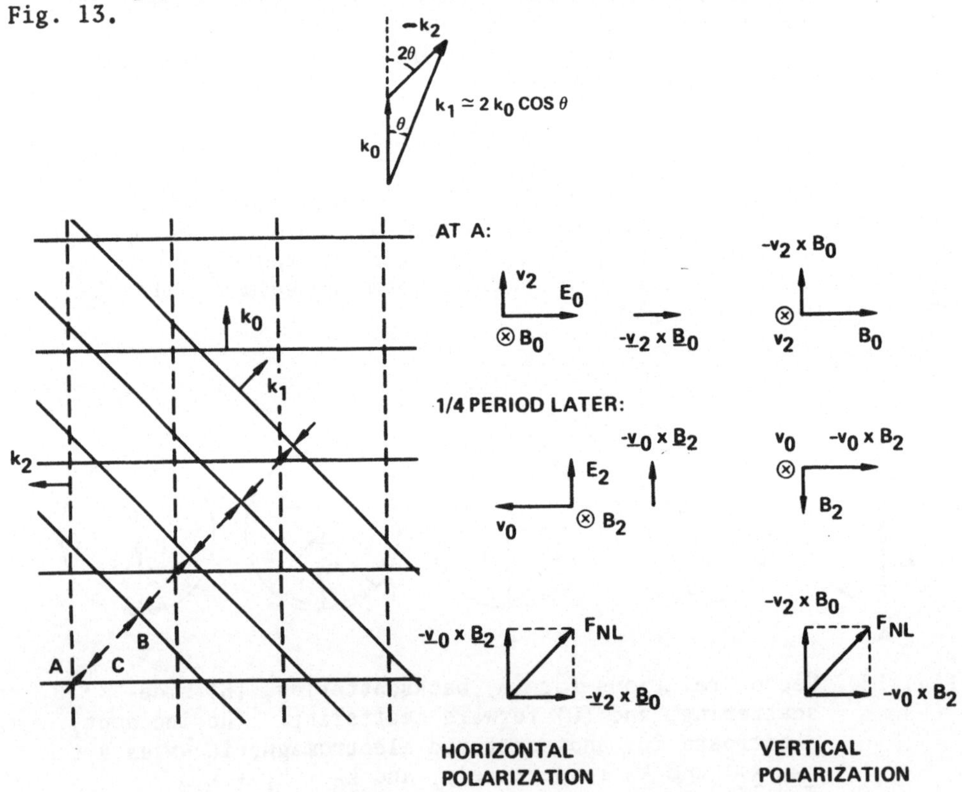

Fig. 13. Physical mechanism of sidescattering.

The wave fronts of the incident wave $\underline{k}_0$, traveling upward, are shown by the solid lines. The dashed lines show the wave fronts of the sidescattered wave $\underline{k}_2$, traveling to the left. For the case of horizontal polarization, $\underline{E}$ is in the plane of the page, and the wave $\underline{k}_2$ is simply rotated by 90° from Fig. 8. At the point A, $\underline{E}_0$ and $\underline{v}_2$ are assumed to be at their maxima. As shown in the center column of Fig. 13, the term $- \underline{v}_2 \times \underline{B}_0$ of the ponderomotive force $\underline{F}_{NL}$, Eq. (17), is then to the right. At the same point A a quarter period later, $\underline{v}_0$ and $\underline{B}_2$ are in phase, and the term $- \underline{v}_0 \times \underline{B}_2$ is upward. Since these two terms are equal in magnitude, the resultant $\underline{F}_{NL}$ is at 45°, as shown. At the point B in Fig. 13, both $\underline{v}$ and $\underline{E}$ are reversed in direction, and $\underline{F}_{NL}$ is the same as at A. At the point C, $\underline{v}$ and $\underline{E}$ are shifted 90° from A in opposite directions; their product, therefore, is the negative of that at A, and $\underline{F}_{NL}$ is in the opposite direction. The resulting pattern of the ponderomotive force causes density striations at $\theta = 45°$, in agreement with Fig. 12B. The wavelength is $\lambda_0/2 \cos\theta$, or $k_1 = 2k_0 \cos\theta$-- also in agreement with the geometrical construction of Fig. 12.

This polarization is not the optimum for excitation, however, because of the $\underline{E}\cdot\nabla\underline{E}$ terms in Eq. (17), which we have neglected. These terms oppose the $\underline{v} \times \underline{B}$ terms in the $\theta = 45°$ case and diminish the net $\underline{F}_{NL}$. The $\underline{E}\cdot\nabla\underline{E}$ terms identically vanish by geometry in the case of vertical polarization, in which both $\underline{E}_0$ and $\underline{E}_2$ are out of the page. In that case, $\underline{B}_0$ - $\underline{B}_2$ are in the plane of the page, and the vectors in the third column of Fig. 13 indicate the total ponderomotive force.

Note that the regular spacing of the density ripples in Fig. 13 is possible in SRS only if $T_e = 0$, so that $\omega = \omega_p$. For finite T_e, the Bohm-Gross dispersion relation requires k_1 to vary with ω_p. The wave fronts of $\underline{k}_1$ must then be curved, and a plane wave $\underline{k}_2$ cannot exist without phase mismatches even in sidescattering.

We wish now to explain the angular dependence of stimulated Brillouin scattering, whose growth rate[7] varies as $\cos\phi(\cos\theta)^{\frac{1}{2}}$ where ϕ is the polarization angle between $\underline{E}_0$ and $\underline{E}_2$. The $\cos\phi$ factor is easily understood, since the entire expression of Eq.(17) for $\underline{F}_{NL}$ is proportional to $\nabla \langle\underline{E}_0\cdot\underline{E}_2\rangle$. The $\cos\theta$ factor is best seen from the $\underline{v} \times \underline{B}$ terms in the case $\phi = 0$, when the $\underline{E}\cdot\nabla\underline{E}$ terms vanish. From Eq. (9) it is clear that $\underline{F}_{NL}$ is proportional to $\langle\underline{E}_0 \times (\underline{k}_2 \times \underline{E}_2)\rangle + \langle\underline{E}_2 \times (\underline{k}_0 \times \underline{E}_0)\rangle$. These terms are of equal magnitude and are in the directions of $\underline{k}_2$ and $\underline{k}_0$, respectively. The phases are such that the averages are of opposite sign. The total F_{NL} is, therefore, proportional to $|\underline{k}_0-\underline{k}_2|$. From Fig. 12, we see that $|\underline{k}_0-\underline{k}_2| = |k_1| \simeq 2k_0 \cos\theta$. The growth rate γ_0 is proportional to F_{NL}, and hence to $\cos\theta$. In Eq. (18), however, we see that γ_0 is also proportional to $\omega_i^{-\frac{1}{2}} = (k_1 c_s)^{-\frac{1}{2}} \simeq (2k_0 c_s \cos\theta)^{-\frac{1}{2}}$. This additional factor is due to the fact that the electric field in the ion wave depends on its wavelength. The growth rate γ_0, therefore, varies as $(\cos\theta)^{\frac{1}{2}}$. This weak dependence on θ, which is valid only for linear ion waves, is in disagreement with the experimental observations of Eidmann and Sigel[15], which show strongly collimated backscattering. If the ion waves are nonlinear, however, it is possible that the fastest growing mode--that with $\theta=0$--is the only one that survives.

FILAMENTATION

Filamentation, or self-focussing, occurs when the incident wave vector lies along incipient density striations (Fig. 14). The refraction caused by the density perturbations of wave number k_1 channels the light beam into the less dense regions. The resulting ponderomotive force pushes plasma away from such regions and increases the density perturbations. In steady-state, F_{NL} balances the electron pressure $\nabla(nKT_e)$, giving rise to the Boltzmann-like relation

$$n = n_0 \exp - \left[\frac{\omega_{p0}^2}{\omega_0^2} \frac{\langle E^2 \rangle}{8\pi n_0 KT_e}\right] \tag{33}$$

which is independent of gradient scale length.

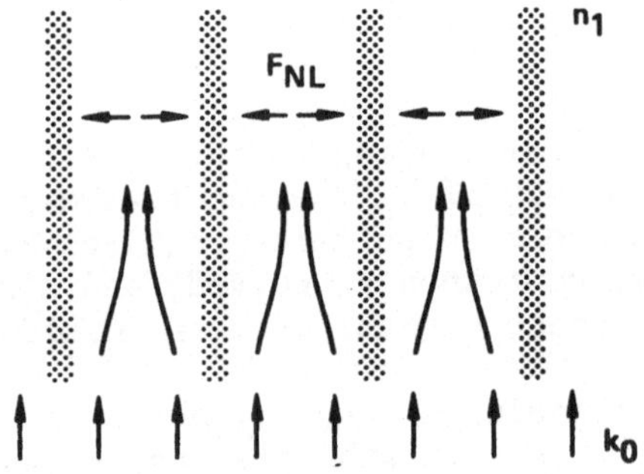

Fig. 14. Physical mechanism of filamentation.

If the light waves are not damped and Debye shielding were perfect, this equilibrium would occur with any intensity; and there would be no threshold for filamentation. The deviation from strict neutrality due to finite Debye length, however, means that F_{NL} must overcome a small electric field proportional to $k_1^2\lambda_D^2$. The filamentation threshold given by Kaw et al.[7,16] is

$$\frac{v_0^2}{c^2} > 4\left(1 + \frac{T_i}{T_e}\right)k_1^2\lambda_D^2 \tag{34}$$

The first density perturbations that grow will be of smallest k_1 and longest wavelength, comparable to the plasma or beam radius. Since v_0^2 is proportional to intensity, or power divided by beam radius, the threshold for self-focussing depends only on total beam power. For a Gaussian beam, this threshold is

$$P_c = 8500(\omega_c/\omega_p)^2 T_{eV} \quad \text{watts} \quad . \tag{35}$$

For $k_1 \approx r^{-1}$, Eqs. (34) and (35) are identical to within a numerical factor. Although the threshold for filamentation is lower than that for absolute parametric instabilities, the growth rate is slower [7,16]:

$$\gamma \approx \frac{1}{2}\frac{v_0}{c}\,\omega_{pi} \quad . \tag{36}$$

This is a factor $(\omega_i/\omega_0)^{1/2}$ smaller than that given in Eq. (18) for SBS.

The preceding picture of self-focussing is well-known; indeed, it was the first application of the ponderomotive force. This phenomenon, however, can also be considered a parametric instability in which the ion wave $\underline{k}_1$ is at right angles to $\underline{k}_0$, as in Fig. 12C. Since $\underline{k}_1$ is small in this limit, the anti-Stokes process $\underline{k}_2 = \underline{k}_0+\underline{k}_1$ is indistinguishable from the Stokes process $\underline{k}_2 = \underline{k}_0-\underline{k}_1$; and both must be considered. Filamentation is, therefore, a four-wave, rather than a three-wave, interaction. [The anti-Stokes diagrams corresponding to Fig. 12 are obtained by reversing the directions of the $\underline{k}_1$ arrows.] The forward-scattered waves $\underline{k}_2$ interfere with the incident wave to produce the refraction and self-focussing in the usual physical picture. The geometry of this instability is identical with that of the OTS instability (Fig. 3), and both are non-convective. However, Since $\underline{k}_2$ is a light wave, the simplification $k_0 \ll k_1$ cannot be made in filamentation.

OPTICAL MIXING AND CASCADING

By using two lasers with a frequency difference, it is possible to couple strongly to an underdense plasma by making $\omega_0-\omega_1 = \omega_p$. This was first suggested by Kroll, Ron, and Rostoker[17] and experimentally tested by Stansfield, Nodwell, and Meyer[18]. Recent work[19,20,21] treats nonlinear saturation and plasma inhomogeneity. Fig. 15 illustrates the basic idea. In the Raman process, two electromagnetic pump waves ω_0 and ω_2 beat with each other to

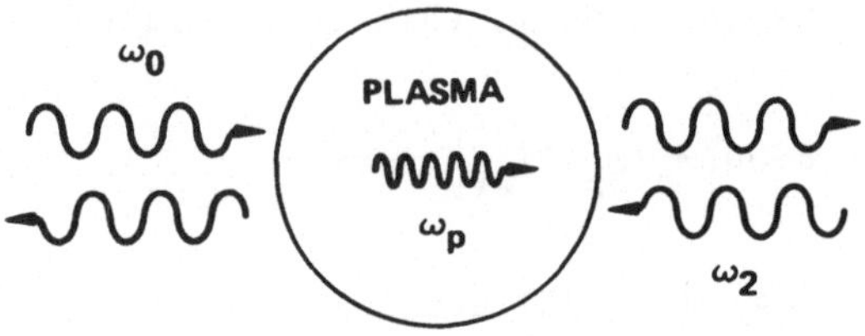

Fig. 15. Laser-beating at $\Delta\omega = \omega_p$.

generate a plasma wave ω_1, such that $\omega_0-\omega_2 = \omega_1 \approx \omega_p$, $\underline{k}_0-\underline{k}_2 = \underline{k}_1 \approx 2\underline{k}_0$. The plasma wave, in turn, interacts with the incident waves to change their amplitudes: the higher-frequency wave ω_0 is damped, while the lower-frequency wave ω_2 is enhanced. The interaction is strongest with $\underline{k}_2 = -\underline{k}_0$, as shown. This process will be recognized as just stimulated backscattering: the backscattered wave ω_2, with the proper frequency shift, is imposed on the plasma, so that it does not have to grow from thermal noise. The threshold, therefore, is lower than in SRS. The theory differs only in that the amplitude of ω_2 is assumed fixed, with the consequence that the ω_p oscillation grows linearly with time rather than exponentially.

It is tempting to try to heat a plasma by this anomalous absorption process. For instance, the 10.6μ and 10.2μ lines of the CO_2 laser can beat to couple with a plasma of density 1.5×10^{16} cm^{-3}, or the 10.6μ and 9.6μ lines with $n = 10^{17}$ cm^{-3}. Unfortunately, the energy given to the wave ω_1 is much less than the energy exchange between ω_0 and ω_2. This is a consequence of the conservation of action. From a quantum mechanical viewpoint, the energy of a wave is $W = N\hbar\omega$, where N is the number of quanta. Since N is conserved, one has

$$\frac{W_0}{\omega_0} = \frac{W_1}{\omega_1} = \frac{W_2}{\omega_2} \quad , \tag{37}$$

where W_1 is small because $\omega_1 \ll \omega_0, \omega_2$. However, the wave momenta $\underline{p} = N\hbar\underline{k}$ are such that they add rather than subtract, and electromagnetic momentum can be transferred to the plasma to achieve low-frequency coupling[22].

To circumvent the restriction on energy coupling, Kaufman et al.[23] suggested a cascade process (Fig. 16). Two incident beams

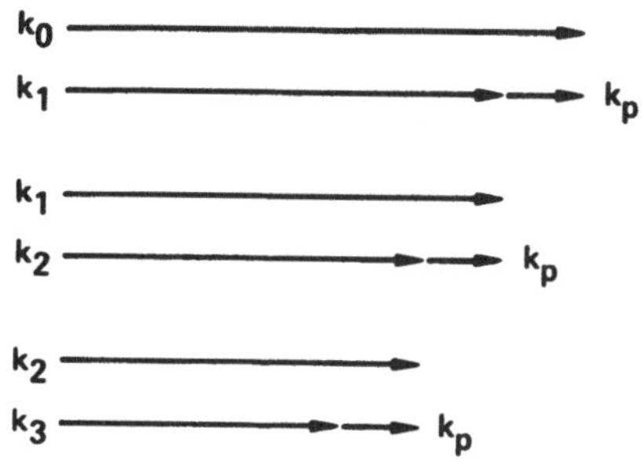

Fig. 16. Decay of wave vectors in cascading.

$\omega_0, \underline{k}_0$ and $\omega_1, \underline{k}_1$ undergo forward scattering to produce a plasma wave $\omega_p = \omega_0 - \omega_1$, $\underline{k}_p = \underline{k}_0 - \underline{k}_1$. The plasma wave then interacts with $\underline{k}_1$ to produce $\omega_2 = \omega_1 - \omega_p$, $\underline{k}_2 = \underline{k}_1 - \underline{k}_p$, and so forth until the laser energy is almost all converted into plasma waves. Repeated k-matching, unfortunately, works only for forward scattering, which is ω_p/ω_0 times less efficient than backscattering. Because of this, the threshold for cascading is rather high.

KINETIC INSTABILITIES

There are a number of parametric instabilities involving resonant particles which cannot be treated by fluid theory. One of these is nonlinear Landau growth, discussed previously. Another is a kinetic modulational instability, which is driven by particles resonant with the group velocity v_g of a large amplitude wave. The physical mechanism is made clear in Fig. 17. The ponderomotive

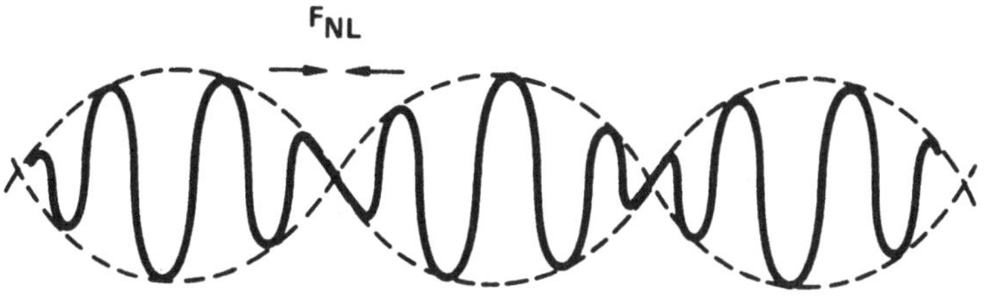

Fig. 17. Mechanism of kinetic modulational instabilities.

force associated with the envelope of a modulated wave interacts with particles traveling near v_g to produce Landau damping or growth, depending on the sign of $f_0'(v_g)$.

The plasma laser is an interesting idea suggested by J.M. Dawson. Let a density ripple of wavenumber k_i exist in a plasma, and let a light wave $\underline{E}_0$ with frequency ω_0 and $k_0 << k_i$ be imposed (Fig. 18). As in the OTS instability, a high-frequency field $\underline{E}_1$ is set up by charge separation. The field $\underline{E}_1$ is the sum of two waves with phase velocities $\pm\omega_0/k_i$ and can therefore interact with particles at these velocities. If the electron distribution is

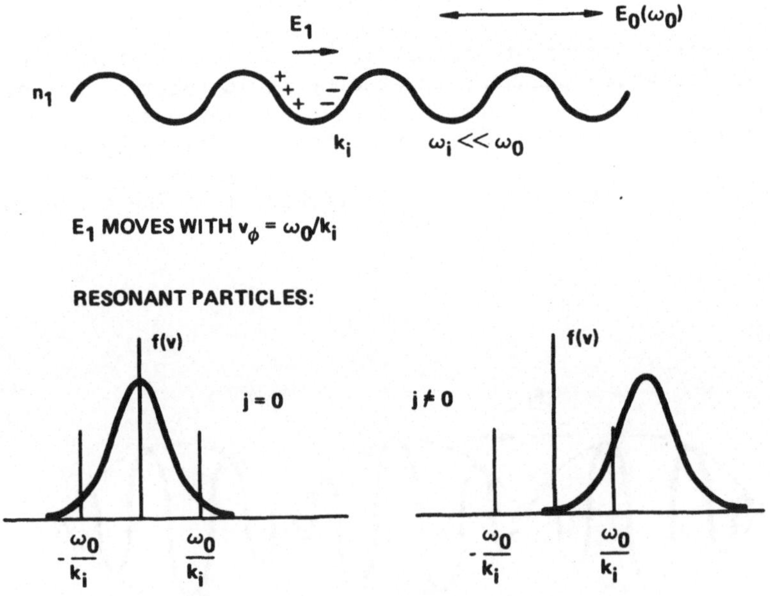

Fig. 18. Mechanism of the plasma laser.

symmetric, no instability occurs. However, if a current exists in the plasma, so that f(v) is shifted as in the lower right-hand part of Fig. 18, the resonant particles can feed energy into $\underline{E}_1$, and hence $\underline{E}_0$. Thus, lasing action occurs without molecular transitions.

The author is indebted to Prof. George Schmidt for a large number of clarifying discussions.

REFERENCES

1. H. Hora, Laser Interaction and Related Plasma Phenomena, ed. by H.J. Schwarz and H. Hora (Plenum Press, New York, 1971), p. 383 ff.
2. G. Schmidt, Physics of High Temperature Plasmas (Academic Press, New York, 1966), p. 47 ff.
3. J.W. Shearer and J.L. Eddleman, Lawrence Livermore Laboratory UCRL-73969 (1972).
4. V.L. Ginzburg, Propagation of Electromagnetic Waves in Plasmas (Pergamon Press, New York, 1964), p. 365.
5. A.Y. Wong and G. Schmidt, UCLA PPG-151 (1973).
6. K. Nishikawa, J. Phys. Soc. Japan 24, 916 (1968).
7. J. Drake, P.K. Kaw, Y.C. Lee, G. Schmidt, C.S. Liu, and M.N. Rosenbluth, UCLA PPG-158 (1973).
8. M.N. Rosenbluth and R.Z. Sagdeev, Comments on Plasma Physics and Controlled Fusion 1, 129 (1972).
9. D.W. Forslund, J.M. Kindel, and E.L. Lindman, to be published; also Phys. Rev. Lett. 30,739 (1973).
10. C.S. Liu and M.N. Rosenbluth, Inst. for Advanced Study COO 3237-11 (1973).
11. D. Pesme, G. Laval, and R. Pellat, Phys. Rev. Lett. 31, 203 (1973); also D. Pesme, Thesis, Univ. of Paris (1973).
12. M.N. Rosenbluth, Phys. Rev. Lett. 29, 565 (1972).
13. W. Kruer, Lawrence Livermore Laboratory, private communication.
14. M.A. Mostrom, D.R. Nicholson, and A.N. Kaufman, Lawrence Berkeley Laboratory LBL-2032 (1973).
15. K. Eidmann and R. Sigel, Inst. f. Plasmaphysik, Garching, Germany, IPP IV/46 (1972).
16. P.K. Kaw, G. Schmidt, and T. Wilcox, UCLA PPG-140 (1972).
17. N.M. Kroll, A. Ron, and N. Rostoker, Phys. Rev. Lett. 13, 83 (1964).
18. B.L. Stansfield, R. Nodwell, and J. Meyer, Phys. Rev. Lett. 26, 1219 (1971).
19. M.N. Rosenbluth and C.S. Liu, Phys. Rev. Lett. 29, 701 (1972).
20. G. Schmidt, UCLA PPG-133 (1972).
21. A.N. Kaufman and B.I. Cohen, Phys. Rev. Lett. 30, 1306 (1973).
22. F.F. Chen, Comments on Plasma Physics and Controlled Fusion 1, 81 (1972).
23. B.I. Cohen, A.N. Kaufman, and K.M. Watson, Phys. Rev. Lett. 29, 581 (1972).

LASER-INDUCED IMPLOSION AND THERMONUCLEAR BURN†*

John H. Nuckolls

University of California Lawrence Livermore Laboratory

Livermore, California 94550

ABSTRACT, INTRODUCTION

In high density laser induced fusion, the key idea is laser implosion of hydrogen isotope microspheres to approximately 10,000 times liquid density in order to initiate efficient thermonuclear burning[1]. Fusion yields 50-100 times larger than the laser energy for laser energies of 10^5-10^6 joules have been achieved in sophisticated computer simulation calculations. Most of the dense pellet is isentropically compressed to a high density Fermi-degenerate state, while thermonuclear burn is initiated in the central region. A thermonuclear burn front propagates radially outward from the central region heating and igniting the dense fuel.

The laser fusion implosion system consists of a tiny spherical pellet of deuterium-tritium surrounded by a low density atmosphere extending to several pellet radii, located in a large vacuum chamber, and a laser capable of generating an optimally shaped pulse of light energy. Figure 1. The atmosphere may be produced by ablating the pellet surface with a laser prepulse.

---

†Presented at the Third Workshop on "Laser Interaction and Related Plasma Phenomena" at Rensselaer Polytechnic Institute, Troy, New York, August 13-18, 1973.

*Work Performed Under the Auspices of the U.S. Atomic Energy Commission.

Reprinted from: *Laser Interaction and Related Plasma Phenomena,* Vol. 3B, H. Schwarz and H. Hora (eds.), Plenum, New York, 1974, pp. 399–425.

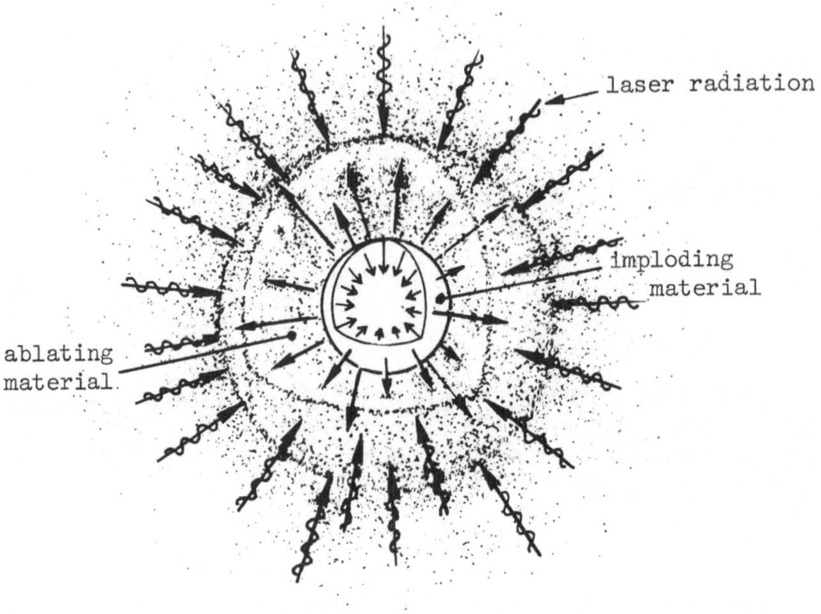

Fig. 1 - Multiple laser beams are focussed on the atmosphere ablated from the surface of the imploding pellet. Laser heated electrons generated in the outer atmosphere transport energy inward through the atmosphere to heat and ablate the pellet surface. Ablation reaction forces implode the pellet.

Mirrors (or lenses) are used to focus the laser light on the atmosphere more or less uniformly from all sides. Absorption of the laser light in the outer atmosphere generates hot electrons. The atmosphere and the pellet surface are heated by electron diffusion and transport. As the electrons move inward through the atmosphere, scattering and solid angle effects greatly increase the spherical symmetry. Violent ablation and blowoff of the pellet surface generates the pressures which implode the pellet. This is essentially a spherical rocket. The laser pulse is shaped (in time) to achieve ultra-high compression of the pellet while fully exploiting electron degeneracy and thermonuclear propagation to maximize the fusion yield.

THERMONUCLEAR BURN

The feasibility of thermonuclear explosions was demonstrated by Los Alamos scientists in the early 1950's[2].

Thermonuclear micro-explosions scale as the density-radius product ρR. The rates of burn, energy deposition by charged reaction products, and electron-ion heating are proportional to the density, and the inertial confinement time is proportional to the radius. Consequently the burn efficiency, self-heating, and feasibility of thermonuclear propagation are determined by ρR. In spherical compression, ρR increases (because $\rho \propto R^{-3}$) and is proportional to $(M\rho^2)^{1/3}$, where M is the mass. Compression by 10^4 reduces the mass--and laser energy--required to initiate an efficient thermonuclear micro-explosion by up to 10^8-fold, depending on the efficiency of the compression process.

The burn efficiency, ϕ, is proportional to the product of the burn rate, $\overline{\rho\sigma v}$, and the inertial confinement time, R/4C, where $\overline{\sigma v}$ is the Maxwell velocity-averaged reaction cross section and C is the sound speed. The factor of 4 arises because in a sphere half the mass is beyond 80% of the radius. Both $\overline{\sigma v}$ and C depend on temperature, but their ratio is approximately constant in the 20-50 keV temperature range characteristic of efficient deuterium-tritium micro-explosions. Using $\overline{\sigma v}/C$ evaluated at 20 keV, and correcting for depletion, it follows that[3]

$$\phi \approx \frac{R\rho}{6+R\rho} \quad .$$

At $\rho R = 3$ g/cm^2, $\phi \approx 1/3$, corresponding to a fusion energy release of $\sim 10^{11}$ joules/g.

The average specific heat energy required for ignition is

$$C_P \, \theta_{ign} \, \beta$$

where C_P is the heat capacity, θ_{ign} is the ignition temperature, and β is a correction for the self-heating by the 3.6 MeV DT alpha particles, and for thermonuclear propagation. If $\rho R >> 0.3$ g/cm^2, then only about 0.3 g/cm^2 in the central region need be heated to approximately 10 keV in order to initiate a radially propagating burn front which ignites the entire pellet. In this case 1.6×10^{10} joules/g of fusion energy will be released from the central region; one fifth of the energy is in alpha particles, sufficient to heat 3 times more DT to 10 keV. The alpha particles will deposit their energy in approximately this mass since their range in 10 keV DT is about 0.3 g/cm^2.

Due to the effects of shock convergence and pulse shaping during implosion, the DT temperature just prior to ignition may be made to vary approximately R^{-2}. Then β is proportional to $(\rho R)^{-2}$. Because of practical limitations on implosion symmetry, a minimum of ~ 0.03 is imposed on β, which occurs for $\rho R \gtrsim 3$ g/cm^2. Then the average ignition energy is 3×10^7 joules/g. This is also the minimum compressional energy of DT at a density of 1000 g/cm^3, and this minimum occurs if the DT electrons are Fermi-degenerate[4]:

$$\varepsilon = \varepsilon_F\left[\frac{3}{5} + \frac{\pi^2}{4}\left(\frac{\theta_e}{\varepsilon_F}\right)^2 + \ . \ . \ .\right]$$

where ε_F, the Fermi-energy, is $\frac{h^2}{8m}\left(\frac{3}{\pi} n_e\right)^{2/3}$, h is Planck's constant, m is the electron mass, θ_e is the electron temperature, and n_e is the electron density. For $\rho = 1000$ g/cm^3 ($n_e \approx 2.5 \times 10^{26}$/cm^3), $\varepsilon_F \approx 1$ keV, and the DT electrons are degenerate if the implosion is carried out so that $\theta_e \ll 1$ keV (except in the central region where ignition occurs).

The minimum average energy of ignition and compression with $\rho R \sim 3$ g/cm^2, assuming propagation and degeneracy, is roughly the sum of the compression and ignition terms, or $\sim 6 \times 10^7$ joules/g. Since the fusion energy produced at this ρR is 10^{11} joules/g, the gain is ~ 1500. Approximately 95% of the laser energy absorbed by the pellet during implosion is lost to kinetic and internal energy of the blowoff. Consequently the energy gain relative to the laser light employed is about 75 fold. This is sufficient for CTR applications with a 10% efficient laser, a 40% thermal-to-electric efficiency, and about 30% of the electrical energy circulated internally.

Figure 2 shows the variation of gain (relative to laser light energy) with compression and laser light energy[1,16]. The curves have been slightly normalized to computer calculations of the implosion and burn. Gains approaching 100 are predicted for laser energies of 10^6 joules. The calculations indicate that less than 1 kJ of laser light may be sufficient for breakeven (gain ≈ 1) and 10^5 joules may be sufficient to generate net electrical energy with a 10% efficient laser. These predicted gains are probably upper limits to what can be achieved. Unforeseen difficulties may cause significant performance degradations when the predictions are experimentally tested.

Similar gain curves may be generated for D_2, DHe^3, and $B^{11}p_x$ pellets seeded with a small percentage of tritium to facilitate ignition. Since these reactions have smaller $\overline{\sigma v}$'s than DT, higher ρR's are required for efficient burn. This may be

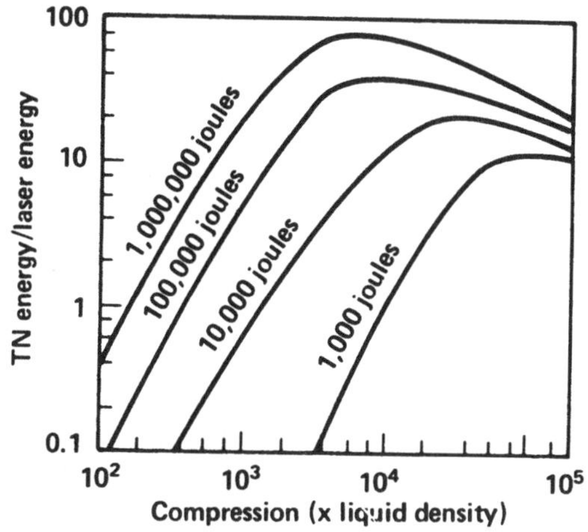

Fig. 2 - Calculated gain (ratio of fusion energy output to laser light energy input) as a function of fuel compression for several different laser light energy inputs. At compressions less than about 10^3, the gain increases strongly with increasing compression because of increasing burn efficiency and propagation. The gain decreases with compressions much greater than 10^4 because the energy of compression (against degeneracy pressures) becomes larger than the energy of ignition.

achieved either by use of larger pellets and higher energy lasers, or by compressing the pellet to higher densities ($\sim 10^4$ g/cm^3). The D_2 and DHe^3 fuels produce fewer neutrons than DT, and the $B^{11}p_x$ fuel produces essentially no neutrons[5].

IMPLOSION

Conditions involving pressure, symmetry, and stability must be satisfied in order to implode a DT sphere to a state at 10^4 times liquid density, in which both Fermi-degeneracy and thermonuclear propagation can be exploited to achieve maximum gain[1]. A pressure of at least 10^{12} atmospheres must be generated. Much higher pressures are required if the electrons in the high density DT are not Fermi-degenerate, i.e. if the implosion is not essentially isentropic. The pressures applied to implode the pellet must be uniform spatially and temporally to less than one part in twenty in order to preserve effective spherical symmetry. During compression the radius decreases by about 20-fold ($10{,}000^{1/3}$), and it is necessary to maintain sufficient symmetry within the central region in order to utilize central ignition and propagation. The hydrodynamic Rayleigh-Taylor instability must be controlled. Otherwise the pellet surface cannot be relatively gradually accelerated during the implosion as required by the optimum pulse shape. These conditions may be satisfied by the characteristics of the laser pulse shape and of the laser implosion scheme.

Pressure

The minimum pressure of DT at 10^4 times liquid density is 10^{12} atmospheres, and occurs if the electrons are Fermi-degenerate. If $\theta = 5$ keV, the ideal ignition temperature for DT, the electrons are not degenerate, and the pressure is 10^{13} atmospheres. The required 10^{12} atmosphere pressure is generated by focussing the high power laser light to high intensities on the outer atmosphere, multiplying the absorbed energy flux by electron transport in the spherically convergent density gradient from the outer atmosphere to the pellet surface, ablating the pellet surface, and multiplying the resulting ablation pressures by the concentration of energy density which occurs in the implosion process.

Absorption. Laser light is focussed to intensities as high as 10^{15} watts/cm^2 on the outer atmosphere. Absorption occurs at densities less than the critical density (where the plasma frequency equals the optical frequency) by inverse bremsstrahlung and by plasma instabilities if the intensity exceeds the instability thresholds. Figure 3 from Shearer[6] shows the inverse bremsstrahlung opacity in matter at one keV temperature for Nd laser light ($\lambda \approx 10^{-4}$cm). The light slows down as it approaches the critical density (10^{21} electrons/cm^3 for Nd light), the electromagnetic field increases, and at high intensities the quivering velocity of the electrons may exceed their thermal velocity. The inverse bremsstrahlung opacity scales as follows:

$$K_{\text{inverse brem}} \sim \frac{Z^3}{A^2} \frac{n^2}{(1-n/n_c)^{1/2}} \lambda^2 \theta_e^{-3/2} \text{ cm}^{-1}$$

where θ_e is the electron temperature, λ is the laser wavelength, n is the electron density, n_c is the critical electron density ($\sim \lambda^{-2}$), A is the atomic weight, and Z is the effective charge. Since in typical laser fusion pellets the scale height of density is comparable to one mm, efficient absorption via inverse bremsstrahlung is not possible with Nd at electron temperatures $\gtrsim$ 10 keV, or with CO_2, at electron temperatures $\gtrsim$ 1 keV. Plasma simulation calculations show that the instability absorption length near the critical density is approximately ten wavelengths[7], or 10^{-3}cm for Nd light. The threshold for the modified two-stream

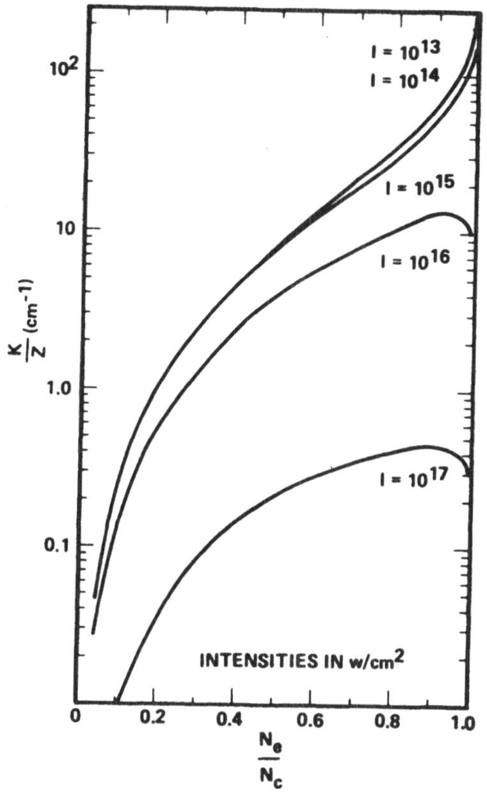

Fig. 3 - Inverse bremsstrahlung opacity for laser light absorption, 1 μm wavelength, one keV electron temperature[6].

instability is approximately[7]

$$\text{Instability threshold} \sim \left(\frac{n_c^3}{\theta_e}\right)^{\frac{1}{2}} Z \times S$$

where the swelling factor, S, corrects for the increasing electromagnetic field as the light approaches the critical density. For 1 keV DT, this threshold is approximately 10^{12} watts/cm^2.

Electron Spectrum. Absorption of intense laser light generates hot electrons. If absorption is by inverse bremsstrahlung the electron spectrum is near Maxwellian

$$f(v_e)_{\text{inverse brem}} \sim v_e \exp\left(-\frac{mv_e^2}{2\theta_e}\right)$$

where v_e is the electron velocity. If absorption is by plasma instabilities the electron spectrum may be non-Maxwellian[8]. Figure 4 shows an electron spectrum calculated by a plasma

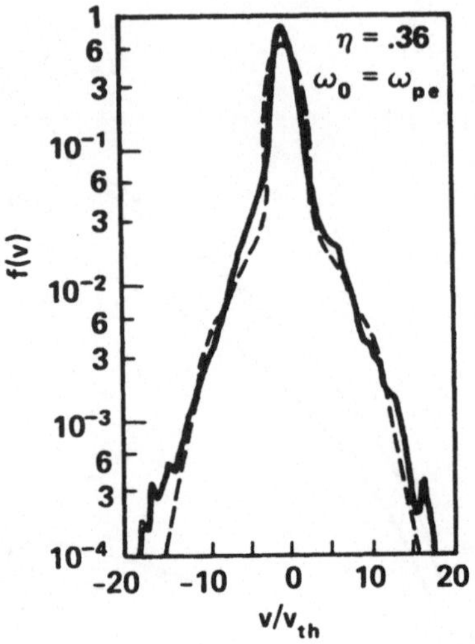

Fig. 4 - Calculated electron spectra generated by absorption of intense laser light via plasma instabilities[8]; v is the electron velocity, and v_{th} the thermal electron velocity. The central core $(v/v_{th} \approx 0)$ is near Maxwellian and the wings are non-Maxwellian. (Dashed is theory).

simulation computer program[8]. An exponential-like tail extends to 100 times the thermal electron energy.

The laser heated electrons heat the pellet atmosphere by electron-electron collisions. The mean free path, λ_{e-e}, is

$$\lambda_{e-e} \sim E_e^2 \, n^{-1}$$

$$\sim 1 \text{ cm}, \; E_e \sim 60 \text{ keV}, \; n \sim 10^{21} \text{ cm}^{-3} \; .$$

where E_e is the electron energy. After a tiny fraction of the electrons escape into the vacuum chamber, electrostatic potentials comparable to θ_e are generated and the remaining electrons are trapped within the atmosphere which expands at ion velocities. Sufficiently high energy electrons will reflect back and forth across the atmosphere many times before losing a significant fraction of their energy.

Ablation Pressure. Typically the density falls off inversely as radius cubed from the ablating radius to the absorption radius. So long as electron-electron collisions are sufficiently rapid energy is strongly coupled by electron transport in this spherically convergent gradient from the absorbing surface to the much smaller ablating surface. Then

$$\dot{E} \approx 2 \, P_A C A_P + \text{smaller terms}$$

where P_A is the ablation pressure, A_P is the ablating surface area, the factor 2 corrects for heating of the blowoff--through which energy is transported--and $\dot{E}$ is the absorbed laser power. The blow-off velocity, C, is approximately sound speed in the hot atmosphere

$$C \approx \sqrt{\frac{P}{\rho}} \sim \theta_e^{1/2} \qquad (\theta_e >> \theta_i)$$

where θ_e is the electron temperature. This temperature is determined largely by the equation for flux limited electron transport

$$n_c\, v_e\, \Delta E\, A_c \approx \dot{E}$$

where $n_c\, v_e$ is the incident electron flux, ΔE is the net energy carried by each electron from the absorption region, and A_c is the absorption area. Then it follows that the ablation pressure is approximately

$$P_A \approx \frac{A_c}{A_p} \left(\frac{\dot{E}}{A_c}\right)^{2/3} \rho_c^{1/3}$$

where ρ_c is the critical density in g/cm^3. If $A_c = A_p$ (plane geometry) this is the equation derived by Kidder[9]. The pressure is multiplied by a factor A_c/A_p by electron transport in the convergent density gradient.

The ablation coupling efficiency is

$$\frac{P_A A_p}{\dot{E}} \approx \frac{v}{2C}$$

where v is the implosion velocity. Typically v varies from 10^6 to 3×10^7 cm/s while C varies from 2×10^7 to 10^8 cm/s, so that the coupling efficiency varies from 2 to 15 percent. Most of the energy is coupled near the end of the implosion, so that the average coupling efficiency is 5-10 percent.

Decoupling. If the hot electron range becomes comparable to the absorption radius, the electron may cross the pellet, re-enter the region in which laser light is absorbed and be further heated before its energy is lost to colder electrons. Then the electrons are heated to even higher energies, have a larger range, and lose energy even more slowly to the colder electrons[10]. The threshold for onset of this decoupling determines a maximum hot electron energy and laser intensity, and thereby a maximum ablation pressure. If self-generated electric fields and anomalous effects are neglected, then the maximum ablation pressure is

$$P_{max} \approx 10^2 \frac{A_c}{A_p} A_c^{1/4} \rho_c^{3/2} .$$

However if the density gradient is steeper than R^{-3}, (as in implosion of hollow pellets), higher pressures may be generated. Decoupling may be a serious problem for CO_2 laser implosions (ρ_c is smaller), and weakly affects Nd laser implosions. For larger pellets and laser energies decoupling becomes less severe because the hydrodynamic time is larger. Decoupling can be compensated for if the initial volume is increased by making the pellet hollow. Then the required implosion velocity may be achieved with limited implosion pressures (acting over a larger volume).

Preheat. Electrons with energies of 50 keV have a range of 100 μ in liquid density DT, comparable to the thicknesses of typical laser fusion pellets. Electrons of this energy are generated by long wavelength lasers even if the electron spectrum is assumed to be Maxwellian because of decoupling effects and by lasers of any wavelength via instabilities if non-Maxwellian electron spectra are assumed. If the flux of these electrons is too high the resulting preheating may make compression to a Fermi-degenerate state impossible, and a high compression extremely difficult[1]. The tolerable fraction of electron energy in hot electrons with range comparable to the pellet thickness is exceeded when the preheat energy flux is comparable to the implosion energy flux:

$$f \approx \frac{P_A v}{P_e v_e}$$

where P_A is the ablation pressure, v is the implosion velocity, P_e is the electron pressure, and v_e the hot electron velocity. In the initial stages of a CO_2 implosion P_A is 1 Mb and v is 1 cm/μs, while P_e is 10^{-2} Mb ($\theta_e \sim 1/2$ keV) and v_e is 10^4 cm/μs, so that f is only one percent. For Nd f is initially 10^{-2}%. Because of the preheat problem, absorption of the laser light by plasma instabilities may not be feasible for some laser fusion pellet designs. Since high intensity CO_2 light must be absorbed by instabilities, it is possible that CO_2 is not suitable for high density laser fusion. In order to increase the instability threshold and the inverse bremsstrahlung opacity, seeding of the pellet with higher Z material, and use of short wavelength lasers is advantageous. The instability threshold may also be increased by frequency modulation of the laser[20]. [However recent studies at Livermore by J. Lindl show that with 1 μ light if $\lesssim$ 20% of the

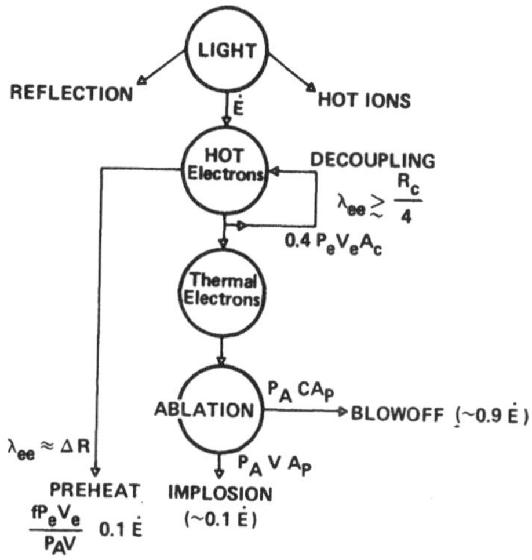

Fig. 5 - Modes of energy transfer and coupling in laser implosion. In the preferred mode ablation occurs via hot and thermal electrons. The implosion may be perturbed or destroyed by the other channels, i.e. reflection, anomalous ion heating, decoupling, and preheat.

electron spectra is non-Maxwellian then the laser implosion may not be strongly degraded.]

Magnetic Field. Spontaneous magnetic fields are generated due to the non-uniform irradiation of the atmosphere with the laser[11]. Gradients in the electron pressure and in the electron density are generated which have a non-zero cross product (via the electron temperature). Then[12]

$$\dot{B} \sim \nabla \times \left[\vec{v} \times \vec{B} + \frac{c}{en_e} \nabla P_e + \cdots\right] .$$

Implosion calculations with long wavelength lasers (e.g. CO_2) show magnetic fields sufficiently large to inhibit electron transport. However the magnetic energy density is much smaller than the electron energy density. The effect of the magnetic field may be reduced by increasing the collision frequency (via short wavelength lasers and seeding with small amounts of high Z material).

Implosion. In the implosion the applied ablation pressure first generates kinetic energy

$$\overline{P}_A\,(V_o - V_f) \sim \frac{1}{2} Mv^2$$

where $\overline{P}_A$ is the average ablation pressure, V_o is the initial volume, and V_f is the final volume. At maximum compression most of this kinetic energy is internal, so that the pressure multiplication ratio is proportional to the compression

$$P_{max} \approx (\gamma-1) \frac{\frac{1}{2}Mv^2}{V_f} \approx (\gamma-1) \frac{\overline{P}_A(V_o-V_f)}{V_f}$$

$$\therefore \quad \frac{P_{max}}{\overline{P}_A} \approx (\gamma-1) \frac{V_o}{V_f}, \quad V_o \gg V_f \quad .$$

Convergence transforms the kinetic to internal energy nearly isentropically--except for a small region near the implosion center. If the matter is compressed to a Fermi-degenerate state (the initial implosion velocity is near sonic and the subsequent ablation pressure increases sufficiently gradually) then

$$P \sim \rho^{\gamma}, \quad \gamma = \frac{5}{3}$$

and

$$\left(\frac{P_{max}}{\overline{P}_A}\right) \approx \left(\frac{\overline{P}_A}{P_{Ao}}\right)^{3/2} \left(\frac{2}{3} V_o \frac{\rho_o}{M}\right)^{5/2}$$

where P_{Ao} is the initial ablation pressure. Large pressure multiplications may be achieved even if $\overline{P}_A$ is limited (e.g. by decoupling) by use of hollow pellets (V_o term). However V_o cannot be made indefinitely large because of symmetry and stability constraints. The multiplication ratio is also increased by making the average ablation pressure large compared to the initial pressure (via pulse shaping). Because of the magnitude of the Fermi-energy in matter under ordinary conditions, the entropy is not significantly decreased by making P_{Ao} much less than one half megabar.

Pulse Shape. The optimum laser pulse shape--which maximizes the fusion yield for a given laser energy--satisfies five conditions.

(1) Densities greater than 1000 g/cm^3 are achieved.

(2) The implosion occurs in less than one sonic transit time so that long wavelength Taylor unstable growth is tolerable.

(3) In most of the compressed pellet the electrons are Fermi-degenerate. This minimizes the laser energy required for compression.

(4) The central core of the compressed pellet--having a radius comparable to the DT alpha range at ignition temperatures--reaches temperatures of $\sim$ 10 keV, initiating rapid thermonuclear burn and propagation. This minimizes the laser energy required for ignition.

(5) The implosion and blowoff velocities are increased together so as to maximize the ablation efficiency.

These five conditions are satisfied by a pulse shape having the following two properties.

(1) The initial implosion velocity is $\sim$ 1 cm/μs, slightly larger than sound speed. This shock is sufficiently weak so that most of the pellet can be compressed to a Fermi-degenerate state, but--with convergence--is strong enough so that the center is significantly shock heated. The peak implosion velocity is sufficiently large ($\gtrsim 3 \times 10^7$ cm/s) so that densities of 1000 g/cm^3 are reached, and central ignition occurs.

(2) The implosion pressure is increased with time so that the hydrodynamic characteristics coalesce within most of the compressed mass, but beyond one alpha particle range from the center. This insures isentropic Fermi-degenerate compression of most of the pellet while igniting the central region.

By means of hundreds of implosion/burn computer calculations, the optimum pulse shape (laser power history) has been determined. In these calculations the pulse shape was represented by an eight element histogram. The amplitudes and durations of the elements were varied to optimize the ratio of fusion energy to laser light energy. For solid DT pellets in which preheat, decoupling, and other effects are not important, the optimum histogram can be approximated by the following equation:

$$\dot{E} = \dot{E}_o \, \tau^{-s}, \text{ where } \tau = 1 - \frac{t}{\tau'}, \; t < t'$$

$$s \sim \frac{3\gamma}{\gamma+1} \sim 2 \; (\text{if } \gamma = \frac{5}{3}), \text{ and } \gamma = \frac{PV}{\varepsilon} + 1.$$

where $\dot{E}_o$ is the initial laser power (which generates an initial implosion velocity of $\sim$ 1 cm/μs), and t' is the implosion time.

If the pellet is hollow or seeded, or if preheat, decoupling, or other effects are important, the optimum pulse shape is significantly modified. Symmetry requirements also affect the optimum pulse shape.

Symmetry

In compression of a sphere by 10^4-fold, the radius decreases somewhat more than 20-fold. If after compression spherical symmetry is required to within 1/2 the compressed radius (1/40 the initial radius) then the implosion velocity (and time) must be spatially uniform (and synchronized) to ∿ one part in 40, or a few percent[1]. In general, for a spherical implosion in which the ratio of initial to final volumes is η, and in which the tolerable error in the final radius, R, is wR, the tolerable fractional error in the integral of velocity and time is approximately

$$\frac{\Delta \int v dt}{\int v dt} \simeq \frac{w}{\eta^{1/3}}, \eta >> 1.$$

Implosion errors may be reduced to about 10% by a many-sided irradiation system, consisting of a laser, beam splitters, mirrors, lenses, and other optical elements. This error can then be reduced to less than 1% by means of electron transport in the atmosphere surrounding the pellet.

Multiple Beam Irradiation. If a sphere is irradiated from all sides with many laser beams focussed to the diameter of the sphere, intensity variations over the spherical surface as small as 10% may readily be achieved. However light which is not incident perpendicular to the spherical surface is refracted in the density gradient[6]. If the incident angle is more than 30° from normal, the light trajectory in the density gradient does not approach the critical density surface. Then absorption by plasma instabilities is not possible. Also absorption by inverse bremsstrahlung is reduced and occurs at low densities, so that the electron decoupling problem is more severe.

Using f/1 optics the entire surface of a sphere cannot be illuminated by as few as six circular laser beams focussed on the center of the sphere. However if each of six beams is focussed to a point one spherical radius beyond the center (along the axis of each beam) then the entire surface is illuminated, and the maximum angular deviation from normal incidence is less than twenty degrees. If the overlapping edges of the beam are blocked (so that each beam is four sided instead of circular) then the maximum error in intensity of illumination is 10-20%. This error may be reduced to less than 10% by using more laser beams, e.g. 12 beams with pentagonal cross section.

Complex lenses and mirrors may also be used to increase the symmetry of irradiation.

Electron Transport in Atmosphere. The region in the atmosphere where the laser light is absorbed has a radius several times that at which ablation occurs. Consequently each point on the ablating surface is heated by nearly 2π steradians of the absorbing region. If many laser beams are used, e.g. 12 pentagons, then each beam occupies a small fraction of 2π steradians, and each point on the ablating surface is effectively heated by many beams. These effects reduce the irradiation error.

The range of the laser heated electrons in the low density region where the laser light is absorbed is comparable to the diameter of one of the multiple focussed laser beams. An initial prepulse may be used to satisfy this condition at the beginning of the implosion (after the atmosphere is formed). Hence non-uniformities which occur over small areas are strongly smoothed. In addition there are many electron scattering mean free paths through the density gradient between the absorbing and ablating surfaces. The smoothing due to scattering during electron transport may be estimated by use of spherical harmonic analysis if a steady state, a short electron mean free path, and a uniform density are assumed[13]. This analysis shows that scattering in the atmosphere reduces the heating error from 10-20% at the absorbing surface to less than 1% at the ablating surface with heating by eight laser beams. Solution of this problem with the real (non-uniform) density gradient would reduce the heating errors even further.

Stability

The unstable growth of small surface perturbations during acceleration of a fluid interface is described by the Rayleigh-Taylor theory[14].

$$A = A_o e^{\int \sigma dt}, \quad \sigma^2 = aK$$

where A_o is the initial amplitude of a perturbation with wave-number K, and a is the acceleration. Unless some stabilization mechanism can be utilized, a fluid shell can be accelerated through only about five thicknesses before it is essentially destroyed by this instability. However in the laser fusion implosion scheme described here the optimum pulse shape may accelerate shells through many tens of thicknesses. Stabilization is achieved by several means.

(1) Growth of perturbations with wavelength less than $\sim 2\pi$ times the shell thickness is prevented because the implosion pressure is generated by surface ablation driven by diffusing electrons. Under these conditions

$$\sigma^2 = aK - \frac{P_A K^2}{\rho}$$

where P_A is the ablation pressure[15]. This stabilizing effect occurs largely because the temperature gradient near the peak of the perturbation is steeper than the gradient near the valley. Consequently the rate of transport of energy is higher to the peak than to the valley, and higher ablation rates and ablation pressures are generated. This reduces the amplitude of the perturbation.

(2) Growth of long wavelength perturbations--which have a slower growth rate--is sufficiently limited by imploding the pellet as rapidly as possible consistent with compression to a Fermi-degenerate state. Consequently the initial implosion velocity is somewhat supersonic.

(3) Pellets are used which have an initial ratio of shell thickness to radius of more than 2%. If thinner shells are used, perturbations with wavelengths greater than 2π times the shell thickness (which are not stabilized by ablation) grow so large that the shell is destroyed.

Because the implosion stability is affected by the symmetry of the pellet irradiation, and because of other complex factors (convergence, compressible DT, finite sound speed, long range electrons, etc.) the theory is uncertain and stability has been demonstrated with two-dimensional computer calculations.

COMPUTER CALCULATIONS

The compression and burn processes which have been described are illustrated in results of a typical computer simulation calculation of the implosion of a fusion pellet to 10,000 times liquid density, and of the resulting thermonuclear micro-explosion. This calculation was carried out at the Livermore Laboratory by Albert Thiessen on the CDC 7600 computer with the LASNEX program developed by George Zimmerman[16]. LASNEX is a two-dimensional (axially symmetric) finite difference code which includes coupled non-linear partial differential equations for the following physical processes (Figure 6)

Hydrodynamics--Lagrangian; real and generalized Von Neumann artificial viscosities; pondermotive, electron ion, photon, magnetic, and alpha particle pressures.

Laser light--absorption via inverse bremsstrahlung and plasma instabilities; reflection at critical density.

Coulomb coupling of charged particle species.

Suprathermal electrons--Multigroup flux-limited diffusion with self-consistent electric fields; non-Maxwellian electron spectra determined by results of plasma simula-

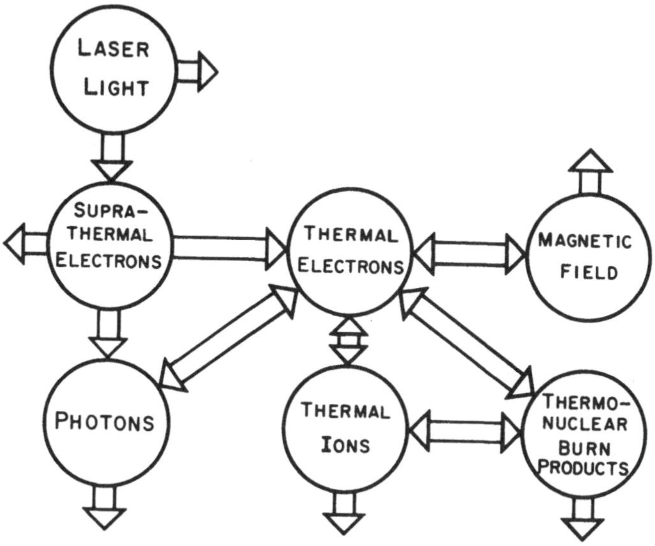

Fig. 6 - Physical processes included in the LASNEX laser implosion /fusion computer program. Arrows represent coupling and possibilities for diagnostics.

tion calculations for laser light absorption by plasma instabilities; inverse bremsstrahlung electron spectrum for classical absorption.

Thermal electrons and ions--flux-limited diffusion.

Magnetic field--includes modification of all charged particle transport coefficients, as well as most of the equilibrium MHD effects described by Braginskii[12].

Photonics--Multigroup flux-limited diffusion[17]; LTE non-LTE average-atom opacities for free-free, bound-free and bound-bound processes[18]; Fokker-Planck treatment of Compton scattering[19].

Fusion--Maxwell velocity-averaged reaction rates; the DT alpha particle is transported by a one group flux-limited diffusion model with appropriate energy deposition into the electron and ion fields; one group transport of the 14 MeV neutron.

Material properties--opacities, pressures, specific heats, and other properties of matter are used which take into account nuclear Coulomb, degeneracy, partial ionization, and other significant effects.

In this implosion calculation, a 54 KJ pulse of laser light is focussed symmetrically onto a low density atmosphere (generated by a laser prepulse) surrounding a 400 μm radius spherical pellet of liquid deuterium-tritium. The applied laser power is optimally increased from $\sim 10^{10}$ to $\sim 10^{15}$ watts in 24 nanoseconds, while the laser wavelength is varied from 4 μm initially to 1/2 micron at the end of the implosion. Figure 7. The initial spike in the pulse shape increases the implosion symmetry by preheating the atmosphere. Long wavelength light also minimizes asymmetries in the early stages of the implosion and short wavelength light avoids decoupling in the late stages. (Recent calculations show that much less frequency variation is required than was used in this calculation.) The laser light is absorbed via inverse bremsstrahlung and plasma instabilities near the critical density in the atmosphere, generating hot electrons with an assumed near-Maxwellian spectra. Coupling between the hot and cold electron distributions due to self-generated electric fields is assumed to be strongly degraded by runaway and other effects, and coupling due to anomalous effects is assumed to be negligible.

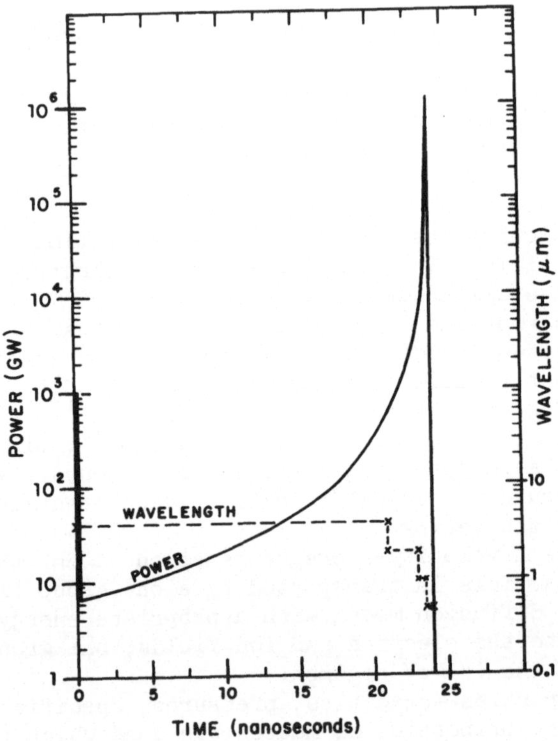

Fig. 7 - Pulse shape: power and wavelength histories. This pulse shape achieves a symmetric implosion in which degeneracy and thermonuclear propagation are fully exploited to maximize the gain.

If these pessimistic assumptions are relaxed, up to 20% of the hot electrons may be generated with a non-Maxwellian (i.e. exponential) distribution without significantly degrading the implosion. If the electron spectrum is hotter than this, then seeding, hollow pellets, frequency modulation, and short wavelength light may be utilized[1]. The atmosphere is heated by the hot electrons to electron temperatures which increase in time from $\sim 10^6$ to 10^8 $^{\circ}$K at the absorption radius. The surface of the pellet is heated and ablated by electron thermal conduction through the hot atmosphere, generating implosion pressures which optimally increase from $\sim 10^6$ to $\sim 10^{11}$ atmospheres. Figure 8. This five order of magnitude increase in implosion pressure occurs at an optimal rate during transit of the initial shock to the center. Consequently the outer part of the pellet is isentropically compressed into a high

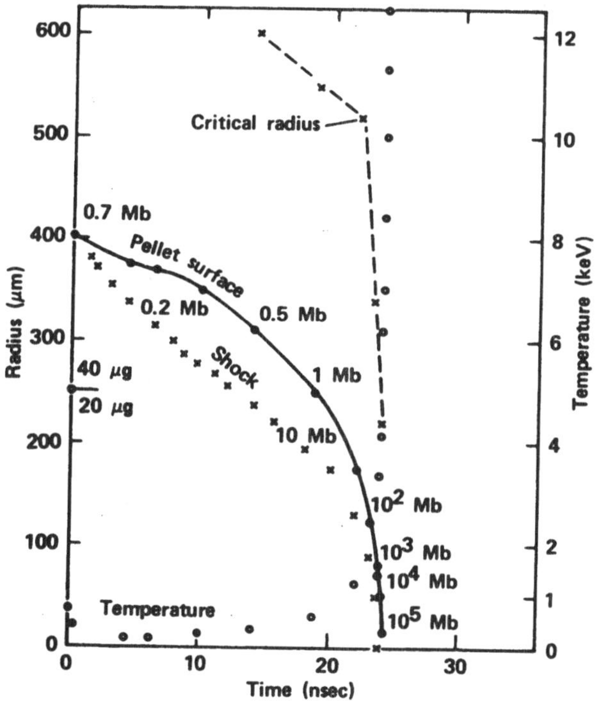

Fig. 8 - Implosion history. The solid line is the surface of the dense pellet and the "x's" inside of this line show the trajectory of the initial shock. Numbers along the solid line indicate the ablation pressures. Circles show the electron temperatures at the radius where the critical density occurs. The dashed line indicates the variation in the critical radius during the implosion.

density spherical shell ($\rho > 100\ g/cm^3$) while at the same time this shell is inwardly accelerated to velocities which increase in time from 10^6 to 3.5×10^7 cm/s. As the internal pressure becomes larger than the ablation pressure the rapidly converging shell slows down and is compressed near isentropically, at sub-Fermi-temperatures, to densities greater than 1000 g/cm^3. Figure 9. The inner region is compressed by the outer shell to densities approaching 1000 g/cm^3, and heated to ion temperatures greater than 10^8 ^{0}K, initiating thermonuclear burn. A thermonuclear burn front then propagates outward. Figures 10, 11. About 1800 KJ of fusion energy is produced in less than 10^{-11} seconds. The energy gain is about 35 fold.

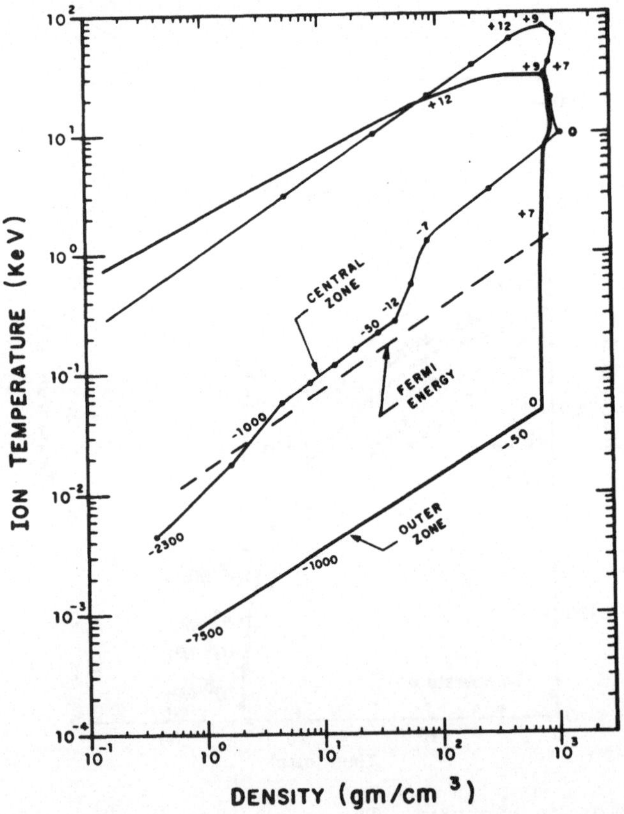

Fig. 9 - Thermodynamic histories of the central and outer regions of the dense pellet. Dashed line shows the Fermi energy. The numbers along curves are times relative to ignition (t=0 ps). The outer material is isentropically compressed to a highly degenerate state while the central material is compressed to thermonuclear ignition conditions.

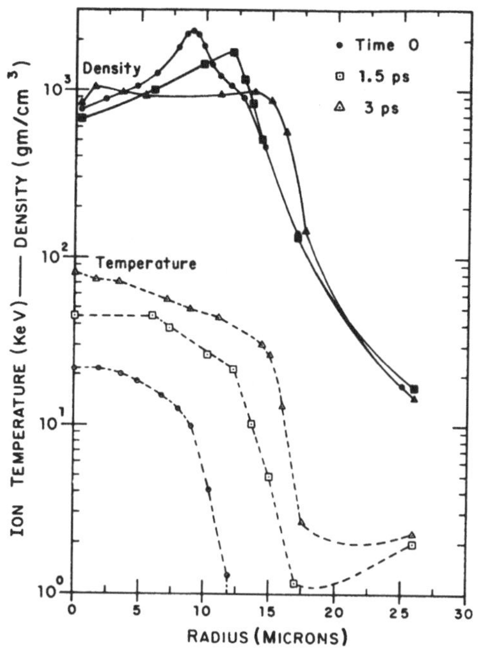

Fig. 10 - Propagation. Density and ion temperature profiles at three times spanning 3 ps during thermonuclear burn.

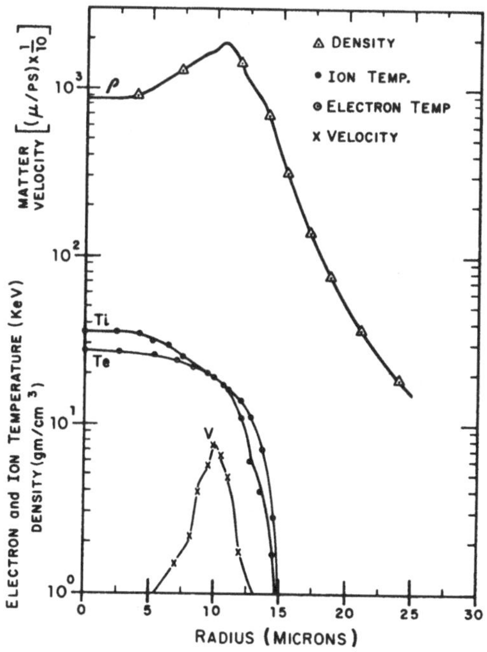

Fig. 11 - Structure of thermonuclear burn front. An electron temperature precursor is caused by deposition of most of the DT alpha particle energy in the electrons at temperatures around 10 keV as well as by thermal conduction.

A two-dimensional calculation has been made of this same pellet configuration to study symmetry and Taylor instability. The pellet was irradiated with ten laser beams consisting of adjacent equiangular (18°) conical shells symmetric about a common axis. The laser intensity varied sinusoidally ± 10% across each beam. Selected frames from a computer movie of the resulting implosion and explosion are reproduced in Figures 12 and 13. A cross section of one quadrant is shown (the horizontal axis is a line of symmetry and the vertical axis is a plane of symmetry). The lines are the Lagrangian mesh and are attached to the matter. The counters show the maximum electron and ion temperatures (TE,TI) and density (D), the inputted laser energy (EL), fusion energy (EP), time, and scale size. The thermonuclear yield produced in this two-dimensional calculation was 1750 KJ, only a few percent less than that in the corresponding one-dimensional calculation. For reasons discussed earlier, the magnetic field physics was not used in this calculation. If the calculation is run with magnetic fields, decoupling effects are observed but the symmetry is not strongly perturbed.

A two-dimensional calculation was made of the same pellet in which the thickness of the initial atmosphere was reduced by a factor of two, the laser wavelength was 1 μ (and was not varied), and the pulse shape did not contain the initial spike. The central region of the pellet is destroyed by a severe axial jet when the radius is 100 μm and the pellet surface is strongly perturbed. Figure 14. Negligible thermonuclear yield is produced.

CTR APPLICATION

Cheap fuel pellets are required for commercial power production. Ten million joules of electrical energy is currently worth roughly one cent, and most of this cent must be used to pay capital and operating costs of the power plant. In the scheme described here the pellet may be a droplet of DT (or deuterium, using multi-megajoule lasers) ejected from a sophisticated eye dropper and spherized by surface tension and viscosity effects, while freely falling in an evacuated drop tower. The pellet may also be a hollow sphere of fusion fuel. The high cost of tritium is obviated by regenerating the consumed tritium by means of neutron reactions in lithium blankets.

Thermonuclear microexplosions producing of the order of 10^{7}-10^{8} joules (5-50 pounds TNT equivalent) may be suitable for commercial power production. Then multigigawatt-electric average power levels could be achieved by burning of the order of 100 pellets per second, perhaps 10 per second in each of 10 explosion

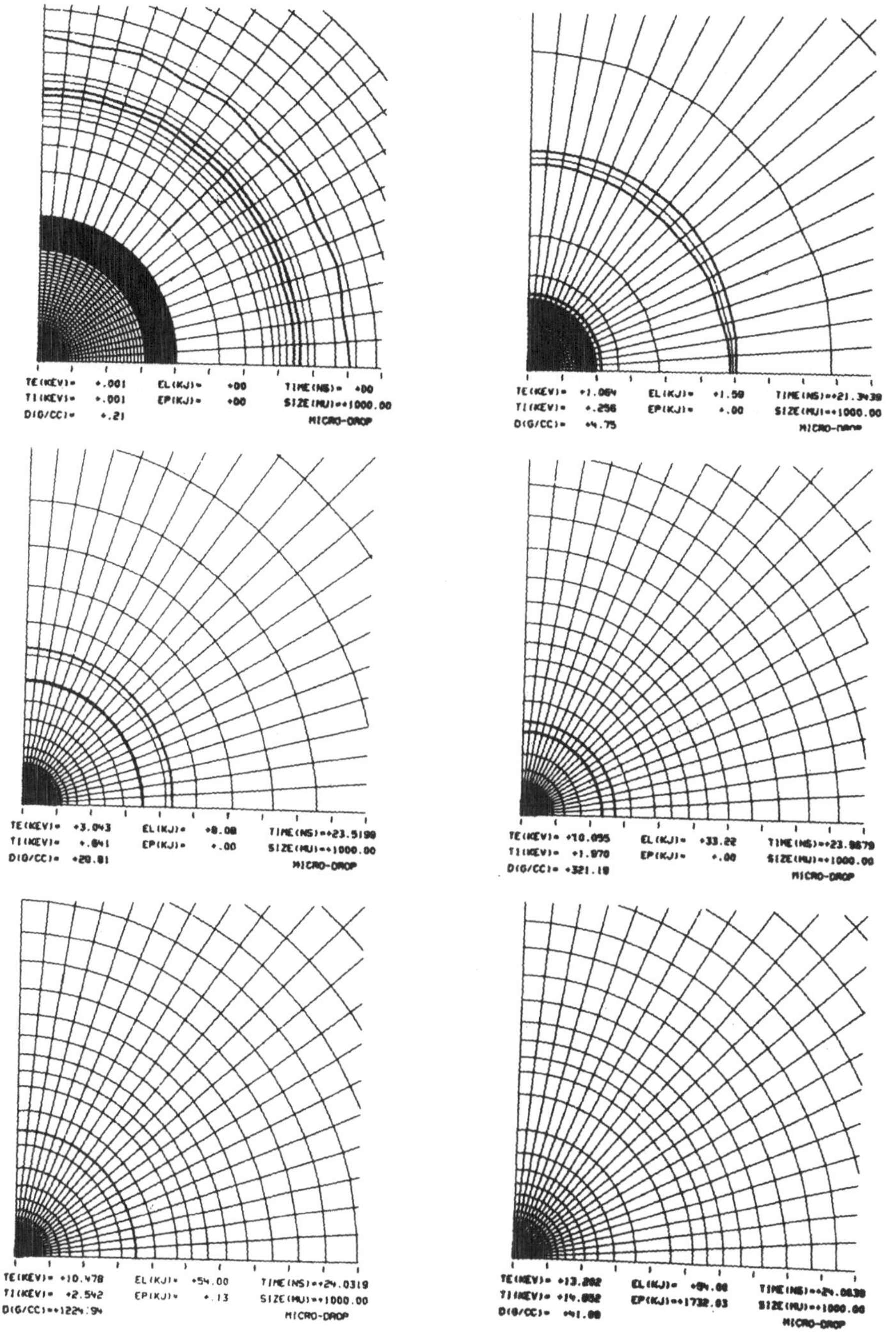

Fig. 12 - Frames from computer movie of two-dimensional implosion of a laser fusion pellet with initial radius 400 μm surrounded by an atmosphere extending beyond 1000 μm. Lagrangian mesh and electron temperature isotherms (dark arcs) are shown (0.1, 0.3, 1, 3, and 10 keV).

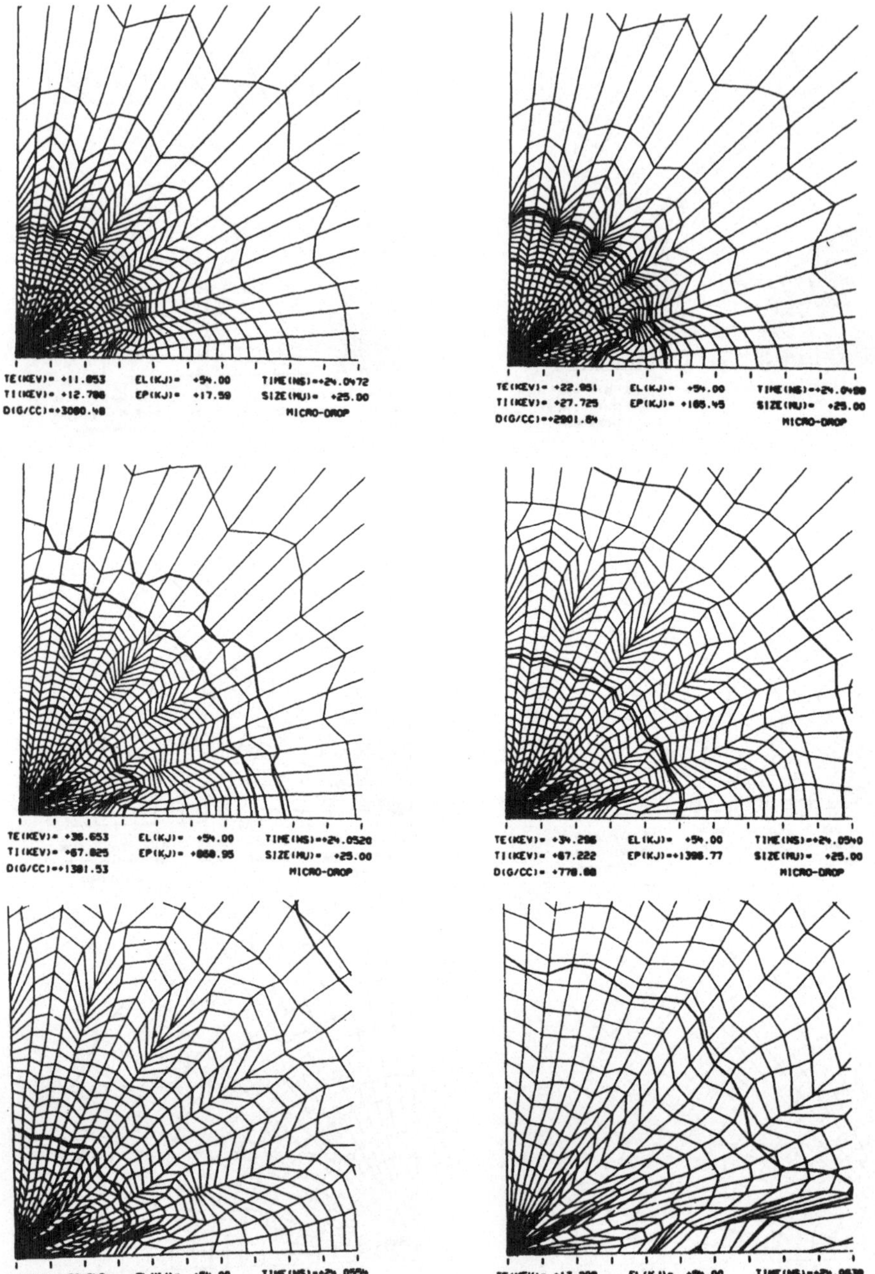

Fig. 13 - Frames from computer movie of two-dimensional calculation of thermonuclear explosion of pellet imploded in Figure 12. Dark arcs are 10, 20, and 50 keV ion temperature isotherms.

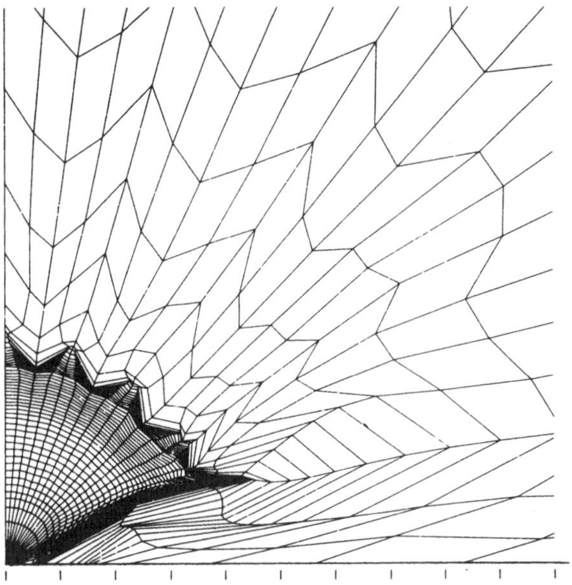

Fig. 14 - Two-dimensional failure of implosion calculation in which atmosphere was too thin and was not preheated, and in which initial laser wavelength was too short.

chambers.

The electrical power, P, is related (via energy conservation) to the laser energy, L, explosion frequency, N, gain, G_L, and thermal-electric efficiency ϕ_L by

$$P = LN\,(G_L\,\phi_t - \phi_L^{-1}) \;.$$

where it is assumed the laser is electrically pumped and the hot gas exhausted from the laser is not used to generate electricity. Less than one third of the electrical energy circulates internally if $G_L\ \phi_t \geq 3\ \phi_L^{-1}$. Then if $\phi_t = 0.4$, $\phi_L = 0.1$, G_L must be 75 or more, which occurs for $L \sim 10^5$-10^6 joules--see Figure one. Consequently if $P = 10^9$ watts, $L = 1/2 \times 10^6$ joules and $N = 100\ sec^{-1}$.

The explosive impulse from a fusion energy pulse is much smaller than that from a chemical explosion of the same energy, because the fusion pellet weight is $< 10^{-6}$ that of the chemical, and the impulse generated is proportional to the square root of the mass. However, a specially designed combustion chamber is required to withstand the neutrons, x-rays, and hot plasma. A chamber several meters in diameter can be designed to withstand the neutron flux. The walls may be protected from x-rays and plasma by a thin film of Li[21].

Conventional thermal cycles may be used to generate electricity if the fusion neutrons are absorbed in lithium blankets. If relatively large DT pellets or essentially pure deuterium pellets are burned (requiring a multi-megajoule laser energy), the fusion neutrons will deposit much of their energy in the fuel plasma, and this plasma may be expanded against a magnetic field, and much of the fusion energy directly converted to electricity[22]. Then higher electrical generating efficiencies may be achieved. With deuterium pellet burning, lithium utilization and tritium storage and cycling may be greatly reduced.

Hybrid reactors--in which the 14 MeV DT neutrons which escape from the combustion chamber are used to fission natural uranium (or thorium)--may be capable of generating more energy than used to pump the laser even with low efficiency, low energy lasers (e.g. 1% efficient, 10 KJ lasers).

A high energy, high efficiency, possibly short wavelength laser which is sufficiently cheap and capable of firing up to 100 times per second must be developed for CTR applications. This laser, the associated electrical power source, and the explosion chamber must be designed to function up to 10^{10} times without replacement. The overall capital cost of this equipment must be less than about $100 per installed kilowatt (electrical). Although these are formidable technological problems, it is believed that they are soluble.

Acknowledgments

I would like to acknowledge important and useful discussions with S. Bodner, J. DeGroot, J. Katz, and W. Kruer. The calculations were made by A. Thiessen and the LASNEX computer program was developed by G. Zimmerman. This work was done in collaboration with my colleague L. Wood who has made many key contributions.

REFERENCES

1. J. Nuckolls, L. Wood, A. Thiessen, G. Zimmerman, Nature 239, 139, (1972).
2. E. Teller, Science 121, 267 (1955).

3. J. Nuckolls, et al., Livermore report UCRL-74116 (1972).
4. J. Mayer, M. Mayer, Statistical Mechanics, Wiley (1940), p. 385.
5. T. Weaver, G. Zimmerman, L. Wood, Livermore report UCRL-74191/UCRL-74352 (1972).
6. J. Shearer, UCRL-72400 (1970).
7. W. Kruer, private communication.
8. J. Katz, J. Weinstock, W. Kruer, J. DeGroot, R. Faehl, Livermore report UCRL-74334 (1972).
9. R. Kidder, Physics of High Energy Density, (1969), Academic Press, p. 315.
10. R. Kidder, J. Zink, Nucl. Fusion 12, 325 (1972).
11. J. Stampen, et al., Phys. Rev. Lett. 26, 1012 (1971).
12. S. Braginskii, Reviews of Plasma Physics 1, 2-5 (1965).
13. L. Wood, et al., Livermore report UCRL-74115 (1972).
14. G. Taylor, Proc. Royal Society 201A, 192 (1950).
15. C. Leith, LLL internal report (1962).
16. G. Zimmerman, Livermore report UCRL 50021-72-1, 107 (1972).
17. D. Post, J. Wilson, LLL internal document (1972).
18. W. Lokke, J. Ramus, LLL internal document (1972).
19. J. Chang, G. Cooper, Jour. of Comp. Phys. 6, No. 1, 1 (1970).
20. S. Bodner, Livermore report UCRL-74074 (1972).
21. L. Booth, et al., Los Alamos report LA-4858-MS, Vol. 1 (1972).
22. B. Freeman, L. Wood, J. Nuckolls, Livermore report UCRL-74486 (1971).

Exact Steady-State Analogy of Transient Gas Compression by Coalescing Waves

L. L. LENGYEL*
Max-Planck-Institut für Plasmaphysik, Garching bei München, Germany

THE possibility of compressing gases to very high densities by converging spherical shock waves was first noted by Guderley.[1] Nuckolls and co-workers[2] suggest compressing matter to extreme densities in very short times (to 10^3 to 10^4 times solid density in 20 to 30 nsec) by a preprogramed train of isentropic spherical compression waves produced by a suitably shaped laser pulse and coalescing at the center of the target. Extreme densities are required for reducing laser power requirements, just as high fluxes are required for efficient radiation absorption and are thus a prerequisite for laser induced thermonuclear fusion. At the temperatures inherent in this process, the matter (deuterium-tritium mixture) behaves like a gas. While in an ideal gas subject to shock compression the attainable density ratio is limited, no theoretical upper bound exists for density changes corresponding to isentropic compression.

The existence of steady-state analogies to certain transient state processes is well-known in continuum mechanics (see, for example, Zel'dovich and Raiser,[3] Zierep,[4] etc.). The respective discussions are, however, usually restricted to the qualitative aspects of the observed similarity. Zierep[4] doubts that quantitative analogies, in general, could be found.

It shall be shown here that the one-dimensional plane variant of the coalescing wave compression has an exact steady-state analogy manifested by the compression of a steady-state two-dimensional supersonic flow over a suitably shaped concave wall. The properties of the steady-state flow may thus be

* Research Scientist.

Reprinted from: *AIAA Journal*, 11, 1347–1349 (1973).

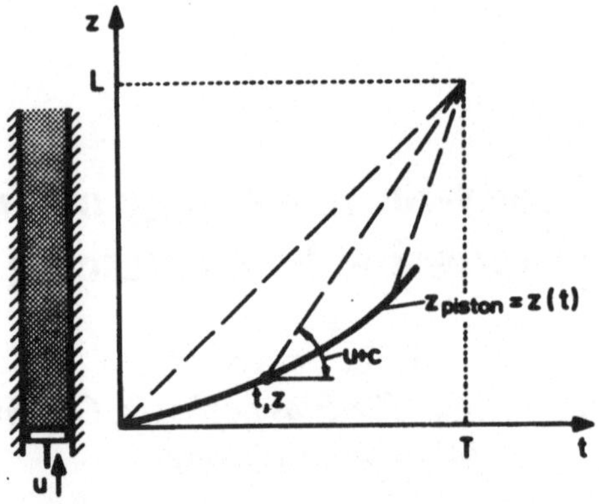

Fig. 1 Transient state compression by coalescing waves.

indicative of the qualities of the respective transient state process. Moreover, analogies sometimes provide simple means for experimental study of some complex phenomena.

A. Transient State Compression

Consider the one-dimensional compression process associated with the accelerating motion of a piston in an ideal gas. The piston path is to be programed in such a way that all compression waves sent forward coalesce at a given distance L measured from the initial piston position. Hence the respective characteristics intersect at a single point of the z, t plane (see Fig. 1). This type of problem involving simple waves can be solved analytically,[5] and analytic solutions have indeed been obtained by various authors.[6,7] For further use, we shall present here the main features of such a solution.

Assume a set of initial conditions: $z = 0$, $u = u_0$, and $c = c_0$ at $t = 0$; where t, z, u, and c denote time, piston position, piston velocity (equal to local gas velocity), and sonic velocity, respectively. The slope of a characteristic (simple wave) originating at z, t and passing through the point L, T can be written as follows:

$$(dz/dt)_{\text{char}} = (L-z)/(T-t) = u_p + c \tag{1}$$

where $L/T = u_0 + c_0$, and $u = u_p = dz/dt$ is the local gas velocity equal to the piston velocity. Since the process is isentropic, u and c are related by the respective Riemann invariant (the specific heat ratio γ is assumed to be 5/3 in this analysis)

$$u - 2c/(\gamma - 1) = \text{const} = u_0 - 2c_0/(\gamma - 1) \tag{2}$$

Eliminating c from Eq. (1) by means of Eq. (2) and introducing the dimensionless variables $z^* = z/L$, $t^* = t/T$ the following first-order differential equation is obtained for the piston path:

$$\frac{dz^*}{dt^*} - \frac{2}{\gamma+1}\left[\frac{1-z^*}{1-t^*} - \frac{1}{M_0+1}\left(1 - \frac{\gamma-1}{2}M_0\right)\right] = 0 \qquad (3)$$

whose solution satisfying the initial conditions is

$$z^* = 1 - \frac{1}{M_0+1}\cdot\frac{2}{\gamma-1}(1-t^*)\left[\frac{\gamma+1}{2}(1-t^*)^{-r} - 1 + \frac{\gamma-1}{2}M_0\right] \qquad (4)$$

where M_0 is the initial Mach number and $r = (\gamma-1)/(\gamma+1)$. Note that $z^* = 1$ at $t^* = 1$, i.e., the piston path also terminates at point L, T (the piston velocity is infinite there). The variation of the velocity, Mach number, pressure, flow work rate, etc. can readily be found from the previous expression and the isentropic flow relations

$$\begin{aligned} p^* &= p/p_0 = (1-t^*)^{-\gamma s}, \qquad s = 2/(\gamma+1) \\ \rho^* &= \rho/\rho_0 = (1-t^*)^{-s} \to \infty \quad \text{as} \quad t^* \to 1 \end{aligned} \qquad (5)$$

B. Steady-State Analogy

Consider now the two-dimensional plane supersonic flow of an ideal gas over a concave surface. The wall shape is to be defined in such a manner that all compression waves originating at the wall intersect at a single point whose coordinates X, Y are given (see Fig. 2). The initial gas state (M_0, etc.) is known.

Prescribing zero wall inclination $\theta = 0$ and parallel flow $\mathbf{u} = (u, 0)$ at $x = 0$, the equation of characteristics intersecting

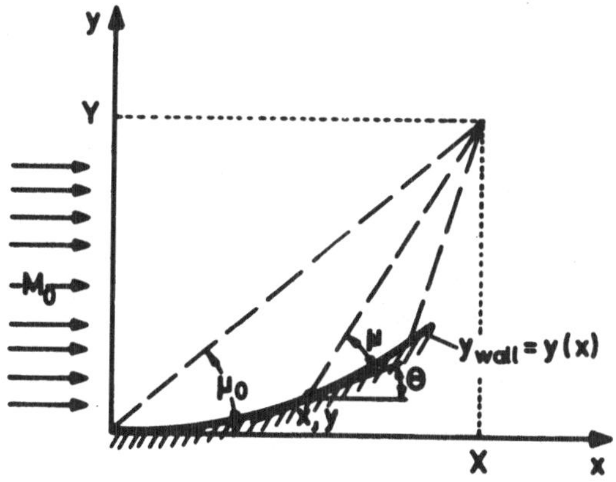

Fig. 2 Steady-state compression of a supersonic flow.

at X, Y can be written at once

$$(dy/dx)_{\text{char}} = (Y-y)/(X-x) = \tan(\theta+\mu) \qquad (6)$$

where $\mu = \arcsin(1/M)$ is the Mach wave angle and $\tan\theta = dy/dx$ represents the wall slope. Note that $X/Y = (M_0{}^2-1)^{1/2}$. Introducing the dimensionless variables

$$y^* = y/Y, \quad x^* = x/X = (M_0{}^2-1)^{-1/2}x/Y$$

the above expression yields the following first-order differential equation for the wall contour line:

$$\frac{dy^*}{dx^*} = \frac{(1-y^*)(M^2-1)^{1/2}-(1-x^*)(M_0{}^2-1)^{1/2}}{(1-x^*)(M^2-1)^{1/2}+(1-y^*)(M_0{}^2-1)^{-1/2}} \qquad (7)$$

Note that in this representation $y^* = 1$ at $x^* = 1$ and, in accordance with the boundary conditions, $dy^*/dx^* = 0$ at $x^* = 0$. The variation of the Mach number as a function of the isentropic pressure change (p_{stg} = const) is given by the expression

$$\{1+[(\gamma-1)/2]M^2\}/\{1+[(\gamma-1)/2]M_0{}^2\} = (p^*)^{(1-\gamma)/\gamma} \qquad (8)$$

where subscript "0" denotes the value at $x^* = 0$. Equations (7) and (8) in combination with Eqs. (4) and (5) are sufficient for showing the analogy proposed.

C. Calculations and Results

The two processes described are analogous both kinematically and dynamically if equal changes of the reduced coordinates $z^* = z^*(t^*)$ and $y^* = y^*(x^*)$ produce equal changes of the gas-dynamic variables p^*, ρ^*, etc. in both coordinate systems; and vice versa. Hence the computation can be carried out in the following manner.

For making the two models consistent, zero initial piston velocity ($u_0 = M_0 = 0$), i.e., zero initial slope of the $z^* = z^*(t^*)$ curve, is assumed for the transient compression case. Selecting now equal fractional increments for the respective independent variables ($\Delta x^* = \Delta t^*$), one computes first the functions $z^*(t^*)$ and $p^*(t^*)$ for a given Δt^* increment. For the pressure ratio thus found one determines, by means of Eq. (8), the Mach number change corresponding to this pressure change in the x^*, y^* plane. In other words, taking equal fractional changes of the independent variables x^* and t^*, i.e., $x^* = t^*$, the combination of Eqs. (5) and (8) yields an expression for the Mach number as a function of the x^* coordinate

$$M^2 = [2/(\gamma-1)](\{1+[(\gamma-1)/2]M_0{}^2\}(1-x^*)^{2r}-1) \qquad (10)$$

Knowing the Mach number change over the distance Δx^*, one determines Δy^* by integrating numerically Eq. (7), and so on. The steady-state freestream Mach number $M_0 = M(x^* = 0)$ is a free parameter in these calculations. As can be seen from the

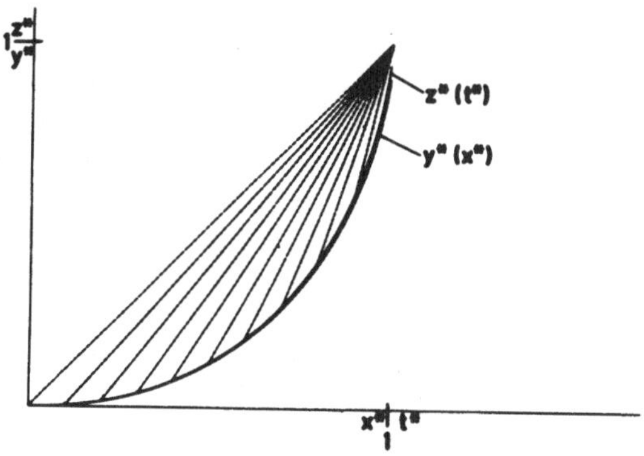

Fig. 3 Reduced piston path $z^*(t^*)$ **and wall contour** $y^*(x^*)$ **for** $M_0 = 10, \gamma = 5/3$.

integrated Prandtl-Meyer expression $\theta - \theta_0 = f(M) - f(M_0)$, where $\theta_0 = 0$ and

$$f(M) = \frac{1}{2}\left[\left(\frac{\gamma+1}{\gamma-1}\right)^{1/2} \arcsin\left(\frac{2\gamma-(\gamma-1)M^2}{2+(\gamma-1)M^2}\right) + \arcsin\left(1 - \frac{2}{M^2}\right)\right] \tag{11}$$

the larger M_0 the better is the correspondence between the unsteady and steady-state processes (with $\gamma = 5/3$ a total turning angle of $\Delta\theta = \pi/2$ corresponds to $M_0 = \infty$ and $M(x^* = 1) = 1$).

Machine plots of z^* (solid line), y^* (broken line), and the respective characteristics computed for $M_0 = M(x^* = 0) = 10$ are shown in Fig. 3. The rather good correspondence is evident.

If the integrated Prandtl-Meyer expression is included in the system of steady-state equations considered, an algebraic solution can be constructed for $y^* = y^*(x^*)$ without integrating Eq. (7).† Indeed, solving Eq. (6) for y^* (or, equivalently, replacing dy^*/dx^* in Eq. (7) by $(M_0{}^2 - 1)^{1/2} \tan\theta$) the following expression is obtained for y^*:

$$y^* = 1 + (M_0{}^2 - 1)^{1/2}(1 - x^*) \cot\left[\theta - \arctan(M^2 - 1)^{1/2}\right] \tag{12}$$

where $\theta = \theta(M)$ and $M = M(x^*)$ are given by Eqs. (11) and (10), respectively. Note that for a given finite M_0 value the Mach number values computed by Eq. (10) become inaccurate as one approaches a limiting x^* value: $x^*_{\text{lim}} = 1 - [(\gamma+1)/(2 + (\gamma-1)M_0{}^2)]$. x_{lim}, of course, approaches unity as $M_0 \to \infty$.

The analogy thus shown is in fact quite natural: the displacement of the piston in an unperturbed gas is replaced in the

† This possibility was noted by the editor reviewing this Note.

steady state case by an equivalent displacement of the wall in the direction normal to the unperturbed motion. Hence some of the known features of supersonic flows over curved surfaces may be generalized and applied to the transient state case as well. It is known, for example, that a supersonic compression flow may be considered as a nearly isentropic one only in regions where the compression waves are spread over a sufficiently large distance and the associated gradients remain sufficiently small. As one approaches the point where the compression waves (or weak oblique shock waves) intersect, the gradients become increasingly steeper and the very condition of isentropicity—vanishingly small gradients—is violated. The compression becomes increasingly nonisentropic in the direction of the intersection point and the flow is compressed with different efficiencies, depending upon the location of the respective streamtubes. For equalizing the pressures downstream from the compression zone, wave or shock reflection takes place at the point of intersection of the compression waves and a slipstream (shear layer) is generated at the point itself.

A similar situation should also be expected in the transient compression case: as time goes on, the spatial extent or spread of the compression wave packet decreases, and the associated gradients rapidly increase in magnitude. As $t^* \rightarrow 1$ the flow conditions deviate more and more from the isentropic ones. From the moment collision of the compression waves, i.e., for $t^* > 1$ an expansion wave must propagate backwards to the piston to adjust the pressure to the shocked value (this applies, of course, only if the piston does not follow the path prescribed by Eq. (4) up to the $z^* = 1$ end position). Hence in the usual gasdynamic approximation a compression by coalescing waves cannot be considered as an isentropic process.

This does not mean, however, that high compression cannot be obtained by coalescing waves. The process can be made more efficient by approximating an isothermal compression,[8] i.e., by heat removal ("entropy removal") from the hot zone during the critical compression phase. Thus the attainable compression ratios depend to a great extent upon the energy exchange processes prevailing in the final compression phase. Detailed numerical calculations performed for spherical and hollow DT targets[2,8] indicate the possibility of reaching density ratios of the order of 10^4 with laser pulses of the order of 10 kjoule output and some nsec duration.

References

[1] Guderley, G., "Strong Spherical and Cylindrical Compression Shocks in the Vicinities of the Sphere Center and the Cylinder Axis" (in German), *Luftfahrtforschung*, Vol. 19, No. 9, 1942, pp. 302–312.

[2] Nuckolls, J., Wood, L., Thiessen, A., and Zimmerman, G., "Laser Compression of Matter to Super-High Densities: Thermo-Nuclear (CTR) Applications," *Nature*, Vol. 239, Sept. 1972, pp. 139–142.

[3] Zel'dovich, Ya. B. and Raiser, Yu. P., *Physics of Shock Waves and High Temperature Phenomena*, Vol. 1, Academic Press, New York, 1966.

[4] Zierep, J., *Lectures on Theoretical Gas Dynamics* (in German), G. Braun Verlag, Karlsruhe, 1963.

[5] Courant, R. and Friedrichs, K. O., *Supersonic Flow and Shock Waves*, Interscience, New York, 1956.

[6] v. Hagenow, K. U., oral presentaticn, July 1972, Max-Planck-Institut für Plasmaphysik, Garching, Germany.

[7] Anisimov, S. I., "On the Transition of Hydrogen into Metallic State in a Laser Pulse Induced Compression Wave" (in Russian), *Letters to the Soviet Journal of Experimental and Theoretical Physics*, Vol. 16, No. 10, 1972, pp. 570–572.

[8] Clarke, J. S., Fischer, H. N., and Mason, R. J., "Laser Driven Implosion of Spherical DT Targets to Thermonuclear Burn Conditions," *Physics Review Letters*, Vol. 30, No. 3, Jan. 1973, pp. 89–92.

Laser-Driven Implosion of Spherical DT Targets to Thermonuclear Burn Conditions*

J. S. Clarke, H. N. Fisher, and R. J. Mason
University of California, Los Alamos Scientific Laboratory, Los Alamos, New Mexico 87544

Calculations predict that carefully timed laser pulses can implode small DT spheres and shells to extreme densities such that thermonuclear burn ensues. We characterize the implosion quality as a function of the pulse time scale, growth law, and initial intensity. Phenomenological rules for mass scaling, input energy threshold, and yield ratio $Y_R = E_{out}/E_{in}$ are presented. We find that $Y_R = 4.7$ for only 1.9 kJ of CO_2 laser input energy to a 3-μg shell. The performance of shells is compared to spheres.

Short-pulse CO_2 and Nd lasers which deliver up to 8 kJ should soon be available.[1] Anticipating this, we discuss the possibility of initiating thermonuclear fusion in small DT targets exposed to properly timed pulses from such lasers.

Theoretical considerations.—At 8 keV the rate of thermonuclear energy production exceeds the bremsstrahlung rate in DT at equilibrium, provided the α-particle reaction products are recaptured by the plasma.[2] This temperature represents 600 J/μg. Thus, conceivably, the proposed lasers could suffice to ignite, say, a 0.75-μg sphere of DT fuel. At solid density, $n = 5 \times 10^{22}$ cm^{-3}, the sphere has a radius $R = 1.13 \times 10^{-2}$ cm. The 8-keV ions have a mean thermal speed $v_{th} = 5.6 \times 10^7$ cm/sec. So the sphere will disassemble by expansion on a time scale $\tau_e \simeq R/v_{th} = 1.7 \times 10^{-10}$ sec. Energy producing reactions occur with the characteristic time $\tau_r = 1/n\langle\sigma v\rangle = 2.9 \times 10^{-7}$ sec, in which the averaged cross-section term $\langle\sigma v\rangle$ increases monotonically with temperature for $T < 65$ keV.[3] Thus, $\tau_e/\tau_r \ll 1$ and any thermonuclear burn in the ambient solid is quenched long before completion. Also, the α-particle mean free path,[4] $\lambda_\alpha = 0.032T^{3/2}$ (keV) = 1 cm, far exceeds the pellet dimensions. Under compression, however, these conditions can be markedly improved since $\tau_r/\tau_e \sim \lambda_\alpha/R \sim n^{-2/3}$. With an extreme 10^4-fold increase in density, for example, $\lambda_\alpha/R = 0.7$ and $\tau_e/\tau_r = 0.27$. α particles from the center of the sphere redeposit their energy, and roughly 27% of the DT fuel is consumed. Complete burn yields 326 kJ/μg, so ~60 kJ should be produced in the sphere.

The utility of high compression is not a new concept. Early work on fission weapons at Los Alamos in 1943 by S. Neddermeyer[5] explored the possibility of improving the neutron economy and enhancing the rate of fission energy release by compressing a solid ball of fissile material to high density using a spherical implosion driven by detonating high explosives. This technique did indeed achieve high compressions and as a consequence the technique has been studied and used a great deal since. It is for this reason that the concept of using the laser energy to compress a DT pellet to improve the retention of α-particle energy and to increase the rate of thermonuclear energy release occurred naturally to workers at Los Alamos and Livermore. The maturing of this idea over the past ten years coupled with the development in laser technology and the experimental and theoretical investigation of the interaction of laser energy with matter has led to the serious consideration of this concept as an approach parallel to that of magnetic confinement for the production of electrical energy for utility use by the fusion process. The first discussion of these ideas in the journal literature appears in an article by Nuckolls *et al.*[6] This Letter presents Los Alamos work on the laser-driven ablative implosion of DT spheres and shells.

Figure 1 demonstrates this process for a shell. The laser heats electrons on the outside of the shell by inverse bremsstrahlung and anomalous mechanisms.[7-10] This heating travels inward as a thermal wave in competition with an outward-

Reprinted from: *Physical Review Letters*, **30**, 89–92 (1973).

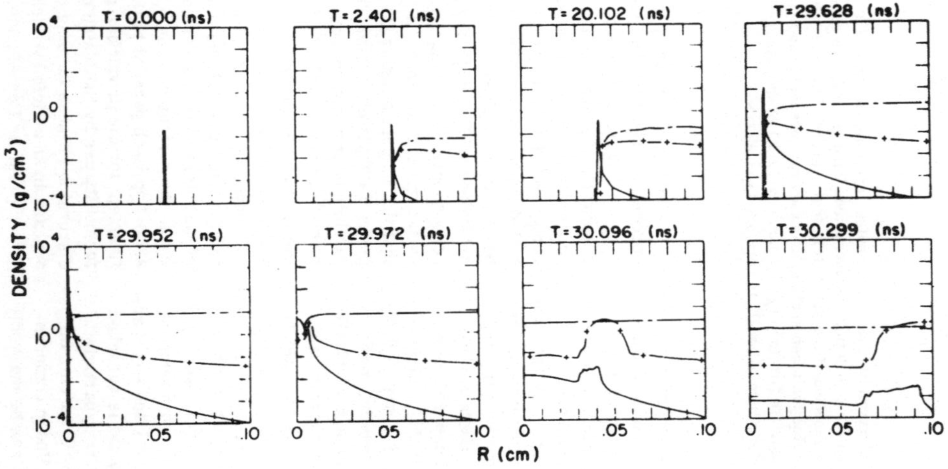

FIG. 1. Time history of the density and temperatures of a 7.5-μg DT shell subjected to 5.3 kJ of CO_2 laser energy; plusses, ion, and dashes, electron temperatures in keV.

moving expansion of ablating ions.[11] The ions are thermalized by electron-ion collisions. The inner ~10% of the shell (its core) is compressed and accelerated[12] by the gradient of electron and ion pressures, and by the reactive force to the ablating ions.

Optimized timing of the laser pulse can significantly improve the quality of implosions. Compression is easiest when the DT core remains cool. In spheres some heating will derive from a first shock generated at the low initial laser intensity. Thereafter, the laser exposure profile can be tailored to keep the subsequent compression of the target core adiabatic to a maximal degree. This is accomplished when the rising laser intensity continuously generates weak, overtaking shocks which first coalesce to a strong shock just before the center. Then, only, as this shock collapses at the center are the high temperatures produced which are required for ignition.

In the adiabatic compression of a sphere or shell, the core temperature obeys $T \sim n^{\gamma-1} \sim R^{-3\gamma+3}$, where γ is the effective ratio of specific heats. The shock overtaking process keeps the mean speed of the compressed portion of the core following $v \sim T^{1/2} \sim R^{(-3\gamma+3)/2}$. Since $v \equiv -dR/dt$ with $R \rightarrow 0$ at time τ, we can therefore solve for the velocity dependence, obtaining $v \sim (1-t/\tau)^{-q}$, $q = (3\gamma-3)/(3\gamma-1)$. Finally, because work is done on the core at a total rate $\dot{W} = 4\pi R^2 P v$ with $P \sim nv^2$, and since we expect this to be proportional to the energy input rate from the laser $\dot{E}(t)$, we conclude that an optimized laser exposure profile is

$$\dot{E}(t) = \begin{cases} \dot{E}_0(1-t/\tau)^{-p}, & E \le E_{in}, \\ 0, & E > E_{in}, \end{cases} \tag{1}$$

where $p = (9\gamma-7)/(3\gamma-1)$ ($=2$ for $\gamma = \frac{5}{3}$), $\dot{E}_0$ is some appropriate initial input power, and E_{in} is the chosen total laser input energy.

Numerical results.—We have made extensive studies of the implosion of DT spheres and shells by computer simulation. The calculations have been performed with a one-dimensional, Lagrangian hydrodynamic, three-temperature (electron, ion, and Planckian radiation) code, that includes energy exchange among the fields (Coulomb, bremsstrahlung), radiation diffusion, classical ion and electron conductivity, and time-dependent nonlocal α-particle energy deposition, with Fermi degeneracy effects included in the equations of state. For the results presented, CO_2 laser light was absorbed by inverse bremsstrahlung up to the critical density n_{crit}, with the remainder deposited at n_{crit} by assumed anomalous mechanisms.

Figure 1 describes the implosion of a 7.5-μg DT shell of initial inner radius $R = 5.4\times10^{-2}$ cm and thickness $\Delta R = R/56$. The laser input energy is 5.3 kJ. The pulse parameters are $\tau = 30$ nsec, $p = 1.875$, and $\dot{E}_0 = 5.7\times10^{8}$ W. The peak power is 3.85×10^{13} W. With the assumed energy dump-all and classical conductivity, this raises the electron temperature at the critical density to a 10-keV maximum. A highly optimized compression results bringing the inner 0.75-μg core of the shell to 10^4 times solid density with a central peak density at 6×10^3 g/cm^3. The mean ion temperature in the core under full compression is 8 keV, although a 30-keV temperature is recorded at the pellet's center just after the shock collapse. The burn yields 58 kJ, in accordance with our theoretical considerations. Thereafter, a blast wave is seen to develop in the density coincident with a region of elevated ion temperature. The higher conductivity of the electrons rapidly flattens their temperature profile.

In Fig. 2 we have collected the data from numerous runs. We have found the quality of implosion to be relatively insensitive to the choice of p and τ in (1) over the ranges $1.5 < p < 2.5$ and $0.5\tau_0 < \tau < 3\tau_0$, where τ_0 is the transit to the center of spheres of the first shock launched with $\dot{E}(t) = \dot{E}_0$. The yield ratio, $Y_R = E_{out}/E_{in}$, is, however, quite sensitive to $\dot{E}_0$, as apparent from Fig. 2(a), which gives results for shells of differing mass. Clearly, a proper choice for $\dot{E}_0$ is more essential at smaller masses. For a given mass m a threshold energy E_{in}^* is required for good returns; more input energy than this is wasted. In general, $E_{in}^*/m = 0.7$ kJ/μg seems to apply. Figure 2(b) shows this result for a 60-μg sphere. The optimal initial intensity $\dot{E}_0^*$ has been found to obey the strict mass dependences (over 3 orders of magnitude) $\dot{E}_0^* \sim m^{1.50}$ for spheres, and $\dot{E}_0^* \sim m^{1.38}$ for the shells, as derived from Fig. 2(c). For a fixed mass, inner radius, and choice of p, the time dependence $\dot{E}_0^* \sim \tau^{-p}$ is obeyed. From this, integrating (1), we can conclude that with fixed $\dot{E}_{in}^*$ the optimal pulse closes at a fixed power level, $\dot{E}^*(t = t_{in})$, independent of the time scale.

Figure 2(d) shows the mean temperature versus mass-density behavior of the 6-μg core of a 60-μg spherical pellet under (1), employing $\tau = 20$ nsec, $p = 2$, and $\dot{E}_0^* = 1.3\times10^{10}$ W. The DT is first shock heated to 10^{-2} keV, then rides up the $\gamma = \frac{5}{3}$ adiabat to 2 keV, where the final shock collapse and burn raises T_i to 95 keV, prior to the expansive disassembly of the core. Clearly, the expansion pursues a similar adiabat down to $\rho = 0.1$ g/cm^3.

Optimal yield ratios (from pulses using $\dot{E}_0^*$) are given versus $\dot{E}_{in}^*$ in Fig. 2(e). Here the cal-

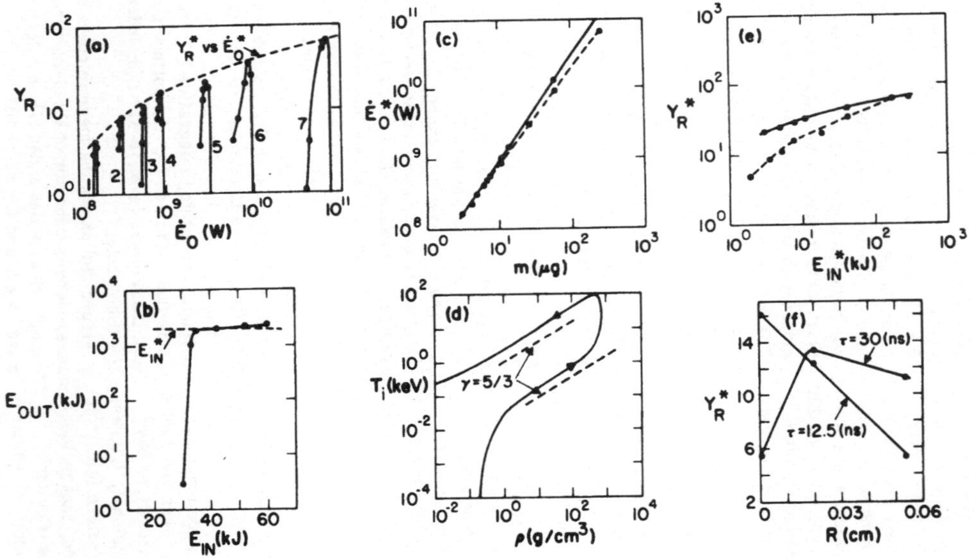

FIG. 2. DT shell and sphere performance characteristics: (a) Y_R versus $\dot{E}_0$ for (1) 3-μg, 2.2-kJ, (2) 5-μg, 3.5-kJ, (3) 7.5-μg, 5.3-kJ, (4) 10.8-μg, 7.5-kJ, (5) 26-μg, 18.2-kJ, (6) 60-μg, 43-kJ, and (7) 250-μg, 178-kJ shells of initial inner radius $R = 5.4\times 10^{-2}$ cm under profile (1) with $\tau = 30$ nsec and $p = 1.875$. (b) Input energy threshold, $\dot{E}_{in}^*$, determination. (c) Optimal $\dot{E}_0$ versus mass: solid line, for spheres ($\tau = 20$ nsec, $p = 2$); dashed curve, for the shells of (a). (d) 6-μg core trajectory, T_i versus ρ, for a 60-μg sphere during its implosion, thermonuclear burn, and expansion. (e) Optimal yield ratios versus input energy: dashed curves, for the shells [in (a)]; solid curves, for spheres. (f) Y_R^* versus inner radius for 7.5-μg shells imploded with the optimized (1) profile for $\tau = 30$ and 12.5 nsec.

culations for spheres employed local α deposition, which artificially improves their performance at small m. For the shells with nonlocal, time-dependent deposition, the rules $Y_R^* \sim (E_{in}^*)^{0.45}$, $E_{in}^* > 8$ kJ, and $Y_R^* \sim (E_{in})^{0.90}$, $E_{in} \leq 8$ kJ, are observed. We calculate $Y_R^* = 4.7$ for only 1.9 kJ delivered to a 3-μg shell. Thus, breakeven with only 350 J of input energy is implied. With nonlocal deposition and parameters other than R and $\dot{E}_0^*$ held constant, our studies show degraded performance for shells relative to spheres under the exposure profile (1), as documented by Fig. 2(f).

Thus, within the constraints of the physical model described, our calculations imply the feasibility of controlled thermonuclear burn by inertial confinement. Scaling laws for a broad range of potential target designs have been presented. Still numerous questions remain, as to the three-dimensional stability of these designs, the details of anomalous light absorption,[13] hyperthermal electron production and deposition,[14] and the effects of spontaneously generated magnetic fields.[15]

We acknowledge valuable assistance from Dr. R. S. Cooper, Dr. R. L. Morse, and Dr. R. Pollock and helpful discussions with Dr. F. Evans, Dr. H. H. Rogers, and Dr. R. N. Thorn at Los Alamos.

*Work performed under the auspices of the U. S. Atomic Energy Commission.

[1]K. Boyer, Bull. Amer. Phys. Soc. 17, 1019 (1972).

[2]S. Glasstone and R. H. Lovberg, *Controlled Thermonuclear Reactions* (Van Nostrand, Princeton, N.J., 1960), p. 33.

[3]J. L. Tuck, Nucl. Fusion 1, 202 (1961).

[4]M. S. Chu, Phys. Fluids 15, 413 (1972).

[5]As described in D. Hawkins, LASL Report No. LAMS-2532, 1961 (unpublished), Vol. I, p. 23.

[6]J. Nuckolls, L. Wood, A. Thiessen, and G. Zimmerman, Nature (London) 239, 139 (1972).

[7]J. P. Silin, Zh. Eksp. Teor. Fiz. 48, 1679 (1965) [Sov. Phys. JETP 21, 1127 (1965)].

[8]P. K. Kaw and J. M. Dawson, Phys. Fluids 12, 2586 (1969).

[9]J. P. Freidberg and B. M. Marder, Phys. Rev. A 4, 1549 (1971).

[10]J. P. Freidberg, R. W. Mitchell, R. L. Morse, and L. I. Rudinski, Phys. Rev. Lett. 28, 795 (1972).

[11]J. L. Bobin, Phys. Fluids 14, 2341 (1971).

[12]R. S. Cooper, LASL Report No. LA-DC-72-104, 1972 (unpublished).

[13]D. W. Forslund, J. M. Kindel, and E. L. Lindman, Bull. Amer. Phys. Soc. 17, 1044 (1972).

[14]R. L. Morse and C. W. Nielson, "Occurrence of High-Energy Electrons and Surface Expansion in Radiantly Heated Target Plasmas" (to be published).

[15]J. A. Stamper, K. Papadopoulos, R. N. Sudan, S. O. Dean, E. A. McLean, and J. M. Dawson, Phys. Rev. Lett. 26, 17 (1971).

Nuclear Fusion Reactions in Solid-Deuterium Laser-Produced Plasma

F. Floux, D. Cognard, L-G. Denoeud, G. Piar, D. Parisot,
J. L. Bobin, F. Delobeau, and C. Fauquignon
Commissariat à L'Energie Atomique, Centre D'Etude de Limeil, 94 Limeil-Brevannes, France

When focusing a 4-GW, fast-rise-time, nsec-range laser onto a solid deuterium target, neutron production is observed. We give evidence for nuclear fusion reactions, measure the electronic temperature, and estimate the number of neutrons for each laser shot.

Nanosecond laser pulse heating of solid-state deuterium targets has been carried on in our laboratory for three years. Evidence for electron temperatures of up to 500 eV was reported earlier.[1–3]

In order to deal with experimental results, a tractable theoretical model was devised.[4] It predicts a plasma temperature proportional to the two-thirds power of the incoming light flux, which is in fair agreement with electron-temperature measurements. Since 500 eV were found with only 1 GW, it was expected that a higher power would lead to plasmas in which a significant amount of nuclear DD reactions would occur.

Our approach to the problem of kilovolt plasmas produced by laser irradiation of a solid target is then quite different from that of Basov's and his coworkers, who focus powerful picosecond pulses onto DLi targets.[5] Another important feature of the interaction process is the coupling between the plasma dynamics and the laser pulse itself: The beam is focused inside the ice (0.7 cm in depth, say), and, indeed, one has to manage in order that the deflagration front reaches the focal volume

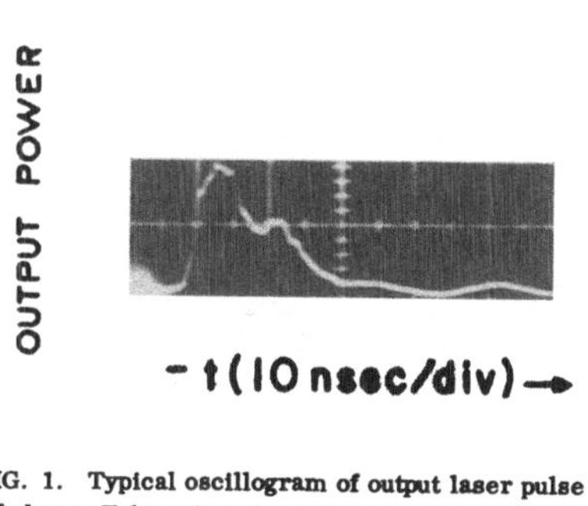

FIG. 1. Typical oscillogram of output laser pulse recorded on a Tektronix 519 oscilloscope. Signal is given by an ITT fast-rise-time photodiode.

when the laser pulse is at its peak value.[4] The increase of the maximum power should be completed by a suitable tailoring of the pulse shape.

In the previous experiment,[3] the Nd glass laser's maximum output power was 1 GW, and the main limitation to the increase of temperature was the too-long rise time of the pulse. Then the following trick was used. A Pockels's cell, made of a potassium dihydrogen phosphate crystal inserted between two Glan prisms, was located after the first amplifier rod. Triggering is provided by part of the 30-nsec pulse deviated towards an 8-atm argon-filled spark gap. By applying to the cell a voltage pulse of 12-kV amplitude and a rise time of 2 nsec, the laser pulse half-width is reduced to 15 nsec, and its energy is about 2 J. It is then amplified by four rods up to 80 J; diameter of the end rod is 45 mm. Figure 1 shows a typical pulse. The rise time is less than 5 nsec. An average output energy of 40 J was used, and the peak power varied from 3–5 GW, depending on each shot. The beam is focused by a 50-mm focal length f/1 aspherical lens.[6] The target is a solid deuterium stick, 1 mm square in cross section. Optimum focusing conditions range from 100–300 μm inside the ice. Light is partly reflected from the target, is amplified backwards through the cascade, and is eventually shuttered by the Glan prism, thus preventing the first stages of the laser from being broken down. The system is now operating well, and can sustain about 3 or 4 J of reflected energy, measured at the Glan-prism level, without any damage.

Electron-temperature recording and neutron detection are used as diagnostics. The "plasma thermometer" uses nickel and beryllium foils. The setup and results were extensively described in Ref. 2 and Ref. 3. Neutron detection is secured by two large, plastic phosphors connected through light pipes with XP 1040 Radiotechnique photomultipliers. Each phosphor is 450 mm in diam. From calibration measurements made by means of a DD

Reprinted from: *Physical Review*, **A1**, 821–824 (1970).

pulsed neutron source, the efficiency of detection was found to be one pulse for about 40 neutrons emitted in a solid angle of 4π sr by a source located 50 cm from the center of the phosphor external boundary. The quantum efficiency of the phosphor is about 50%. The detectors are completely shielded by lead sheets 5 mm thick. They are located 50 cm from the plasma in opposite directions. One of them can be moved from 50 to 150 cm in order to make time-of-flight measurements. X rays and neutron signals are recorded by Tektronics 585

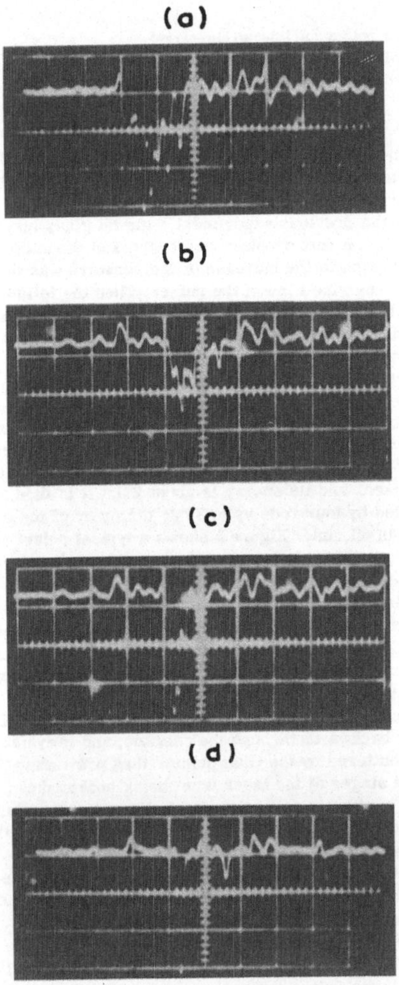

FIG. 2. Typical oscillograms of neutron signals. Detectors are located, respectively, at (a) 50 cm and (b) 150 cm from the target. The recording is made with 200 mV/div and 40 nsec/div. Cases (c) and (d) correspond to a less efficient laser shot, with the two detectors located at 150 cm from the target.

cathode-ray oscilloscopes. A time recorder, whose steps are registered with each track to be analyzed, provides a standard time scale. Then, by taking into account the photomultiplier time delay – which turned out to be close to 36 nsec – a chronology of laser pulse, x rays, and neutron signals can be made with an accuracy better than 10 nsec. More than 150 shots have been performed. Energy varies from shot to shot in the range 30–50 J, and the peak power ranges from 3 to 5 GW. Since the focal-spot diameter inside the ice is about 100 μm, the maximum flux density is a few times 10^{13} W/cm.^{2} Shots are carried on in series of 15 and, within each series, signals were recorded from neutron detectors 10 times over. After improving the focusing accuracy, 90% of the shots yielded signals from neutron detectors.

To determine the origin of these signals, tests were made under the following conditions.
(a) The lead shield also protects the detector against the laser flash lamp light. When several shots were made without target, no signals were detected. Thus, our neutron detectors are not sensitive to spurious electromagnetic noise. (b) To determine if these signals could be caused by hard x rays (>120-keV energy), the lead shield was taken off one of the detectors. The signals recorded on the two detectors were of the same amplitude. (c) Our experimental setup allows us to use, alternatively, deuterium and hydrogen targets; this is quickly effected by merely changing the gas entering the cryogenic device. Thus, other experimental conditions remain identical. When shining on hydrogen targets, the electron temperature was recorded and was found to be the same as those recorded in deuterium plasmas, within the experimental uncertainties.

In all tests, signals from neutron detectors were recorded with deuterium targets: No signal occurred when replacing deuterium by hydrogen; and signals were recorded again when returning to the deuterium ice. Thus, we got evidence for believing that the recorded signals coming from the plasma were due to the presence of deuterium atoms inside the target. Further information is derived from signal chronology. It was checked that within a 10-nsec accuracy: (a) Soft x rays are contemporary to the laser peak power; (b) shots were made, locating one of the phosphors at 50 cm and the other one at 150 cm from the plasma. Typical signals are displayed in Figs. 2(a) and 2(b).

Taking into account the first spikes, these are found to be about 45 nsec apart, which is consistent with the time of flight of 2.45-MeV neutrons produced by DD reaction (46 nsec). Further spikes occur randomly up to 120 to 150 nsec after the first: They are due to scattering, as will be shown later. Figures 2(c) and 2(d) correspond to a less efficient laser shot; with both detectors 150 cm apart from the target, no further spikes occurred. With respect to the laser peak power, the first neutron spikes occur 20 or 25 nsec later, when detectors are located at 50 cm from the target. When this distance is varied to 75, 100, and 150 cm, the signals appear 35, 50, and 70 nsec after the laser peak value. For these measurements, the measured photomultiplier time delay of 36 nsec has been taken into account. Now if one subtracts the time of flight of DD neutrons from the experimentally recorded chronology, and compares it with the laser pulse time history, it is found that most of the detected signals are consistent with neutrons emitted when the laser power is maximum, as shown in Fig. 3. The number of neutrons for each shot is estimated on the oscillogram by studying the maximum amplitude of signals and by counting the different pile-up spikes. For one laser shot, the number of neutrons varies from 100–500 emitted in the total solid angle. This number strongly depends on the laser power and the focusing conditions. The tail observed for later times is consistent with neutron paths larger than direct trajectories by about 1 m, which is most likely due to scattering on nearby items, such as the helium bath, gas pipes and tanks, and iron structures. In another test, the most sensitive detector was entirely shielded by a 3-mm-thick cadmium foil and by 25 cm of paraffin. In this case, no signal occurred, whereas, for the same shot, a signal was recorded from the bare detector. Taking off the shield, signals occurred again on the two detectors. Finally a BF_3 detector, using three counters inserted into paraffin with an over-all efficiency of one pulse for 100 neutrons emitted at 15 cm in 4π sr, was set up and one to five pulses were recorded for each laser shot. Connected with these measurements, electron temperatures were derived from x-ray signals. Since we concentrated on neutron detection, the results are not at this time very accurate. However, average electron temperatures range between 500–700 eV. For these values, provided ions are at the same temperature as electrons, rough calculations show that some DD fusion reactions are likely to occur.

We seem to have gotten definite evidence for a neutron yield from a laser-produced deuterium plasma. These neutrons are most likely due to DD fusion reactions. This result, which was obtained by carefully tailoring the pulse shape of an otherwise-classical, nanosecond giant pulse laser, appears promising for the quest of very-high-temperature plasmas. In this respect, nanosecond lasers are expected to be very convenient and efficient tools, since neutrons were obtained with a peak power of only 4 GW.

The authors are greatly indebted to J. L. Bocher and D. Meynial who, over a period of two years, contributed substantially to the success of this experiment. We are deeply grateful to P. Langer, E. Laviron, and C. Delmare for laser and electronic support. We thank, especially, J. de Metz for the very fine focusing lens he devised for us; C. Faure, A. Perez, and P. Meunier, who achieved counter calibration and neutron detection; and H. Guillet, C. Meunier, D. LeGoff, and J. Osmalin, of the C. G. E. Co., who devised the laser-sharpening system and participated in the experiment. Finally, P. Nelson and P. Veyrié are thanked for their crucial contribution to the beginning of this experiment.

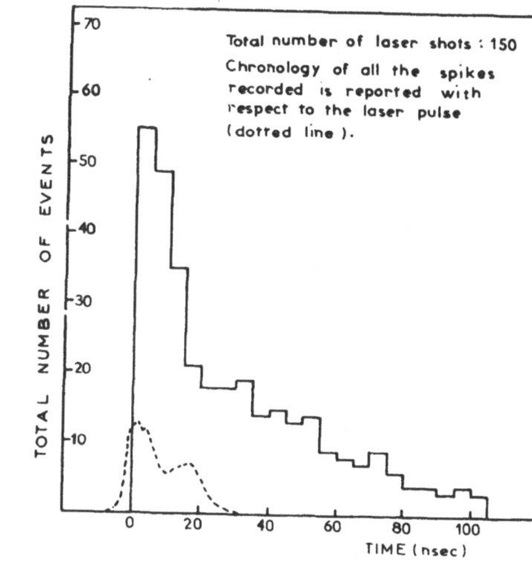

FIG. 3. Histogram of up-to-date results. Neutron spikes are numbered with amplitude and time for each laser shot. The total number of so-recorded events is plotted in a time scale given by laser pulse. Chronology takes account of times of flight corresponding to distances between plasma and phosphors.

[1]C. Colin, Y. Durand, F. Floux, D. Guyot, P. Langer, and P. Veyrié, J. Appl. Phys. 39, 2991 (1968).

[2]J. L. Bobin, F. Floux, P. Langer, and H. Pignerol, Phys. Letters 28A, 398 (1968).

[3]J. L. Bobin, F. Delobeau, C. Fauquignon, G. De Giovanni, and F. Floux, Nucl. Fusion 9, 115 (1969).

[4]C. Fauquignon and F. Floux, Phys. Fluids (to be published). A brief account of this work is given in Ref. 3.

[5]N. G. Basov, S. D. Zakharov, P. G. Kryukov, Yu V. Senatskii, and S. V. Chekalin, I. E. E. E. J. Quant. Elec. QE4, 855 (1968) Zh. Eksperim. i Teor. Fiz. Pis'ma v Redaktsiyu 8, 26 (1968) [English transl.: Soviet Phys. – JETP Letters 8, 14 (1968)].

[6]J. L. Champetier, J. P. Marioge, J. de Metz, F. Millet, and A. Terneaud, C. R. Acad. Sci. Paris, 266, 838 (1968).

HEATING OF LASER PLASMAS FOR THERMONUCLEAR FUSION*

N. G. Basov, O. N. Krokhin, G. V. Sklizkov

Lebedev Physical Institute, Academy of Sciences

Leninsky Prospect 53, Moscow, USSR

ABSTRACT

In this survey the experimental results of the investigation of the laser produced plasmas in a wide interval of energy and light pulse duration are reported. The processes taking place at a dense plasma heated by laser radiation for the obtaining of a thermonuclear yield have been analyzed. The directional motion for heating in the various schemes using cumulative effects is discussed.

The experimental results of plasma heating investigation are described. The laser description and the methods of investigation of plasma parameters are given. The results of measurements of neutron emission from deuterized polyethylene plasma are considered.

Spectroscopic and interferometric studies of gas-dynamical parameters in the process of heating are reported. The interfermetric method of measurement of plasma pressure and momentum carried away by the heated plasma during laser pulse has been carried out. It has been observed that for a conventional laser the maximum of the pressure coincides with the beginning of a pulse.

---

*Presented at the Second Workshop on "Laser Interaction and Related Plasma Phenomena" at Rensselaer Polytechnic Institute, Hartford Graduate Center, August 30-September 3, 1971.

Reprinted from: *Laser Interaction and Related Plasma Phenomena*, Vol. 2, H. Schwarz and H. Hora (eds.), Plenum, New York, 1972, pp. 389–408.

The application of multi-beam lasers for the target heating is discussed. The laser having a rectangular shape of light pulse with energy of 10^3 J within 2-16 nsec has been made. The efficiency of the beam energy absorption by the spherical target and the influence of laser beam parameters on this efficiency are determined.

1. INTRODUCTION

At present the idea of laser application for plasma heating seems to be prospective and is universally accepted. In Reference 1 it has been noted that in order to obtain a useful yield of a d-t thermonuclear reaction an amount of $\sim 10^5 - 10^6$ J of laser energy is required, and it seems to be quite a reasonable value. The same energy values are also presented in References 2 and 23. The same value was mentioned at the International Conference on Laser Plasma held in Moscow, November 1970. The obtaining of thermonuclear fusion by means of lasers can be realized nowadays only with the help of a Nd-glass Q-switched laser. One should remember that the efficiency of such lasers equals to about 1 per cent. Such a low efficiency will require a development of a new type of laser to be used in a real thermonuclear arrangement.

Now the investigators face two problems. On the one hand the enhancement of laser energy, and on the other hand the increase of neutron yield due to the improvement of the target and the elucidation of the phenomena taking place at laser interaction plasma. The up-to-date experimental studies of plasma heating are performed with the energy of several tens of Joules, see References 3-7, 22, and 27. Some works reported the neutron emission from the LiD target[5], solid deuterium[6], and deuterized polyethylene[7], the maximum number of neutrons equals to 10^4 - 10^5 neutrons per pulse. Thus, laser energy increase from 10^2 J to 10^6 J is required to enhance the neutron yield up to 10^{16} - 10^{17} neutrons per pulse in order to achieve a thermonuclear reaction. Speaking of the laser energy, we mean the radiative energy at the optimal duration (for various targets it is within 10^{-10} - 10^{-9} sec).

The usage of complex targets with mixtures having the atomic weight $A \sim 250$ can be promising when lasers of supershort duration are employed[8].

The present paper deals with the phenomena that take place in the heated plasma during a laser pulse. The usage of a multi beam laser resulted in the measurement of the heating efficiency of the spherically irradiated target at a level of 10^3 J energy.

2. TARGET AND LASER RADIATION PARAMETERS

In the case of the pulsed plasma heating the condition for a thermonuclear fusion is

$$n\tau \sim n \frac{r}{v_s} K \qquad (1)$$

where n is plasma density, r is the characteristic radius of a target for spherical irradiation, and v_s is thermal velocity. Hence, for $N\tau \geq 10^{14}$, $r = 10^{-2}$ cm, $v_s = 10^8$ cm/sec the required density of the heated substance should be more than 10^{24} cm^{-3}. One of the ways to achieve such densities is to use spherical hydrodynamic plasma implosion[24] or a statical compression of substance. The experiments on collisions of two laser flares showed that in the plane of the flare convergence an increase of plasma density and temperature is observed[9, 10, 25]. With the spherical geometry it seems to be reasonable to expect the density of two or three orders of magnitude. Such a substance compression is experimentally observed in the devices such as a plasma focus[11]. However, the use of targets with a high density requires the preliminary studies of the heating efficiency of plasma, the density of which is higher than the critical one.

The laser radiation parameters (divergence, duration, energy) are determined on the one hand by the target dimensions, and on the other, by the optical quality of laser elements and focusing system. The beam divergence α should be determined by an inequality

$$\alpha \leq \frac{r}{f}$$

(at $r \approx 10^{-2}$ cm and focus $f \approx 10^2$ cm, $\alpha \leq 10^{-4}$ rad). The use of objectives with $f \approx 10$ cm is possible only at the energy of about 10^3 J.

3. THE MEASUREMENT OF PLASMA PRESSURE

To understand the processes of plasma heating one should know the value of plasma pressure and its variation in the process of heating. In this section the value of pressure depending on time is determined using the data of high-speed interferometric measurements[4].

The experimental arrangement (Fig. 1) consisted of an Nd-glass laser used for plasma heating and a ruby laser that served as a pulse source for the illumination of an interferometer.

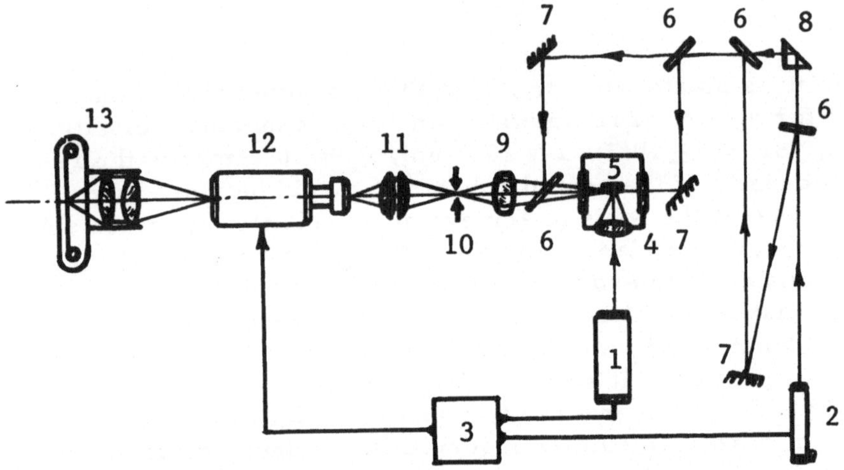

Fig. 1. Scheme of the arrangement for a time scanning of a flare interferogram using an image converter camera in streak mode. 1 - Nd-glass laser; 2 - ruby laser; 3 - controlling device; 4 - focusing objective; 5 - target; 6 - splitters; 7 - mirrors; 8 - full reflective prism; 9 - objective that is projecting the image of the interferogram on the slit 10; 11 - objective that is projecting the image of the slit on the photocathode of the electro-optical tube; 12 - image converter camera; 13 - photocamera.

The carbon target was placed in vacuum at pressure not higher than 10^{-5} mmHg. Nd-glass laser energy was equal to 8 J at pulse duration of 80 nsec at the level of 0.1 amplitude. Pulse oscillogram is shown in Fig. 2. The maximum beam divergence that corresponded to the maximum intensity was equal to 2×10^{-3} rad. During the rise time the divergence was increasing linearly with the intensity, and the focal spot radius varied from 0.05 mm to 0.2 mm (the radius of the beam at focal area due to abberation was equal to 0.05 mm).

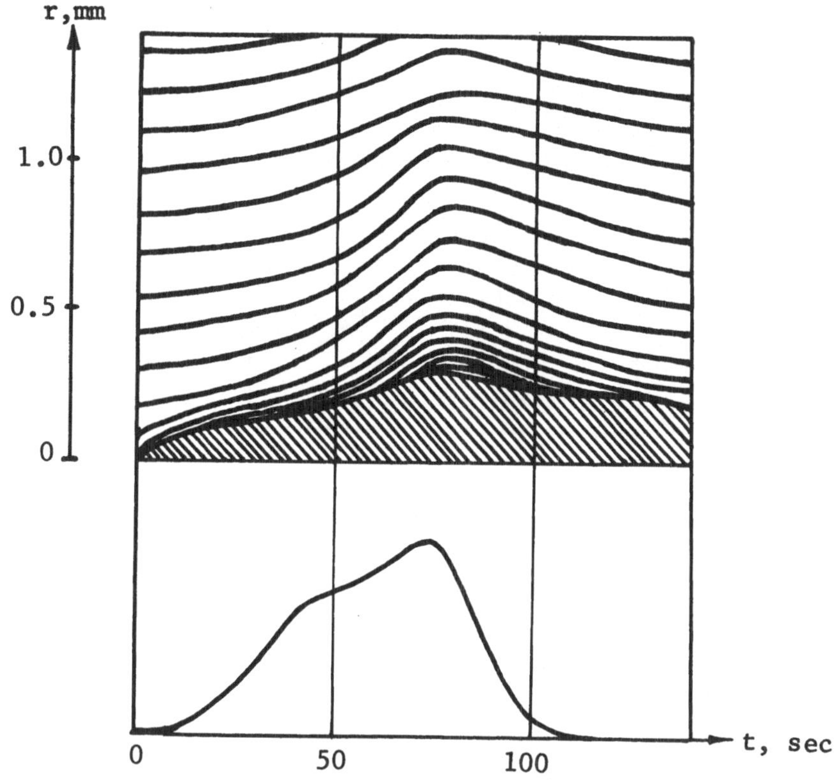

Fig. 2. Time scanning of the carbon interferogram which is correlated with the laser pulse (bottom curve). The slit of the camera coincides with the Nd-glass laser beam axis and is perpendicular to the target surface. Interference fringes in the zero position are parallel to the target. r - distance from the target, r = 0 - corresponds to the target surface. Dotted region is an opaque area.

The split image of the interferogram was scanned in time with an image converter camera in streak mode.

A typical streak interferogram is shown in Fig. 2. It is seen that an opaque zone is broadening with time, and it reaches a maximum value of 0.25 mm in 76 nsec after the beginning of the heating pulse. The electron density in the zone edge was equal to $5 \times 10^{19} cm^{-3}$. The profiles of the electron density for various time intervals have been determined by using the interferogram (Fig. 3). Interferograms have been unfolded by the parabolic method assuming an axis symmetry [21].

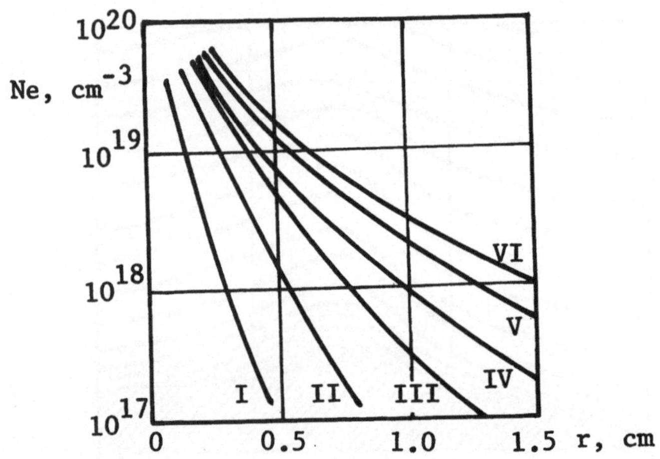

Fig. 3. Distribution of the electron density in the direction of the laser beam axis in various time intervals. Figures above the curves indicate the time from the beginning of a laser pulse in ns: I - 10; II - 16; III - 23; IV - 36; V - 56; VI - 76.

The full mass of plasma heated to a given moment was determined by integrating the density curves. Here the missing parts of curves were approximated by an expression:

$$n_e = \frac{10^{17}}{R^2} \frac{(1-\xi)^{\alpha}}{\xi^2} \quad (2)$$

where $\xi = (r + 0.01)R^{-1}$, r - distance from the focal point, $R = v_u t$ - radius of the plasma edge, v_u - asymptotic velocity of plasma particles. Plasma expansion is assumed to be spherically symmetrical. The best approximation of all the curves is achieved at index $\alpha \approx 16$. The gradient of plasma density in the heated region is seen to be the largest and is reducing with time $\sim t^{-3}$.

The data of the density profile and of the velocity of various regions of plasma make it possible to calculate the momentum that is carried away by the plasma. Since the plasma acceleration occurs near the target the momentum value determines the plasma pressure on the target. Here we assume that plasma expands spherically symmetrically. During the laser pulse the plasma edge expands far away in comparison with the focal spot diameter. For the plasma momentum projection normal to the target surface we have

$$F(t) = \int \rho(r,t) v_n(r,t) dV \quad (3)$$

where $\rho(r,t) = n_e m_i z^{-1}$, m_i - ion mass, z - effective charge, $v_n(r,t)$ - projection of the velocity to the normal, r - distance from the centre of the focal spot. Integration is over the whole of volume.

Hence the pressure in the hot part of a flare will be

$$p(\zeta) = (\pi r_o^2 \tau)^{-1} \frac{dF}{d\zeta} \quad (4)$$

where r_o - radius of the focal spot, τ - time of heating, $\zeta = t\tau^{-1}$ - dimensionless time.

The profile of the velocity v(r) used in the calculations is presented in Fig. 4. The choice of such a profile is conditioned by the experimental data on the measurement of plasma velocity according to the Doppler shift[12], as well as on the results of time scanning of ion expansions of various changes in the light of corresponding lines[13]. Here v_o - sound velocity, r_1 - 0.06 cm. Taking into account that $v_o \sim q^{1/4}$(q - radiation flux density), $r_o = r_{o\tau}(x_1 + x_2\zeta)$, $r_{o\tau}$ - maximum value of the focal spot at the moment of time; τ, x_1 and x_2 equal 0.25 and 0.75, respectively.

For the case of an Nd-glass laser with the radiation pulse shown in Fig. 2, and 8 J energy to the end of a pulse (τ = 76 nsec), the temperature has been measured.

In our case we have temperature of $T_m \approx 50$ eV and an effective ion charge $z \approx 5$. Here $A_1 = 0.5 \times 10^7$ cm/sec, $v_{umax} = 2.5 \times 10^7$ cm/sec, $r_o = 2 \times 10^{-2}$ cm. Laser beam divergence increases linearly with time during the largest part of the pulse front. Angular and space characteristics of the light beam were investigated by a slit-optical converter camera. To calculate the pressure the growing part of the pulse was approximated by a linear function. Noting the mentioned conditions the pressure is calculated by Formula (4). A force, acting on the target can be derived from the momentum

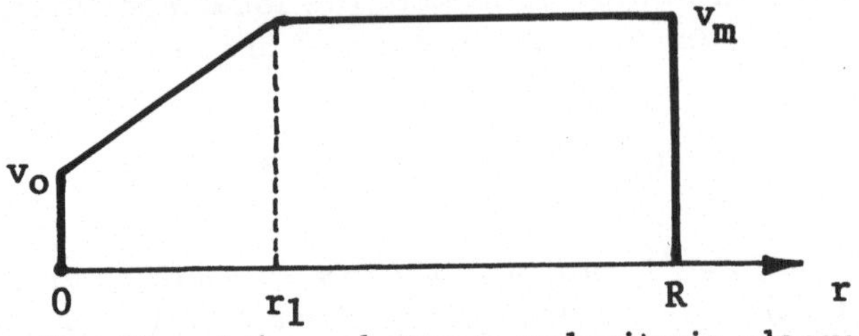

Fig. 4. Profile of the substance velocity in plasmas along the laser beam axis.

linear function. Noting the mentioned conditions the pressure is calculated by Formula (4). A force, acting on the target can be derived from the momentum

$$f(\zeta) = \frac{1}{\tau}\frac{dF(\zeta)}{d\zeta} \qquad (5)$$

Figure 5 illustrates the curves of the dependence in time of the force, momentum and plasma pressure on the target. The maximum pressure is seen to act at the beginning of a pulse due to the fact that the divergence is small at small intensity and the focal spot diameter is determined in this experiment only by the lens parameters. At later times (as is seen from Fig. 3) the density profile is rising. The plasma region where the absorption occurs, is shifting from the target and is growing. The mass of gas heated by laser is also increasing. The decrease of pressure indicates an attenuation of laser radiation. The temperature in the hot region falls and the larger part of the radiative energy is transformed into the kinetic energy of the expanding substance.

Thus, changing the time dependence of the radiation divergence one can shift the maximum pressure and use the laser energy optimally for the plasma heating in real conditions. The presence of the peak of pressure and hence, the maximum temperature at the beginning of the laser pulse can explain the experiments of Reference 14, the authors of which have observed the intensive peak of x-ray emission from the flare in the very beginning of a laser pulse with the radiative power much less than the maximum one.

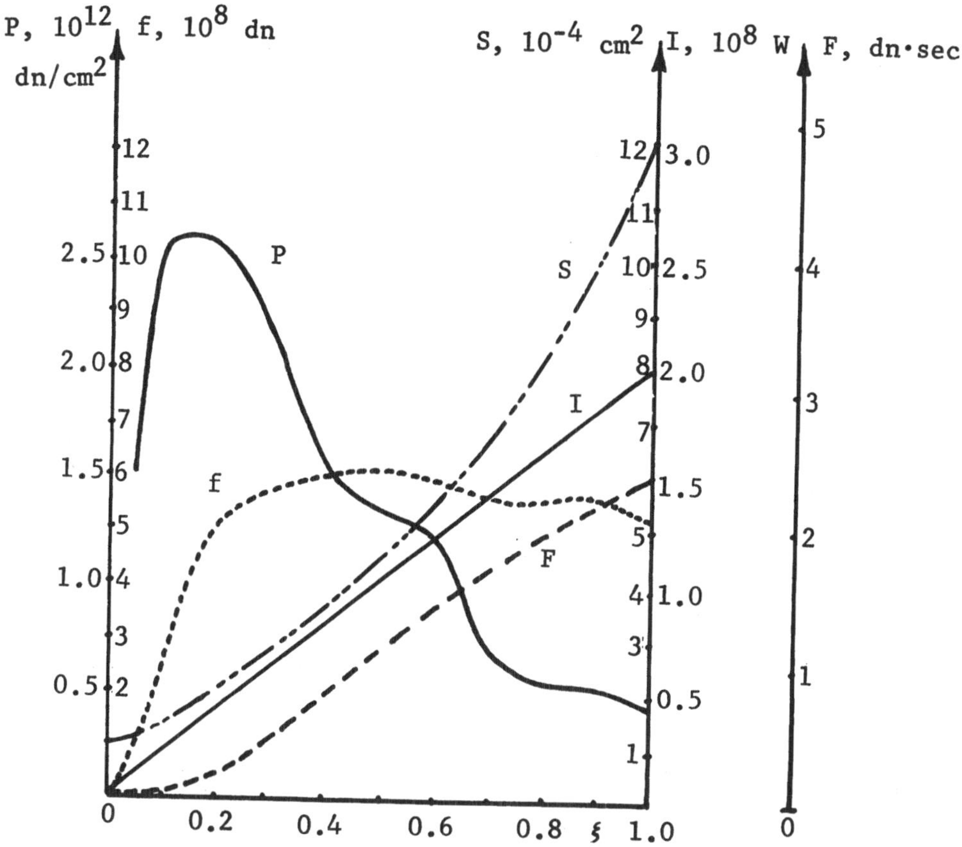

Fig. 5. Time variations of plasma pressure p in the "hot" region and momentum F carried away by plasma during the growing part of the laser pulse, the intensity of which is approximated by the linear function (line I). f is force, applied to the target, S is a focal square.

4. NEUTRON GENERATION

The improvement of an Nd-glass laser has made it possible to record neutron emission from the deuterized polyethylene target[7]. The presence of heavy ions in the deuterium plasma should lead to the increase of temperature and decrease of density[15] (in comparison to pure deuterium)

$$T = T_D\, Z^{2/3} \left(\frac{2}{Z+1}\right)^{2/3} \left(\frac{A}{2}\right)^{2/9}$$

$$N = N_D Z^{-1} \left(\frac{2}{Z+1}\right)^{1/2} \left(\frac{A}{2}\right)^{1/6} \qquad (6)$$

where Z, A - average charge and ion mass, respectively, $T_D N_D$ - temperature and pressure in the case of pure deuterium.

In the experiment we used a laser with five amplifiers. The laser beam was focused by a lens f = 100 mm onto the massive target prepared from the powder-like polyethylene. At the beam divergence of 1.2×10^{-4} rad the heated area was equal to 10^{-4} cm^2. Maximum laser energy was of 80 J and the duration did not exceed 3.5 nsec at the 1% level.

Neutron emission was recorded by a photomultiplier with a plastic scintillator. The photomultiplier recorded the light flash of recoil photons.

Figure 6 illustrates the pulses from the neutron detector at distances from the scintillator to the target of 10 cm and 60 cm. Laser pulse from the coaxial photodiode recorded at the same trace served as a time mark. The minimum number of neutrons is easily determined because at the distance of 60 cm the neutrons are always recorded in each experiment (at energy not less than 14 J). It makes not less than 10^4 neutrons for the total number.

5. SPECTRAL MEASUREMENTS

To measure electron temperature T_e of the dense laser plasma there have been used the relative intensities of the spectral lines in the vacuum ultraviolet spectrum region, in the range 100 - 200 Å. They correspond to the transitions $2s2p^n - 2s2p^{n+1}$ of ions of the elements with charge of Z = 20-30, that are stripped at temperatures of 100 - 1000 eV[16].

Stark line broadening of the hydrogen-like ion CVI has been used to measure the electron density. The choice of the ion CVI of a rather high ionization potential (498 eV) permits the

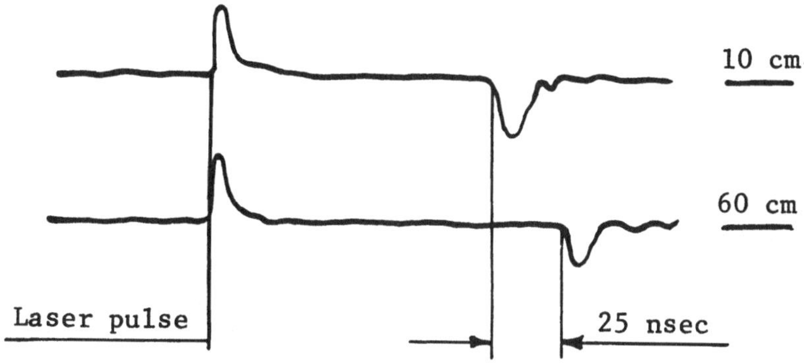

Fig. 6. Oscillograms of the pulses from the neutron detector distanced 10 cm and 60 cm from the target. A light pulse from the coaxial photodiode is recorded at the beginning of the traces.

evaluation of the density distribution in the hot flare region during the action of the laser pulse[19] by the integrated spectrograms.

The measurement of the profile of plasma velocity was performed by the analysis of data on time scanning of CVI ion lines in the visual spectrum region[13] and on the Doppler shift of absorption lines[17] of the resonance doublet of CIV ion (Fig. 7). The hot dense flare nucleus radiates in a continuous spectrum through an expanding cloud of the rarefied cold plasma. It leads to the appearance of the absorption lines similar to the "Fraunhofer lines" in the solar spectrum. The shift of absorption lines relative to the emission lines to the blue side of $\lambda \sim 1\text{Å}$ permits the evaluation of the velocity of the directed motion of the "cold" envelope. Spectrogram of Fig. 7 shows that the multi-charged laser plasma that is leaving a dense "hot" region, accelerates due to the thermal electron energy up to the velocity $\sim 3 \times 10^7$ cm/sec.

6. MULTIBEAM LASER[26]

With the help of a powerful laser arrangement there has been measured the efficiency of radiation absorption in the target with spherical irradiation. The arrangement consisted of a successive-parallel system of amplifiers and it provided the obtaining of

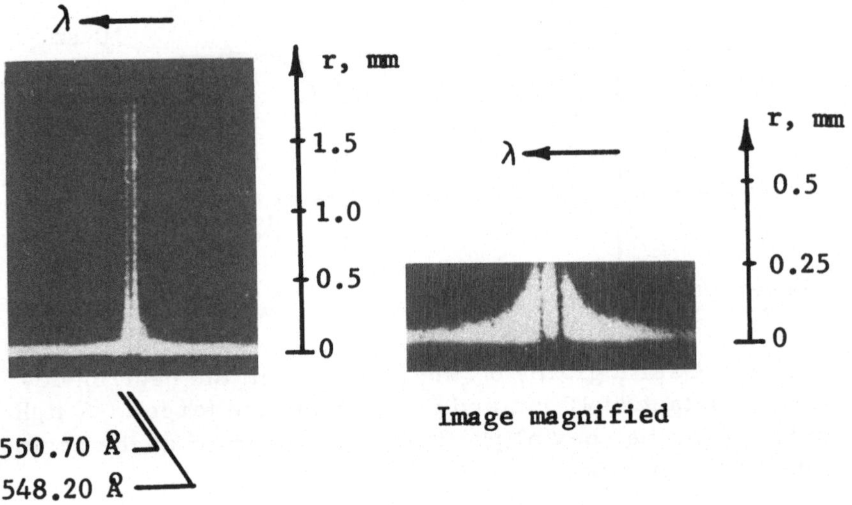

Fig. 7. Spectrograms of the lines λ = 1550.77 Å, λ = 1548.20 Å, r - distance from the target. Absorption lines of the cold envelope are shifted on a Δλ = 1.2 Å relatively to the emission line.

about 10^3 J energy at the controllable pulse duration from 2 to 16 nsec. Nine beams were formed at the output. Divergence of each beam did not exceed 2×10^{-4} rad. The background energy did not exceed 0.1 percent of the useful energy. The energetic characteristics of the laser arrangement are listed in Table I. E - total output energy in Joules (with an accuracy of about 10%), B - radiative brightness, q - flux density on the target, η - efficiency relating to the electric energy of the condensor. From Table I one can see that with the pulse reduction from 16 to 2 nsec the energy falls about a factor of two. Accordingly, the space distribution of intensity and coherent properties of the beams do not vary, in fact.

A preamplified laser beam was directed to the powerful amplifying cascade (Fig. 8) where the light beam obtains the whole energy despite a rather small amplification coefficient (~10). Amplification proceeds into two stages. In the first stage a beam is split into three parts, each of which is amplified by a rod after the corresponding isolating shutter. In the second stage each beam is again divided into three, and each is again amplified. As the

τ, nsec	E, J	$B \cdot 10^{-16}$, $W/cm^2 \cdot sr$	q, W/cm^2	η, %
2	600	4.3	I.5	0.I5
4	800	2.9	I	0.2
8	I000	I.8	0.6	0.24
I6	I300	I.2	0.4	0.3

TABLE I

active elements Nd-rods are used, 4.5 cm diameter and 60 cm pumping part length.

The arrangement provided a flux density on the target that equaled 10^{16} W/cm^2 (for a flat target) and 2 x 10^{15} W/cm^2 (for a spherical target). In the case of spherical irradiation the time derivative of the specific energy yield in the heated plasma can achieve 10^{18} W/cm^3, in the condition of full radiation absorption by the plasma. For r_o = 2.5 x 10^{-2} cm an energy yield of 5 x 10^{15} W/cm^3 has been experimentally achieved. The latter value can be substantially increased, using heavy targets of smaller dimensions[8].

Compensation of the optical paths of each beam was realized by optical compensators that provided an accuracy of synchronization of light pulses coming to the target not greater than 3 x 10^{-11} sec.

The formation of a rectangular light pulse is performed by means of a shutter controlled by a spark gap with laser triggering[18]. The shutter provided a contrast of 0.5 x 10^5. The oscillograms of light pulses after the shutter are given in Fig. 9.

After amplification all nine beams are directed to the focusing objectives mounted in the walls of a vacuum chamber. The focusing system provided a practically uniform irradiation of the

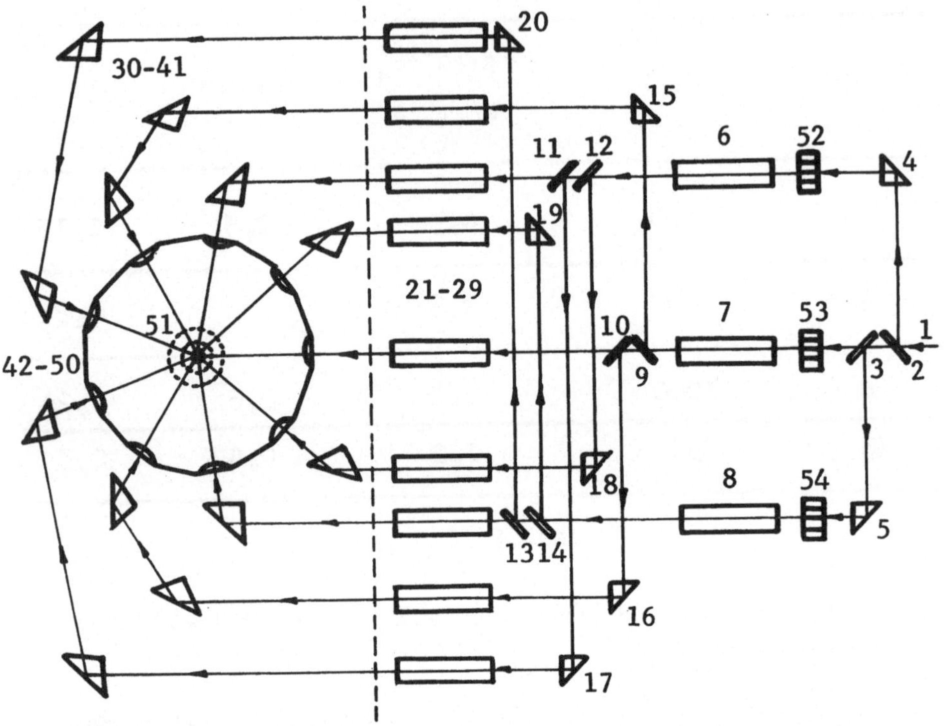

Fig. 8. Scheme of the powerful amplifying system. 2, 3, 9, 10, 11 + 14 - dividing elements; 4, 5, 15-20, 30-41 -- prisms of the full internal reflection; 6-9, 21 + 29 - Nd-rods; 42-50 -- focusing objectives; 51 - spherical target; 52-54 -- isolating shutters. Calorimeters and coaxial photodiodes are not shown in the scheme. In some beams the light delays are used to compensate optical ways (they are not shown in the scheme).

spherical target, 0.1 - 0.5 mm diameter.

To photograph high temperature processes that occur at the target a system of a high speed laser photography has been used[19]. Frame exposure equaled 0.4 ns, a number of frames - 7.

7. EFFICIENCY OF THE TARGET HEATING

Light radiation was focused on the spherical target dimensioned from the condition $r_0 \approx v\tau$, where v - thermal velocity of expansion at temperature T. At $T \approx 1$ KeV and pulse duration $\tau \approx 2 \times 10^{-9}$ sec we have $r_0 \approx 2.5 \times 10^{-2}$ cm.

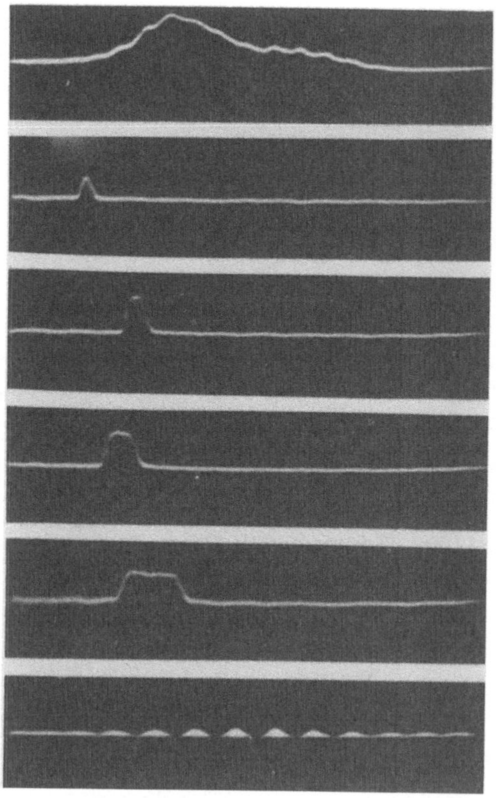

a b c d e f

Fig. 9. Radiation light pulses of the powerful laser. a - a pulse after the oscillator; b, c, d, e - radiative pulses after a forming shutter and a first three passes amplifier for the regimes, respectively: 2, 4, 8 and 16 nsec; f - time marks of 10 nsec.

The energy, absorbed in the target, has been measured by observation of a gasdynamic expansion of the spherical shock wave in the residual gas. The motion of the wave is described by formulae of the spherical instant point explosion[20] when the mass of the shock driven gas is sufficiently greater than the mass of the heated target. In our case it occurs when the radius of the shock wave equals $R \approx 20\ r_o$.

Figure 10 illustrates a seven-frame shadow photograph of the shock wave formed in air at pressure of 17 torr. Due to the spherical irradiation of the target the shock wave has good spherical symmetry. The long duration of the shock wave formation (first

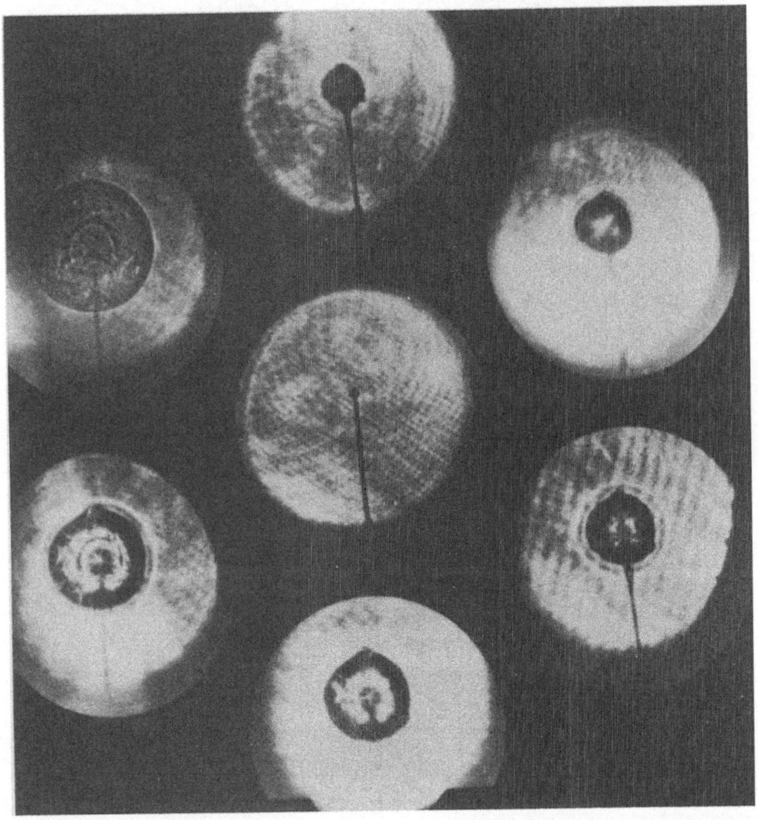

Fig. 10. Shadowgram of the spherical shock wave. Time intervals correspond to the following delays from the beginning of heating. 1 - 2.0; 2 - 20; 3 - 45; 4 - 70; 5 - 140; 6 - 210; 7 - 460. Still diameter equals 5.0 cm. A polyethylene target has $r_o = 2.5 \times 10^{-2}$ cm. Laser energy is equal to 560 J. Pulse duration - 2 nsec.

three frames) compared to the time of heating, 2 nsec, is explained by high target density versus the density of the surrounding gas.

The law of the shock wave front motion is defined by the expression

$$R = \left[\frac{a(\gamma)E_e t^2}{\rho_o}\right]^{1/5} \tag{7}$$

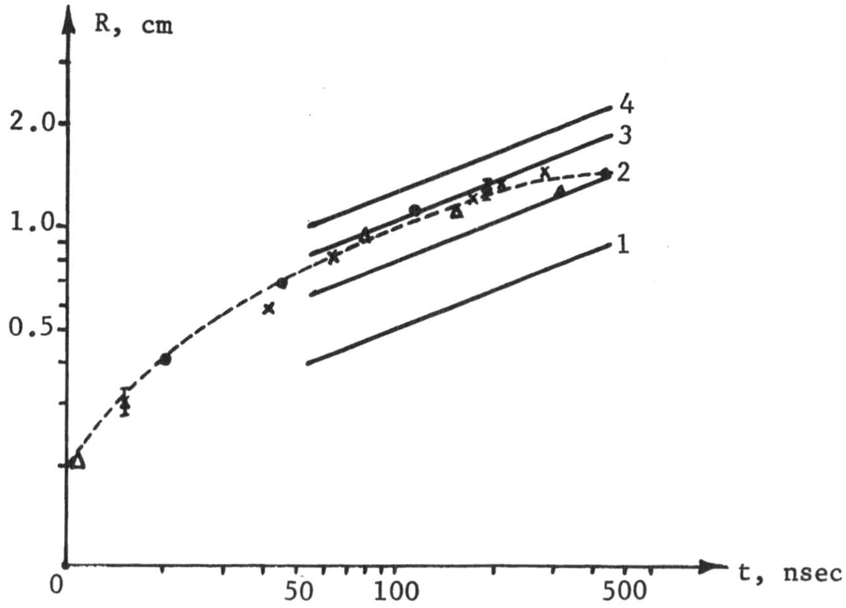

Fig. 11. Experimental dependence of the shock wave radius on time - a dotted curve. The marks o, x, Δ correspond to the three separate experiments in the mode of $\tau = 2$ nsec. Theoretical curves correspond to the explosion energies: 10, 10^2, 3.5×10^2; and 10^3 J. Gas pressure (air) before the wave equals 15 mm Hg.

where $a(\gamma)$ - coefficient that depends on the adiabatic index, which in our case is equal to 1, E_e - energy of the explosion, which equals the absorbed laser light energy neglecting the plasma radiation, ρ_0 - density of unperturbed gas.

Figure 11 shows the dependence of R(t) for three experiments at the same conditions. The same figure illustrates R-t diagrams that are calculated by Formula (7) at the explosion energies of 10, 100, 350, 1000 Joules. Comparing theoretical and experimental curves one can conclude that the explosion energy equals 300 J. This is the minimal value for the laser radiation energy absorbed by the plasma. Taking into account the energy losses on the optical elements of the converging and focusing systems one can derive that a light energy of about 450 J falls on the target surface, more than 70 percent of which is absorbed in the target.

After 250 nsec the dependence of radius on time changes from $R \sim t^{0.4}$ to $R \sim t^{0.2}$. This can be explained by the losses due to recombinative radiation for which the plasma is transparent behind the shock wave front in the considered time interval.

Note that the reduction of the energy yield (see the Table) for shorter pulse length cannot yet be explained. It can turn out to be an important point when designing a thermonuclear system, because for a laser with 10^6 J energy the optimal pulse duration is equal to 10^{-10} - 10^{-9} sec. If the energy reduction is connected with the saturation of the lower nonradiating transition then for the decrease of the pulse to 10^{-10} sec the energy reduction won't be more than about two times. However, this assumption needs a special experimental verification.

REFERENCES

1. N. G. Basov, O. N. Krokhin, "Laser Application for Thermonuclear Fusion", Vestnik Ac. Sci. USSR, N 6, 55 (1970).

2. H. Hora, "Application of Laser Produced Plasmas for Controlled Thermonuclear Fusion", p. 427 in Laser Interaction and Related Plasma Phenomena, Plenum Press, New York-London, 1971.

3. N. G. Basov, V. A. Gribkov, O. N. Krokhin, G. V. Sklizkov, "Investigation of High-Temperature Phenomena Induced by Powerful Laser Radiation Focused onto the Solid Target", JETP 54, N 4, 1073 (1968).

4. N. G. Basov, V. A. Boiko, V. A. Gribkov, S. M. Zakharov, O. N. Krokhin, G. V. Sklizkov, "Gasdynamics of Laser Plasma During Heating", JETP 61, N I, 154 (1971).

5. N. G. Basov, P. G. Kriukov, S. D. Zakharov, Yu. V. Senatsky, S. V. Tschekalin, IEEE J. Quant. Electron., QE-4 864 (1968). G. W. Gobeli, J. C. Bushnell, P. S. Peercy, E. D. Jones, Phys. Rev. 188, N I, 300 (1969).

6. F. Floux, D. Cognard, L. Denoeud, G. Piar, D. Parisot, J. Bobin, F. Delobeau, C. Fauquignon, "Nuclear Fusion Reactions in Solid Deuterium Laser Produced Plasma", Phys. Rev. A (General Physics), I, N 3, 821 (1970).

7. N. G. Basov, V. A. Boiko, S. M. Zakharov, O. N. Krokhin, G. V. Sklizkov, "Neutron Emission from Laser Plasma Heated by Nano-second Pulses", JETP Letters, 13, N 12, 691 (1971).

8. Yu. V. Afanasiev, E. M. Belenov, O. N. Krokhin, I. A. Poluektov, "About Possibility of Obtaining of the Intense Neutron Source at Laser Plasma Heating", JETP Letters, 13, N 5, 257 (1971).

9. N. G. Basov, O. N. Krokhin, G. V. Sklizkov, "Investigation of Dynamics and Heating and Expansion of Plasma Created by Laser Radiation Focused onto the Substance", Trydi FIAN, 52, 171 (1970) "Quantum Radiaphysics".

10. V. A. Gribkov, V. Ya. Nikulin, G. V. Sklizkov, "Plasma Density Increase at Laser Flare Collision", Short Communications in Physics, FIAN, N 2, 45-49 (1971).

11. N. V. Filippov et al, "Experimental and Theoretical Investigation of Pinch Discharge Like 'Plasma Focus'", Proc. of the 4th Conference on Controlled Thermonuclear Fusion, 1971, USA, Madison, Report CN-28-D6. N. J. Peacock, M. G. Hobby and P. O. Morgan, "Measurements of the Plasma Confinement and Ion Energy in the Dense Plasma Focus", Report CN-28-D3.

12. E. V. Aglitzky, V. A. Boiko, S. M. Zakharov, G. V. Sklizkov, "Determination of Electron Density Profile in Laser Plasma by Stark Spectral Line Broadening", Preprint N 143, FIAN, Moscow, 1970.

13. N. G. Basov, V. A. Boiko, Yu. A. Drozhbin, S. M. Zakharov, O. N. Krokhin, G. V. Sklizkov, V. A. Yakovlev, "Investigation of Initial Stages of Gasdynamical Laser Flare Plasma Expansion", DAN, 192, N 6, 1248 (1970).

14. G. L. Bobin, F. Floux, P. Langer, M. Pignerol, "X-rays from a Laser Created Deuterium Plasma", Physics Letts., 28A, 398 (1968).

15. O. N. Krokhin, "High-Temperature and Plasma Phenomena Induced by Laser Radiation", Proc. of the Intern. School of Physics, Enrico Fermi, Course XLVIII, Academic Press NY-L, 1971, 278-305.

16. N. G. Basov, V. A. Boiko, Yu. P. Voinov, E. Ya. Kononov, S. L. Mandelshtam, G. V. Sklizkov, JETP Letters, 5, 179 (1967); 6, 849 (1967). V. A. Boiko, Yu. P. Voinov, V. A. Gribkov, G. V. Sklizkov, Optics and Spectroscopy, XXIX, 1023 (1970).

17. E. V. Aglitzky, V. A. Boiko, S. M. Zakharov, G. V. Sklizkov, "Determination of Velocity and Electron Density Profile in Laser Plasma by Measurements in UV Spectral Region", Short Communications in Physics, FIAN, N 6, p.3 (1971).

18. A. H. Guenther, J. R. Bettis, "Laser Triggered Switching", p. 131-172 in Laser Interaction and Related Plasma Phenomena,' ed. by H. Schwarz and H. Hora, Plenum Press, NY-L, 1971. N. N. Zorev, G. V. Sklizkov, S. I. Fedotov, A. S. Shikanov, "Investigation of Spark Channel Triggered by Laser Radiation", Preprint FIAN, N 56, Moscow 1971.

19. N. G. Basov, O. N. Krokhin, G. V. Sklizkov, "Laser Application for the Production and Diagnostics of Pulse Plasma", Appl. Optics 6, N 11, 1814 (1967).

20. Korobeinikov, N. S. Melnikov, E. V. Ryazanov, "Theory of the Point Explosion", Fizmatgiz, Moscow (1961).

21. V. A. Gribkov, V. Ya. Nikulin, G. V. Sklizkov, "Interferometric Investigation of Laser Plasma Collision", Preprint FIAN N 153, Moscow (1970).

22. G. V. Sklizkov, "Kinetic and Ionization Phenomena in Laser Produced Plasmas", p. 235-257 in Laser Interaction and Related Plasma Phenomena, Plenum Press, NY-L, 1971 (ed. by H. Schwarz and H. Hora).

23. P. P. Pashinin, A. M. Prokhorov, "Obtaining of High Temperature Dense Plasma at Laser Heating of a Special Gas Target", JETP, 60, 1630 (1971).

24. J. W. Daiber, A. Hertzberg, C. E. Wittliff, "Laser Generated Implosions", Phys. of Fluids, 9, N 3, 617 (1966).

25. H. Puell, H. Opower, H. G. Neusser, "Experiments with Two Laser Produced Interpenetrating Plasmas", Phys. Letts. 31A, N 1, 4 (1970).

26. N. G. Basov, O. N. Krokhin, G. V. Sklizkov, S. I. Fedotov, A. S. Shikanov, "Powerful Laser Installation with a Successive-Parallel System of Amplifiers for Plasma Heating", JETP, N 1, (1972).

27. M. P. Vaniukov, V. Venchikov, V. I. Isaenko, P. P. Pashinin, A. M. Prokhorov, JETP, To be translated.

NEUTRON GENERATION IN SPHERICAL IRRADIATION OF A TARGET BY HIGH-POWER LASER RADIATION

N.G. Basov, Yu.S. Ivanov, O.N. Krokhin, Yu.A. Mikhailov, G.V. Sklizkov, and S.I. Fedotov
P.N. Lebedev Physics Institute, USSR Academy of Sciences

One of the essential problems in the use of lasers for thermonuclear purposes is the determination of the dependence of the temperature and of the neutron yield on the laser radiation energy and on the heating conditions. A yield of not less than 10^4 neutrons from a CD_2 target heated by a sharply-focused nanosecond laser beam, at 50 J energy, was registered in [1]. In [2], where a solid D_2 target was used, the maximum neutron yield was $\sim 5 \times 10^4$ at a laser energy up to 100 J and a duration 3.5 nsec at half-altitude.

At large values of the light-pulse energy, sharp focusing of the radiation on the target surface is not the decisive factor, owing to the smearing of the high-temperature zone by the heat-conduction and gasdynamic mechanism of energy redistribution.

The development of a multiple-beam laser has made it possible to experiment with heating a spherical target by spherically symmetrical irradiation [3]. In this case the neutron yield from a heated solid deuterated-polyethylene target greatly exceeded the results obtained with sharp focusing. In contrast to the known experiments, the dimension of the heated target was approximately equal to the diameter of the focal spot, and the heated mass was determined by the mass of the particle.

The scheme used to focus nine laser beams on the target was analogous to that used in [3]. The focusing was with two-element lenses of f = 6 cm producing a focal-spot diameter of 20 μ. Thus, taking into account the divergence of the laser radiation, the latter was focused into a spot of 50 μ diameter. Unlike in [3], a divergent beam with angle 10^{-2} rad was used in the preamplifier system. Before reaching the splitting system, this beam was made parallel, with simultaneous compensation of the astigmatism in the laser system. The energy of all nine beams, for a duration of 6 nsec at the base, was 214 J, and the average temperature was 840 eV. To reduce reflection and for more uniform target irradiation, the focal plane of the objective was located a distance of 200 μ away from the target surface. The target diameter was 110 μ.

The neutrons were registered by three scintillation detectors located at different distances from the target [1]. To measure the time of flight, a time marker from a laser beam was aimed on each detector scintillator with the aid of a light pipe. The neutron pulses were thus identified by the delay time of the neutron signal relative to the laser signal, which was proportional to the time of flight of the neutrons from the target to the detector.

The quantitative measurements were performed with nuclear emulsions, using the recoil protons. A type NIKFI-R emulsion 300 μ thick was placed 6 cm from the target. A control emulsion from the same batch was located in a different room. After three experiments in which neutrons were recorded by the scintillation counters, both emulsions were developed simultaneously under the same conditions. The irradiated emulsion had 87 tracks per cm^2, corresponding to the recoil protons from the neutrons of the D-D reaction, while the control emulsion had 48 tracks. To verify that the photographic properties of both emulsions were identical, the number of stars in the same square centimeter was measured and turned out to be 45 and 49 respectively. It follows therefore that if the

Reprinted from: *JETP Letters*, 15, 417–419 (1972).

neutrons are assumed to be isotropically emitted from the plasma, the number of neutrons per flash is 3×10^6. The table lists the results of the temperature measurements for spherical targets with different radii. At a radius of 30 μ, the temperature reached 4 keV.

Target radius cm	Laser energy J	Average temp., eV	Neutron yield per pulse	
			exper.	calc.
$2.50 \cdot 10^{-2}$	600	40	–	–
$1.25 \cdot 10^{-2}$	202	120	–	10^2
$5.50 \cdot 10^{-3}$	214	840	$3 \cdot 10^6$	$8 \cdot 10^7$
$3.00 \cdot 10^{-3}$	232	$4 \cdot 10^3$	–	$1 \cdot 10^{10}$

It is interesting to estimate the plasma temperature and the parameter $n\tau$ under the conditions of our experiment. Since the laser-pulse parameters correspond to an intermediate heating regime (between the thermal-conductivity and gasdynamic stages), we present here two independent estimates. According to [5] we have for CD_2

$$T_{ther} = 6.5 \cdot 10^9 \frac{Q^{4/9} \tau^{2/9}}{r^{8/9} n^{2/9}} \text{ eV} = 3.6 \cdot 10^3 \text{ eV},$$

$$T_{gas} = 3.8 \cdot 10^2 \frac{Q^{4/9}}{r^{2/3}} \text{ eV} = 2.15 \cdot 10^3 \text{ eV},$$

$$n\tau_{ther} = 9.2 \cdot 10^6 \, T_{eV}^{3/2} = 2.4 \cdot 10^{12},$$

$$n\tau_{gas} = 0.6 \cdot 10^{11} \left(\frac{10^4}{T_{eV}} \right)^{1/2} Q^{1/9} = 2 \cdot 10^{11},$$

where Q is the radiation power in GW, r is the target radius in 10^{-2} cm, and n is the total ion density. The numerical values are given for Q = 8 GW, r = 30 μ, $n = 10^{23}$, and $\tau = 10^{-9}$ sec.

One cannot exclude the possibility that a certain role is played in our experiments by the cumulation effect. Estimates of the pressure, using the results of [4], give a value not less than 10^8 bar. It should be noted, however, that the laser-pulse duration used in the present study was much longer than the optimal value determined by the gasdynamic lifetime of the heated target, which was ∼0.5 nsec in our case.

The authors thank V.G. Larionova and L.I. Ivanova for help with the reduction of the emulsion data and also V.M. Groznov, A.A. Erokhin, N.N. Zorev, and N.V. Novikov for help with the work.

[1] N.G. Basov, V.A. Boiko, S.M. Zakharov, O.N. Krokhin, and G.V. Sklizkov, ZhETF Pis. Red. 13, 691 (1971) [JETP Lett. 13, 489 (1971); S.W. Mead, R.E. Kidder, and J.E. Swain, Preliminary Measurements of X-ray and Neutron Emission from Laser-produced Plasmas, Lawrence Radiation Laboratory (University of California), Preprint UCRL-73356 (1971).

[2] C.Yamanaka, T. Yamanaka, T. Sasaki, K. Yoshida, M. Waki, and H. Kang, Plasma Generation and Heating to Thermonuclear Temperature by Lasers, Institute of Plasma Physics (Nagoya University). Preprint IPPJ-117 (1972); F. Floux, D. Cognard, L. Denoeud, G. Pior, D. Parisot, J. Bobin, F. Delobeau, and C. Fauguignon, Phys. Rev. A (General Physics) 1, 821 (1970).

[3] N.G. Basov, O.N. Krokhin, G.V. Sklizkov, S.I. Fedotov, and A.S. Shikanov, Zh. Eksp. Teor. Fiz. 62, 203 (1972) [Sov. Phys.-JETP 35, No. 1 (1972)].

[4] N.G. Basov, V.A. Boiko, V.A. Gribov, S.M. Zakharov, O.N.Krokhin, and G.V. Sklizkov, ibid. 61, 154 (1971) [34, 81 (1972)].

[5] N.G. Basov and O.N. Krokhin, Vestnik AN SSSR, No. 6, 55 (1970); in: Intern. School of Physics Enrico Fermi, Corso 48, Acad. Press, 1971.

Anomalous Heating of a Plasma by a Laser

C. Yamanaka,* T. Yamanaka,* T. Sasaki, K. Yoshida, and M. Waki
Faculty of Engineering, Osaka University, Osaka 565, Japan

and

H. B. Kang
Institute of Plasma Physics, Nagoya University, Nagoya 464, Japan

As lasers have an ability to deliver a large amount of energy very rapidly to matter, one can produce a plasma of thermonuclear temperature by laser bombardment of matter. We observed a neutron yield from a solid deuterium target irradiated by the beam of a glass laser, which had a power of 20 GW for 2 nsec. The theoretically estimated threshold laser power for anomalous heating owing to the parametric instability was in agreement with the experimental result. Above this threshold, an increase in the electron temperature, the appearance of a fast-ion group, and an anomaly in the reflection of light from the plasma were observed. These phenomena were closely correlated with the neutron yield. At the high-temperature region above a few hundred electron volts, the anomalous heating plays an essential role in reaching the fusion-reaction temperature.

I. INTRODUCTION

With the development of high-intensity lasers, one can produce plasmas with high density and high temperature which are relevant to thermonuclear-fusion research.[1] The Limeil group[2] has reported substantial neutron emission from a solid deuterium target irradiated by a 7-nsec 4-GW Nd^{3+} laser. (Recently they used a 3.5-nsec 24-GW glass-laser system.)

In order to investigate the potential of laser produced plasma for a future fusion reactor, a detailed investigation of the properties[3] of these plasmas is very important. In particular the laser heating process is one of the most important topics to study. Classical absorption is weak at high temperatures and becomes inefficient as the temperature increases. The absorption length l near the cutoff density of glass-laser light is

$$l = 5 \times 10^{35}\, T_e^{3/2} n^{-2}\ (\mathrm{cm}),$$

where T_e is the electron temperature in eV and n is the density in cm^{-3}. For a density of $10^{21}\ \mathrm{cm}^{-3}$ and a temperature of 1 keV, the absorption length is 150 μ. Therefore the absorption is small within a focal spot of 100 μ in diameter with a plasma of this temperature and density. According to the classical process, the laser beam would be perfectly reflected or transmitted with very little absorption, depending on whether the laser frequency ω_0 is less than or greater than the plasma frequency ω_{ep}. Considering our experimental results, it seems that anomalous absorption due to nonlinear effects induced by the high field strength of the laser beam is important in heating the plasma.

In this paper we report the properties of laser plasmas from solid deuterium and LiH targets, measured by various kinds of diagnostics, and we discuss the correlation of the neutron yield to the anomalous effects.

II. EXPERIMENTAL ARRANGEMENT

The laser was composed of an oscillator and five amplifiers[4,5] constructed in our laboratory. Three alternative oscillators could be used. Two of these were glass lasers, operating in the nanosecond and picosecond ranges, respectively. The third was a YAG oscillator which had a narrow spectrum beam especially suitable for scattering spectroscopy. The nanosecond oscillator delivered a pulse whose duration could be varied[6] from 2 to 10 nsec with a rise time of 1 nsec. A pulse-shaping system consisting of a laser-triggered spark gap, a potassium diphosphate (KDP) Pockels cell, and Glan prisms was employed. The oscillator pulse could be amplified up to more than 40 J, corresponding to 20 GW for a 2-nsec pulse. The energy of the pulse was kept below its maximum attainable value as a precaution against damage to the glass lasers by the beam reflected from the target. The damage to the glasses has been com-

Reprinted from: *Physical Review*, **A6**, 2335 2342 (1972).

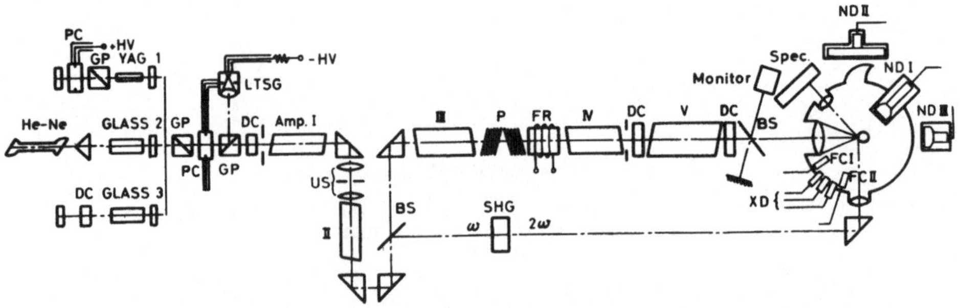

FIG. 1. Experimental arrangement of laser plasma research. 1: Pockels cell Q-switched yttrium-aluminum-garnet (YAG) laser, 2: rotating prism Q-switched glass laser, 3: mode locked glass laser, PC: Pockels cell, DC: saturable dye cell, XD: x-ray detector, ND: neutron detector, FC: Faraday cage, Spec: spectrometer, US: uniguide slit, A: attenuator.

pletely studied and has been reported elsewhere.[7] To protect the glasses from the reflection, we used a Faraday rotator, dye cells, and a special uniguide slit which was provided with an afocal lens system and a pin-hole slit as shown in Fig. 1. In this system the weak forward laser beam passes through the slit but the more intense backward one produces a plasma at the slit, which prevents the reflected beam passing through the slit. By these methods the reflected beam could be attenuated by more than a factor of 10^4.

The picosecond oscillator could deliver a single pulse selected from a train of picosecond mode locked pulses by the pulse transmission mode method, which has been described in a previous paper.[3] To amplify the picosecond pulse, a $POCl_3$-Nd^{+3} liquid[8] laser seemed to be most suitable since it has a larger transition cross section than that of a glass amplifier.

The yttrium-aluminum-garnet (YAG) laser was Q switched by a Pockels cell and fed to the glass preamplifier through the pulse-forming system. Then the beam was separated into two, one of which was used to generate second harmonics of a few MW by a KDP crystal and the other was sent to the main amplifier (see Table I).

The divergence of the output beam was less than 1 mrad. It was focused onto the target using an aspherical lens of focal length 50 mm and $f = 0.83$.

The experimental arrangement is shown schematically in Fig. 1. A cryostat of liquid helium produced a solid deuterium stick, the diameter of which was 2 mm, in a vacuum of 10^{-7} Torr.

III. EXPERIMENTAL RESULTS

A. Preliminary Results for Laser Plasma

To identify the properties of the plasma that depend on the type of laser target, we performed experiments on a small suspended particle and on a solid thick plate, each made of LiH. Streak photographs of a LiH particle 50 μ in diameter irradiated by a glass-laser beam were taken using an STL streak camera. The expansion velocity was about 5×10^7 cm/sec as shown in Fig. 2. At a distance of 30 cm from the irradiated point, H^{1+}, Li^{1+}, and Li^{2+} ions were resolved by an energy analyzer, and their velocities were found to be 2.4×10^7, 2.7×10^7, and 4×10^7 cm/sec, respectively.[9] This is in agreement with the well-known characteristics[10–14] of multicharged laser plasmas.

From our simple gas-dynamical computation,[15,16] when the glass-laser rise time is 2 nsec and the power is 10 GW, the temperature reaches a maximum of 300 eV within 1 nsec and then decreases. Applying a magnetic field to suppress the expansion, the decrease in the temperature reverses and the temperature can rise again to 600 eV after the expansion. Without a magnetic field, inertial confinement seems to finish after 1 nsec. The experimental values of electron temperature are in

TABLE I. Parameters of the glass amplifier.

	Osc.	Amp. I	Amp. II	Amp. III	Amp. IV	Amp. V
Rod dimension (mm)	$10^{\phi} \times 150^{l}$	$20^{\phi} \times 320^{l}$	$20^{\phi} \times 320^{l}$	$30^{\phi} \times 320^{l}$	$30^{\phi} \times 320^{l}$	$40^{\phi} \times 600^{l}$
Nd_2O_3 (wt%)	3.5	3.5	3.5	3.5	3.5	3.5
Flash lamp	2	4	4	6	6	10
Pumping energy (kJ)	0.8	9	9	13.5	13.5	60
Output power (MW)	4.5	31.5	160	720	2900	10^4
Gain		7	5	4.5	4	3.5

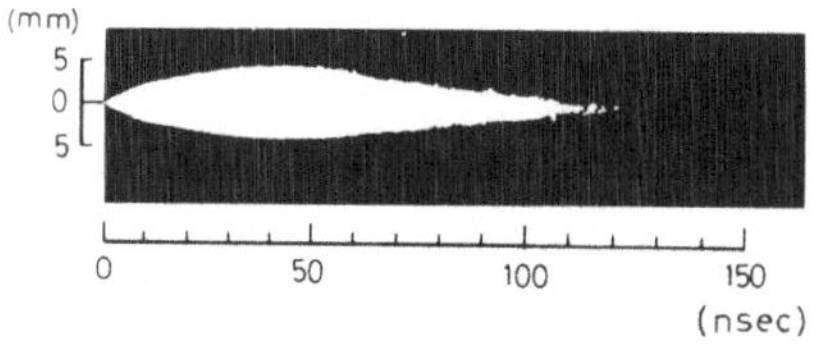

FIG. 2. Streak photograph of suspended LiH particle 50 μ in diameter irradiated by glass laser.

good agreement with the calculated values, which are shown in Fig. 3.

When we used a thick target, the plasma formed had some spatial structure[10,17] which was probably caused by the self-induced field.[18] After the appearance of the plasma, very-high-density neutral gas appears. The electron and neutral gas densities were measured[19] using Thomson and Rayleigh scattering of the laser light as shown in Fig. 4. The appearance of this neutral gas is a drawback for the magnetic confinement of a laser plasma.

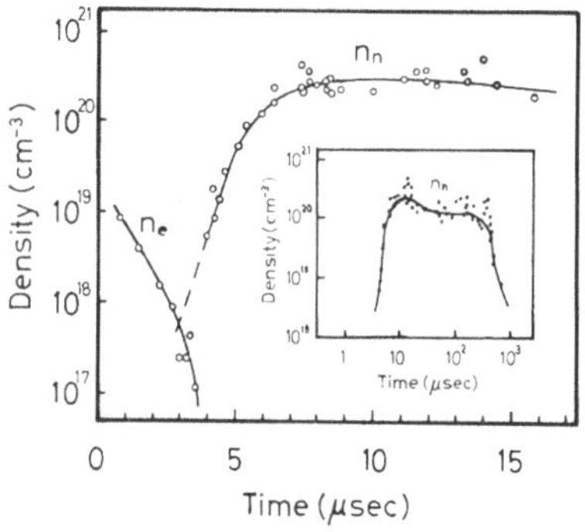

FIG. 4. Evaporation of dense neutral atoms from laser-irradiated solid target measured by light scattering.

B. Neutron Yield from Deuterium Targets

To investigate the laser-heating process for a deuterium plasma we examined the correlation between the plasma parameters and the neutron yield.

1. Electron Temperature

The electron temperature was measured from the absorption of soft x rays from the plasma by plastic scintillators with beryllium windows of different thickness (25, 50, and 100μ). The small type of photomultiplier HTV-R292 has a time resolution of 10 nsec. The experimental results for solid deuterium are shown in Fig. 5. The laser-pulse[13] duration did not seem to influence the electron temperature for the cases where the duration was 2, 4, and 10 nsec. The dependence of the electron temperature on the laser power seems to have an abrupt change at a power of 2 GW. Below this point the electron temperature had a dependence on laser power P of $P^{2/3}$, while above this point the dependence was $P^{1.2}$, and also the data began to scatter. The above point of discontinuity corresponded to a laser intensity of about 10^{13} W/cm^2; this will be referred to as the critical intensity.

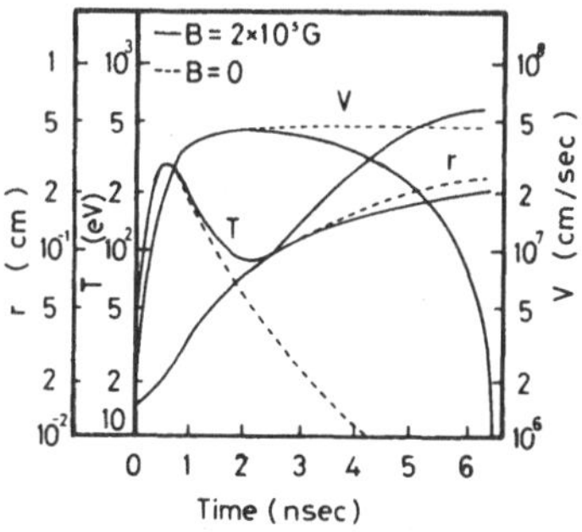

FIG. 3. Gas-dynamical behavior of laser-irradiated LiH particle, laser power: 10 GW, rise time: 2 nsec, focal spot size: 100 μ, particle size: 50 μ, magnetic field: 2×10^5 G.

2. Laser-Beam Reflection

A fairly large amount of the input laser energy was reflected by the target. The pulse form of the input and reflected laser light was measured by the same biplanar photodiode, type HTV-R317. The ratio of reflected energy to incident energy changed suddenly from 4 to 20% at a laser intensity of 10^{13} W/cm^2, i.e., at the critical intensity.

At the lower side of this intensity where the electron temperature was about 200 eV, the reflected laser light[20] was often observed to oscillate with a frequency of about 10^9 Hz.

3. Ion Collection and Time of Flight

As previously reported,[17,21] the ejection of the laser plasma from the target had some directional

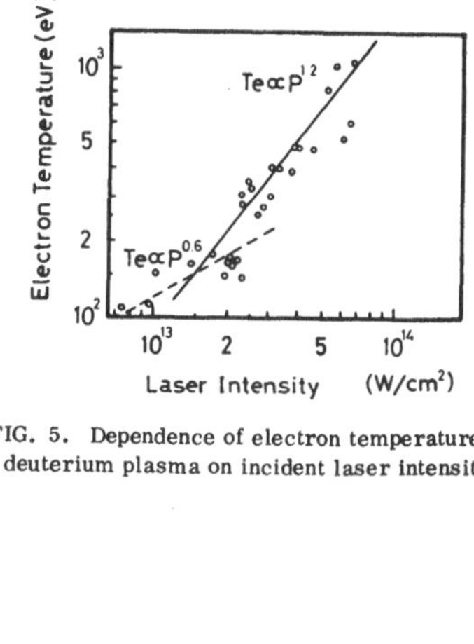

FIG. 5. Dependence of electron temperature of deuterium plasma on incident laser intensity.

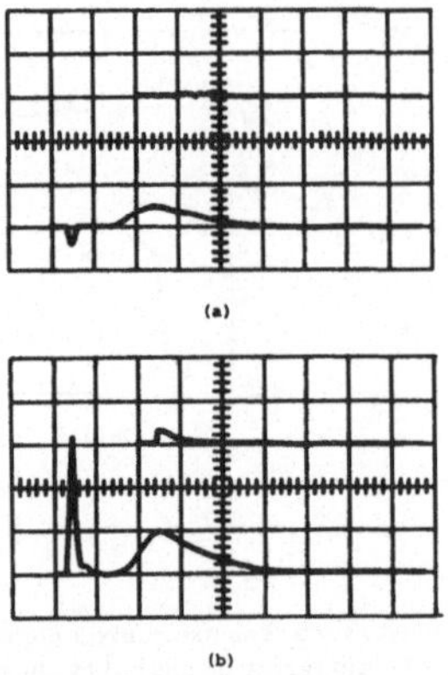

FIG. 6. Time of flight of ions from deuterium solid target. (a) laser power 1.1 GW, (b) laser power 3 GW; sweep time: 200 nsec/division. Upper trace: 60° from the incident beam, lower trace: 30° from the incident beam.

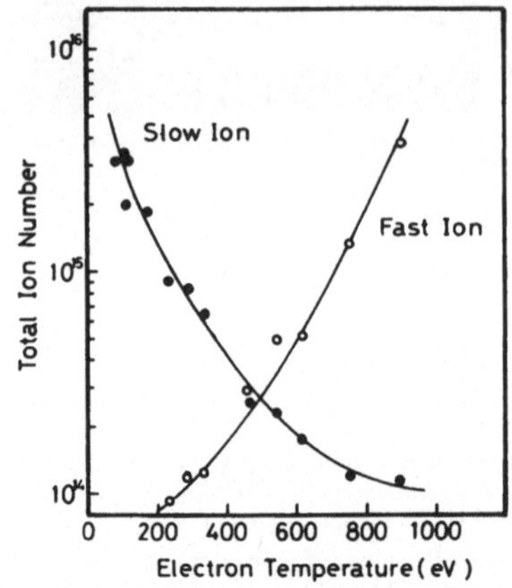

FIG. 8. Ion numbers of fast and slow components vs the electron temperature.

dependence. We measured the ion flux at 30° and 60° to the incident beam as shown in Fig. 6. The time-of-flight measurements showed the existence of two groups of ions, fast and slow, above the critical laser intensity. Below the critical point only the slow-ion component was observed. The dependence of the mean ion velocity V_k on the electron temperature is shown in Fig. 7. The number of ions in the slow component decreased with the increasing of electron temperature, while the number in the fast component increased as shown in Fig. 8. The ion temperature T_i [21,22] was estimated from the velocity profiles of the ions, which are shown in Fig. 9. The temperature of the fast component had a dependence on laser power of $P^{1.2}$, while the dependence of the slow component was $P^{0.3}$.

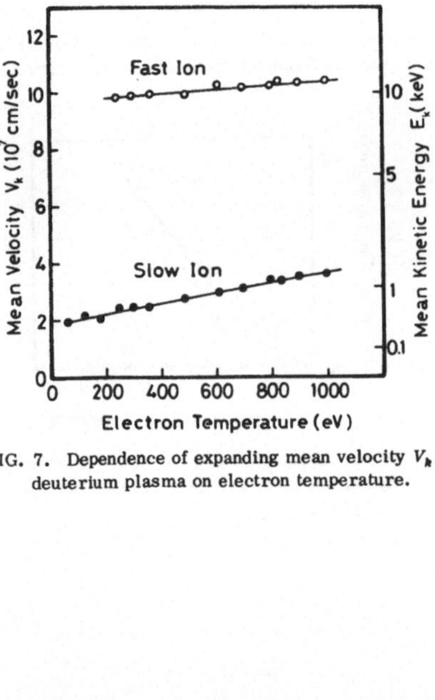

FIG. 7. Dependence of expanding mean velocity V_k of deuterium plasma on electron temperature.

The appearance of these fast ions exactly corresponded to the appearance of neutrons from the plasma.

The ratio of the energies of the fast ions and the slow ions was 4 to 1 at a laser intensity of 3×10^{13} W/cm^2, i.e., roughly at the critical point.

4. Neutron Yield

We used three plastic scintillators which were set at distances of 5, 10, and 40 cm from the target to detect the neutrons. The detectors were calibrated using an Am-Be neutron source.

The threshold laser energy for neutron emission was 5 J for a 2-nsec pulse, corresponding to just above the critical intensity of 10^{13} W/cm^2, and the electron temperature was about 500 eV. With a laser energy of 10 J neutrons were observed in 40% of the shots, and neutrons appeared in every shot when the laser energy was more than 15 J.

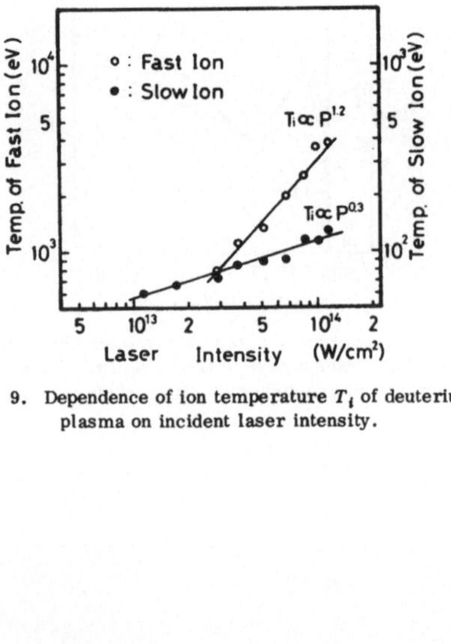

FIG. 9. Dependence of ion temperature T_i of deuterium plasma on incident laser intensity.

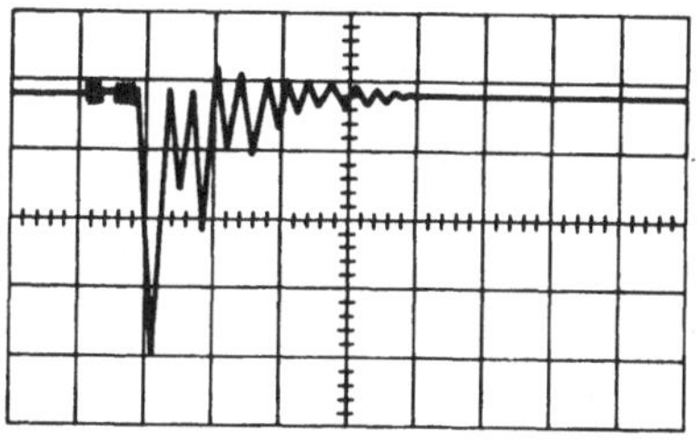

FIG. 10. Neutron yield from solid deuterium irradiated by laser beam. 50 nsec/division. Absorbed laser energy is 5 J.

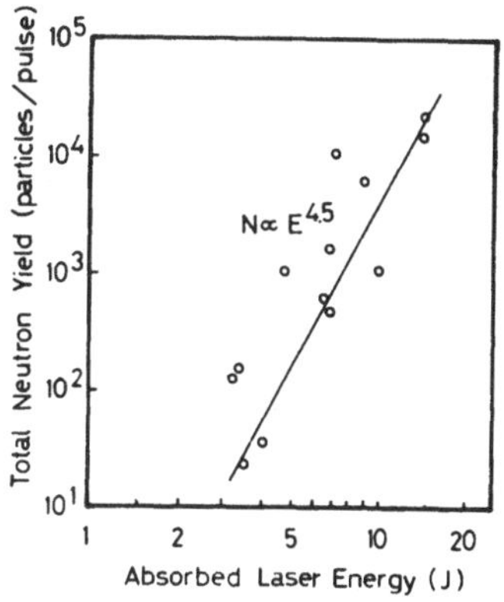

FIG. 11. Dependence of neutron yield on absorbed laser energy.

The position of the focal spot on the target was very critical for the production of neutrons. Figure 10 shows the response of a plastic scintillator for the neutron flux at the threshold. The dependence of the neutron yield on the absorbed laser energy is shown in Fig. 11. The number of neutrons N was strongly dependent on the laser energy ϵ, varying as $\epsilon^{4.5}$. When the number of collected fast ions was about 10^{15}, assuming the ion temperature to be 2.5 keV, then the calculated neutron yield is 10^4, which agrees with the experimental result. There was a close correspondence between the abrupt change of the electron temperature dependence on P, the appearance of fast ions, and the threshold of the neutron yield. Table II shows a sample of the experimental results leading to this conclusion.

IV. DISCUSSION

When a laser pulse, whose duration is larger than a few nsec, is used for plasma production, a hydrodynamic expansion takes place and the energy of the laser is diffused outwards. If a shorter pulse can complete the energy injection into the plasma before the development of the expansion, then we can expect effective heating. The laser energy is mainly supplied to electrons either by inverse bremsstrahlung in a collisional plasma or by an anomalous effect in the collisionless state.

In the region of electron temperature less than a few hundred electron volts and for a laser intensity of 10^{12} W/cm^2, the energy relaxation between electrons and ions is affected by electron-ion collisions. According to the classical theory, the electron-electron relaxation time τ_{ee} is 10^{-13} sec and the electron-ion relaxation time τ_{ei} is given by

$$\tau_{ei} = \frac{3}{8\sqrt{2\pi}} \; \frac{m_i T_e^{3/2}}{m_e^{1/2} e^4 z^2 n_e \ln\Lambda} , \tag{1}$$

and the ion transit time τ_s is given by

$$\tau_s \approx \frac{x_0}{v_{ac}} = x_0 \left(\frac{5zT_e}{3m_i}\right)^{-1/2} , \tag{2}$$

where x_0 is the plasma dimension, which is of the order of 10^{-2} cm, m_e and m_i are the electron and ion masses, z is the charge, n_e is the electron density, and v_{ac} is the sound velocity. When we put $T_e = 200$ eV, $n_e = 10^{21}$ cm^{-3}, then the values are $\tau_{ei} \sim 10^{-10}$ sec and $\tau_s \sim 10^{-9}$ sec. If the initial electron temperature increases, τ_{ei} becomes large and and relaxation decreases. The electron temperature T_e^* at $\tau_{ei} = \tau_s$ gives a condition for relaxation; i.e., for $T_e^* > T_e$ the ion temperature will approach the electron temperature. For a deuterium plasma T_e^* has been estimated[23] to be 400 eV. The dependence of the electron temperature on the laser power can be estimated by considering the energy relaxation in the hydrodynamic expansion of the

TABLE II. Properties of deuterium plasma produced by laser.

		Fast ion		Slow ion		
Input energy (J)	Electron temperature T_e (keV)	Mean energy E_k (keV)	Temperature (estimated from velocity spread) T_i (keV)	Mean energy E_k (keV)	Temperature (estimated from velocity spread) T_i (eV)	Total neutron yield
3	~0.2	...	...	0.7	70	...
5	~0.5	10	1.1	0.9	80	~300
12	~2	11	3.2	1.5	90	~5000
20	~4	~17	~7	2	140	~20000

plasma[24] and has been found to be $T_e \propto P^{2/3}$. Our experimental results show that this law holds up to a laser intensity of 10^{13} W/cm^2.

The maximum incident laser intensity was 10^{14} W/cm^2. At this intensity we may expect anomalous absorption of the laser energy in our experiments. According to the theory of anomalous absorption[25-30] predicted by Nishikawa,[25] two types of instability are expected near $\omega_0 = \omega_{ep}$. One is the oscillating two-stream instability, which appears at $\omega_{ek} > \omega_0$, where ω_{ek} is the Bohm–Gross frequency. The threshold P_T is given by the following equation:

$$P_T = 2\gamma_e(\omega_{ek}\,\nu_{ek}/\omega_{ep}^2)\,n_0\,c\kappa T_e \ (\mathrm{W/cm^2}), \qquad (3)$$

where γ_e is compression ratio of the electrons, ν_{ek} is the plasma wave damping decrement including electron Landau damping, and c is the velocity of light.

The other instability is the parametric instability which appears for $\omega_0 > \omega_{ek}$. The threshold intensity P_p is given by the following equation:

$$P_p = \begin{cases} \dfrac{2\sqrt{3}\,\gamma_e}{9}\,\dfrac{\nu_{ek}^2}{\omega_{ep}}\,\dfrac{\nu_{ik}}{\Omega_k^2}\,n_0 c\kappa T_e \ (\mathrm{W/cm^2}) & (\nu_{ek} > \Omega_k) \qquad (4) \\[2ex] \dfrac{\gamma_e\,\nu_{ek}\,\nu_{ik}}{\omega_{ep}\,\Omega_k}\,n_0 c\kappa T_e \ (\mathrm{W/cm^2}) & (\Omega_k > \nu_{ek}), \qquad (5) \end{cases}$$

where ν_{ik} is the ion acoustic wave damping decrement including ion Landau damping and Ω_k is the frequency of ion acoustic wave. We have estimated[26] the threshold for these instabilities in the range of our experiments and these are shown in Fig. 12. For our experimental conditions $n_e \approx 10^{21}$ cm^{-3}, $\omega_0 = 1.8\times10^{15}$ rad/sec, and Eq. (4) is satisfied in the low-temperature region. This expression has a minimum at $k\lambda_D = 0.2$, where k is the wave number of the instability and λ_D is the Debye shielding distance. The threshold laser intensity decreases with increasing T_e, varying approximately as $T_e^{-7/2}$, since the effective collision frequency of the Landau damping of an ion wave is smaller than the ion-ion collision frequency.

In the high-temperature region, Eq. (5) is satisfied and has a minimum at $k\lambda_D \approx 0.15$. In this region the effect of the Landau damping of the ion wave is dominant, the effective collision frequency of the ion wave is independent of temperature (the value of ν_{ik} is about 4×10^{12} Hz), and the threshold laser intensity vary as $T_e^{-1/2}$. The threshold of the parametric instability is smaller than that of the oscillating two-stream instability when the electron temperature is larger than 200 eV.

In our experimental conditions the electron temperature was 300 and 400 eV when the laser intensity at the focal point was 2×10^{13} and 3×10^{13} W/cm^2, respectively. These values almost agree with the calculated value 10^{13} W/cm^2. Up to an electron temperature of 200 eV, the heating process is mainly controlled by classical absorption,

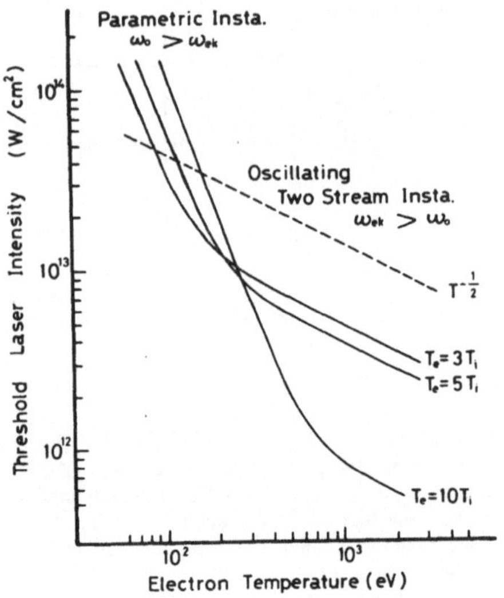

FIG. 12. Estimated threshold laser intensity of oscillating two-stream instability and parametric instability. D^+ plasma, $n_0 = 10^{21}$ cm^{-3}, laser frequency: $\omega_0 = 1.8\times10^{15}$.

but beyond this temperature the anomalous heating due to the parametric instability seems to dominate. The growth rate of this instability is known to be very fast above threshold[31] and the induced plasma waves tend to saturate and heat the plasma. As shown before, the electron temperature measurement of soft x ray verifies these conclusions. This anomalous effect seems to be very effective in the high-temperature range as shown in Fig. 5. At the threshold of neutron[32] yield we observed the appearance of a fast-ion component as well as the abrupt change of the electron temperature dependence on P. Concerning the sudden change of the reflection of the laser radiation by the plasma at this critical point, it seems possible that above the threshold the strong laser radiation produces a sharp boundary of very dense plasma, and this induces the strong reflection. While below the threshold the diffused plasma attenuates the reflected laser beam. The oscillation of reflected light just below the threshold may be due to a macroscopic variation of the plasma surface.

The incident laser energy is mainly taken up by the fast-ion component. The pulse duration of the nanosecond laser has no appreciable influence on the plasma properties at high temperatures.

At the present time experiments are being performed using a picosecond pulse to heat the plasma.

In conclusion, the plasma heating in the keV region, which is a very important subject for laser plasma research, appears to be mainly due to the anomalous absorption of laser light. Also we can

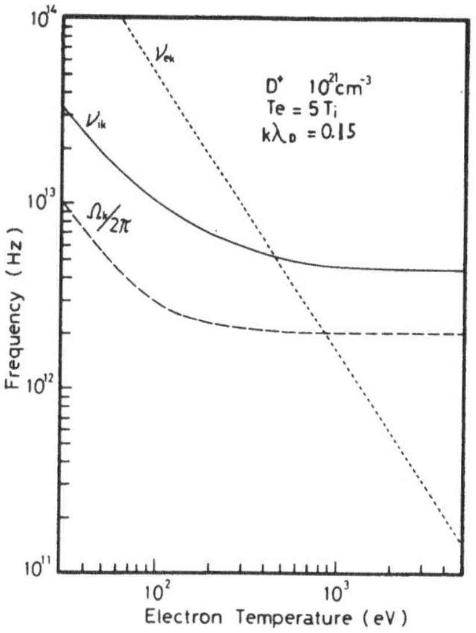

FIG. 13. Wave damping decrement of D⁺ plasma and ion acoustic frequency vs electron temperature. Plasma density is 10^{21} cm^{-3}, $T_e = 5T_i$, and $k\lambda_D = 0.15$.

say that the threshold intensity of the anomalous heating agrees very well with the theoretical prediction of the onset of the parametric instability of laser radiation and plasma in the hot-electron regime. However, the quantitative distribution of the energy between the electrons and the ions needs to be investigated more thoroughly.

ACKNOWLEDGMENTS

The authors wish to thank Dr. K. Nishikawa for stimulating discussion and Dr. Y. Sakagami and N. Miyajima for assistance in performing the experiments. The authors are very grateful to the Toray Science Promotion Society and also to the Ministry of Education for scientific research funds. Also the authors express their sincere thanks to Professor K. Husimi and Professor K. Takayama of the Institute of Plasma Physics, Nagoya University, for their encouragement.

APPENDIX A

In the estimation at high β of the plasma heating, we use the following equations[16]:

$$\left(P - \frac{B^2}{8\pi}\right) 4\pi r^2 \frac{dr}{dt} = \frac{1}{2}\bar{M} \frac{d}{dt}\left(\frac{dr}{dt}\right)^2 , \tag{A1}$$

$$P = \frac{3(N_i + N_e)}{4\pi r^3} \kappa T , \tag{A2}$$

$$\frac{3}{2}(N_i + N_e)\kappa \frac{dT}{dt} + \left(P - \frac{B^2}{8\pi}\right) 4\pi r^2 \frac{dr}{dt} = W , \tag{A3}$$

$$W = \begin{cases} \Phi(r/R_0)(1 - e^{-K_\nu r}) & (R_0 > r) \\ \Phi(1 - e^{-K_\nu r}) & (R_0 < r) , \end{cases} \tag{A4}$$

where N_i and N_e are the total number of ions and electrons; B is the applied magnetic field; $\bar{M}$ is the average plasma mass; r and R_0 are, respectively, the radius of plasma at time t and the focal spot of the laser beam; P is the plasma pressure; κ is the Boltzmann constant; T is the plasma temperature; W is the rate of energy absorption by the plasma; Φ is the input laser power as a function of time; and K_ν is the classical plasma absorption coefficient. A typical solution of these equations is shown in Fig. 3. We assumed that the initial temperature, electron density, and velocity of expansion due to precursor radiation were 10 eV, 10^{22} cm^{-3}, and 2×10^6 cm/sec, respectively. The laser light was simply absorbed by the classical process, and this solution shows the highest attainable temperature with classical absorption.

APPENDIX B

If the plasma has a Maxwellian distribution in energy, the wave damping decrement due to collisional and Landau damping of plasma waves and ion acoustic waves is approximately

$$\nu_{ek} = \nu_{ei} + \frac{1}{2}\pi \frac{\omega_{pe}}{k^3\lambda_D^3} e^{-1/2k^2\lambda_D^2} , \tag{B1}$$

$$\nu_{ik} = \nu_{ii} + \frac{1}{8}\pi\Omega_k\left(\frac{zT_e}{T_i}\right)^{3/2} e^{-zT_e/2T_i} \qquad (T_e \gg T_i) , \tag{B2}$$

where ν_{ei} and ν_{ii} are the collision frequency between electron-ion and ion-ion, respectively, T_e and T_i are the electron and ion temperature, and z is the average charge of the ions. In equation (B2), the Landau damping caused by the electrons has been neglected. The values of these under our experimental condition are shown in Fig. 13. The solid and dotted curves are, respectively, the wave damping decrement of ion acoustic waves and that of plasma waves. The dashed curve is the frequency of ion acoustic waves.

---

*Guest staff of Institute of Plasma Physics, Nagoya University, Nagoya, Japan.

[1]N. G. Basov, P. G. Kriukov, S. D. Zakharov, Yu. V. Senatsky, and S. V. Tchekalin, IEEE J. Quantum Electron. QE-4, 864 (1968).

[2]F. Floux, D. Cognard, L.-G. Denoeud, G. Piar, D. Parisot, J. M. Bobin, F. Delobeau, and C. Fauquignon, Phys. Rev. A 1, 821 (1970).

[3]C. Yamanaka, T. Yamanaka, T. Sasaki, H. Kang, K. Yoshida, and M. Waki, in International Quantum Electronics Conference, Kyoto, 1970, Digest of Technical Papers, p. 16 (unpublished); Proceedings of the International Conference on Laser Plasma, Moscow, 1970 (unpublished).

[4]T. Sasaki, T. Yamanaka, G. Yamaguchi, and C. Yamanaka, Japan J. Appl. Phys. 8, 1037 (1969).

[5]C. Yamanaka, T. Yamanaka, and T. Sasaki, in International Quantum Electronics Conference, Kyoto, 1970, Digest of Technical Papers, p. 404 (unpublished).

[6]K. Yoshida, T. Yamanaka, T. Sasaki, H. Kang, M. Waki, and C. Yamanaka, Japan J. Appl. Phys. 10, 1643 (1971).

[7]C. Yamanaka, T. Sasaki, M. Hongyo, and Y. Nagao, in Proceedings of the Conference on Damage in Laser Materials, Boulder, 1971 (unpublished); *ASTN Damage in Laser Materials*, edited by A. J. Glass and A. Guenther (National Bureau of Standards, Boulder, Colo., 1971), p. 104.

[8]C. Yamanaka, T. Sasaki, and M. Hongyo, IEEE J. Quantum Electron. QE-7, 291 (1971).

[9]M. Ohnishi and C. Yamanaka, Tech. Rept. Osaka Univ. 20, 121 (1970).

[10]P. Langer, G. Tonon, F. Floux, and A. Ducauge, IEEE J. Quantum Electron. QE-2, 499 (1966).

[11]B. C. Boland, F. E. Irons, and R. W. P. McWhirter, J. Phys. B 1, 1180 (1968).

[12]B. E. Patron and N. R. Isenor, Can. J. Phys. 46, 1237 (1968).

[13]M. Mattioli and D. Véron, Plasma Phys. 11, 684 (1969).

[14]W. Demtröder and W. Jantz, Plasma Phys. 12, 691 (1970).

[15]C. Yamanaka and T. Yamanaka, Progress Reports of Plasma Electronics, Osaka University, 1968 (unpublished).

[16]A. F. Haught and D. H. Polk, Phys. Fluids 9, 2047 (1966).

[17]T. Yamanaka and C. Yamanaka, Tech. Rept. Osaka Univ. 18, 155 (1968).

[18]J. A. Stamper *et al.*, Phys. Rev. Letters 26, 1012 (1971).

[19]Y. Izawa and C. Yamanaka, Japan J. Appl. Phys. 7, 954 (1968).

[20]M. Waki, T. Yamanaka, H. Kang, K. Yoshida, and C. Yamanaka, Japan J. Appl. Phys. 11, 420 (1972).

[21]H. Kang, T. Yamanaka, K. Yoshida, M. Waki, and C. Yamanaka, Japan. J. Appl. Phys. 11, 765 (1972).

[22]F. J. Allen, J. Appl. Phys. 14, 3048 (1970).

[23]N. G. Basov *et al.*, Quantum Radiophysics Laboratory Preprint No. 60, 1970 (unpublished).

[24]C. Fauquignon and F. Floux, Phys. Fluids 13, 386 (1970).

[25]K. Nishikawa, J. Phys. Soc. Japan 24, 916 (1968); 24, 1152 (1968).

[26]C. Yamanaka *et al.*, Phys. Letters 38A, 495 (1972).

[27]E. A. Jackson, Phys. Rev. 153, 235 (1967).

[28]P. K. Kaw, E. Valeo, and J. M. Dawson, Phys. Rev. Letters 25, 430 (1970).

[29]E. Valeo, C. Oberman, and F. W. Perkins, Phys. Rev. Letters 28, 340 (1972).

[30]D. F. DuBois and M. V. Goldman, Phys. Rev. Letters 28, 218 (1972).

[31]P. Kaw, J. Dawson, W. Kruer, C. Oberman, and E. Valeo, Princeton University, Institute of Plasma Physics, Report No. Matt-817, 1970 (unpublished).

[32]C. Yamanaka, T. Tamanaka, T. Sasaki, K. Yoshida, M. Waki, and H. B. Kang, Institute of Plasma Physics Nagoya University Research Report No. IPPJ-117, 1972 (unpublished).

Neutron Emission from Laser-Produced Plasmas*

G. H. McCall, F. Young, A. W. Ehler, J. F. Kephart, and R. P. Godwin
Los Alamos Scientific Laboratory, University of California, Los Alamos, New Mexico 87544

Neutron emission from laser-produced plasmas is shown, experimentally, to be adequately explained by electron heating and electrostatic ion acceleration. The absence of anomalous ion-heating mechanisms is not proved, but these mechanisms, if they exist, may be unimportant in reported experiments. It is concluded that no neutrons of thermonuclear origin are necessary to explain the results obtained thus far.

Because of the possible application of lasers for the production of thermonuclear fusion, much attention has focused on the generation of neutrons in laser-produced plasmas of deuterium or deuterated polyethylene (CD_2). Inverse bremsstrahlung and other mechanisms which selectively heat only electrons are discounted as ways of heating ions to thermonuclear temperatures because of the long electron-ion equilibration time in these plasmas.[1-4] Various anomalous ion-heating mechanisms are invoked to explain the neutron emission.[5-8] In the work described below, we show that anomalous ion heating is unnecessary and possibly incorrect as a description of the neutron-generation mechanism.

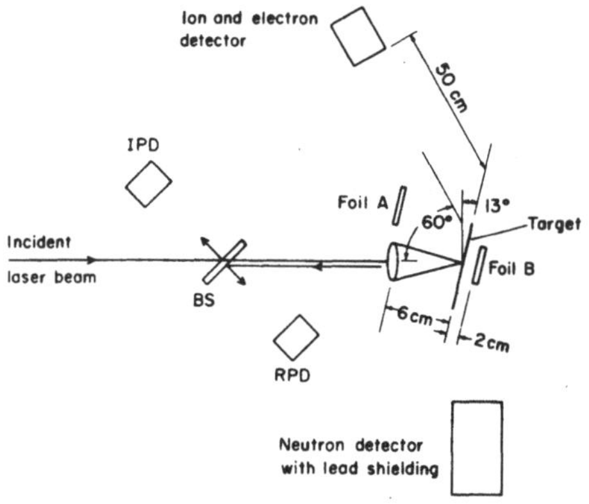

FIG. 1. Experimental setup. IPD and RPD are photodiodes for measuring incident and reflected laser energy, respectively. BS is a beam splitter.

It is well known that neutrons are observed in plasmas heated with 2- to 10-nsec pulse lengths,[9-12] but not to any great extent in plasmas heated with picosecond pulses.[10,13,14] We have investigated the problem experimentally using a mode-locked Nd:YAlG (yttrium aluminum garnet), Nd:glass laser capable of delivering up to 17 J, in a bandwidth-limited pulse approximately 25 psec in length, to a solid target. These experiments employed an $f/3.5$ aspheric lens which produced a spot diameter of 50 μm or less on a 125-μm-thick target of CH_2 or CD_2, giving a peak intensity of approximately 3×10^{16} W/cm^2 at the target. The experimental setup is indicated in Fig. 1. No attempt has been made to show all of the extensive set of diagnostics, but the relevant ones are indicated.

The neutron detector was a cylindrical Pilot-F scintillator 15 cm in diameter and 12.5 cm in length viewed by an RCA 4522 photomultiplier. The detector was calibrated using a known flux of 2.45-MeV neutrons from an accelerator, so its sensitivity and efficiency were known. One count would appear in the detector with 50% probability for isotropic emission by the plasma of 100 neutrons. Typical signals encountered corresponded to 1 to 20 neutrons detected. The rise time of the output pulse was approximately 5 nsec with a full width at half-maximum of 14 nsec. The neutron time of arrival was calibrated by removing the 3-cm-thick lead shielding between the detector and the target and observing the >100-keV x rays passing through the brass target chamber. Time was measured from an incident-light photodiode signal used to trigger the scope. This signal was also displayed on the

Reprinted from: *Physical Review Letters*, 30, 1116–1118 (1973).

Tektronix type-454 oscilloscope. A target of CH_2 was used to confirm the effectiveness of the lead shielding used to eliminate x rays in the detector.

Initially, no neutrons were observed. X rays having energies above 100 keV were detected, and a fast-ion component as shown in Fig. 2 was observed. This is identical to the x-ray and ion results for nanosecond plasmas except that for longer pulses the fast ions are accompanied by neutrons. To confirm that the fast ions were deuterons, a CD_2 foil (2 cm×2.5 cm) was placed 6 cm from the target in the position indicated as foil *A* in Fig. 1. A neutron signal was then observed with a delay time corresponding to the velocity of the fast-ion component ($\sim 2\times10^8$ cm/sec) observed at the ion detector. Assuming an ion energy of 65 keV and a *d-d* cross section given by the Gamow formula,[15] with a range over which significant neutron production occurs of 2.5×10^{-5} g/cm^2, one obtains approximately 2×10^{12} ions striking foil *A*. No neutrons were detected from foil *B* of Fig. 1, which was placed 3 cm behind the target (although for some shots a factor of 5 lower yield would have been undetectable), so the fast ions are assumed uniformly distributed in the forward hemisphere. This assumption gives agreement between the foil and ion detector measurements and yields a total fast-ion number of 1×10^{14}. It is interesting that before foils *A* and *B* were inserted, a few neutrons were observed after approximately ten shots had been fired, with a delay time corresponding to generation at the chamber walls. Because of the short interaction range of a 65-keV deuteron, only a few monolayers of deuterium on the chamber walls are required to give the same result as a solid foil. This problem should be especially troublesome in experiments with solid deuterium targets. For example, the assumption of wall generation for at least part of the neutrons is not inconsistent with the data shown in Fig. 10 of Ref. 10. Thus, it is possible to observe fast ions in the absence of neutron production.

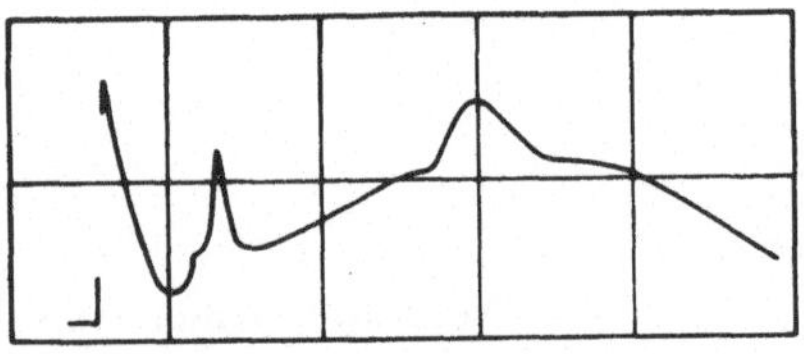

FIG. 2. Typical ion detector signal. Distance to target is 50 cm. Horizontal scale 200 nsec/div. Vertical scale 100 mV/div. Initial pulse is produced by x-ray emission from target. The second pulse is the fast-deuteron peak.

The fast-electron theory of Morse and Nielson[16] provides a possible explanation of the observed phenomena. X-ray spectral measurements[17] indicate a fast-electron component which gives a slope of 30 keV on an intensity-versus-energy plot. According to the calculation of Morse and Nielson, one should observe ion energies of the order of or, perhaps, twice that of the fast electrons. The calculation of the electron energy is based on flux balance between the laser input and the electron conduction out of the absorption region and is insensitive to the details of the electron heating mechanism. Using only flux balance, one obtains an electron energy which is in agreement with the x-ray spectral measurements. It should be noted that the ions are accelerated away from the target only, and no neutrons were observed from ion collisions in the target region. This is in agreement with a fast-electron model which predicts that the electric field on the target side of the absorption region should be canceled by the flow of cold electrons from the dense region into the absorption region; it also is in agreement with the absence of anomalous ion heating. Recent measurements with an ion spectrometer indicate that the ratio of the maximum fast-ion energy to the ion charge is a constant independent of the ion species. These measurements will be reported in detail later.

The differences in neutron-emission results for nanosecond and picosecond pulses can now be understood. The ion expansion, reflected-light threshold, and x-ray emission are similar for the two cases. This indicates, at least, a nonthermal mechanism for electron heating which may be only resonance absorption,[1,2] with the enhanced reflectivity being due to stimulated Brillouin scattering.[7] For nanosecond pulses, however, a low-density, low-temperature plasma is generated in front of the target before significant fast-electron production occurs, and when electrostatic acceleration occurs near the critical surface (where the plasma frequency and laser frequency are equal), the fast ions pass through the cold plasma generating neutrons. For picosecond pulses, no low-density region exists and the fast ions produce no neutrons. The neutron yield then depends on the time characteristics of the pulse, and this mechanism may explain the difference in results obtained by different investigators.

To test this hypothesis, the CD_2 target was irradiated by two pulses of approximately 6 J each separated in time by about 4 nsec. Neutrons were produced in the target region at a time corresponding to the second pulse on each of two shots. The ion energies and x-ray spectrum were the same as for the single pulse case. (No further data were taken.)

Thus, we believe that electrostatic acceleration of a relatively small number of ions is the dominant mechanism responsible for neutrons which have been observed in laser-produced plasmas. This is not to say that the proposed instabilities are entirely absent or that they will be unimportant as laser energies are increased.

---

*Work performed under the auspices of the U. S. Atomic Energy Commission.

[1]J. P. Freidberg, R. W. Mitchell, R. L. Morse, and L. I. Rudsinski, Phys. Rev. Lett. 28, 795 (1972).
[2]R. P. Godwin, Phys. Rev. Lett. 28, 85 (1972).
[3]L. Spitzer, *Physics of Fully Ionized Gases* (Interscience, New York, 1956).
[4]J. Dawson and C. Oberman, Phys. Fluids 5, 517 (1962).
[5]D. F. Dubois and M. Y. Goldman, Phys. Rev. Lett. 14, 544 (1965).
[6]K. Nishikawa, J. Phys. Soc. Jap. 24, 1152 (1968).
[7]D. W. Forslund, J. M. Kindel, and E. L. Lindman, Los Alamos Scientific Laboratory Report No. LA-DC-72-1355 (to be published), and Bull. Amer. Phys. Soc. 17, 1044 (1972).
[8]J. P. Freidberg and B. M. Marder, Phys. Rev. A 4, 1549 (1971).
[9]F. Floux, D. Cognerd, L. G. Denoeud, G. Pier, D. Parisot, J. L. Bobin, R. Delobeau, and C. Fauquignon, Phys. Rev. A 1, 821 (1970).
[10]C. Yamanaka, T. Yamanaka, T. Sasaki, K. Yoshida, M. Waki, and H. Kang, Phys. Rev. A 6, 2335 (1972).
[11]J. W. Shearer, S. W. Mead, J. Petruzzi, F. Rainer, J. E. Swain, and C. E. Violet, Phys. Rev. A 6, 764 (1972).
[12]K. Büchl, K. Eidmann, P. Mulser, H. Salzmann, and R. Sigel, Max-Planck-Institute für Plasmaphysik Report No. IPP IV/28, 1971 (unpublished).
[13]N. G. Basov, P. G. Krinkov, S. D. Zakharov, Yu. V. Senatsky, and S. V. Tchekalin, IEEE J. Quantum Electron. 4, 864 (1968).
[14]G. W. Gobeli, J. C. Bushnell, P. S. Peerey, E. D. Jones, Phys. Rev. 188, 300 (1969).
[15]W. R. Arnold, J. A. Phillips, G. A. Sawyer, E. J. Stovall, and J. L. Tuck, Phys. Rev. 93, 483 (1954).
[16]R. L. Morse and C. W. Nielson, Los Alamos Scientific Laboratory Report No. LA-4986-MS, July 1972 (to be published).
[17]J. F. Kephart, R. P. Godwin, and G. H. McCall, Bull. Amer. Phys. Soc. 17, 971 (1972).

Saturation of Stimulated Backscattered Radiation in Laser Plasmas*

L. M. Goldman,† J. Soures, and M. J. Lubin‡
Laboratory for Laser Energetics, College of Engineering and Applied Science, University of Rochester, Rochester, New York 14627

Measurements of the back reflection from laser-produced plasmas have been carried out for laser intensities of up to 1.5×10^{16} W/cm^2 with pulse durations of 120 psec using spherical low-Z targets. The reflectivity of the plasma increases with incident intensity but saturates at a value of 10%. The intensity-dependent reflectivity is attributed to stimulated Brillouin scattering.

Over the past several years theoretical calculations[1-3] and experimental measurements[4-6] have indicated that nonlinear processes such as stimulated Brillouin, Raman, or Compton backscattering might prove to be a limiting factor in the heating of a dense plasma by laser radiation. The experiments reported here indicate that for intensities below 1.5×10^{16} W/cm^2 the measured reflectivity (into the focusing system) of the plasma appears to saturate at a value close to 10%. The experimental intensities were above the expected threshold for Brillouin scattering, and near the threshold for stimulated Raman and Compton scattering.

The experiments were conducted with a Nd-glass laser producing 120-psec (full width at half-maximum) pulses varying in energy from 5 to 40 J. In all cases, a prepulse of approximately 0.5% of the total laser energy preceded the main pulse by 6 nsec. The laser beam was focused on spherical targets of lithium deuteride or $C_{36}D_{74}$ with a 20-cm focal length $f/2$ lens. From far-field measurements of the beam divergence, the spot size was determined to be less than 60 μm. The targets varied from 150 to 250 μm in diameter; the beam size was always small compared to the target dimension. The spherical shape of these targets should be noted since most of the previously reported experiments at high intensities[4-6] have used planar targets.

Measurements were made of the incident and reflected laser energy, the expansion energy in the plasma, the electron temperature of the plasma, the ionic species produced, the neutron yield, and the spectral characteristics of both the incident and reflected light. Details of all these measurements will appear in a forthcoming publication.[7]

The energy absorbed was deduced from the expansion energy in the plasma as determined by a set of charged-particle collectors and a time-of-flight mass spectrograph.[8] Figure 1 shows the measured spatial distribution of the ion density and the energy flux for a typical plasma. Integration of the energy flux over all angles gives the total expansion energy. This represents a lower limit on the absorbed energy as it does not include the energies of evaporation, ionization, or reradiation.

In Fig. 2(a) the dependence of reflectivity on incident light intensity is shown for two different experimental conditions. Reflectivity is defined as the ratio of light intensity reflected back through the focusing optics to the incident light intensity. In approximately half the experiments a single 120-psec pulse was used. In the remainder of the experiments there were two 120-psec pulses separated by 10 nsec, both having approximately the same amplitude.

In the single-pulse experiments the reflectivity increases with incident intensity. In the two-pulse studies the reflectivity rapidly reaches a constant level independent of incident intensity.

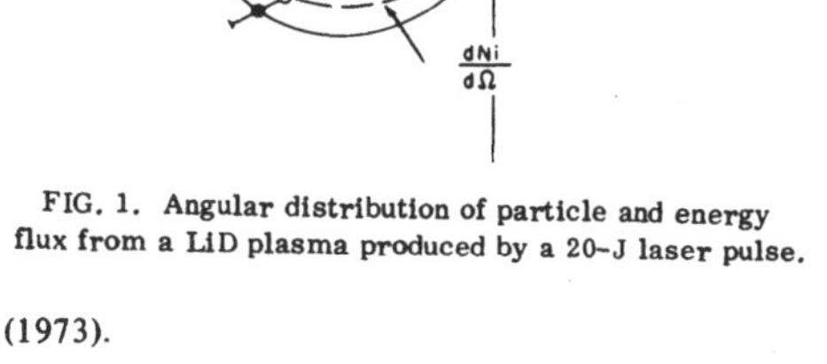

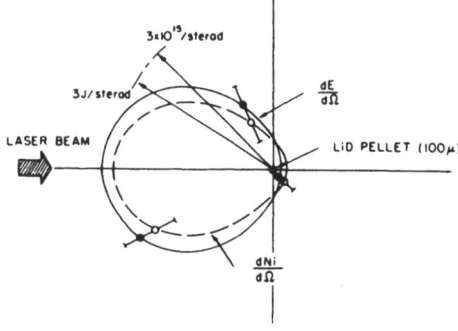

FIG. 1. Angular distribution of particle and energy flux from a LiD plasma produced by a 20-J laser pulse.

Reprinted from: *Physical Review Letters*, **31**, 1184–1187 (1973).

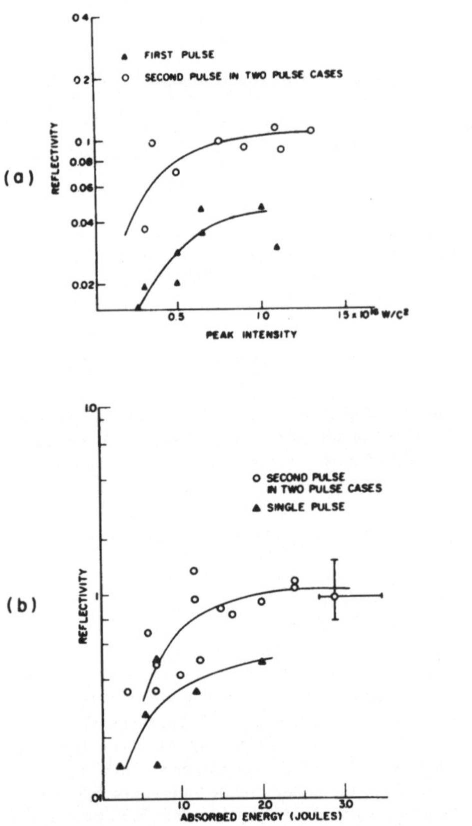

FIG. 2. Measured reflectivity (a) versus incident intensity and (b) versus absorbed energy.

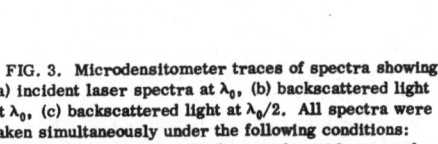

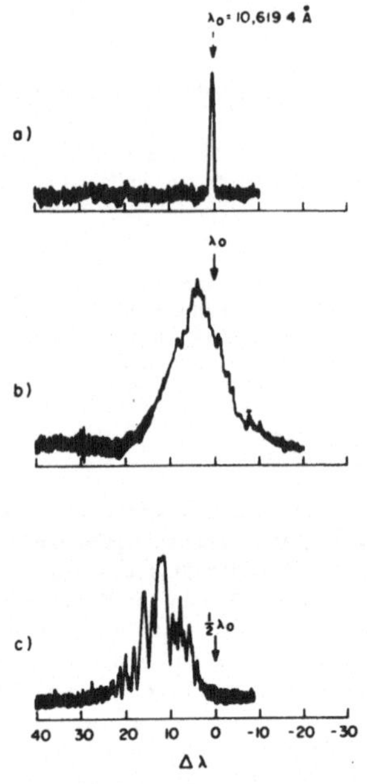

FIG. 3. Microdensitometer traces of spectra showing (a) incident laser spectra at λ_0, (b) backscattered light at λ_0, (c) backscattered light at $\lambda_0/2$. All spectra were taken simultaneously under the following conditions: single 30-J, 120-psec-wide laser pulse with a prepulse incident on a 150-μm LiD target.

An alternative method of presentation shows the variation of reflectivity with absorbed energy E_A, Fig. 2(b). The reflectivity is essentially constant from 1 to 3 J absorbed, indicating a rather complete saturation. It is of interest to note that the linearly polarized incident beam was depolarized by less than 10% upon reflection.

The spectrum of the backscattered radiation has been measured at both the incident wavelength and twice the indicent wavelength. The spectrum of the incident beam was typically less than 2 Å wide, centered around $\lambda_0 = 10\,620$ Å. A sample of all three spectra is shown in Fig. 3. In almost all cases the peak of the reflected radiation at λ_0 was shifted to the red by 2–6 Å and was broadened by amounts which ranged from 2 to 15 Å. At $\lambda_0/2$ the shifts were increased to 10–15 Å, with a broadening similar to that of the reflection at λ_0. The intensity of backscattering at $\lambda_0/2$ was only 10^{-3} of the backscatterer at λ_0, i.e., a maximum of 10^{-4} of the incident intensity.

There are a number of possible mechanisms contributing to the backscatter, such as reflection from the critical density interface, and stimulated Raman, Brillouin, and Compton scattering. A portion of the measured backscattered light is due to reflection from the critical interface. However, if one assumes that the plasma acts as a diffuse body reflecting 90% of the incident energy, then a maximum of 4% of that energy is reflected into the $f/2$ collecting optics. If one assumes that the reflection from the critical surface is specular, it is very difficult to estimate the fraction of light reflected back along the incident beam as this is a sensitive function of the shape and position of the reflecting surfaces. It is possible that this sensitivity contributes to the large scatter in the measured reflectivity at the lower intensities. The additional intensity-dependent reflectivity may be accounted for by stimulated mechanisms, and/or density modulations.[9] The Brillouin scattering threshold for an inhomogeneous plasma is given by Liu and Rosenbluth[1]:

$$P_B > (m^2c^4\omega_0/\pi e^2L)(v_e/c)^2 \text{ ergs/cm}^2\text{ sec}, \quad (1)$$

where L is a density scale length associated with the density profile $n(x) = n_c(1 + x/L)$, v_e is the electron thermal velocity, and ω_0 is the laser frequency. The measured electron temperature in these experiments is approximately 1.0 keV. The density scale length L is taken to be 3×10^{-3} cm, as determined from calculations of the density profiles produced by the prepulse.[10] The same scale length is used for the second-pulse case, although a longer scale length might be more appropriate. This leads to a threshold for stimulated Brillouin scattering of 3.0×10^{14} W/cm^2. The expected threshold for Raman scattering will be larger by a factor of c/v_e, leading to a threshold of 3×10^{16} W/cm^2 under the conditions described above. Using the measured values of the pulse energy, pulse width, and focal spot size one may calculate an average (rms) power on the target surface assuming the intensity is uniform across the focus. From such averaged intensities we derive peak intensities which range from 0.25×10^{16} to 1.5×10^{16} W/cm^2, and thus stimulated Brillouin rather than Raman scattering should be dominant. The shift in the backscattered spectrum is in the correct direction and of the proper magnitude for stimulated Brillouin scattering. This shift, as determined by the resonant conditions on ω and k, may be written as

$$\Delta\lambda/\lambda = (2n/c)(kT_e/m_i)^{1/2}, \qquad (2)$$

where n is the index of refraction in the plasma. At an electron temperature of 1 keV for the deuterium component of the plasma,

$$\Delta\lambda/\lambda = (1.33 \times 10^{-3})n. \qquad (3)$$

To account for the observed shift at 1.06 μm of 2–6 Å the refractive index would take on values from 0.14 to 0.42, representing densities varying from 0.98 to 0.90 of n_c, the critical density for 1.06-μm radiation.

An estimate of the effect of stimulated Compton scattering may be made following the analysis of Krasyuk, Pashinin, and Prokhorov.[11] The reflectivity due to Compton scattering can be written as

$$R = R_0 \exp[\beta I_0 L(1 - R)], \qquad (4)$$

where I_0 is the incident power density, R_0 is the reflectivity in the absence of Compton scattering, L is the effective plasma thickness, and

$$\beta = \frac{c r_0^2}{\pi \nu_0^3} \frac{v_0 n_e}{kT_e} \left(\frac{mc^2}{2\pi k T_e}\right)^{1/2} \exp\left(-\frac{m v_0^2}{2kT_e}\right), \qquad (5)$$

with v_0 the expansion velocity of the plasma and r_0 the classical electron radius. For the conditions of these experiments, namely, $v_0 \simeq 2 \times 10^7$ cm/sec, $T_e \simeq 1$ keV, $n_e \simeq 4 \times 10^{20}$, $L \simeq 3 \times 10^{-3}$ cm, and $\beta \simeq 1.4 \times 10^{-14}$ cm/W, we find

$$R \simeq R_0 \exp[(4.7 \times 10^{-17}) I_0 (1 - R)].$$

For $R_0 = 0.01$ this expression predicts a reflectivity of $R \simeq 0.013$ at 5×10^{15} W/cm^2 and $R \simeq 0.03$ at 2×10^{16} W/cm^2. This estimate indicates that stimulated Compton scattering could be contributing to the observed backscatter. One point which appears to argue against Compton effects is the spectral width, which is large compared to the width of the initial spectrum.[12]

It is our best estimate that the intensity-dependent portion of the observed reflection from the plasma is due to stimulated Brillouin scattering. Higher reflectivities reported in the literature may be due to differing target geometries, laser pulse width, and prepulse configuration.[4,6] The significant point is that for some geometries the effect saturates at a reflectivity of 10% and thus should not seriously limit energy absorption in the plasma at these power densities.

---

*Work supported by the Laser Fusion Feasibility Project.

†Permanent address: General Electric Company Corporate Research and Development, Schenectady, N.Y. 12301.

‡Department of Mechanical and Aerospace Sciences, and The Institute of Optics.

[1]C. S. Liu and M. N. Rosenbluth, Princeton Institute for Advanced Study Report No. COO 3237-11, 1972 (unpublished).

[2]D. W. Forslund, J. M. Kindel, and E. L. Lindman, Phys. Rev. Lett. 30, 939 (1973).

[3]A. A. Galeev, G. Laval, T. M. O'Neil, M. N. Rosenbluth, and R. Z. Sagdeev, Pis'ma Zh. Eksp. Teor. Fiz. 17, 48 (1973) [JETP Lett. 17, 35 (1973)].

[4]F. Floux, J. F. Bernard, D. Cognard, and A. Sahers, in *Laser Interaction and Related Plasma Phenomena*, edited by H. Hora and H. J. Schwartz (Plenum, New York, 1972), p. 409.

[5]C. Yamanaka, Ref. 4, p. 481.

[6]K. Eidmann and R. Sigel, Max-Planck-Institute für Plasmaphysik Report No. IPP IV/46, 1972 (unpublished).

[7]J. Soures, L. M. Goldman, and M. J. Lubin, to be published.

[8]M. Oron and Y. Paiss, to be published.

[9]A. V. Vinogradov, B. Y. Zel'dovich, and I. I. Sobelman, Pis'ma Zh. Eksp. Teor. Fiz. 17, 271 (1973) [JETP Lett. 17, 195 (1973)].

[10]A. Brauer, thesis, University of Rochester (unpublished).

[11]I. K. Krasyuk, P. P. Pashinin, and A. M. Prokhorov, Pis'ma Zh. Eksp. Teor. Fiz., 17, 130 (1973) [JETP Lett. 17, 92 (1973)].

[12]J. Peyraud, J. Phys. (Paris) 29, 872 (1968).

Pair Production by Relativistic Electrons from an Intense Laser Focus*

J. W. Shearer, J. Garrison, J. Wong, and J. E. Swain
Lawrence Livermore Laboratory, University of California, Livermore, California 94550

A preliminary discussion is given of electron-positron pair production by means of electrons accelerated to relativistic velocities at the focus of a laser beam. First, the pair-production cross section was numerically evaluated near its energy threshold. Then, two methods of relativistic electron production by focused laser light were considered: the coherent oscillation of electrons, and the acceleration of a few high-velocity electrons by plasma waves excited by laser-driven instabilities. It was found that the first method would produce pairs for neodymium-laser light at intensities close to the achievable limit ($10^{19}-10^{20}$ W/cm^2). The second method, however, may be capable of producing pairs at lower intensities.

I. INTRODUCTION

In recent years the brightness of pulsed laser light sources has increased from 10^{12} W/cm^2/sr up to the range of $10^{17}-10^{20}$ W/cm^2/sr at the neodymium-laser wavelength of 1.06 μ.[1,2] With a well-designed short-focus lens, such laser pulses can be focused to corresponding intensities of $10^{17}-10^{20}$ W/cm^2. At these intensities electron-positron pair production by the strong electromagnetic field at the focus might be possible either by direct vacuum pair production or indirectly from relativistic electrons accelerated by these strong fields.

Two calculations of the probability of vacuum pair production by multiphoton absorption have been published[3,4]; both of them indicate that the production of a detectable number of pairs from this process would require many more orders of magnitude of intensity than contemporary lasers provide. A simple order-of-magnitude estimate can show this. For an appreciable pair-production probability, the pair energy $2m_0c^2$ should be of the order of the electric-field potential energy at one Compton wavelength $\bar{\lambda}$. That is,

$$eE_l \bar{\lambda} \simeq 2m_0c^2, \tag{1}$$

where E_l is the average electric field of the focused laser light. Therefore, the intensity I_l of the laser light in vacuum can be written

$$I_l = c\epsilon_0 E^2 \simeq \frac{m_0c^3}{\pi r_0 \bar{\lambda}^2}, \tag{2}$$

where r_0 is the classical radius of the electron:

$$r_0 = e^2/4\pi\epsilon_0 m_0 c^2. \tag{3}$$

Numerically, this is an intensity of 2×10^{30} W/cm^2, which is many orders of magnitude greater than available intensities, but is in rough agreement with values previously cited.[3,4]

However, it is well known that focused laser pulses create hot plasmas in matter. It has been pointed out that there exists the "wholly real possibility" of observing pairs produced in such laser-plasma experiments by means of the excitation of high-energy electrons.[5] When the electron kinetic energy E_v exceeds the pair-production threshold $2m_0c^2$, the fast electron can produce an electron-positron pair by scattering in the Coulomb potential of a nucleus as first calculated by Bhabha.[6] This is often called the "trident" process (see Fig. 1). In this paper we shall discuss some mechanisms by which energetic electrons can be created in the laser-plasma focus, and we shall attempt to evaluate the number of electron-positron pairs which can be produced by the subsequent trident process.

II. TRIDENT CROSS SECTION AT ELECTRON ENERGIES NEAR THE PAIR-PRODUCTION THRESHOLD

We have calculated the cross section σ_T for the trident pair-production process by two different

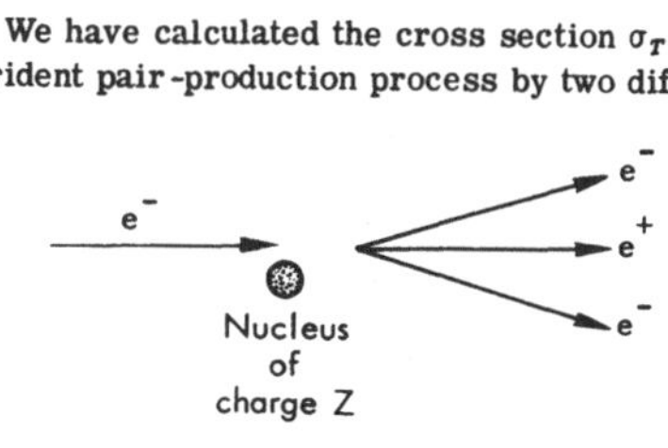

FIG. 1. Schematic of trident process of pair creation.

Reprinted from: *Physical Review*, **A8**, 1582–1588 (1973).

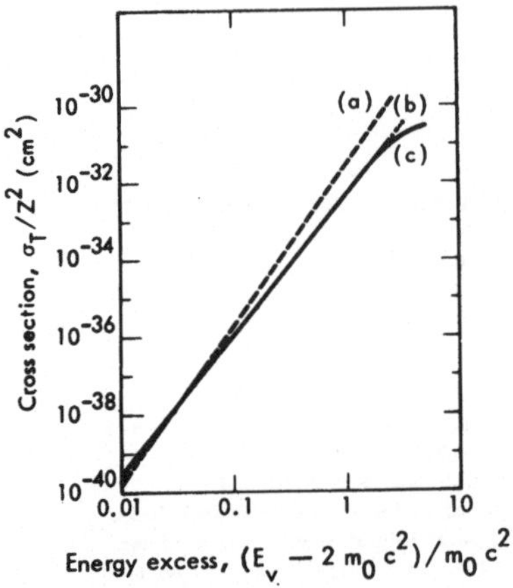

FIG. 2. Total cross section σ_T of trident process plotted vs the dimensionless energy excess above threshold, where E_v is kinetic energy of the incident electron. Curve (a) is Bhabha's analytical calculation (Ref. 6) for this range [see Eq. (4)]. Curve (b) is our computer calculation based on Brodsky and Ting's method (Ref. 7). Curve (c) is our approximate fit to the computer calculation.

methods; the results are plotted in Fig. 2. In the first method we integrated Eq. (30) of Bhabha's paper,[6] and obtained

$$\sigma_T = \frac{(\alpha r_0 Z)^2}{128}\left(\ln\frac{1}{\gamma_B^2} - \frac{161}{60} + c_1 + c_2 + c_3\right)\left(\frac{E_v}{m_0c^2} - 2\right)^4, \tag{4}$$

where α is the fine-structure constant, r_0 is the classical electron radius, Z is the nuclear charge, and γ_B is defined by

$$\frac{1}{\gamma_B} \equiv 1 + \frac{E_v}{m_0c^2}. \tag{5}$$

The symbols c_1, c_2, and c_3 in Eq. (4) are lengthy algebraic functions of γ_B which are given in Eq. (28) of Bhabha's paper. Numerical calculations of Eq. (4) are plotted in Fig. 2.

Bhabha's approach has the merit of yielding a convenient analytical expression for the cross section; however, he makes two approximations having attendant errors that are difficult to estimate. In the first place, the initial electron is described by a classical straight-line trajectory; secondly, the interference effects between the two electrons in the final state are neglected. Both approximations are avoided by evaluating the lowest-order Feynman diagrams which are applicable to the trident process. Our numerical calculation uses a program for evaluating these diagrams which was written by Brodsky and Ting.[7] The program numerically evaluates the necessary products of Dirac matrices and takes the trace to get the differential cross section. This procedure is more accurate than the usual one, which involves algebraic reduction of the matrix products. The differential cross section is then numerically integrated to obtain the total cross section as a function of incident-electron energy. The results of these calculations are also plotted in Fig. 2.

It is seen that the second, more exact calculation gives lower cross-section values over most of the energy range of interest—near the trident-production threshold. The reduction in cross section is presumably due to inclusion of the interference between direct and exchange scattering, an effect which was omitted in Bhabha's calculation. For the subsequent portions of this paper we use the equation

$$\sigma_T \cong 9.6\times10^{-4}(\alpha r_0 Z)^2\left(\frac{E_v}{m_0c^2} - 2\right)^{3.6}, \tag{6}$$

which approximates the more exact cross-section curve. In Fig. 2 we see that the fit of the approximation is good up to kinetic energies of $\simeq 4m_0c^2$, which is a sufficient range for our problem. The possibility of pair production by bremsstrahlung from the electrons, mentioned in a footnote of Ref. 7, is not as likely as trident production in this energy range because it is a two-step process.

III. HIGH-INTENSITY CIRCULARLY POLARIZED LIGHT

We consider first the coherent motion of plasma electrons in the electromagnetic wave of the laser light. Because the pair-production threshold is twice the rest energy, this motion must be treated relativistically. This treatment is more difficult than the well-known nonrelativistic treatments.[8] We have not treated the linearly polarized wave, because it has been shown to be coupled to a longitudinal plasma wave.[9] However, the circularly

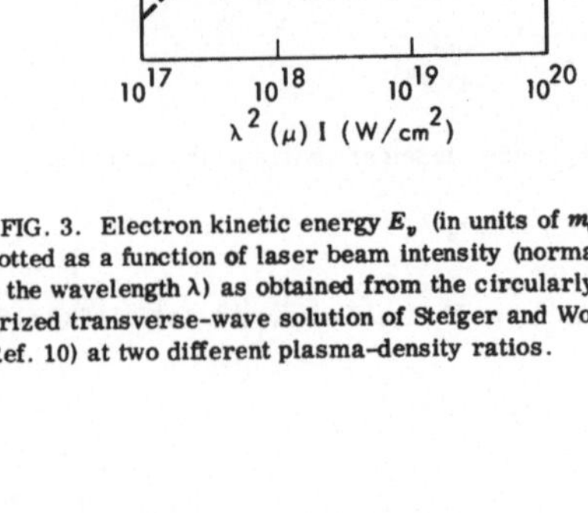

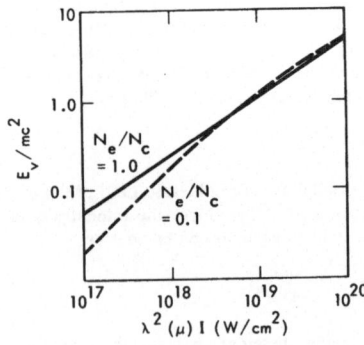

FIG. 3. Electron kinetic energy E_v (in units of m_0c^2) plotted as a function of laser beam intensity (normalized to the wavelength λ) as obtained from the circularly polarized transverse-wave solution of Steiger and Woods (Ref. 10) at two different plasma-density ratios.

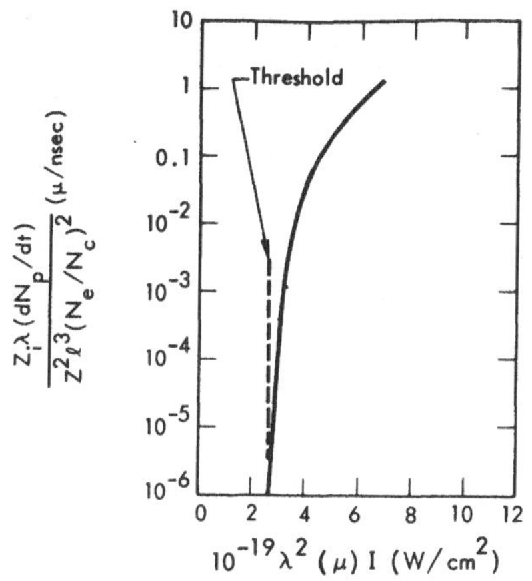

FIG. 4. Density-normalized pair-production rate plotted vs laser intensity, as computed from Eqs. (7) and (13).

polarized wave is a simple transverse wave at all intensities, and a convenient solution for this case has been obtained by Steiger and Woods.[10] We shall use these authors' results for the case which includes the relativistic mass change and the inverse Faraday effect, but we neglect energy losses caused by radiation by the electron.

Steiger and Woods[10] found that at high circularly polarized laser-beam intensities where the coherent electron-orbit velocity is relativistic, the electron kinetic energy E_v is a strong function of laser intensity, but is almost independent of plasma density. We replotted this result (Fig. 3) and found that a useful approximation is given by the following simple power law:

$$\frac{E_v}{m_0c^2} \cong \left(\frac{\lambda^2 I}{\lambda_0^2 I_0}\right)^n = \left(\frac{\lambda^2 I}{9\times 10^{18}}\right)^{0.657}, \qquad (7)$$

where the laser wavelength λ is measured in μ and the intensity I is in W/cm^2.

An equation that describes the rate dN_p/dt of creation of pairs by means of the trident process in a volume whose characteristic dimension is approximately l times the wavelength λ is

$$\frac{dN_p}{dt}\ (\text{pairs/s}) = (l\lambda)^3 N_i N_e \sigma_T v_e\ , \qquad (8)$$

where N_i is the ion (nucleus) density and v_e is the velocity of the electron. Assuming that the thermal velocity can be neglected in comparison with the relativistic coherent velocity of gyration in the intense electromagnetic field, we have

$$v_e \equiv \beta c = (c/\gamma)(\gamma^2-1)^{1/2}, \qquad (9)$$

where the coefficient γ is the normalized total electron energy

$$\gamma = 1 + \frac{E_v}{m_0c^2} = 1 + \left(\frac{\lambda^2 I}{9\times 10^{18}}\right)^{0.657}, \qquad (10)$$

where we have substituted from Eq. (7).

Consider a plasma containing ions of charge $Z_i = N_e/N_i$, where Z_i is not necessarily the nuclear charge Z. Substitute Eqs. (6) and (9) into Eq. (8) to obtain the result

$$\frac{dN_p}{dt} = \frac{9.6\pi^2}{10^4}\left(\frac{c}{\lambda}\right) l^3\alpha^2\frac{Z^2}{Z_i}\left(\frac{N_e}{N_c}\right)^2(\gamma-3)^{3.6}\frac{(\gamma^2-1)^{1/2}}{\gamma}, \qquad (11)$$

where the density N_c,

$$N_c \equiv \epsilon_0 m\left(\frac{\omega_L}{e}\right)^2 = \frac{\pi}{r_0\lambda^2}, \qquad (12)$$

is the nonrelativistic cutoff-density parameter at which the plasma frequency ω_p equals the laser frequency ω_L. In convenient units, Eq. (11) becomes

$$\frac{dN_p}{dt}\ (\text{pairs/ns}) = \frac{0.15}{\lambda(\mu)}\frac{Z^2}{Z_i} l^3\left(\frac{N_e}{N_c}\right)^2(\gamma-3)^{3.6}\frac{(\gamma^2-1)^{1/2}}{\gamma}. \qquad (13)$$

Equations (7) and (13) are presented in graphical form in Figs. 4 and 5.

The pair-production rate threshold ($\gamma = 3$) corresponds to a threshold laser beam intensity I_T of

$$I_T\left(\frac{\text{W}}{\text{cm}^2}\right) = \frac{2.6\times 10^{19}}{[\lambda(\mu)]^2}. \qquad (14)$$

The pair-production rate then rises steeply to interesting values at somewhat higher intensities.

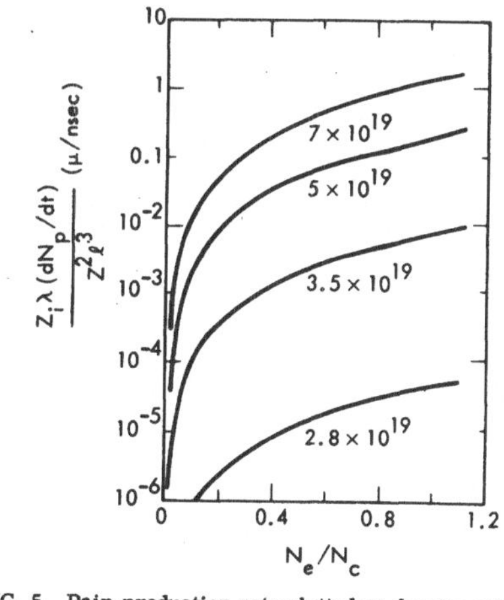

FIG. 5. Pair-production rate plotted vs density ratio for several values of the laser intensity. The figures on the curves are $\lambda^2 I\,[(\mu)^2$ (W/cm^2)]. The curves continue beyond cutoff [$(N_e/N_c) > 1$] because of the possibility of relativistic beam penetration (Refs. 9 and 11).

These intensities (greater than 10^{19} W/cm^2 for neodymium lasers) are at the upper limit of current laser practice.[1,2]

Of course, this is an idealized single-particle calculation which neglects unstable collective effects in the plasma. One should regard it mainly as an order-of-magnitude estimate that trident-process pair production is hard to produce by means of the coherent electron motion alone.

When the electron motion is relativistic the laser beam can penetrate an overdense plasma,[9,10] because the plasma current is limited to the value $N_e ec$, instead of increasing with increasing beam intensity.[11] In such cases the pair-production rate will be enhanced, provided the intensity is above the threshold.

IV. DISCUSSION OF PLASMA INSTABILITIES AT LOWER LIGHT INTENSITIES

Anomalous heating of plasmas by intense electromagnetic waves is now recognized as a significant effect both at radio-wave frequencies and at laser frequencies.[12,13] The incident wave energy is coupled into plasma waves by plasma instabilities.[14] One of these is the ion-acoustic decay instability[15] and the other is the oscillating two-stream instability.[16]

According to numerical computations of these processes,[13,17] the growth of the amplitude of the plasma waves is sufficiently rapid for the instability mechanism to saturate, causing the plasma to become turbulent. Many such calculations show a high-velocity, high-energy group of electrons, which is called a "suprathermal electron tail" on the Maxwellian electron-velocity distribution. One physical picture of this process is that some of the electrons are trapped by high-phase-velocity, high-amplitude plasma waves.[13,17]

Although these numerical calculations were nonrelativistic, the same qualitative arguments would be expected to hold for plasma waves whose phase velocity v_p is nearly the velocity of light ($v_p \to c$). Thus we must examine the plausibility of relativistic electron production by these plasma instabilities. If relativistic electrons ($E_v > 2m_0c^2$) are produced, the trident mechanism for pair production is possible.

The dispersion relation for longitudinal plasma waves can be written in the form[18]

$$v_p^2 = (\omega_p^2/K^2) + 3v_t^2 , \tag{15}$$

where K is the wave number and v_t is the thermal velocity of the electrons. Let us examine the wave whose phase velocity v_p is equal to the velocity of light, because waves which trap relativistic particles will have velocities which closely approach this limiting solution. Then for $v_p = c$, we find

$$(K_c \lambda_D)^2 = v_t^2/c^2 - 3v_t^2 , \tag{16}$$

where K_c is the corresponding wave number and λ_D is the Debye length of the plasma ($\lambda_D\omega_p = v_t$). From Eq. (16) we find that waves which are potentially capable of trapping electrons in the relativistic energy range tend to have small values of K_c, corresponding to longer-wavelength waves. As $v_t \to 0$, $K_c \to \omega_p/c$, as can be seen by eliminating λ_D from Eq. (16). Thus in the low-temperature limit, the wavelength of the longitudinal wave of velocity c is equal to the vacuum wavelength of an electromagnetic wave whose frequency is equal to the plasma frequency.

Now we ask whether plasma waves of wave number K_c are likely to be excited in the plasma by laser-driven instabilities. Consider first the threshold conditions for the ion-acoustic decay instability. The most unstable wave number K_p can be written[19]

$$K_p^2 = \frac{\omega_L^2 - \omega_p^2}{3v_t^2} . \tag{17}$$

If we set $K_p = K_c$, and make the approximation $v_t \ll c$, we find from Eqs. (16) and (17) that

$$1 - \frac{\omega_p^2}{\omega_L^2} = 3\left(\frac{\omega_p^2}{\omega_L^2}\right)\frac{v_t^2}{c^2} . \tag{18}$$

Here Eq. (12) can be used to find the plasma density

$$\Delta N/N_c \cong 3(v_t^2/c^2) , \tag{19}$$

where ΔN is the difference between the cutoff density N_c and the electron density N_e and where $\Delta N \ll N_c$.

This result shows that electron plasma waves of velocity $v_p \simeq c$ can be excited by the parametric ion-acoustic instability at plasma densities close to the cutoff density. This is just the plasma-density regime where the threshold intensity for this instability is low.[20] For example, if the electron temperature T_e is 1 keV, we find that $N_e = 0.994\ N_c$. Also, we find that in this case the wavelength of this longitudinal wave is approximately the same as the vacuum wavelength of the laser radiation. Because most focal spots used in practice are at least several wavelengths in diameter, several wavelengths can build up inside the focus.

At incident laser intensities high above threshold, the ion-acoustic decay instability will be excited at lower plasma densities, where the most unstable mode will have a phase velocity less than the velocity of light. However, other modes will also be excited, particularly after saturation of the initial growth of the instability. Plasma-simulation calculations[17] show that the wave-number spectrum is enhanced in the low-K regime after saturation. Thus, long-wavelength plasma waves whose phase velocity is comparable to c should be excited over a wider range of densities in the plasma.

Another requirement for relativistic electron production is that the plasma-wave amplitude be sufficiently high to provide the necessary acceleration electric field. In order to reach the threshold energy for trident pair production ($2m_0c^2$), we must have

$$2m_0c^2 \cong eE_p(\lambda_p/2) \pm \tfrac{1}{2}m_0v_{ep}^2 , \tag{20}$$

where E_p is the average electric field of the plasma wave, λ_p is the wavelength, and v_{ep} is the velocity component of the individual electron in the direction of the electric field E_p. The number of electrons that will be accelerated thus depends not only on the wave intensity, but also on the shape of the velocity distribution on the "tail" of the distribution function. If the suprathermal tail is sufficiently large, an appreciable number of electrons can be accelerated to the threshold for trident pair production.

We conclude that it may be possible to obtain pair production from electrons accelerated in the turbulent plasma environment of anomalous absorption instabilities. However, we have not been able to estimate the probability of this process, or to make quantitative estimates of the production rate of the relativistic electrons. Such a capability awaits development of a relativistic theory or a relativistic plasma-simulation numerical code which can be applied to the inhomogeneous plasmas which are produced within the small dimensions of the focal-spot region of the focused laser light.

V. EXPERIMENTAL EVIDENCE FOR RELATIVISTIC ELECTRONS

In Sec. IV we have shown the plausibility of relativistic electron production by the plasma instabilities associated with the anomalous absorption of laser radiation. Experimental indications of such electrons have recently been seen at this laboratory.

These experiments were done with our "long-path" neodymium-glass-disk laser system.[21,22] A 6-ns double pulse was used whose intensity peaks were separated by 3 ns; the pulse shape was as shown in Ref. 21, not Ref. 22. The total output energy in each pulse was 80± 10 J. One difference from the older work was that the light path through the system made only five passes through the disks, rather than nine passes. Optical spectrometer measurements of this new pulse indicated a narrower over-all spectral width (≃ 60Å) than the previous pulse (≃ 100Å). A detailed account of the substructure of the output pulses from the long-path laser is available.[23]

Although the narrowing of the over-all spectral width seemed like a minor change, an unusually penetrating hard component of the x rays was seen when this new pulse was incident on our standard polyethylene target [approximate composition $(CH_2)_n$]. The detector was a plastic fluor (15-cm diam × 20 cm long) originally intended for neutron measurements with $(CD_2)_n$ targets; its front face was located 22 cm from the target. The absorption curve obtained from these measurements is plotted in Fig. 6.

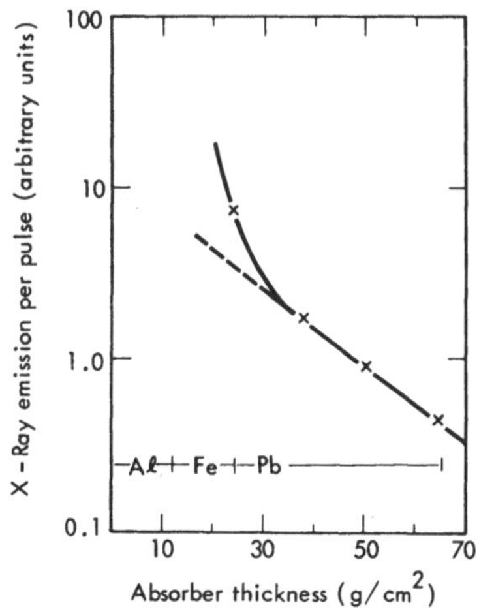

FIG. 6. X-ray absorption curve for laser-target experiment described in text. The ordinate scale is based upon the energy absorbed by the detector. The abscissa represents absorber thickness; the aluminum and iron were never removed.

The absorption coefficient μ of the hard component in Fig. 6 is approximately $\mu = 0.05\ cm^2/g$, corresponding to x rays in the 1-10-MeV range (independent of the absorbers used).[24] Such extremely hard x rays were not seen in such abundance in earlier experiments[22]; the reasons for the difference are not known. Whatever the reasons, it appears that in at least one set of experimental conditions hard x rays were seen which are best interpreted as bremsstrahlung created by relativistic electrons of energy greater than $2m_0c^2$ (~1 MeV).

However, we are unable to estimate the absolute number of relativistic electrons produced, because we do not know where the x rays were produced. One-MeV electrons have a range in cold material of 0.4 g/cm² or more.[24] The range in hot plasma would be higher, but even the cold-material range is already much greater than the dimensions of the laser focal spot. Thus, relativistic electrons created at the target would be expected to escape into the vacuum chamber, where they would travel to the walls, creating bremsstrahlung x rays at many locations. Until further experimental work is done to isolate and measure these effects, it is not possible to estimate the absolute number of electrons produced.

These considerations also cast doubt on whether pair production takes place at the focus, since in some directions the distance from the focus that the relativistic electron travels is much greater than the focal-spot dimensions.

VI. COMPARISON OF PAIR PRODUCTION AND BREMSSTRAHLUNG BY FAST ELECTRONS

For experimental purposes it is of interest to compare the average energy loss of fast electrons by pair production to the average energy loss by bremsstrahlung x radiation. To do this, consider the probability dP that pairs are produced in distance dx:

$$dP = \sigma_T N_n \, dx, \tag{21}$$

where σ_T is the trident cross section and N_n is the density of nuclei per unit volume. Near the threshold we can assume that approximately all of the kinetic energy E_v is lost when the trident process occurs, so that the averaged energy loss (over many electrons) can be written

$$\left(\frac{dE}{dx}\right)_{\text{pairs}} \cong -E_v \frac{dP}{dx} = -9.6\times 10^{-4}\alpha^2 r_0^2 Z^2 N_n E_v \times\left(\frac{E_v}{m_0c^2}-2\right)^{3.6}, \tag{22}$$

where we have substituted from Eq. (6).

It is well known that a similar equation exists for the radiative energy loss due to bremsstrahlung,[25]

$$\left(\frac{dE}{dx}\right)_{\text{rad}} = -4\alpha r_0^2 Z^2 N_n E_v \ln\left[\left(\frac{183}{Z}\right)^{1/3}\right]. \tag{23}$$

This expression is the average energy loss at all x-ray frequencies. For comparison with the experiment described in Sec. V, however, we want to know the average radiation loss by emission of hard x rays ($E \gtrsim 2m_0c^2$). Because the x-ray spectrum is almost constant, this can be written approximately as

$$\left(\frac{dE}{dx}\right)_{\text{hard x}} \simeq -21\alpha r_0^2 Z^2 N_n (E_v - 2m_0c^2), \tag{24}$$

where we have put $Z = 1$ in the slowly varying logarithmic term.

The approximate ratio of average pair-production energy loss to average hard x-ray emission is then found from Eqs. (22) and (24):

$$\frac{(dE)_{\text{pairs}}}{(dE)_{\text{hard x}}} \simeq 4.6\times 10^{-5}\alpha\left(\frac{E_v}{m_0c^2}\right)\left(\frac{E_v}{m_0c^2}-2\right)^{2.6}. \tag{25}$$

This result can be used to make an estimate of the possibility of pair production in an experiment where bremsstrahlung x rays have been produced. In our earlier experiment,[22] 10^{-9} J of x-ray energy was emitted by the source as x rays of 100 keV or greater. One would expect much less energy to have been emitted as x rays of $2m_0c^2$ or greater. For electrons near threshold $[(E_v/m_0c^2 - 2) \ll 1]$, Eq. (25) predicts that the ratio of pair energy to hard x-ray energy would be much less than 10^{-6}. So one concludes that the average pair-production energy loss in this example was much less than 10^{-15} J. However, the threshold energy $2m_0c^2$ for production of a single pair is of the order of 10^{-13} J, which is still greater than this extreme upper limit. Thus we conclude that no pairs were produced in our early experiment,[22] and that one should not expect to see pair production in similar experiments unless orders-of-magnitude-greater x-ray bremsstrahlung intensities are detected.

In the more recent experiment, described above, in which more hard x rays were detected, we cannot tell whether pairs were produced, because we could not estimate the absolute x-ray intensity for that experiment.

VII. SUMMARY

We have examined various mechanisms for electron-positron pair production by intense, focused laser light pulses. Vacuum pair production was estimated to be unobservable, in agreement with previous authors. The trident process of pair production by high-energy electrons was then considered, and the cross section was calculated. The remaining question is how the high-energy (kinetic energy $> 2m_0c^2$) electrons can be created at the laser focus.

In one case, that of the coherent "quivering velocity" of an electron in a circularly polarized beam, we were able to obtain a result for the threshold of pair production which was at the extreme upper end of contemporary feasible focused intensities.

In another case, that of the parametric ion-acoustic instability which is excited in laser-produced plasmas, we have given qualitative arguments for the plausibility of production of at least a few relativistic electrons by the longitudinal plasma waves. On the basis of available experimental information, it seems unlikely that electron-positron pairs have been produced at experimental intensities of 10^{12}–10^{15} W/cm². When new experiments are done in the intensity range 10^{16}–10^{18} W/cm², however, a few pairs may possibly be produced. Thus, an experimental search for positron-electron pairs need not wait for a focused laser intensity as high as 10^{19}–10^{20} W/cm², as previously estimated.[5]

ACKNOWLEDGMENTS

We would like to acknowledge interesting discussions of these problems with Ray E. Kidder, Stephen Bodner, and James Eddleman. We are indebted to Arno Steiger and C. H. Woods for discussions and access to some of their calculations. We wish to thank Stanley J. Brodsky for sending us his calculation program.

---

*Work performed under the auspices of the U.S. Atomic Energy Commission.

[1]W. F. Hagen, J. Appl. Phys. **40**, 511 (1969).
[2]P. G. Kryukov and V. S. Letokhov, Usp. Fiz. Nauk **99**, 169 (1969) [Sov. Phys.-Usp. **12**, 641 (1970)].
[3]F. V. Bunkin and I. I. Tugov, Dokl. Akad. Nauk SSSR **187**, 541 (1969) [Sov. Phys.-Dokl. **14**, 678 (1970)].
[4]E. Brezin and C. Itzykson, Phys. Rev. D **2**, 1191 (1970).
[5]F. V. Bunkin and A. E. Kazakov, Dokl. Akad. Nauk SSSR **193**, 1274 (1970) [Sov. Phys.-Dokl. **15**, 758 (1971)].
[6]H. J. Bhabha, Proc. R. Soc. **152**, 559 (1935).
[7]S. J. Brodsky and S. C. C. Ting, Phys. Rev. **145**, 1018 (1966).
[8]V. L. Ginzburg, *The Propagation of Electromagnetic Waves in Plasmas*, translated by J. B. Sykes and R. J. Taylor (Pergamon, New York, 1964).
[9]P. Kaw and J. Dawson, Phys. Fluids **13**, 472 (1970).
[10]A. D. Steiger and C. H. Woods, Phys. Rev. A **5**, 1467 (1972).
[11]C. Max and F. Perkins, Phys. Rev. Lett. **27**, 1342 (1971).
[12]P. Kaw, J. Dawson, W. Kruer, C. Oberman, and E. Valeo, Sov. J. Quantum Electron. **1**, 205 (1971).
[13]W. L. Kruer and J. M. Dawson, Phys. Fluids **15**, 446 (1972).
[14]K. Nishikawa, J. Phys. Soc. Jap. **24**, 1152 (1968).
[15]D. F. DuBois and M. V. Goldman, Phys. Rev. Lett. **14**, 544 (1965).
[16]P. K. Kaw and J. M. Dawson, Phys. Fluids **12**, 2586 (1969).
[17]J. S. DeGroot and J. I. Katz, Phys. Fluids **16**, 401 (1973).
[18]L. Spitzer, Jr., *Physics of Fully Ionized Gases*, 2nd ed. (Interscience, New York, 1962), p. 56.
[19] J. W. Shearer and J. J. Duderstadt, Nucl. Fusion 13, 401 (1973).
[20]C. Yamanaka, T. Yamanaka, H. Kang, K. Yoshida, M. Waki, and T. Shimamura, Phys. Lett. A **38**, 495 (1972).
[21]S. W. Mead, R. E. Kidder, J. E. Swain, F. Rainer, and J. Petruzzi, Appl. Opt. **11**, 345 (1972).
[22]J. W. Shearer, S. W. Mead, J. Petruzzi, F. Rainer, J. E. Swain, and C. E. Violet, Phys. Rev. A **6**, 764 (1972).
[23]L. W. Coleman, J. E. Swain, F. Rainer, and R. A. Saroyan, University of California Lawrence Livermore Laboratory Report No. UCRL-74626, 1973 (unpublished).
[24]*American Institute of Physics Handbook*, 2nd ed. (McGraw-Hill, New York, 1963), pp. 8-45–8-98.
[25]E. Fermi, *Nuclear Physics*, revised ed. (University of Chicago Press, Chicago, 1950), p. 47.

Estimates for the efficient production of antihydrogen by lasers of very high intensities

HEINRICH HORA
Rensselaer Polytechnic Institute, Graduate Center Hartford, Conn., USA

Starting from the non-linear relativistic equations of motion for charged particles in the very high intensity fields of laser radiation, the maximum kinetic energy ϵ_{kin} of the resulting oscillation is derived exactly. In non-relativistic conditions ϵ_{kin} agrees with the well-known value $e^2 E_v^2/(2m_0 \omega^2 |n|)$, showing a dependence on the rest mass m_0 of the particle. In the relativistic case, the mass dependence vanishes. The multipole radiation is calculated on the basis of Sommerfeld's formula for relativistic conditions. It is shown that this radiation is not important for oscillation energies up to $\epsilon_{kin}^{mr} = 70\, m_0 c^2$ for electrons in neodymium glass laser radiation and up to higher values for CO_2 lasers and for protons. With the limitation $m_0 c^2 < \epsilon_{kin} < \epsilon_{kin}^{mr}$, the formula for ϵ_{kin} is used to calculate the pair production (a) for singly oscillating particles in vacuum without collisions and (b) for plasmas with collisions. Taking into account the local increase of the effective electric laser field near the cut-off density due to the decrease of $|n|$ (n is the complex refractive index), there is the possibility of efficient proton pair production at intensities of 10^{19} W cm^{-2} for neodymium glass lasers and of 10^{17} W cm^{-2} for CO_2 lasers, besides electron pair production.

1. Introduction

With high laser intensities, the motion of singly charged particles or of plasma particles in the laser fields becomes important. If the oscillating particles reach relativistic energies, the question arises of producing antiparticles, as studied by Prokhorov *et al.* and Bunkin *et al.* [1, 2]. These treatments used the total relativistic energy of the particles. The total energy splits into that of oscillation and of translation [3], determined by the time dependence of the pulse and the initial conditions of the particles. A further question is the generation of standing waves, and the dielectric properties with respect to this translation. To solve for the dielectric properties simply, only the oscillation of the particles in their co-ordinate system is considered. The problem of electron pair production in vacuum [3] due to vacuum polarization [4] is excluded, because this occurs at intensities exceeding 10^{26} or 10^{24} W cm^{-2} for Nd: glass or CO_2 lasers, while the following treatment is for smaller intensities.

From an analytical treatment of the relativistic equation of motion we shall derive the general value of the maximum kinetic enegy ϵ_{kin} during the oscillation. With this value the multipole radiation of a charge will be calculated on the basis of the derivation given

Reprinted from: *Opto-Electronics*, 5, 491–501 (1973).

by Sommerfeld [5]. Because the duration is limited by emission of multipole radiation at much higher laser intensities than the relativistic threshold, the oscillation energy can be used for deriving the equations defining the probability of pair production. Finally collective effects are considered in a plasma with such high densities that the plasma frequency ω_p is equal to the light frequency ω. The thresholds of pair production are then decreased by similar changes of the refractive index as these are the origin of non-linear forces of direct collisionless interaction of laser radiation with plasmas [6].

2. Relativistic motion

In the non-relativistic case, the forces acting on a free particle (electron, proton, etc.) of rest mass m_0 and (single) charge e in a laser field are due to the electric field E, resulting in a velocity

$$\boldsymbol{v} = \frac{e}{m_0} \boldsymbol{E} = \frac{e}{m_0} \boldsymbol{i}_2 E_0 \cos \omega t \; ; \quad E_0 = \frac{E_v}{\sqrt{|n|}} , \tag{1}$$

where E_0 is the amplitude of a plane, linearly polarized wave propagating in the direction of the unit vector $\boldsymbol{i}_1$ with an orientation of $\boldsymbol{E}$ parallel to the $\boldsymbol{i}_2$-unit vector and where a complex refractive index n is present. E_v is the corresponding amplitude in the vacuum if the wave penetrates with negligible loss into the plasma. The maximum kinetic energy during the oscillation of the particle is

$$\epsilon_{\text{kin}} = \frac{1}{2} \frac{e^2 E_0^2}{m_0 \omega^2} , \tag{2}$$

to which nothing has to be added from the motion of the particle due to the magnetic field

$$\boldsymbol{H} = \boldsymbol{i}_3 |E_v| \frac{\sqrt{|n|}}{c} \cos(\omega t + \phi) \tag{3}$$

because a magnetic field does not change the energy of a moving particle. In Equation 3, the phase ϕ is effective only if $n \neq 1$.

The relativistic equation of motion without the retardation of the field is defined by the velocity $\boldsymbol{v}$

$$\frac{\mathrm{d}}{\mathrm{d}t} \frac{m_0 v_y}{\{1 - (v_x^2 + v_y^2)/c^2\}^{\frac{1}{2}}} = e \frac{E_v}{\sqrt{|n|}} \cos \omega t \tag{4a}$$

$$\frac{\mathrm{d}}{\mathrm{d}t} \frac{m_0 v_x}{\{1 - (v_x^2 + v_y^2)/c^2\}^{\frac{1}{2}}} = \frac{e}{c} v_y E_v \sqrt{(|n|)} \cos(\omega t + \phi) , \tag{4b}$$

where only the components v_y and v_x exist parallel to $\boldsymbol{E}$ and $\boldsymbol{E} \times \boldsymbol{B}$ respectively. Using the substitution

$$z = \{1 - (v_x^2 + v_y^2)/c^2\}^{\frac{1}{2}} = \{1 - \boldsymbol{v}^2/c^2\}^{\frac{1}{2}} , \tag{5}$$

we integrate Equation 4a:

$$v_y = \frac{e E_v z}{m |n| \omega} \sin \omega t . \tag{6}$$

Using Equations 5 and 6 we find from Equation 4b

$$\frac{d}{dt} \frac{\{c^2 - z^2(c^2 + e^2 E_v^2 \sin^2 \omega t/(m_0^2 |n| \omega^2)\}^{\frac{1}{2}}}{z}$$

$$= z \frac{e^2 E_v^2}{m_0^2 c\omega} [\tfrac{1}{2} \sin 2\omega t \cos \phi - \sin^2 \omega t \sin \phi] . \tag{7}$$

Obviously, the expression under the radical cannot become negative, therefore z can vary only between

$$0 \leqslant z^2 \leqslant z^{*2} , \tag{8}$$

where

$$z^{*2} \left(1 + \frac{e^2 E_v^2}{c^2 m_0^2 |n| \omega^2}\right) = 1 . \tag{9}$$

The maximum value of z^*,

$$z^* = \{1 - v_{max}^2/c^2\}^{\frac{1}{2}} = 1 \Big/ \left\{1 + \frac{e^2 E_v^2}{m_0^2 \omega^2 c^2 |n|}\right\}^{\frac{1}{2}} \tag{10}$$

defines the maximum kinetic energy of the oscillation, using the definition of the relativistic kinetic energy

$$\epsilon_{kin} = m_0 c^2 \left(\frac{1}{\{1 - v^2/c^2\}^{\frac{1}{2}}} - 1\right) = \left[\left\{1 + \frac{e^2 E_v^2}{m_0^2 \omega^2 c^2 |n|}\right\}^{\frac{1}{2}} - 1\right] m_0 c^2. \tag{11}$$

For $E_v \ll E_v^r$ we reproduce the non-relativistic value in agreement with Equation 1:

$$\epsilon_{kin} = \frac{1}{2} \frac{e^2 E_v^2}{m_0 \omega^2 |n|} , \tag{12}$$

where E_v is the relativistic limit

$$\frac{E_v^2}{|n|} = m_0 c^2 \frac{m_0 \omega^2}{e^2} . \tag{13}$$

This limit for neodymium glass lasers with a wavelength $\lambda = 2\pi c/\omega = 1.06$ µm(Nd) and for CO_2 lasers ($\lambda = 10.6$ µm) in the case of electrons (e) and protons (p) is:

$$\frac{E_v^r}{\sqrt{|n|}} = \begin{cases} 3 \times 10^{10} \text{ V cm}^{-1} & \text{(e)} \quad \text{(Nd)} \\ 5.5 \times 10^{13} \text{ V cm}^{-1} & \text{(p)} \quad \text{(Nd)} \\ 3 \times 10^{9} \text{ V cm}^{-1} & \text{(e)} \quad (CO_2) \\ 5.5 \times 10^{12} \text{ V cm}^{-1} & \text{(p)} \quad (CO_2) . \end{cases} \tag{14a}$$

The corresponding laser intensities $I_v = I_v^r$ in vacuum are

$$\frac{I_v^r}{|n|} = \begin{cases} 1.1 \times 10^{18}\ \mathrm{W\,cm^{-2}} & \text{(e)}\quad \text{(Nd)} \\ 3.7 \times 10^{24}\ \mathrm{W\,cm^{-2}} & \text{(p)}\quad \text{(Nd)} \\ 1.1 \times 10^{16}\ \mathrm{W\,cm^{-2}} & \text{(e)}\quad (\mathrm{CO_2}) \\ 3.7 \times 10^{22}\ \mathrm{W\,cm^{-2}} & \text{(p)}\quad (\mathrm{CO_2}), \end{cases} \tag{14b}$$

where the refractive index n is to be used at the actual location r_0 within the plasma neglecting absorption between the vacuum and r_0.

For the pure relativistic case $E_v \gg E_v^r$ we find from Equation 11:

$$\epsilon_{kin} = \frac{e\,c\,E_v}{\omega\sqrt{|n|}} \sim \sqrt{I}\,. \tag{15}$$

This result demonstrates that there is no dependence of the oscillation energy of relativistic particles on their rest mass m_0, and the increase is proportional to the square root of the intensity I only. The consequent use of the refractive index $|n|$ diminishes the effect of neglect of the retardation of the laser field at the cut-off densities of plasmas.

3. Multipole radiation

The oscillation of particles follows Equations 4 as long only as the emission of multipole radiation can be neglected. Here we like to know only the upper limit of laser intensities beyond which the re-emission is very strong. Weak effects like frequency shifts can occur at much lower intensities, indeed [1], but this is of less importance when we wish to know the energy of the particles with respect to collision induced pair production.

The energy re-emitted per oscillation from a particle of one elementary charge is [5]

$$S = \frac{e^2\,\omega^3\,x_0^3}{3c^3(2\pi)^3\,(1 - v^2/c^2)^2}\,, \tag{16}$$

where x_0 is the amplitude of the oscillation. In the non-relativistic case ($v^2 \ll c^2$) with $x_0 = e^2\,E_0^2/(m_0^2\,\omega^4)$ we find

$$S^{NR} = \frac{e^4\,E_0^2}{3(2\pi\,c)^3\,m_0^2\,v}\,. \tag{17}$$

Using ϵ_{kin} from Equation 12 with $n = 1$, we find a negligible ratio $S^{NR}/\epsilon_{kin} = 2e^2\,\omega/(24\pi^3\,c^3\,m) = 4.49 \times 10^{-11}$ for neodymium glass lasers and 4.49×10^{-12} for CO_2 lasers.

For the relativistic case we over estimate S a little by putting $x_0 = 2\pi\,c/\omega = c\lambda$ and using $\epsilon_{kin}/(m_0\,c^2) = (1 - v^2/c^2)^{-1/2}$ from Equation 11, we find from Equation 16 with Equation 15:

$$S^R = \frac{e^2\,\omega}{96\pi\,c}\left(\frac{\epsilon_{kin}}{m_0\,c^2}\right)^6 = \frac{e^8\,E_v^6}{96\pi\,c^7\,m_0^6\,|n|^3\,\omega^5}\,. \tag{18}$$

The upper limit for neglecting multipole radiation (mr) is that laser field strength E_v^{mr} or kinetic energy ϵ_{kin}^{mr} of the particle for which $\epsilon_{kin} = S^R$, following from Equations 15 and 18:

$$E_v^{mr}/\sqrt{|n|} = (96\pi\,c^8\,m_0^6\,\omega^4/e^7)^{1/5}\,, \tag{19}$$

$$\frac{\epsilon_{\text{kin}}^{\text{mr}}}{m_0 c^2} = \frac{96\pi\, m_0\, c^3}{e^2\, \omega}\,. \tag{20}$$

Using the rest masses for electrons (e) and protons (p) and the frequencies for neodymium glass (Nd) and CO_2 lasers again, we find from Equations 19 and 20:

$$\frac{E_v^{\text{mr}}}{\sqrt{|n|}} = \begin{cases} 2.15 \times 10^{12}\ \text{V cm}^{-1} & \text{(e)} \quad \text{(Nd)} \\ 3.95 \times 10^{15}\ \text{V cm}^{-1} & \text{(p)} \quad \text{(Nd)} \\ 2.15 \times 10^{11}\ \text{V cm}^{-1} & \text{(e)} \quad (\text{CO}_2) \\ 3.95 \times 10^{14}\ \text{V cm}^{-1} & \text{(p)} \quad (\text{CO}_2) \end{cases} \tag{21}$$

$$\frac{I^{\text{mr}}}{|n|} = \begin{cases} 6.34 \times 10^{21}\ \text{W cm}^{-2} & \text{(e)} \quad \text{(Nd)} \\ 2.14 \times 10^{28}\ \text{W cm}^{-2} & \text{(p)} \quad \text{(Nd)} \\ 6.34 \times 10^{19}\ \text{W cm}^{-2} & \text{(e)} \quad (\text{CO}_2) \\ 2.14 \times 10^{26}\ \text{W cm}^{-2} & \text{(p)} \quad (\text{CO}_2) \end{cases} \tag{22}$$

$$\frac{\epsilon_{\text{kin}}^{\text{mr}}}{m_0 c^2} = \begin{cases} 71 & \text{(e)} \quad \text{(Nd)} \\ 321 & \text{(p)} \quad \text{(Nd)} \\ 112.5 & \text{(e)} \quad (\text{CO}_2) \\ 505.9 & \text{(p)} \quad (\text{CO}_2) \end{cases} \tag{23}$$

Because of the increase of the multipole radiation with the sixth power of E_v in the relativistic region, the limits E_v^{mr} are very strongly pronounced, implying the multipole radiation is accurately negligible for $E_v < E_v^{\text{mr}}$. The large numbers of Equation 23 demonstrate that there is a very wide relativistic range where the multipole radiation is negligible.

The fact that I^{mr} for electrons is always less than the relativistic threshold I^{r} (Equations 14) for protons demonstrates that in the relativistic range for protons, the electrons will not have the kinetic energy ϵ_{kin} of Equation 15 but a smaller value due to the quadrupole radiation. For the mechanism of pair production of relativistic protons, it is of less importance what energies the colliding electrons have.

4. Conditions for pair production

Besides of the concept of interaction of a high intensity laser field with the vacuum for producing electron proton pairs by vacuum polarization [3, 4] which has a threshold near an intensity of 10^{26} W cm^{-2} for neodymium glass laser, another mechanism was discussed by Erber [7] where a free electron is oscillating in the laser field without any collision. The pair production is given by a transition rate

$$\exp\,(-\ 8/3\gamma)\,, \tag{24}$$

where [2]

$$\gamma = \frac{\epsilon_{\text{kin}}}{m_0\, c^2} \frac{E_0}{E_{\text{cr}}} \text{ and } E_{\text{cr}} = m_0^2\, c^3/(e\, \hbar)\,.$$

Using $E_0 = E_v$ (putting $|n| = 1$) and ϵ_{kin} from Equation 15 we arrive at

$$\gamma = \frac{e^2 \hbar}{\omega m_0^3 c^3} E_v . \tag{25}$$

With a value $\gamma = 0.1$, resulting in an exponential function before Equation 24 of 2.6×10^{-12} we find field strength E_v^v and intensities I_v^v:

$$E_v^v = \begin{cases} 4.4 \times 10^{13}\ \text{V cm}^{-1} & \text{(e)} \quad \text{(Nd)} \\ 2.7 \times 10^{23}\ \text{V cm}^{-1} & \text{(p)} \quad \text{(Nd)} \\ 4.4 \times 10^{12}\ \text{V cm}^{-1} & \text{(e)} \quad (\text{CO}_2) \\ 2.7 \times 10^{22}\ \text{V cm}^{-1} & \text{(p)} \quad (\text{CO}_2) \end{cases} \tag{26}$$

$$I_v^v = \begin{cases} 2.7 \times 10^{24}\ \text{W cm}^{-2} & \text{(e)} \quad \text{(Nd)} \\ 1.03 \times 10^{44}\ \text{W cm}^{-2} & \text{(p)} \quad \text{(Nd)} \\ 2.7 \times 10^{22}\ \text{W cm}^{-2} & \text{(e)} \quad (\text{CO}_2) \\ 1.03 \times 10^{42}\ \text{W cm}^{-2} & \text{(p)} \quad (\text{CO}_2) . \end{cases} \tag{27}$$

which are similar to the result with the uncorrected value [2] of ϵ_{kin} and which are higher than the laser intensities considered below.

A third mechanism is the pair production due to collisions resulting at lower thresholds by the condition only of $\epsilon_{kin} \gg m_0 c^2$ or $I_v > I_v^r$ or $E_v > E_v^r$ of Equations 14, given by an order of magnitude for the cross-section of pair production [2]:

$$\sigma = \frac{e^8}{\pi \hbar^2 m_0^2 c^6} \ln^3 \left(\frac{\epsilon_{kin}}{m_0 c^2} \right) . \tag{28}$$

The number N_p of pairs produced in a plasma volume V during a time τ and a density n_e of electrons is

$$N_p = \frac{e^8 n_e^2}{\pi \hbar^2 m_0^2 c^5} V \tau \ln^3[\epsilon_{kin}/(m_0 c^2)] , \tag{29}$$

or by using Equation 15

$$N_p = \frac{e^8 n_e^2}{\pi \hbar^2 m_0^2 c^5} V \tau \ln^3[e E_v/(\omega \sqrt{|n|}\, m_0 c)] . \tag{30}$$

Because of the logarithmic function in Equation 30, the difference of our result to that with the uncorrected value [2] of ϵ_{kin} is indeed not very strong, at least it is not a question of many orders of magnitude.

Because the following considerations are also directed to questions of pair production of protons besides of that of electrons, the use of Equation 30 is then a very rough approximation, because the interaction of the protons with other particles and quanta is neglected. The mass m_0 in the factor before V appearing in Equation 30 was derived from electrodynamic quantities only expressed by the fine structure constant, therefore at this point the electron mass will remain. The only condition for proton pair production is that of intensities I_v and fields E_v exceeding the relativistic thresholds of Equation 14 for the protons.

5. Collective effects

The result of pair production in plasmas due to collisions, Equation 30, has a remarkable rate, if the particle energy exceeds the energy of the rest mass by a factor of 3 or more. For electrons, the needed laser intensities I_v and field strength E_v are then indeed not too far from the present state of technology if thin plasmas with $n = 1$ are presumed. But the results of Equation 14 indicate intensities exceeding strongly the present aspects of laser technology, if the pair production of protons is desired. In this case we have a chance only by assuming collective effects at such electron densities n_e of the plasma that the laser frequency ω is equal to the plasma frequency ω_p.

$$\omega_p^2 = \frac{4\pi e^2 n_e}{m_0} \{1 - v^2/c^2\}^{\frac{1}{2}}, \tag{31}$$

where the relativistic correction of the electron mass was used as indicated by Kidder [8]. The refractive index n,

$$n^2 = 1 - \frac{\omega_p^2}{\omega^2} \cdot \frac{1}{1 - i\nu/\omega}, \tag{32}$$

is determined by the collision frequency ν defining at high laser intensities an absorption constant K_{NL} if the electron density n_e is much less than the cut-off density n_{eco} derived from Equation 31 for $\omega = \omega_p$. In this case $|n| = 1$ and we find for a hydrogen plasma

$$K_{NL} = \frac{\nu}{c} \frac{\omega_p^2}{\omega^2} \tag{33}$$

$$= \frac{e^6 n_e^2 \pi^{3/2} \ln \Lambda}{c\,\omega^2 (2m_0 k T)^{3/2}} = \frac{e^6 n_e^2 \pi^{3/2} \ln \Lambda}{c\,\omega^2 (2m_0 e c E_v/3\omega)^{3/2}}. \tag{34}$$

Here the collision frequency,

$$\nu = \frac{e^4 n_e \ln \Lambda}{(2kT/\pi)^{3/2} m_0^{\frac{1}{2}}}, \tag{35}$$

was used for describing the electron-ion collisions. We have to neglect the correction for electron-electron collisions in this case because of the coherent motion of the electrons in the laser field. $\ln \Lambda$ is the Coulomb logarithm where, with E_v in V cm^{-1},

$$\Lambda = \frac{c}{2e^2 \omega(\pi n_e)^{\frac{1}{2}}} \cdot \frac{E_v}{\sqrt{|n|}} = \begin{cases} 3.9 \times 10^{-5} E_v/\sqrt{|n|} & (\text{Nd}) \\ 3.9 \times 10^{-3} E_v/\sqrt{|n|} & (\text{CO}_2) \end{cases} \text{ if } E_v > E_v^r. \tag{36}$$

For the following relativistic cases of cut-off density we shall use $\ln \Lambda = 17$. Here the electron energy $\epsilon_{kin} = 3\,kT$ was substituted in the well known [9] expression of $\ln \Lambda$ taking into account that ϵ_{kin} is the maximum value of the oscillating electron and the average value

$$\langle \epsilon_{kin} \rangle = \epsilon_{kin}/2 .$$

For the condition $\omega = \sqrt{(\omega_p^2 + \nu^2)}$ from Equation 32, there follows the minimum value of

$$|n| = \sqrt{(\mathrm{Re}(n)^2 + I_m(n)^2)} = \frac{1}{2}\sqrt{\frac{\nu}{\omega}} = \frac{1}{2}\left(\frac{e^4\, n_e \ln \Lambda}{(kT)^{3/2}\,\omega\, m_0^{\frac{1}{2}}}\right)^{\frac{1}{2}}, \tag{37}$$

from which we find for $\ln \Lambda = 17$ and T in eV,

$$|n| = \frac{a}{T^{3/4}}\,; \quad a = \begin{cases} 2.26 & (\mathrm{Nd}) \\ 0.715 & (\mathrm{CO_2}) \end{cases} \tag{38}$$

Substituting again for T by ϵ_{kin} in Equation 37, we find

$$|n| = \frac{1}{2}\left(\frac{e^4\, n_e (\ln \Lambda)\, 3^{3/2}\, \omega^{\frac{1}{2}}}{(2m_0\, e\, c\, E_v)^{3/2}}\right)^{\frac{1}{2}}. \tag{39}$$

The physical meaning of this value is the effective increase of the vacuum field E_v by the denominator $\sqrt{|n|}$, resulting in $E_v^2/\sqrt{|n|} = E_v^2\, T^{3/4}/a$ producing an increase of E_v by factors of 100 and more, if the laser light penetrates into the plasma with a negligible reflection of light and a negligible absorption. This increase of the actual field strength E^{act} causes non-linear forces in the plasma of a ponderomotive kind, as were first described in 1967 [10] and later on more generally [6]. A case of an exact evaluation avoiding the WBK-approximation was given by Lindl and Kaw [11], while the predominance of the non-linear force over the thermal forces was shown by Steinhauer and Ahlstrom [12] at high temperatures for all intensities. Their evaluation could be used to prove the predominance for laser intensities exceeding 10^{14} W cm^{-2} (Nd) and 10^{12} W cm^{-2} (CO_2) generally [13]. Also numerical evaluations demonstrated the predominance in a WBK-case for 10^{15} W cm^{-2} (Nd) immediately [14] and a non-WBK-case including the reflection of light by Mulser and Green [15] showed the larger non-linear force for the global (net) acceleration over the thermokinetic force even for intensities of 5×10^{14} W cm^{-2} (Nd). It does not cancel these results, for a dynamic calculation with laser pulses incident on a plasma demonstrated at selected conditions of long laser pulses that the forces in the shocked plasma below the absorption region can determine the expansion to a greater degree [16] than the mechanisms within the interaction region of the laser light with the plasma. The influence of the non-linear force also on the whole dynamical numerical description was demonstrated by Shearer, Kidder, and Zink [17]. If the derivation of the non-linear force includes some approximations, the resulting expression for the force only in certain conditions [18] has the same value as that from a general derivation without the approximations. The non-linear forces caused by changes of n for cases of infra-red lasers and microwaves were treated by Chen [19] and Klima [20] and were included in the calculations of laser compression of fusion plasmas by Nuckolls *et al.* [21].

In Equation 39 it was assumed that the absorption in the plasma with relativistically oscillating electrons is due only to collisions even at densities where $\omega_p \approx \omega$. The knowledge of instabilities [22] or anomalous absorption [23] at these densities may modify the values given in Equations 38 and 39. These instabilities are still not known in the pure relativistic conditions $E_v \gg E_v^r$, therefore the following considerations lack this knowledge and should be considered very preliminary.

6. Conclusions of proton pair production

Considering a hydrogen plasma with densities exceeding n_{eco} (10^{21} cm^{-3} or a little less due to relativistic effects) where neodymium glass laser pulses of intensities of 10^{19} W cm^{-2} corresponding to field strengths of $\sim 10^{11}$ V·cm^{-1} are incident, the pulse length being assumed sufficiently long ($> 10^{-10}$ s) to build up a plasma region of cut-off density of a thickness exceeding many vacuum wave lengths with little absorption, we shall expect electron pair production in all regions of the interaction zone. But in the cut-off-region we find by a first iteration from $E^{act} = E_v T^{-3/8}$, with T in electron-volts, values with $T = 10^6$ eV, and therefore $E^{act} = 1.7 \times 10^{13}$ V cm^{-1}. This causes again an effective increase of the electron energy accompanied by a new density profile and/or conditions of reflectivity for electron energies of 1.7×10^8 eV. This would again increase E^{act} within the next step of iteration with this last energy, we shall exceed the field strength at 10^{14} V cm^{-1} which will cause pair production of protons following Equation 30.

Though these conclusions are very preliminary (with respect to the lack of detailed knowledge of the instabilities in relativistic conditions [24] and of the processes of multipole radiation emission of electrons beyond the thresholds of Equations 19 to 23, and of following processes of re-absorption and perhaps meson generation etc.) we can expect the production of proton pairs even with such laser intensities which are needed for electron pair production only. The only condition is the use of cut-off plasma densities and the described decrease of the refractive index n in the cut-off region as known from the mechanisms of the non-linear force.

Finally, we can estimate very roughly the size of an apparatus for most optimistic conditions, resulting in an effective conversion of laser energy into proton pairs. Assuming the logarithmic function in Equation 30 has a value 25, we arrive at

$$N_p = 10^{24}\, V \tau = 10^{24}\, F\, l\, \tau\,, \tag{40}$$

if F is the cross-section of the laser bundle and l is the length of the reaction zone. We compare the total energy E_p of the pairs ($E_p = 2m_{0\,\text{proton}}\, c^2 N_p$) with that of the incident laser energy E_L,

$$E_L = 2F \tau I_v{}^r{}_{\text{electron}}, \tag{41}$$

being a little above the relativistic threshold for electrons in the case of neodymium glass lasers.

We arrive at

$$\frac{E_p}{E_L} = \frac{2m_0{}^r{}_{\text{proton}}\, c^2 N_p}{2F \tau I_{v\,\text{electron}}} = 5.45 \times 10^{-5}\, l\,. \tag{42}$$

Neglecting the details of absorption, Equation 42 results in a length of the reaction zone of $l = l_0$, if a conversion efficiency $E_p/E_L = 1$ should be reached, of

$$l_0 = 73\ \text{m}\,. \tag{43}$$

The use of a self-focused filament of a diffraction limited beam will be preferable because of stabilizing quasistationary conditions of the reaction. A diffraction-limited beam from a Nd laser ($F = 10^{-7}$ cm^2) as described in Equation 41 would result, using Equation 14b, in

$$E_L = 2.2 \times 10^{-12} \tau \text{ J}, \tag{44}$$

which corresponds to a feasible laser power of 2×10^{12} W.

In our gedankenexperiment we have assumed a value $|n| = 10^{-6}$ which causes an effective velocity of light of 3×10^2 m s^{-1} in the reaction zone, resulting in a minimum time of operation of 0.25 s. This comes to a laser pulse energy of 5×10^{11} J.

The conditions of a 73 m long reaction zone and laser energies of 500 GJ of 0.25 s length are not too fantastic conditions for an efficient production of antiprotons, taking into account that the energy to be stored within 1 s in the magnetic fields of a future tokamak device for nuclear fusion is of the same order of magnitude.

Acknowledgements

The author gratefully acknowledges stimulating discussions with Academician Basov, Professor Krokhin, Professor Bunkin (Moscow) and Dr Mulser (Garching).

References

1. F. V. BUNKIN and A. M. PROKHOROV, Polarization Matiere et Rayounement, Volume Jubilaire en l'Bonneur d'Alfred Kastler, Paris (1969); N. G. BASOV and O. N. KROKHIN, Seminar at Lebedev Inst., 1969.
2. F. V. BUNKIN and A. E. KAZAKOV, *Sov. Phys. Doklady* **15** (1971) 758–759.
3. T. W. KIBBLE, *Phys. Rev.* **138B** (1965) 740–753.
4. W. HEISENBERG, *Z. Physik* **90** (1934) 209–231.
5. A. SOMMERFELD, 'Vorlesungen Vol. III, Elektrodynamik,' Wiesbaden (1948) 294.
6. H. HORA, *Phys. Fluids* **12** (1969) 182–191.
7. TH. ERBER, *Rev. Mod. Phys.* **38** (1966) 626–659.
8. R. KIDDER, Varenna Summer School (July 1969), UCRL-Preprint 71775 (1969).
9. L. SPITZER, JR., 'Physics of Fully Ionized Plasmas', Interscience, New York (1956).
10. H. HORA, D. PFIRSCH, and A. SCHLÜTER, *Z. Naturforsch* **22a** (1967) 278.
11. J. LINDL and P. KAW, *Phys. Fluids* **14** (1971) 371–377.
12. L. C. STEINHAUER and H. G. AHLSTROM, *ibid* **13** (1970) 1103–1105.
13. H. HORA, *Opto-Electronics* **2** (1970) 201–214.
14. H. HORA, 'Laser Interaction and Related Plasma Phenomena', eds. H. Schwartz and H. Hora, Plenum, New York, Vol. II (1972) 341.
15. P. MULSER and B. GREEN – see Fig. 5 of Reference 14.
16. B. GREEN and P. MULSER, 'Laser Interaction and Related Plasma Phenomena,' eds. H. Schwarz and H. Hora, Plenum, New York, Vol. II (1972) 381.
17. J. W. SHEARER, R. E. KIDDER, and J. W. ZINK, *Bull. Am. Phys. Soc.* **15** (1970) 1483.
18. J. W. SHEARER, LLL (Livermore) Report UCRL-51254 (Aug. 1972).
19. F. F. CHEN, *Comments on Plasma Physics* **1** (1972) 81.
20. R. KLIMA, *Plasma Physics* **12** (1970) 123.
21. J. NUCKOLLS, L. WOOD, A. THIESSEN, and G. ZIMMERMANN, Paper presented at the VII International Quantum Electronics Conference, Montreal (May 1972).
22. K. NISHIKAWA, *J. Phys. Soc. Japan* **24** (1968) 916–922.
23. J. F. FREIDBERG and B. M. MARDER, *Phys. Rev.* **A4** (1971) 1549–1553; P. K. KAW and J. M. DAWSON, *Phys. Fluids* **12** (1969) 2586–2591; W. L. KRUER and J. M. DAWSON, 'Laser Interaction and Related Plasma Phenomena', eds. H. Schwarz and H. Hora, Plenum, New York Vol. II (1972) 317.

24. The relativistic instabilities would cause complications if circulary polarized light were used: (C. MAX and F. PERKINS, *Phys. Rev. Letters* **29** (1972) 1731–1734).
For linear polarized light however P. KAW and J. DAWSON, *Phys. Fluids* **13** (1970) 472 demonstrated an increase of the transparency of the plasma. Therefore the 'effective collisions' decrease, losses by absorption diminish, and the desired dielectric effects which decrease $|n|$ and so increase the effective electric field strength became more pronounced.

SUBJECT INDEX